Precision Agriculture Technologies for Food Security and Sustainability

Sherine M. Abd El-Kader
Electronics Research Institute, Cairo, Egypt

Basma M. Mohammad El-Basioni
Electronics Research Institute, Cairo, Egypt

A volume in the Advances in Environmental Engineering and Green Technologies (AEEGT) Book Series

IGI Global
PUBLISHER of TIMELY KNOWLEDGE

Published in the United States of America by
IGI Global
Engineering Science Reference (an imprint of IGI Global)
701 E. Chocolate Avenue
Hershey PA, USA 17033
Tel: 717-533-8845
Fax: 717-533-8661
E-mail: cust@igi-global.com
Web site: http://www.igi-global.com

Library of Congress Cataloging-in-Publication Data

Names: Abd El-Kader, Sherine Mohamed, 1972- editor. | Mohammad El-Basioni, Basma Mamdouh, 1983- editor.
Title: Precision agriculture technologies for food security and sustainability / Sherine Mohamed Abd El-Kader and Basma Mamdouh Mohammad El-Basioni, editors.
Description: Hershey, PA : Engineering Science Reference, an imprint of IGI Global, [2021] | Includes bibliographical references and index. | Summary: "This book compiles a comprehensive, multidisciplinary review of current research in the field of precision agriculture. It also discusses cutting-edge tools and models that can help facilitate and improve the systems implementation"-- Provided by publisher.
Identifiers: LCCN 2020005476 (print) | LCCN 2020005477 (ebook) | ISBN 9781799850007 (hardcover) | ISBN 9781799854371 (paperback) | ISBN 9781799850014 (ebook)
Subjects: LCSH: Agricultural engineering. | Sustainable engineering. | Food security.
Classification: LCC S675 .P66 2021 (print) | LCC S675 (ebook) | DDC 631.3028/6--dc23
LC record available at https://lccn.loc.gov/2020005476
LC ebook record available at https://lccn.loc.gov/2020005477

This book is published in the IGI Global book series Advances in Environmental Engineering and Green Technologies (AEEGT) (ISSN: 2326-9162; eISSN: 2326-9170)

British Cataloguing in Publication Data
A Cataloguing in Publication record for this book is available from the British Library.

For electronic access to this publication, please contact: eresources@igi-global.com.

Advances in Environmental Engineering and Green Technologies (AEEGT) Book Series

Sang-Bing Tsai
Zhongshan Institute, University of Electronic Science and Technology of China, China & Wuyi University, China
Ming-Lang Tseng
Lunghwa University of Science and Technology, Taiwan
Yuchi Wang
University of Electronic Science and Technology of China Zhongshan Institute, China

ISSN:2326-9162
EISSN:2326-9170

Mission

Growing awareness and an increased focus on environmental issues such as climate change, energy use, and loss of non-renewable resources have brought about a greater need for research that provides potential solutions to these problems. Research in environmental science and engineering continues to play a vital role in uncovering new opportunities for a "green" future.

The **Advances in Environmental Engineering and Green Technologies (AEEGT)** book series is a mouthpiece for research in all aspects of environmental science, earth science, and green initiatives. This series supports the ongoing research in this field through publishing books that discuss topics within environmental engineering or that deal with the interdisciplinary field of green technologies.

Coverage

- Air Quality
- Green Technology
- Policies Involving Green Technologies and Environmental Engineering
- Green Transportation
- Renewable Energy
- Cleantech
- Pollution Management
- Sustainable Communities
- Contaminated Site Remediation
- Industrial Waste Management and Minimization

IGI Global is currently accepting manuscripts for publication within this series. To submit a proposal for a volume in this series, please contact our Acquisition Editors at Acquisitions@igi-global.com or visit: http://www.igi-global.com/publish/.

The Advances in Environmental Engineering and Green Technologies (AEEGT) Book Series (ISSN 2326-9162) is published by IGI Global, 701 E. Chocolate Avenue, Hershey, PA 17033-1240, USA, www.igi-global.com. This series is composed of titles available for purchase individually; each title is edited to be contextually exclusive from any other title within the series. For pricing and ordering information please visit http://www.igi-global.com/book-series/advances-environmental-engineering-green-technologies/73679. Postmaster: Send all address changes to above address.

Titles in this Series

For a list of additional titles in this series, please visit:
https://www.igi-global.com/book-series/advances-environmental-engineering-green-technologies/73679

Energy-Efficient Underwater Wireless Communications and Networking
Nitin Goyal (Chitkara University Institute of Engineering and Technology, Chitkara University, Punjab, India) Luxmi Sapra (Chitkara University Institute of Engineering and Technology, Chitkara University, Punjab, India) and Jasminder Kaur Sandhu (Chitkara University Institute of Engineering and Technology, Chitkara University, Punjab, India)
Engineering Science Reference • © 2021 • 339pp • H/C (ISBN: 9781799836407) • US $195.00

Impacts of Climate Change on Agriculture and Aquaculture
Ahmed Karmaoui (Southern Center for Culture and Sciences, Morocco) Kirby Barrick (University of Florida, USA) Michael Reed (University of Kentucky, USA) and Mirza Barjees Baig (King Saud University, Saudi Arabia)
Engineering Science Reference • © 2021 • 333pp • H/C (ISBN: 9781799833437) • US $195.00

Development Challenges and Strategies in Global Arid Regions
Umar Garba Benna (Benna Associates, Nigeria)
Engineering Science Reference • © 2021 • 300pp • H/C (ISBN: 9781799837930) • US $195.00

Molecular Plant Breeding and Genome Editing Tools for Crop Improvement
Pradip Chandra Deka (Sir Padampat Singhania University, India)
Engineering Science Reference • © 2021 • 489pp • H/C (ISBN: 9781799843122) • US $195.00

Recent Advancements in Bioremediation of Metal Contaminants
Satarupa Dey (Shyampur Siddheswari Mahavidyalaya, India) and Biswaranjan Acharya (School of Computer Engineering, KIIT University (Deemed), India)
Engineering Science Reference • © 2021 • 363pp • H/C (ISBN: 9781799848882) • US $195.00

Global Environmental Impacts on Food Security, Nutrition, and Human Health
Hicham Chatoui (Private University of Marrakech, (UPM), Morocco & Cady Ayyad University, Morocco) Mohamed Merzouki (Faculty of Science and Technology, Sultan Moulay Slimane University, Beni Mellal, Morocco) Hanane Moummou (Private University of Marrakech (UPM), Marrakech, Morroco) Mounir Tilaoui (Private University of Marrakech (UPM), Sultan Moulay Slimane University, Beni Mellal, Morocco) and Nabila Saadaoui (Private University of Marrakech (UPM), Marrakech, Morroco)
Engineering Science Reference • © 2020 • 300pp • H/C (ISBN: 9781799837480) • US $195.00

IGI Global
PUBLISHER of TIMELY KNOWLEDGE

701 East Chocolate Avenue, Hershey, PA 17033, USA
Tel: 717-533-8845 x100 • Fax: 717-533-8661
E-Mail: cust@igi-global.com • www.igi-global.com

Editorial Advisory Board

List of Reviewers

Table of Contents

Section 3
Applications and Solutions

Detailed Table of Contents

Section 1
Introduction

This section gives insights on PA from all of the addressed perspectives. Essential information is provided on PA history, world interest, scope, enabling technologies, advantages and disadvantages, threats and challenges, economics, policy in line with UN SDG, adoption in developing countries, and adoption in the horticultural sector.

The global population is increasing at a tremendous speed; thus, the demand for safe and secure food to meet this population is in demand. Therefore, traditional farming methods are insufficient to meet this demand; thus, the next revolution in agriculture is required, which is Precision Agriculture (PA), the Fourth Agriculture Revolution. PA is a technology where the concept of farm management is based on observation, measuring, and responding to inter- and intra-field variability in crops. The technologies used for performing precision agriculture are mapping, global positioning system (GPS), yield monitoring and mapping, grid soil sampling application, variable-rate fertilizer application, remote sensing, geographic information systems (GIS), quantifying on farm variability, soil variation, variability of soil water content, time and space scales, robots, drones, satellite imagery, the internet of things, smartphone, and machine learning. Hence, the current chapter will be emphasizing the overview, concepts, history, world interest, benefits, disadvantages, and precision farming needs.

Precision agriculture is a management system that aims to reduce inputs like seeds, water, and energy; protect the environment; and maximize profitability. Precision agriculture uses advanced technology like positioning technology, geographical information systems, satellite navigation, and remote sensing. There are different factors affect the adoption of precision agriculture like farm size, legal affairs, and social interaction. Under climate change and increases in world population, adoption of precision

MATLAB will be utilized to validate the various irrigation systems and report it; the air temperature, wind, and humidity will be member functions to improve the efficiency of irrigation performance before the irrigation process, and the fuzzy information system consists of fuzzy rules, which are derived from information of experts or from input-output learning of the system. Rules mimic human reasoning. Mamdani method is mostly applied in the fuzzy inference technique, and the generalized bell function is used for both of temperature and wind where the triangular function used for humidity. The principles were based on the last experiments and personal experiences, and the output will change into a crisp value from this procedure of defuzzification. There are many different methods to do defuzzification, but the most common technique is centroid method. The resultant surface graphic is enough to monitor, validate, and report the irrigation system efficiency to execute and schedule the irrigation practice management.

In this chapter, the research theme is focused on the relationship between small farmers and information and communication technologies (ICT). Although there are other previous works that have already analyzed this same relationship, the authors believe that access to information remains a major challenge for farmers. With the application of workshops on agricultural practices of maize, in communities of Oaxaca and Veracruz, they learned about the practices of farmers around the production cycle and applied a survey to find out their opinion regarding the use of ICT. In addition, they used a specialized database to complement the workshops objectives. Next, in collaboration between researchers in the areas of biology and computing, they developed a web platform for access and use of information related to the variables of interest to farmers. Among the main results, they highlight that the community prefers to use cell phones to access such information and that the older generations are looking to transmit experiences and knowledge to the young with the aim of conserving ancestral knowledge.

This research was carried out to study the effect of crop factor, which is represented in some plant roots of crops residues (broad bean and wheat) on soil mechanical properties (cohesion strength, C, and internal friction angle, Φ), and thus, the effect of these roots on the power requirements of chisel plow to face the soil resistance was studied. Soil mechanical properties and power requirements of chisel plow (7 blades) were measured in the field directly at various soil depths of (0.05, 0.1, 0.15, and 0.2 m) with and without roots at constant tractor forward speed of about 4 km/h. Moisture content of the soil, broad bean roots, and wheat roots were 21%, 16%, and 14%, respectively, The results showed that the effect of roots of previous crops residues had a significant effect on the soil mechanical properties and power requirements for chisel plow when using the crop factor, which is represented in characteristic of crop residual roots in terms of root mean diameter.

Section 4
Agricultural Systems Analysis Streamlining, Modeling, and Improvements: Towards Precision Agronomics (Economic and Policy Perspective)

This section addresses the agriculture development and food security from the economic and policy perspective. It presents research on the use of quantitative and qualitative analysis methods to recognize the agricultural export future under the new trends of World Trade Organization taking the Egyptian exports as the case study, and to analyzing the role of institutions, funding/assistance sources, information sources in achieving food security, and extension policy taking the Indonesia food security policy as the case study.

The chapter aims to analyze the role of institutional sources of funding/assistance, source information in term of food security and extension policy. The results of the research show the information on problematic areas in policy implementation with respect to food security and extension policy in term of poverty reduction. The role of the institutional, assistance, and source information is useful in the implementation of other policy sectors in the future, because similar issues or problems may be avoided by taking precautions not to repeat the same mistakes or by applying measures which address the implementation issues. In this way, policy implementation can be carried out effectively, and the outcomes will meet the policy objectives.

Section 5
Agricultural Systems Analysis Streamlining, Modeling, and Improvements: Towards Precision Agronomics (Technological Perspective)

This section takes the technological perspective especially with respect to plant biotechnology subject and its role to secure food and to alleviate the risk of loss of the genetic variability of cultivated plants as a result of environmental changes and human practices. Also, it touches the issue of increasing the efficiency of producing animal-derived food products.

Plant biotechnology (micropropagation, secondary metabolites, etc.) is the science of growing different plant cells under controlled artificial environment conditions. Thence, successful plant biotechnology depends on the techniques of plant tissue culture. Plant tissue culture laboratories have style needs that distinguish them from alternative forms of laboratories and a few wants are distinctive for successfully plant growth as well as production of plant secondary metabolites. The most often manipulated physical

Preface

Food security is one of the important topics of concern to the world, and its achievement represents a great challenge. As defined by the United Nations (UN): food security means that all people, at all times, have physical, social, and economic access to sufficient, safe, and nutritious food that meets their food preferences and dietary needs for an active and healthy life, and it is included as the second goal of the UN seventeen Sustainable Development Goals (SDGs) which need the concerted efforts of all the world to be achieved. SDG 2, Zero hunger: achieve food security and improved nutrition and promote sustainable agriculture. Smartness and precision in agriculture were and still are main pillars in achieving sustainable agriculture.

The Precision Agriculture (PA) management concept, also known as precision farming, or precision agronomy, integrates new technologies with the agronomic experience to intelligently manage the high spatial variability of all agricultural variables and the time-scales at which these variables change in such a way that consolidates the agricultural decision making process to take the right action in the right place at the right time. It is proven that the right application of PA helps to increase the size and quality of the agricultural production, saves resources, improves environmental quality, helps to achieve self-sufficiency, food security and agricultural sustainability, increases exports, in addition to other economic benefits.

The field of PA despite its oldness, it will remain an emerging field bears continuous appearance of new trends, technologies, implementations, and methods especially with the continuous advancements in the supporting technologies, therefore there is a need to periodically issue a reference tries to accommodate the new in this field.

This book addresses these important issues and aims to provide a comprehensive approach introduces PA, food security and agriculture sustainability from the technology, policy, and economic perspectives for a broad range of audience who has a past experience in the field or a small knowledge.

Introductory chapters are provided for giving insights on PA from all of the addressed perspectives. Essential information is provided on PA history, world interest, scope, enabling technologies, advantages and disadvantages, threats and challenges, economics, policy inline with UN SDG, adoption in developing countries, and adoption in the horticultural sector.

The second part of the book talks about the usage of the engineering technologies such as: electronics, information technologies, communication networks, and artificial intelligence in PA systems to collect field data, convey it, record, process and analyze the collected data, extract decisions and take automatic or manual actions. The topics covered by this part include an updated review of the use of the Internet of Things (IoT) in the PA systems showing and approaching its success relation with a concise bibliographic background categorizing and synthesizing the potential of both technologies, overview of the existing technologies of soil sensors for monitoring soil fertility/quality and its integration in smart

Automated Decision-Making Systems, and overview of microwave technologies in agricultural and food processing applications.

The third part of the book considers specific applications and proposed solutions which are, a platform allows to collect fundamental physical phenomena which will be treated to estimate the need for water towards optimal irrigation, an intelligent model to monitor, validate and report the automated irrigation systems efficiency to correct the irrigation practices managements, a demonstration of how machine learning techniques can help in monitoring crop quality and disease identification, a framework of precision management practices for legal cultivation of cannabis by putting a special emphasis on water management, nutrient management, and disease control particularly under greenhouse situations to meet fibrous and medicinal needs of the humankind, a research addresses the intelligent adoption of ICT access facilitation means to achieve the agricultural sector development and that the study of the socio-cultural reality of technology users is a way to achieve this, proposes a conceptual and operational framework for information related to the maize production cycle and obtaining a data set comes from an effort to combine complete and consistent data that will systematize and thereby facilitate the decision-making process of corn farmers in Mexico, and finally enhancing a model to predict power requirements for operating chisel plow.

The fourth part of the book is about agricultural systems analysis streamlining, modeling, and improvements, explaining the sub-disciplines of agronomy towards Precision Agronomics. This part is further divided into two sub-parts, the first one takes the agriculture development from the economic and policy perspective where it presents a model for predicting agricultural exports value, and an analysis on the role of institutions, sources of funding/assistance, information sources in food security and extension policy; the second sub-part takes the technological perspective especially with respect to plant biotechnology subject and its role to secure food and to alleviate the risk of loss of the genetic variability of cultivated plants as a result of environmental changes and human practices.

Sherine Mohammed Abd El-Kader
Electronics Research Institute, Cairo, Egypt

Basma Mamdouh Mohammad El-Basioni
Electronics Research Institute, Cairo, Egypt

Acknowledgment

The editors would like to acknowledge the help of all the people involved in this project and, more specifically, to the authors and reviewers that took part in the review process. Without their support, this book would not have become a reality.

First, the editors would like to thank each one of the authors for their contributions. Our sincere gratitude goes to the chapter's authors who contributed their time and expertise to this book.

Second, the editors wish to acknowledge the valuable contributions of the reviewers regarding the improvement of quality, coherence, and content presentation of chapters.

Sherine Mohammed Abd El-Kader
Electronics Research Institute, Cairo, Egypt

Basma M. Mohammad El-Basioni
Electronics Research Institute, Cairo, Egypt

Section 1

Introduction

This section gives insights on PA from all of the addressed perspectives. Essential information is provided on PA history, world interest, scope, enabling technologies, advantages and disadvantages, threats and challenges, economics, policy in line with UN SDG, adoption in developing countries, and adoption in the horticultural sector.

Chapter 1
Introduction to Precision Agriculture:
Overview, Concepts, World Interest, Policy, and Economics

Akalpita Tendulkar
Malaysia University of Science and Technology, Malaysia

Abstract

The global population is increasing at a tremendous speed; thus, the demand for safe and secure food to meet this population is in demand. Therefore, traditional farming methods are insufficient to meet this demand; thus, the next revolution in agriculture is required, which is Precision Agriculture (PA), the Fourth Agriculture Revolution. PA is a technology where the concept of farm management is based on observation, measuring, and responding to inter- and intra-field variability in crops. The technologies used for performing precision agriculture are mapping, global positioning system (GPS), yield monitoring and mapping, grid soil sampling application, variable-rate fertilizer application, remote sensing, geographic information systems (GIS), quantifying on farm variability, soil variation, variability of soil water content, time and space scales, robots, drones, satellite imagery, the Internet of things, smartphone, and machine learning. Hence, the current chapter will be emphasizing the overview, concepts, history, world interest, benefits, disadvantages, and precision farming needs.

CHAPTER OVERVIEW AND CONCEPT

With the growing population and lifestyle changes the demand for food is on the increase. People are now more inclined in eating healthy organic food, all for the reason for green life. However, arable acreage cannot keep pace with this ever-growing demand as the area for growing crop is decreasing due to climatic and human activity. Hence, the impending food security threat could easily devolve into regional or even global instability. Similarly, there is a growing need to balance the ratio of the cultivable land and consumption rate.

DOI: 10.4018/978-1-7998-5000-7.ch001

The traditional farming methods relies on managing a large field, while planting, harvesting, irrigation and applying pesticides and fertilizers majorly based on the regional conditions and historical data. Therefore, just depending on traditional farming method is not faceable to meet the escalating need. As a result, there is a need to manage the agriculture activity to increase the crop yield by maximizing land use, since the area for arable land is limited and traditional farming method has left certain previously fertile land now barren. Therefore, precision agriculture can be used to overcome the drawbacks faced by traditional agriculture, namely by, maximize land usage, management of time, reduction in the usage of water and chemicals which will benefit in the healthier and higher crop yield. Such farming method will not only benefit the farmers but also the environment by preserving resources.

Precision agriculture in general is an approach to manage farm using information technology (IT) for ensuring optimal health and productivity of the crop. Precision Agriculture also known as PA in acronym, had a major breakthrough in farming due to the investment by AL Myers on the on-the-go crop yield monitor.

PA is also known as satellite farming or site-specific crop management (SSCM), wherein the concept of farm management is based on observation, measuring and responding to inter and intra-field variability in crops. Phytogeomorphology, is an approach in precision agriculture which attributes to topological terrain features to tie multi-year crop growth stability and/or characteristics. An interest in such an approach arises from the fact that the geomorphological component typically dictates hydrology in the farm sector.

Agriculture sector has witnessed transformation from the time human started planting, wherein an increase in mechanized agriculture was observed from 1900 to1930 which considered as the first revolution in agriculture, leading to feeding about 26 people. Followed by Green revolution in the 1960s which led to feeding approximately 155 people using gene modification techniques in plants. Finally, precision agriculture came in as the next revolution in the farming industry, wherein the farmer is expected to feed about 265 people on the same acreage. The first initiation of precision agriculture was using satellite and aerial imagery, weather prediction, variable rate fertilizer application, and indicator of plant health. Next advancement in PA saw the use of machine (data mining) to even more accurately predict precise planting of crops, topographical mapping, and soil data.

In contrast to traditional farming methods, precision farming incorporates sensors, robots, GPS, mapping instruments and data analysis software to adapt based on the plant's requirement without an increase in labour. Information regarding a plant stem size, shape of leaf and soil humidity around a plant is send to a computer via a stationary or robot-mounted sensors and camera-equipped drones in the form of images and data on individual plants for seeking signs of health and stress. These information on individual plants is accessible to farmers in real time as a feedback, so that they can decide on the distribution of water, pesticide or fertilizer in a calibrated dosage on the required region. Thus, also help in estimating the time of planting and harvesting.

The use of unmanned aerial vehicles in precision agriculture such as the DJI Phantom, which are comparatively economical and can be operated by novice pilots. These drones are equipped with hyperspectral or RGB cameras that capture numerous field images which are managed by using ortho-photos and NDVI maps using photogrammetric techniques.

The GPS system in precision agriculture enables farmers in planning and mapping the field, soil sampling, crop scouting and mapping of yields. Conditions such as fog, dust, rain and darkness which are visibly low and does not hinder the farming process. Satellite imaging is also applied in precision-based agriculture, where satellite images are used to characterize a field data in detail. This information in combination with GIS data is often used for more intensive and efficient practice of farming.

The method for precision agriculture is done in four stages to observe spatial variability, where first is data collection, variables, strategies and final stage implementing practices. Data collection is accumulating information by geo-locating field for analysing soil and residual nitrogen, and information on previous crops and soil resistivity. These geolocations are done in two ways, wherein an in-vehicle GPS receiver is used in a tractor and second by using aerial or satellite imagery. Second stage, Variables, is on the intra- and inter- field variability which is based on numerous factors such as climate, soils, cropping practices, weeds and diseases. Third stage is 'Strategies', it uses soil maps, wherein the farmers are bale to pursue two strategies (predictive approach and control approach) for the adjustment of field inputs. The control approach in strategies, is based on the information from static indicator such as sampling, remote sensing, proxy-detection and aerial or satellite remote sensing. Fourth stage is 'Implementing practice', wherein crop management at the field level by using the new ICTs makes it more efficient and easier.

The aim of precision agriculture is to optimize field level management regarding crop science, environmental protection, and economics. The world now is already in the fourth industrial revolution wherein ICT has taken centre stage, thus the world interest in incorporating the information technology to increase per capita of yield has brought a new revolution in agriculture in the form of precision agriculture.

Besides, precision agriculture also has other technological breakthrough wherein robots, the Internet of Things (IoT), smartphone application and machine learning are some of the technologies being used.

Precision agriculture in various way have benefited and affected the way we do agriculture today. The benefits of precision agriculture by using the new technologies has provided us with a wealth of information to build up the record of the farm, improve decision-making, adopt better traceability, improve market value of farm products, increase lease arrangements and relationship with landholders, and enrich quality of farm produce. The effects of precious agriculture as it has high capital cost and requires expert advice before actual implementation since the technology is still in its infant stage. Lack of awareness and knowledge on the technology also makes it difficult for farmers to adopt this farming method.

Therefore, countries can incorporate certain policies related to sustainable farming, wherein precision agriculture technologies can be applied. United nation Sustainable Development Goal Plans 2019 to transform the world, which has about 17 goals wherein goal 2 focuses on 'Zero Hunger' and food security. To achieve this goal one of the ways is have efficient farming method, therefore the use of precision agriculture in farming helps in achieving this goal. Thus, countries can adopt PA in their policies inline to achieve the UN SDG goal number two. Such policy making will have positive impact on the economy and environment, as it will reduce the pressure on the environment by increasing the efficiency of machinery and putting it in use. Thereby, saving cost and decreasing harmful chemical runoffs into the aquatic environment.

HISTORY OF PRECISION AGRICULTURE

Agriculture sector has seen three revolution from the start of the industry. The first agriculture revolution (Neolithic Revolution) stared approximately 10,000 BC, wherein we saw the prehistoric transition from hunting and gathering to a settled agriculture. The second agriculture revolution saw the Arabic agriculture revolution in the 8th- 13th century, followed by the British and Scottish in 17th- 19th century.

The third agriculture revolution also known as the Green Revolution started in early 1930- 1960s which observed an increase in the agriculture production, especially in the developing countries

Green revolution refers to a series of research, development and technology transfer for an initiative to increase the production of high-yielding cereal grains, expansion of irrigation infrastructure, and distribution of hybridized seeds, synthetic fertilizers, and pesticides to farmers (Ameen and Raza, 2017). Green revolution as a term for this industrial expansion was first coined by former USAID director William Gaud. The goal of starting the green revolution is to meet the growing populations need of the developing countries, example can be observed in India wherein the annual wheat production raised from 10 million tons in the 1960s to 73 million in 2006 (BBC News, 2006 and Ameen and Raza, 2017). Green revolution gave an increase in the crop yield, but had some disadvantages as it was expensive for the farmers to have the hybrid variety of seeds which consumed more fertilizers (chemical fertilizers) and water, the start of using new machineries reduced the manual labor leading to un-employment and rural- urban migration, making people to work at lower wages. It was also being said that this type of farming practice also greatly affected the environment as it caused loss of biodiversity and pollution. Green Revolution was criticized for the decline in the agriculture quality, concerns about social implications and sustainability.

Amidst this, the demand for a better technology to meet the need of the growing population has become a necessity. Thus, in early 1990's the use of Geographic Information System (GIS) and Global Positioning System (GPS) were used in Agriculture Technology giving the third way of agriculture revolution which is known as 'Precision Agriculture or Precision farming'. The idea was initiated by John Deere, wherein GPS location data from satellites was introduced. Here the GPS system was installed in the farmer's tractor for automatically steering the equipment based on the co-ordination of a field. Thus, reducing the steering error caused by the drivers. Therefore, resulting in less wastage of seed, fertilizer, fuel, and time.

Mulla, D., & Khosla, R. (2016), gave an in-depth explanation on the evolution and advances in precision agriculture. They explain the evolution and advances in the various farming practices namely, soil sampling, geo-statistics and Geographic Information Systems (GIS), farming by soil, variable rate fertilizer, site-specific farming, management zones, Global Positioning System (GPS), yield mapping, variable rate herbicides, variable rate irrigation, remote sensing, automatic tractor navigation and robotics, proximal sensing of soils and crops, and profitability and adoption of precision farming.

Furthermore, Mulla, D., & Khosla, R. (2016), in the chapter 'Historical Evolution and Recent Advances in Precision Farming' explains on the early advances in the soil sampling, Geo-statistics and GIS in precision farming (table 1.1 is the tabular for the same chapter). They mentioned that in soil sampling, spot applications of fertilizers were encouraged by Linsley and Bauer in early 1920s, due to the combination of cheap labor and fertilizer with increase in farm rea caused the shift in famers' behavior in the user of uniform application while waiting for the revolution in precision farming which took place during the 1980s. Later scientists started recognizing that the sparse soil sampling was a poor representation of average fertilizer requirement as the variability was large in soil fertility. Thus, in 1961 Melsted and Peck designed an intensive grid sampling study with a spacing of 24.3 m, this design was in view of minimizing the cost of determining an average soil fertility (Franzen 2007). Customized fertilizers were recommended by Melsted to be more efficient than a single uniform one, as there was variability in the value of the tested soil (Melsted 1967). Until 1994 the intensive grid sampling method was practiced at a routine time interval in same field. Nevertheless, until several decades later, there was little practical application of this concept. For further reading about this topic, several excellent summaries of soil sampling techniques are given in the literature by Wollenhaupt et al. 1997; Mulla and McBratney 2000.

Furthermore, Pierre Robert actively promoted the idea of precision farming and organizing the first workshop called "Soil Specific Crop Management" during the early 1990s. Thus, often regarded him as the father of precision farming. He used computerized farming method wherein soil database were used. Later, in early 1990s variable rate fertilizer applicators using a control system was developed. This idea was studied by several scientists during the early to mid-1980s, including John Hummel (Hummel 1985) working with the United States Department of Agriculture–Agricultural Research Service (USDA-ARS). This system was first tested in using variable rate lime by Luellen 1985 or fertilizer application by Schmitt et al., 1986 in Minnesota, USA. In the beginning of 1986, site-specific farming was studied in irrigated agriculture where high-cash crops example potatoes were grown (Mulla and Hammond 1988; Hammond et al. 1988, Hammond and Mulla 1989). The profit generated from these crops could justify the intensive grid soil sampling which were used to define the management zones for variable P and K fertilizer applications. Until present day, the use of management zones in precision farming has persisted but the concept and definition has been altered. Site-specific and precision farming initially in term was introduced by John Schueller from the University of Florida in the scientific literature (Schueller 1991, 1992). Gradually, farming by soil and site-specific were replaced by variable rate technology which proved to be both profitable and beneficially because of its efficiency in farm inputs, maintenance i.e. improved crop yield and quality, and protected water quality (Sawyer 1994).

In precision agriculture GPS was of primary interest due to its method of identifying the location on a real-time data on spatial variability in crop grain yield (Auernhammer and Muhr 1991; Schueller and Wang 1994; Auernhammer et al. 1994). Before the availability of GPS, the concept of detail spatial mapping of the crop yield was pioneered by Schueller and Bae in 1987. Here they used variations in motor pace as a surrogate for grain flow to the combine, which was once worked beneath steady throttle and cutting head height. Thus, reducing the error s in measuring grain yield. Further in precision farming could see various area of research Variable rate herbicide and irrigation, Aerial remote Sensing and Proximal sensing of soil and crops wherein technological development and testing were done.

The table 1.1 gives a generalized idea on the research area and nature of contribution from each of the key references that has been thorough been reviewed and tabulated by Mulla and Khosla. For further reading about this topic can refer to the chapter Mulla, D., & Khosla, R., 2016.

WORLD INTEREST TOWARDS PRECISION AGRICULTURE

One of the top ten (10) modern agriculture innovation is Precision farming (Crookston 2006). In the early 1980s, the concept of precision agriculture was first developed in the United States, wherein researchers at the University of Minnesota used varied data inputs in crop fields. At the time grid sampling method i.e. applying fixed grid of one sample per hectare was performed. The techniques for the first input on the recommendation maps for fertilizers and pH correction was observed towards the end of 1980s. Ever since the use of yield sensors developed from new technologies in combination with the GPS receivers' precision farming gained ground in agriculture technology. Thus, covering several million hectares of arced land today.

Table 1. Early Advances in Research on Soil Sampling, Geo-statistics, and GIS Remote Sensing, Proximal Sensing of Soils, and Proximal Sensing of Crops for Precision Farming in Precision Farming (Source: Mulla, D., & Khosla, R., 2016)

Research Area	Nature of Contribution	Key References
Soil sampling	Grid sampling recommendation	Melsted (1967), Dow et al. (1973b), Mulla and Hammond (1988), Wollenhaupt et al. (1994)
Geostatistics and GIS	Map interpolation and reclassification for soil fertility data	Mulla and Hammond (1988), Mulla (1989, 1991, 1993)
Farming by soil	Proposed concept	Robert (1982), Rust (1985)
Variable rate fertilizer	Machinery development and field testing	Hummel (1985), Luellen (1985), Ortlip (1986), Schmitt et al. (1986), Borgelt et al. (1989, 1994), Carr et al. (1991), Mulla et al. (1992)
Site-Specific farming and Management zones	Proposed and tested concept	Mulla (1991, 1993), Mulla et al. (1992), Doerge (1999), Fraisse et al. (2001)
Precision farming	Proposed concept	Schueller (1991, 1992)
GPS	Technology adaptation to farming and testing	Larsen et al. (1988), Auernhammer and Muhr (1991), Tyler (1993)
Machinery navigation and autosteer	Technology development and testing	Reid and Searcy (1987), Palmer (1991), O'Connor et al. (1996), Greatline and Greatline (1999), Keller et al. (2001)
Yield mapping	Technology development and field testing	Bae et al. (1987), Schueller and Bae (1987), Searcy et al. (1989a), Karlen et al. (1990), Stafford et al. (1991)
Variable rate herbicide	Technology development and testing	Haggar et al. (1983), Guyer et al. (1986), Beck and Kinter (1988a,b), Thompson et al. (1990, 1991), Felton and McCloy (1992), Stafford et al. (1993)
Variable rate irrigation	Technology development and testing	McCann and Stark (1993), Fraisse (1994), Evans et al. (1996)
Aerial remote Sensing	Ability to identify spatial patterns, identification of sensitive wavelengths	Robert (1982), Zheng and Schreier (1988), Bhatti et al. (1991), Blackmer et al. (1995), Barnes et al. (1996), Bausch and Duke (1996), Blackmer and Schepers (1996b), Haboudane et al. (2002), Thenkabail (2003)
Proximal sensing of soil	Development and testing of technology	Rhoades and Corwin (1981), Shonk and Gaultney (1988), Colburn (1991), Gaultney et al. (1991), Sudduth et al. (1991), Lesch et al. (1992), Carter et al. (1993), Doolittle et al. (1994), Christy and Lund (1998), Adamchuk et al. (1999)
Proximal sensing of crops	Development and testing of technology	Leamer et al. (1973), Daughtry et al. (1980), Walburg et al. (1982), Chappelle et al. (1992), Solie et al. (1996), Raun et al. (2002)

Precision agriculture is associated with sustainable agriculture but in the American Midwest (US) the farmers are also trying to maximize their income with the aid of spending cash solely in areas that require fertilizer. Hence, such practice allows the farmers to diverse the rate of fertilizer across the field according to the need recognized by way of GPS guided Grid or Zone

Fertilizers thus can be applied to those areas that requires it than that do not, thereby optimizing its use.

Precision agriculture has gained tremendous recognition around the globe, countries such as United states, Canada and Australia are forerunners in this technology. In Europe, the United Kingdom was first to practice precision farming, which was followed closely by France. In France, it started in 1997- 1998.

The adaptation of precision agriculture in Latin America wherein the leading country is Argentina. In Latin America, it began in the middle 1990s with the support of the National Agriculture Technology Institute. Later Brazil established a state-owned enterprise, Embrapa for the research and development of sustainable agriculture in the country. To steady the management practice of precision farming development of GPS and variable-rate spreading techniques very beneficial (Simon Blackmore, 2016). In France, recently less than 10% of the farmers are equipped with variable-rate system. The acceptance of GPS in precision agriculture is more widespread, but this has not stopped the farmers from using other services related to precision agriculture, which supplies field-level recommendation maps (Dwivedi et al., 2017).

The world started looking for an alternate and more sustainable mode of agriculture, since one third of the global population still relies on agriculture produce for their living. However, the use of more advanced technologies in precision farming requires a substantial upfront investment, farmers in developing countries are benefitting from mobile technology. Such service contributes the farmers in mobile payments and receipts to improve productivities. For example, as reported in the 2016 article of Technoserve, around 30,000 framers from Tanzania were using mobile phones for contracts, payments, loans and business organization.

The use of technology has benefited the economic and environment of a country which can also been confirmed in China. The Chinese system is characterized by small-scale family-run farms, this makes it difficult for the country to adopt precision agriculture compared to other western countries like in Europe and United States where it is quite far ahead. Hence, China is attempting to introduce precision agriculture technological to its farmers and reduce some risks, which will eventually pave the way for China's technology to boost precision agriculture in the future (Kendall, et al., 2017).

The global market on precision agriculture has projected a growth from $ 4.84 billion in 2018 to $ 10.16 billion by 2024, presenting a Compound Annual Growth Rate (CAGR) of more than 13 per cent over the forecast period, owing to several government initiatives to implement modern farming practises. Through the implementation of advance agricultural technologies, the immense demand for the food supply for the rising population can be accomplished. In addition, over the forecast period it is expected that the increasing weather conditions and the need to increase crop yields and increase farm productivity will contribute to the global market for precision agriculture. The market for precision agriculture can be segmented according to part, technology and application. The market can be separated into hardware and software primarily based on the component. During the forecast period, the hardware segment dominated the market in 2018 and is predicted to continue its market dominance as farmers undertake automation and control devices to enhance productivity. The market can be divided into yield monitoring, field mapping, crop scouting, weather tracking & forecasting, among others in terms of application.

Precision agriculture is gaining recognition and importance in recent year and is expanding to other parts of the world like Asia-Pacific, North America, Europe, South America and the Middle East & Africa. By 2018, the global precision market is dominated by North America, backed by the presence of leading market players and adoption of advanced technologies by countries across the region. Owing to the rising population in developing countries like India and China, it is predicted that the Asia- Pacific region will gain momentum in the coming year as the demand for food will increase.

Currently, in the global precision agriculture market there are few major industrial players namely, Bayer CropScience AG, Trimble Inc., Deere & Company, Iteris, Inc., AgJunction Inc., Farmers Edge, Inc., Ag Leader Technology, CropMetrics, AGCO Corporation, Teejet Technologies, etc. In order to stay relevant in the market these major companies are developing advanced technologies, launching new products and having strategies involving mergers and acquisitions.

With the advancement of technologies in precision agriculture, it has shown uneven on the adoption in both geographically and temporally. The economic principle of innovation forecasts the development and adoption of new technologies in agriculture industry where they make extra environmentally friendly use of the scarcest productive resources. With the adoption of precision agriculture, it can be observed that the labour cost is expensive while the land and capital are relatively less expensive. The inconsistent adoption rate of precision farming is due to the normal cycles in which it replaces the expensive machinery in which many precision farming innovations are represented, in an area where precision farming is practised. The decisions to upgrade equipment are influenced by many factors that are exogenous to the farm, such as bank interest rates and commodity prices. Adoption in countries wherein there is scarcity of labour but land-abundant is likely to continue, with adoption rates rising when commodity prices are high and interest rates low (Swinton, &Lowenberg-Deboer, 1998).

The uneven adaptation of PA technology can be examined across time or space. Given the rapid growth of global trade and the widespread availability of VRA equipment and yield tracking, adoption rates seem to vary markedly from country to country, at least on the basis of the available informal data (Norton and Swinton, 2001). Compared to Brazil or France Argentina has stared adopting yield monitoring in a rapid pace. Even though there is a rapid adoption of yield monitoring technology, Argentina has shown little use of site-specific fertilizer (Lowenberg-DeBoer, 1999). In south-east Asia, Malaysia, which has been known for its rubber plantation, has adopted site-specific fertilization method, but not to its rice fields.

Also, it is observed that even nations such as the United States PA adoption rates vary by a factor of ten i.e., from 11.3% of farms in the Midwest "Heartland" to only 1.1% in the Southeast Seaboard in 1998 (Daberkow and McBride, 2000). Overall, we note that yield monitoring or VRA fertilization has only surpassed 5 percent in the United States and Canada in preferential areas. It would seem that acceptance rates in Australia, Brazil, Denmark, United Kingdom, and Germany could be within the range of 1-5 per cent (again, only for preferred sub-regions). The implementation of PA technologies is virtually unknown in Africa and Asia, with the exception of a few yield monitors in South Africa and some VRA fertilization in isolated plantation agricultural enclaves. These uneven adoption patterns may seem puzzling given the potential benefits of precision agriculture for farm profitability (and environmental protection) (Swinton and Lowenberg- DeBoer, 1998).

TECHNOLOGIES USED IN PRECISION AGRICULTURE

I. Mapping

The very first and most important step in precision agriculture is mapping of crops and soil properties. Such maps can quantify variability in space and provide the basis for monitoring spatial variability. In field operations mapping can be performed by RS, GIS and manually.

II. Global Positioning System (GPS) Receivers

In Precision farming, it is import to accurately determine of location, principally for mapping the variability in soil fertility or crop yield, and in locating farm machinery that can disperse variable rates of fertilizer relative to the data in these maps. Thus, allowing Global Positioning System satellites broadcast to send signals that permit GPS receivers to compute their location. This information is provided in real time. Having precise location information at any time allows soil and crop measurements to be mapped.

Originally, the primary focus in GPS for precision farming has been to identify the location of a combination that collects real-time information on spatial variability in crop yield (Auernhammer and Muhr 1991; Schueller and Wang 1994; Auernhammer et al. 1994). Later, interest in GPS turned to its use in agricultural navigation and autosteering (Zhang et al. 1999; Keller et al. 2001).

III. Yield Monitoring and Mapping

Yield monitoring is a part of precision farming that helps and offers adequate information for farmers to make intelligent decisions regarding their fields. Grain yield sensors continuously calculate and record the flow of grain in a combine's clean-grain elevator in highly mechanized systems. Yield monitoring is a relatively new concept, enabling agriculture equipment such as combining harvesters or tractors to gather vast amounts of data, such as grain yield, levels of moisture, soil properties, and more. When connected to a GPS receiver, yield monitors could provide information necessary to map yields. Applied appropriately, yield knowledge provides significant insight on the effects of controlled inputs such as modifications to fertilizers, plants, pesticides and cultural practices like tillage and irrigation. Yield monitoring operates in three rather easy steps: the grain is harvested and loaded into the grain elevator which has sensors reading the grain's moisture content. Following the step as the grain is transported to the storage tank, more sensors track the yield of grain. Since both of these sensors operate, the information is sent to the driver's cab and is displayed on the screen, the information is geo-referenced so that it can be mapped as well as studied in detail at a later time or date. Farmers benefit from the use of yield monitoring technology, even though one of the main advantages is that it enables to provide the farmer with accurate and often geo-referenced data on their field. Farmers can learn crop yield and crop-related relevant information to ameliorate potential threats or enhance potential opportunities. Other benefits of a performance monitoring system include the potential of a farmer to export information to a computer, allowing this information to be available in a variety of formats, including displays on the farm, at home or in print. As well, in the home or office, a farmer can use advanced software to evaluate and understand better the recorded information. Yield monitors can control grain by field or load on a load basis, giving the farmer an enormous degree of flexibility and providing with instant information on the load they have collected.

IV. Grid Soil Sampling and Variable-rate Fertilizer (VRT) Application

Soil and grid soil sampling utilizes the same working principle wherein it helps in increasing the sampling intensity. This can be better understood with an example wherein assuming that the sampling area is of around 20-acre and have 10 samples using a grid sampling system in an area of 2-acre (hence each sample will be spaced 300 feet apart) when compared to traditional method using one sample. The aim of grid soil sampling is to produce a graph of nutritional necessity.

V. Remote Sensing

The application of remote sensing in precision agriculture are mostly based on the reflectance of the visible light coming from the sun and the near-infrared light by soils or crops. Usually remote sensing data is executed by the use of cameras installed on satellites, airplanes, towers, or unmanned aerial vehicles instead of the contact between the sensor and the soil or crop.

Earliest known application of remote sensing in farming was focused primarily on estimating the crop yield (Pinter et al., 1981; Wiegand et al., 1991). Remote sensing is a method where field data are collected and these data can be processed using a computer. The data sensors used for collecting the information can simply be portable devices, mounted on aircraft or satellite based. These devices as a tool are used for providing data on crop health, example various physiological stresses to plant like moisture, nutrients, compaction, crop diseases and other plant health issues can be easily detected in an overhead image. Electronic cameras can also capture photographs similar to the infrared, which are highly correlated with healthy plant tissue. Remote sensing can expose variation in weather that affects crop yield and may help to have enough time to make management decisions that improve the current crop's profitability.

VI. Geographic Information Systems (GIS)

Geographic information systems (GIS) are computer hardware and software used to generate maps using function attributes and location data. A significant function of an agricultural GIS is to maintain many levels of information, such as yields, maps of soil surveys, remotely sensed data, reports of crop scouting and soil nutrient levels.

VII. Robots

Over the past three decades, accurate automatic navigation has been one of the most concentrated areas of research and implementation. The use of self- steering tractors is one of the example of this automate technology. These tractors were one of the equipment from the John Deere works, wherein it does all the farm warm, only when there are any technical emergencies that the farmers have to step-in. Technology is evolving towards GPS-programmed driverless machinery for spreading fertilizer or plowing land. The benefits of this practice include decreased operator strain, reduction of overlaps and skips of equipment, and increased fuel use performance and drug application. Recent emerging innovation include the solar-powered equipment's, which detects weeds and destroys them precisely with a dose of herbicide or laser. There are already agricultural robots, also known as AgBots, but sophisticated harvesting robots are being designed to classify ripe fruits, change their shape and size, and carefully pluck them out of branches.

VIII. Drones and Satellite Imagery

The poster-child of smart agriculture technology is 'The Unmanned Aerial Vehicle (UAV). Developments in drone and satellite technology are useful for precision farming because drones take images of high quality, while satellites capture the bigger image. Civil Aviation Authority approved the first unmanned crop-spraying drone in 2015 under new rules, but the potential is clear for a multitude of tasks from farm management to real-time animal tracking on remote paddocks. Pilots of light aircraft can combine aerial photography with data from satellite records to predict future yields based on current field biomass rates. Aggregated images may generate contour maps to track where water flows, assess variable seeding levels and generate yield maps of more or less productive areas. A Chinese-developed drone technology called Agra MG-1 "octocopter" is used to spray pesticides or fertilizers on large areas of farmland. The DJI manufacturer of this drone claims that in 10 minutes it can spray 4,000-6,000m2. Haptly, a New Zealand based drone company is working on technology to gather information about pasture quality and to send it to a farmer's computer via the cloud. Whilst, GPS-ithas based in Tauranga built drones that map farms and orchards, and record crop densities from above.

IX. The Internet of Things

The Internet of Things is the network of physical objects equipped with electronics for the processing and storage of data. With the development of sensors and farm-management software IoT comes in the picture (Sophocleous, M., 2015). As an example, farmers may spectroscopically measure nitrogen, phosphorus, and potassium, which is notoriously incoherent. Now the farmers can even check the fields see where cows have urinated already and add fertilizer to only those places that need it. Thus, the fertilizer utilization cost by up to 30%. To determine the perfect time to water plants can now be determined using s moisture sensor in the soil (Sophocleous M., 2015). It is possible to configure the irrigation systems to adjust which side of the tree trunk they water based on the needs and rainfall of the plant. The use of IoT in precision farming is not just constrain to its use in plant but can also be used in animal husbandry. In animal farming, the livestock can be equipped with internal sensors for tracking stomach acidity and digestive problems. Remote sensors monitor movement patterns to assess the health and fitness of the cow, to sense physical injuries, and to establish optimum breeding times. To track trends and patterns, all of this data from sensors can be aggregated and evaluated.

The monitoring technology can be used as another example to make beekeeping more effective. Honeybees are of substantial economic importance and through the pollination of a variety of crops provide a vital service to agriculture. Tracking the health of a honeybee colony through wireless temperature, humidity, and CO2 sensors help to improve bee productivity and read early warnings in data that could threaten the very existence of a whole hive (IoT ONE).

X. Smartphone Applications

Apps on smartphones and tablets are becoming popular in recent years in precision farming. Smartphones come with numerous practical applications, including the camera, microphone, GPS, and accelerometer. Requirements are also made for various agricultural requirements such as field mapping, animal monitoring, weather and crop information gathering, etc. These are easily portable, inexpensive, and have strong computing (Pongnumkul, et al., 2015).

XI. Machine Learning

Machine learning is widely used in conjunction with computers for things such as drones, robots and the internet. This allows data from each of those sources to be submitted. The system then processes this information and returns those devices with the appropriate actions. It helps robots to deliver the perfect amount of fertilizer or to supply IoT devices directly to the soil with the perfect amount of water (Zadra, et al., 2012). Every year, the future of agriculture shifts more towards an architecture of machine learning.

ECONOMIC AND ENVIRONMENTAL IMPACT

Precision agriculture, as the name implies, means the application of the precise and correct amount of inputs like water, fertilizer, pesticides, etc. at the appropriate time to the plant for expanding its profitability and maximizing its yields. Precision management practices in agriculture will drastically reduce the number of nutrients and certain other crop resources used while boosting yields (Ferrell, 2016). Therefore, farmers get a profit for their investment by saving their expenses on water, pesticides, and fertilizers. The second, wider-scale benefit of feedback targeting relates to environmental effects. Implementing the correct quantity of pesticides in the right location and at the right moment supports crops, soils and groundwater, and thus the whole crop round.

The second, wider-scale benefit of feedback targeting relates to environmental effects. Implementing the correct quantity of pesticides in the right location and at the right moment supports crops, soils and groundwater, and thus the whole crop cycle. Precision agriculture has, therefore, become a keystone of sustainable agriculture, because it understands crops, soils, as well as farmers. Sustainable agriculture aims at ensuring a constant food supply inside the economic and environmental, political and social limits necessary to maintain the crop for a long-term production. An article by Rajvanshi, in 2015 attempted to demonstrate that precision farming can help farmers in developing countries such as India.

Precision agriculture decreases the environmental pressures on cultivation by growing the equipment performance and bringing it into practice. Using remote management systems such as GPS, for example, decreases emissions for agriculture, whereas variable rate application of nutrients or chemicals can vastly decrease the application of these inputs, thus saving costs and lowering hazardous runoff into the waterways (Brevik, et al., 2006).

Precision farming enables variability in the rate of fertilizer, manure, and pesticides delivered to better match geographic trends in soil fertility and pesticide adsorption, and to respond to evolving temporal patterns in crop nutrient stress and weed, insect and disease infestations. However, precision farming decreases the usage of chemical products due to the similarity among successive passes of chemical applicator machinery with auto-steering technologies. These factors contribute, conceptually, to better quality and sustainability of the environment (Larson et al. 1997; Bongiovanni and Lowenberg-DeBoer 2004).

POLICY AND SDG FOOD SECURITY

I. The policy framework for the promotion of farm-level precision agriculture:

Precision agriculture is recently taking a serious recognition, thus countries adopting this new technology should consider a policy framework that can better understand and help in implementation. Following approach was suggested by Manglia, 2010:

- Recognize the specialized areas needed to promote crop-specific precision farming.
- Development of interdisciplinary teams incorporating agricultural scientists in different fields, engineers, manufacturers, and economists, to research the overall scope of precision farming.
- Pursue the progressive farmers who have ample risk-bearing ability for precision farming technology.

- Encourage farmers to use primary field-level data to study the spatial and temporal variability of the input parameters.
- Provide the farmers with complete technical backup support to develop pilots or models that can be reproduced on a large scale.
- The pilot study should be carried out in the field of farmers to show the results of the introduction of precision farming.
- Promote farmers to implement farm-level water budgeting procedures, and then use micro-level drainage systems and water-saving strategies.
- Government regulations banning farmers using indiscriminate farm inputs and thus creating ecological/environmental imbalances will encourage the farmer to go for alternative approaches.
- Having farmers aware of the risks of implementing imbalanced doses of agricultural inputs such as irrigation, fertilizers, insecticides, and pesticides.
- Policy support on procurement cost, efficient technology transfer to farmers, formulation of cooperative groups or self-help groups, because many of the precision farming equipment are expensive (GIS, GPS, RS, etc.).

II. SDG and precision agriculture

United Nation Sustainable Development Goal Plans 2019 to transform the world, with about 17 goals centred on' Zero Hunger' and food security in goal 2. One approach to achieve this aim is to have an efficient method of farming, hence the use of precision farming in agriculture can helps to achieve this goal. The goal 2 in SDG talks about ending hunger for all people regardless of their physics state, race and religion. It also focuses on achieving food security and improve the nutrition impute and promote a sustainable way of agriculture. It has about 7 targets in this goal, wherein first is to end hunger and ensure nutritious, safe and sufficient food access to all people the whole year, particularly for poor and people in vulnerable situations, including infants. Second, to end all form of malnutrition. Third, to increase the agriculture productivity according to demand and surge income of small-scale food producers. Fourth, to ensure sustainable food production systems and introduce robust farming practices that improve productivity and efficiency, help maintain habitats, enhance adaptability to climate change, extreme weather, drought, flooding and other disasters, and gradually improve the quality of land and soil. Fifth, preserve the genetic variety of seeds, cultivated plants and farmed and domesticated animals and their associated wild species. Sixth, increase investment, incorporating thru superior worldwide cooperation. Last, the seventh target talks about taking steps to ensure the proper functioning of food commodity markets and their derivatives and to encourage timely access to market information, including food reserves. The United Nation (UN) is expecting to achieve all this goals by the year 2030.

ADVANTAGES, DISADVANTAGES AND NEED OF PRECISION AGRICULTURE

I. Advantages of precision farming

The principle of "doing the right thing in the right place at the right time" has a strong instinctive allure which provides farmers the capacity to make more effective use of all operations and crop inputs. The efficient use of inputs leads to higher crop yield and/or quality without environmental pollution. Precision farming can tackle environmental and economic problems concerning today's agricultural production.

The objective of incorporating precision agriculture is to improve the crop yield, while reducing potential environmental threats, thus by doing so can have following benefits:

- It helps in monitoring the soil and crops physicochemical parameters
- By the use of various techniques introduced in precision farming we can obtain a real-time data
- It helps the famers to device better judgment on farm management
- It helps to save time and cost; as it reduces the unwanted or over used of fertilizers and pesticides.
- Offer good agricultural records necessary for sale and succession.

II. Drawbacks of precision farming

Even though precision agriculture has great potential, but there are few limitation in this technology such as the use of technologies like GIS, drone, robotics and satellite imagery etc… at the same time is very costly, especially for developing countries. At present, technologies implemented in precision agriculture are in their infancy stage, and it's hard to narrow down the pricing of equipment and services. Scarcity of experts in the technological knowledge also hinders in its application. The sustainability of precision agriculture largely depends on how well and how quickly the requisite expertise can be found to direct the new technologies

III. Need of precision agriculture

Precision agriculture are generally used for assessing and managing the various field variabilities. It is understood that fields, due to variations in management practices, soil properties and/or environmental characteristics, have variable yields across the countryside. One's cognitive knowledge collection about how to handle various areas in a field involves years of trial-and-error study and implementation. Presently, due to the variable farm sizes and changes in farmed areas due to annual shifts in leasing arrangements, that level of knowledge of field conditions is hard to maintain. Precision farming provides the ability to automate and simplify the data collection and evaluation.

Precision agriculture is need by the farmers as the variables obtained from the use of different technologies are later assessed, which eventually enables management decisions to be taken and executed in the right time in small areas inside wider fields in the right places. It is also need for having higher crop productivity, since precision farming suggests prescribing custom-made management practices, yield per unit of land will definitely increase as long as the other uncontrollable factors of nature are in favor of it. Precision farming is needed for increasing the efficiency of inputs, by increasing the productivity per unit of input utilized shows improved input efficiency. It is needed because it helps in maximum use of smaller land unit by a farmer.

CONCLUSION

The chapter gives a brief introduction to the topic of precision agriculture (PA), focusing on the overview, concepts, history, world interest, benefits, disadvantages and precision farming needs. Therefore, based on the information it can be understood that PA is a modern technological advancement in agriculture, wherein various application, namely the GPS system, drones, and satellite, etc… are used to better and increase the production of crops (especially the cash crops). According, to the information, if the right policy is implemented, the countries around the world who have adopted this farming technology can be able to achieve the SDG goal on 'Zero Hunger', making it a topic of world interest. Therefore, a further research and education in this, can help countries economically and also help solve their hungry problem.

REFERENCES

Adamchuk, V. I., Morgan, M. T., & Ess, D. R. (1999). An automated sampling system for measuring soil pH. *Transactions of the ASAE. American Society of Agricultural Engineers*, *42*(4), 885–891. doi:10.13031/2013.13268

Aubert, B. A., Schroeder, A., & Grimaudo, J. (2012). IT as enabler of sustainable farming: An empirical analysis of farmers' adoption decision of precision agriculture technology. *Decision Support Systems*, *54*(1), 510–520. doi:10.1016/j.dss.2012.07.002

Auernhammer, H., Demmel, M., Muhr, T., Rottmeier, J., & Wild, K. (1994). GPS for yield mapping on combines. *Computers and Electronics in Agriculture*, *11*(1), 53–68. doi:10.1016/0168-1699(94)90052-3

Auernhammer, H., & Muhr, T. 1991. GPS in a basic rule for environment protection in agriculture. In G. Kranzler (Ed.), Automated Agriculture for the 21st Century. ASAE Publ.

Bae, Borgelt, Searcy, Schueller, & Stout. (1987). Determination of spatially variable yield maps. *Am. Soc. Agric. Eng*.

Barnes, E. M., Moran, M. S., Pinter, P. J., & Clarke, T. R. (1996). Multispectral remote sensing and site-specific agriculture: Examples of current technology and future possibilities. *Precision Agriculture*, *3*, 845–854.

Bausch, W. C., & Duke, H. R. (1996). Remote sensing of plant nitrogen status in corn. *Transactions of the ASAE. American Society of Agricultural Engineers*, *39*(5), 1869–1875. doi:10.13031/2013.27665

Beck, J. L., & Kinter, M. L. (1998a). *Photodetector circuit for an electronic sprayer*. U.S. Patent No. 5,763,873, issued June 9, 1998.

Beck, J. L., & Kinter, M. L. (1998b). *Agricultural implement having multiple agents for mapping fields*. U.S. Patent No. 5,809,440, issued September 15, 1998.

Bhatti, A. U., Mulla, D. J., & Frazier, B. E. (1991). Estimation of soil properties and wheat yields on complex eroded hills using geostatistics and Thematic Mapper images. *Remote Sensing of Environment*, *37*(3), 181–191. doi:10.1016/0034-4257(91)90080-P

Blackmer, T. M., Schepers, J. S., & Meyer, G. E. (1995). Remote sensing to detect nitrogen deficiency in corn. In P. C. Robert, R. H. Rust, & W. E. Larson (Eds.), *Site-Specific Management for Agricultural Systems* (pp. 505–512). ASA-CSSA-SSSA. doi:10.2134/1995.site-specificmanagement.c35

Blackmer, T. M., Schepers, J. S., Varvel, G. E., & Meyer, G. E. (1996a). Analysis of aerial photography for nitrogen stress within corn fields. *Agronomy Journal*, *88*(5), 729–733. doi:10.2134/agronj1996.00021962008800050008x

Bongiovanni, R., & Lowenberg-DeBoer, J. (2004). Precision agriculture and sustainability. *Precision Agriculture*, *5*(4), 359–387. doi:10.1023/B:PRAG.0000040806.39604.aa

Borgelt, S. C., Searcy, S. W., Stout, B. A., & Mulla, D. J. (1989). *A method for determining spatially variable liming rates.* Paper No. 89-1034, International Summer Meeting, American Society of Agricultural Engineers, Quebec City, Quebec, Canada.

Borgelt, S. C., Searcy, S. W., Stout, B. A., & Mulla, D. J. (1994). Spatially variable liming rates: A method for determination. *Transactions of the ASAE. American Society of Agricultural Engineers*, *37*(5), 1499–1507. doi:10.13031/2013.28233

Carr, P. M., Carlson, G. R., Jacobsen, J. S., Nielsen, G. A., & Skogley, E. O. (1991). Farming soils, not fields: A strategy for increasing fertilizer profitability. *Journal of Production Agriculture*, *4*(1), 57–61. doi:10.2134/jpa1991.0057

Carter, Rhoades, & Chesson. (1993). *Mechanization of soil salinity assessment for mapping*. ASAE Paper 931557.

Chappelle, E. W., Kim, M. S., & McMurtrey, F. E. III. (1992). Ratio analysis of reflectance spectra (RARS): An algorithm for the remote estimation of the concentrations of chlorophyll a, chlorophyll b, and carotenoids in soybean leaves. *Remote Sensing of Environment*, *39*(3), 239–247. doi:10.1016/0034-4257(92)90089-3

Christy, C., & Lund, E. (1998). *Device for measuring soil conductivity.* U.S. Patent No. 5,841,282, issued November 24, 1998.

Colburn, J. W., Jr. (1991). *Soil chemical sensor and precision agricultural chemical delivery system and method.* U.S. Patent No. 5,033,397, issued July 23, 1991.

Crookston, K. (2006). A top 10 list of developments and issues impacting crop management and ecology during the past 50 years. *Crop Science*, *46*(5), 2253–2262. doi:10.2135/cropsci2005.11.0416gas

Daberkow, S. G., & McBride, W. D. (2000). Adoption of Precision Agriculture Technologies by U.S. Farmers. *5th International Conference on Precision Agriculture*, Bloomington, IN.

Daughtry, C. S. T., Bauer, M. E., Crecelius, D. W., & Hixson, M. M. (1980). Effects of management practices on reflectance of spring wheat canopies. *Agronomy Journal*, *72*(6), 1055–1066. doi:10.2134/agronj1980.00021962007200060045x

Doerge, T. (1999). Defining management zones for precision farming. *Crop Insights*, *8*, 1–5.

Doolittle, J. A., Sudduth, K. A., Kitchen, N. R., & Indorante, S. J. (1994). Estimating depths to claypans using electromagnetic induction methods. *Journal of Soil and Water Conservation*, *49*, 572–575.

Dow, A. I., James, D. W., & Russell, T. S. (1973b). *Soil Variability in Central Washington and Sampling for Soil Fertility Tests. Washington Agricultural Experiment Station Bulletin 788.* Washington State University.

Dwivedi, A., Naresh, R. K., Kumar, R., & Yadav, R. S. (2017). *Precision Agriculture. Promoting agri-hortucultural, technological innovations.* Parmar Publishers & Distributors.

Evans, R. G., Han, S., Kroeger, M. W., & Schneider, S. M. 1996. Precision center pivot irrigation for efficient use of water and nitrogen. In P. C. Robert, R. H. Rust, & W. E. Larson (Eds.), *Precision Agriculture. Proc. 3rd Intl. Conf.* ASA-CSSA-SSSA. 10.2134/1996.precisionagproc3.c8

Felton, W., & McCloy, K. (1992). *Controller for agricultural sprays*. U.S. Patent No. 5,144,767, issued September 8, 1992.

Ferrell, S. L. (2016). Legal Issues on the Farm Data Frontier, Part I: Managing First-Degree Relationships in Farm Data Transfers. *The Drake Journal of Agricultural Law, 21*, 13.

Fraisse, C. W. (1994). *Variable water application with moving irrigation systems* (PhD dissertation). Colorado State University, Ft. Collins, CO.

Fraisse, C. W., Sudduth, K. A., & Kitchen, N. R. (2001). Delineation of site-specific management zones by unsupervised classification of topographic attributes and soil electrical conductivity. *Transactions of the ASAE. American Society of Agricultural Engineers, 44*(1), 155–166. doi:10.13031/2013.2296

Franzen, D. W. (2007). Lessons learned from 40 years of grid-sampling in Illinois. *2007 Indiana CCA Conference Proceedings*, 1–11.

Gaultney, L. D., Schulze, D. G., Shonk, J. L., & Van Scoyoc, G. E. (1991). *Real-time soil organic matter sensor.* U.S. Patent No. 5,044,756, issued September 3, 1991.

Greatline, M. W., & Greatline, S. E. (1999). *Method and apparatus for prescription application of products to an agricultural field.* U.S. Patent No. 5,919,242, issued July 6, 1999.

Guyer, D., Miles, G., Schreiber, M., Mitchell, O., & Vanderbilt, V. (1986). Machine vision and image processing for plant identification. *Transactions of the ASAE. American Society of Agricultural Engineers, 29*(6), 1500–1506. doi:10.13031/2013.30344

Haboudane, D., Miller, J. R., Tremblay, N., Zarco-Tejada, P. J., & Dextraze, L. (2002). Integrated narrow-band vegetation indices for prediction of crop chlorophyll content for application to precision agriculture. *Remote Sensing of Environment, 81*(2-3), 416–426. doi:10.1016/S0034-4257(02)00018-4

Haggar, R., Stent, C., & Isaac, S. (1983). A prototype hand-held patch sprayer for killing weeds, activated by spectral differences in crop/weed canopies. *Journal of Agricultural Engineering Research, 28*(4), 349–358. doi:10.1016/0021-8634(83)90066-5

Hammond, M. W., & Mulla, D. J. (1989). Field variation in soil fertility: Its assessment and management for potato production. *Proc. 28th Annual Washington State Potato Conference.*

Hammond, M. W., Mulla, D. J., & Fairchild, D. S. (1988). Development of management maps for soil variability. *Proc. 39th Annual Far West Regional Fertilizer Conference.*

Hummel, J. W. (1985). Monitoring and control of field machines. *Illinois Research, 27*(4), 8–10.

Ameen & Raza. (2017). Green Revolution: A Review. *International Journal of Advances in Scientific Research, 3*(12), 129–137.

Karlen, D. L., Sadler, E. J., & Busscher, W. J. (1990). Crop yield variation associated with Coastal Plain soil map units. *Soil Science Society of America Journal*, *54*(3), 859–865. doi:10.2136ssaj1990.036159 95005400030041x

Keller, R. J., Nichols, M. E., & Lange, A. F. (2001). *Methods and apparatus for precision agriculture operations utilizing real time kinematic global positioning system systems*. U.S. Patent No. 6,199,000, issued March 6, 2001.

Kendall, H., Naughton, P., Clark, B., Taylor, J., Li, Z., Zhao, C., Yang, G., Chen, J., & Frewer, L. J. (2017). Precision Agriculture in China: Exploring Awareness, Understanding, Attitudes and Perceptions of Agricultural Experts and End-Users in China. *Advances in Animal Biosciences*, *8*(2), 703–707. doi:10.1017/S2040470017001066

Larsen, W. E., Tyler, D. A., & Nielsen, G. A. (1988). Field navigation using the global positioning system (GPS). Am. Soc. Agric. Eng.

Larson, W., Lamb, J., Khakural, B., Ferguson, R., & Rehm, G. (1997). Potential of site-specific management for nonpoint environmental protection. In F. Pierce & E. Sadler (Eds.), *The State of Site-Specific Management for Agriculture* (pp. 337–367). ASA-CSSA-SSSA. doi:10.2134/1997.stateofsitespecific.c15

Leamer, R. W., Myers, V. I., & Silva, L. F. (1973). A spectroradiometer for field use. *The Review of Scientific Instruments*, *44*(5), 611–614. doi:10.1063/1.1686195

Lesch, S. M., Rhoades, J. D., Lund, L. J., & Corwin, D. L. (1992). Mapping soil salinity using calibrated electromagnetic measurements. *Soil Science Society of America Journal*, *56*(2), 540–548. doi:10.2136 ssaj1992.03615995005600020031x

Lowenberg-DeBoer, J. (1999). Precision Agriculture in Argentina. *Modern Agriculture*, *2*(2), MA13–MA15.

Luellen, W. R. (1985). Fine-tuned fertility: Tomorrow's technology here today. *Crops Soils*, *38*, 18–22.

Manglia, M. (2010). *Precision Agriculture at a Glance*. Agropedia. Retrieved from: http://agropedia.iitk.ac.in/content/precision-agriculture-glance

McCann, I. R., & Stark, J. C. (1993). *Method and apparatus for variable application of irrigation water and chemicals*. U.S. Patent No. 5,246,164, issued September 21, 1993.

Melsted, S. W. (1967). The philosophy of soil testing. In *Soil Testing and Plant Analysis Part I*. SSSA Special Publication Series No. 2. Soil Science Society of America, Inc.

Mulla, D., & Khosla, R. (2016). *Historical evolution and recent advances in precision farming. Soil specific farming*. CRS Press. https://www. taylorfrancis. com/books/9781482245349

Mulla, D. J. (1989). Soil spatial variability and methods of analysis: I. Kriging soil fertility patterns. In C. E. Whitman, J. F. Parr, R. I. Papendick, & R. E. Meyer (Eds.), *Rainfed Agriculture in the Near East Region: Soil, Water and Crop/Livestock Management Systems* (pp. 185–204). USDA/USAID.

Mulla, D. J. (1991). Using geostatistics and GIS to manage spatial patterns in soil fertility. In G. Kranzler (Ed.), *Automated Agriculture for the 21st Century* (pp. 336–345). American Society of Agricultural Engineers.

Mulla, D. J. (1993). Mapping and managing spatial patterns in soil fertility and crop yield. In P. Robert, W. Larson, & R. Rust (Eds.), *Soil Specific Crop Management* (pp. 15–26). American Society of Agronomy. doi:10.2134/1993.soilspecificcrop.c2

Mulla, D. J., Bhatti, A. U., Hammond, M. W., & Benson, J. A. (1992). A comparison of winter wheat yield and quality under uniform versus spatially variable fertilizer management. *Agriculture, Ecosystems & Environment, 38*(4), 301–311. doi:10.1016/0167-8809(92)90152-2

Mulla, D. J., Gowda, P., Koskinen, W. C., Khakural, B. R., Johnson, G., & Robert, P. C. (2002). Modeling the effect of precision agriculture: Pesticide losses to surface waters. In *Terrestrial Field Dissipation Studies. ACS Symposium Series No.* 842. ACS. 10.1021/bk-2002-0842.ch020

Mulla, D. J., & Hammond, M. W. (1988). Mapping of soil test results from large irrigation circles. *Proc. 39th Annual Far West Regional Fertilizer Conference*.

Mulla, D. J., & McBratney, A. B. (2000). Soil spatial variability. In M. E. Sumner (Ed.), *Handbook of Soil Science* (pp. A321–A352). CRC Press.

Mulla, D. J., Perillo, C. A., & Cogger, C. G. (1996). A site-specific farm-scale GIS approach for reducing groundwater contamination by pesticides. *Journal of Environmental Quality, 25*(3), 419–425. doi:10.2134/jeq1996.00472425002500030006x

Norton, G. W., & Swinton, S. M. (2001). Precision Agriculture: Global Prospects and Environmental Implications. In G. H. Peters & P. Pingali (Eds.), *Tomorrow's Agriculture: Incentives, Institutions, Infrastructure and Innovations*. Ashgate.

O'Connor, M., Bell, T., Elkaim, G., & Parkinson, B. 1996. Automatic steering of farm vehicles using GPS. In P. C. Robert, R. H. Rust, & W. E. Larson (Eds.), *Precision Agriculture. Proc. 3rd Intl. Conf.* ASA-CSSA-SSSA. 10.2134/1996.precisionagproc3.c91

Ortlip, E. W. (1986). *Method and apparatus for spreading fertilizer*. U.S. Patent No. 4,630,773, issued December 23, 1976, Assignee: Soil Teq.

Palmer, R. J. (1991). Progress report of a local positioning system. In G. Kranzler (Ed.), *Automated Agriculture for the 21st Century* (pp. 403–408). American Society of Agricultural Engineers.

Pinter, P. J. Jr, Jackson, R. D., Idso, S. B., & Reginato, R. J. (1981). Multidate spectral reflectance as predictors of yield in water stressed wheat and barley. *International Journal of Remote Sensing, 2*(1), 43–48. doi:10.1080/01431168108948339

Pongnumkul, S., Chaovalit, P., & Surasvadi, N. (2015). Applications of smartphone-based sensors in agriculture: A systematic review of research. *Journal of Sensors*.

Rajvanshi, A. K. (2015). *Is precision agriculture the solution to India's farming crisis" Business Standards, online article*. Retrieved from: https://www.business-standard.com/article/news-ians/is-precision-agriculture-the-solution-to-india-s-farming-crisis-comment-special-to-ians-113101500127_1.html

Raun, W. R., Solie, J. B., Johnson, G. V., Stone, M. L., Mullen, R. W., Freeman, K. W., Thomason, W. E., & Lukina, E. V. (2002). Improving nitrogen use efficiency in cereal grain production with optical sensing and variable rate application. *Agronomy Journal, 94*(4), 815–820. doi:10.2134/agronj2002.8150

Reid, J. F., & Searcy, S. W. (1987). Vision-based guidance of an agriculture tractor. IEEE Control Syst. Mag., 7, 39–43. doi:10.1109/MCS.1987.1105271

Rhoades, J. D., & Corwin, D. L. (1981). Determining soil electrical conductivity-depth relations using an inductive electromagnetic soil conductivity meter. *Soil Science Society of America Journal*, *45*(2), 255–260. doi:10.2136ssaj1981.03615995004500020006x

Robert, P. C. (1982). *Evaluation of some remote sensing techniques for soil and crop management* (PhD dissertation). University of Minnesota.

Rust, R. H. (1985). Computerized soil maps benefit crop management. *Farming with Pride Magazine*, 4–10.

Sawyer, J. E. (1994). Concepts of variable rate technology with considerations for fertilizer application. *Journal of Production Agriculture*, *7*(2), 195–201. doi:10.2134/jpa1994.0195

Schmitt, M. A., Walker, W. G., & Fairchild, D. (1986). Computerized fertilizer application by soil type. In *Proc. Great Plains Soil Fertility Workshop*. Kansas State Univ.

Schueller, J. K. (1991). In-field site-specific crop production. In G. Kranzler (Ed.), *Automated Agriculture for the 21st Century* (pp. 291–292). American Society of Agricultural Engineers.

Schueller, J. K. (1992). A review and integrating analysis of spatially-variable control of crop production. *Fertilizer Research*, *33*(1), 1–34. doi:10.1007/BF01058007

Schueller, J. K., & Bae, Y. H. (1987). Spatially attributed automatic combine data acquisition. *Computers and Electronics in Agriculture*, *2*(2), 119–127. doi:10.1016/0168-1699(87)90022-6

Schueller, J. K., & Wang, M.-W. (1994). Spatially-variable fertilizer and pesticide application with GPS and DGPS. *Computers and Electronics in Agriculture*, *11*(1), 69–83. doi:10.1016/0168-1699(94)90053-1

Searcy, S. W., Schueller, J. K., Bae, Y. H., Borgelt, S. C., & Stout, B. A. (1989a). Mapping of spatially-variable yield during grain combining. *Transactions of the ASAE. American Society of Agricultural Engineers*, *32*(3), 826–829. doi:10.13031/2013.31077

Shonk, J. L., & Gaultney, L. D. (1988). Spectroscopic sensing for the determination of organic matter content. Am. Soc. Agric. Engin..

Solie, J. B., Raun, W. R., Whitney, R. W., Stone, M. L., & Ringer, J. D. (1996). Optical sensor-based field element size and sensing strategy for nitrogen. *Transactions of the ASAE. American Society of Agricultural Engineers*, *39*(6), 1983–1992. doi:10.13031/2013.27700

Sophocleous, M. (2015). *Thick-film underground sensors* (Doctoral dissertation). University of Southampton.

Stafford, J. V., Ambler, B., & Smith, M. P. (1991). Sensing and mapping grain yield variation. In G. Kranzler (Ed.), *Automated Agriculture for the 21st Century* (pp. 356–365). American Society of Agricultural Engineers.

Stafford, J. V., & Miller, P. C. H. (1993). Spatially selective application of herbicide to cereal crops. *Computers and Electronics in Agriculture*, *9*(3), 217–229. doi:10.1016/0168-1699(93)90040-8

Sudduth, K. A., Hummel, J. W., & Cahn, M. D. (1991). Soil organic matter sensing: A developing science. In G. A. Kranzler (Ed.), *Automated Agriculture for the 21st Century* (pp. 307–316). ASAE.

Swinton, S. M., & Lowenberg-DeBoer, J. (1998). Evaluating the Profitability of Site-Specific Farming. *Journal of Production Agriculture, 11*(4), 439–446. doi:10.2134/jpa1998.0439

Swinton, S. M., & Lowenberg-Deboer, J. (2001, June). Global adoption of precision agriculture technologies: Who, when and why. In *Proceedings of the 3rd European Conference on Precision Agriculture* (pp. 557-562). Citeseer.

Thenkabail, P. S., Lyon, J. G., & Huete, A. (2010). *Hyperspectral remote sensing of vegetation and agricultural crops: Knowledge gain and knowledge gap after 40 years of research.* CRC Press.

Thompson, Stafford, & Miller. (1990). *Selective application of herbicides to UK cereal crops.* Paper, American Society of Agricultural Engineers 90-1629.

Thompson, J. F., Stafford, J. V., & Miller, P. C. H. (1991). Potential for automatic weed detection and selective herbicide application. *Crop Protection (Guildford, Surrey), 10*(4), 254–259. doi:10.1016/0261-2194(91)90002-9

Tyler, D. A. (1993). Positioning technology (GPS). In P. C. Robert, R. H. Rust, & W. E. Larsen (Eds.), *Proc. Soil Specific Crop Management* (pp. 159–166). American Society of Agronomy.

Walburg, G., Bauer, M., Daughtry, C. S. T., & Housley, T. L. (1982). Effects of nitrogen nutrition on the growth, yield and reflectance characteristics of corn canopies. *Agronomy Journal, 74*(4), 677–683. doi:10.2134/agronj1982.00021962007400040020x

Wiegand, C. L., Richardson, A. J., Escobar, D. E., & Gerbermann, A. H. (1991). Vegetation indices in crop assessment. *Remote Sensing of Environment, 35*(2-3), 105–119. doi:10.1016/0034-4257(91)90004-P

Wollenhaupt, N. C., Mulla, D. J., & Gotway, C. A. (1997). Soil sampling and interpolation techniques for mapping spatial variability of soil properties. In F. J. Pierce & E. J. Sadler (Eds.), *The State of Site-Specific Management for Agriculture* (pp. 19–54). ASA/CSSA/SSSA. doi:10.2134/1997.stateofsitespecific.c2

Zadra, A., & Robert, G. (2012). Dream recall frequency: Impact of prospective measures and motivational factors. *Consciousness and Cognition, 21*(4), 1695–1702. doi:10.1016/j.concog.2012.08.011 PMID:23092675

Zhang, Q., Reid, J. F., & Noguchi, N. (1999). Agricultural vehicle navigation using multiple guidance sensors. *Proc. Intl. Conf. Field Service Robotics.*

Zheng, F., & Schreier, H. (1988). Quantification of soil patterns and field soil fertility using spectral reflection and digital processing of aerial photographs. *Fertilizer Research, 16*(1), 15–30. doi:10.1007/BF01053311

KEY TERMS AND DEFINITIONS

Compound Annual Growth Rate (CAGR): Is the rate at which the return that would be required for a funding to develop from its initial balance to its closing balance, assuming that the profits were reinvested at the stop of each 12 months of the investment's lifespan.

GIS: Geographic information system is a mechanism for data collection, management, and analysis.

GPS: Global positioning system is a radio-navigation satellite system.

IoT: Stands for internet of things; is a collection of interdependent computing devices, computational and digital machines, artifacts, animals or humans supplied with unique identifiers (UIDs) and the ability to convert information over a system without needing human-to-human or human-to-computer intervention.

Precision Agriculture (PA): Also called as satellite farming or site specific crop management (SSCM) is a next generation farm management technology which is based on the concept of observing, measuring and responding to inter and intra-field variability in crops.

Remote Sensing: Remote sensing is the mechanism by which the phenotypic traits of an environment are measured and tracked by analyzing its transmitted and emitted radiation at a distance (usually from satellite or aircraft).

Chapter 2
Precision Agriculture:
A New Tool for Development

Waleed Fouad Abobatta
Horticulture Research Institute, Agriculture Research Center, Egypt

ABSTRACT

Precision agriculture is a management system that aims to reduce inputs like seeds, water, and energy; protect the environment; and maximize profitability. Precision agriculture uses advanced technology like positioning technology, geographical information systems, satellite navigation, and remote sensing. There are different factors affect the adoption of precision agriculture like farm size, legal affairs, and social interaction. Under climate change and increases in world population, adoption of precision agriculture could assist farmers to face various challenges to achieve ideal production and maximizing profitability. Information, technology, and management are considered the backbone of the precision agriculture system, and combining these elements reduces inputs and maximizes productivity. Different threats attacked precision agriculture including threats to confidentiality, threats to integrity, threats to availability, and crowding of the spectrum signal. This chapter explains the different roles of precision agriculture in developing agricultural production.

INTRODUCTION

The main purpose of precision agriculture (PA) is to support farmers by providing customized information and technologyservices that enhancing productivity, increase profitability and reduce environmentally pollutions, PA working on improving crop production and livestock, and contribution of different factors like soil fertility, water, and pest control in increasing farmer's profits, and protect the environment at the same time (Ess& Morgan, 2003) &(Rains & Thomas, 2009).

PA also called Precision Farming(PF) or Satellite Farming (SF) includesadvanced management technologies like soil sensing and mapping, yield monitoring and mapping, satellite-based positioning, remote sensing, fieldand crop scouting, geographical information systems (GIS), variable rate application (VRA), and automatic steering (Balafoutiset al.,2017).

DOI: 10.4018/978-1-7998-5000-7.ch002

Adoption of the PA system is affected by various elements like personality characters of farmers, farm size, features of the machines, characteristics of the technology, legal affairs, and social interaction, etc. (Tey&Brindal, 2012).

PA plays a direct role in developed country-specific intelligent platforms that provide farmers with context-relevant and personalized agricultural recommendations through their mobile phones, also, PA helps render these advisory services more customizable and intelligent with time, by evaluates and enhances the existing agricultural systems (Paustian&Theuvsen 2017). PA is becoming an interesting scope for managing natural resources like water, soil, and seeds and applying modern sustainable agricultural development, it is bringing agriculture into the digital information era,also, PA in the animal production sector use to improve meat and milk productivity (Kutter, Tiemann, Siebert, &Fountas, 2011).

This chapter addressees the role of the PA system to increase agricultural production, increasing farmers' profitability, sustain natural resources and protect the environment,exploring different technologies used in PA and the threats which attacked the PA system.

The work was arranged by focusing on the objective of the importance of precision agriculture, exploring the various technologies used, explaining the advantages of precision agriculture, adopting to the PA system in developing countries and the challenges for the P A system, ending with the potential threats to P A.

Searching for the literature published from 2000 to 2019 was conducted, through Egyptian knowledge Bank(https://www.ekb.eg/), Academia (https://www.academia.edu/), Researchgate (https://www.researchgate.net), Google scholar (https://scholar.google.com/), Springer- (https://www.springer.com/), Elsevier-(https://www.elsevier.com/), IEEE (https://ieeexplore.ieee.org/) and The Egyptian Journal of Remote Sensing and Space Sciencehttps://www.sciencedirect.com/science/journal/11109823using "precision agriculture", "soil mapping", "variable rate application", "threats to precision agriculture " as key terms. 87 papers out of a total database of 252 papers were shortlisted after a thorough reading, all selected papers were analyzed, discussed, and classified into a specific category. The literature was distributed on the basis of related to the subject of the work, whereas (Table 1) highlights various e-resources accessed for acquiring in-depth knowledge of the subject.

Table 1. Summary of data of papers selected on the basis of search string "Precision Agriculture, soil mapping, variable rate application, and threats to precision agriculture"

No	E-resource	Content
1	https://www.ekb.eg/	Journals, Proceedings, books
2	https://www.academia.edu/	Conference, Journals, Proceedings
3	https://www.researchgate.net	Journals, Proceedings, books
4	https://scholar.google.com/	Conference, Journals, Proceedings
5	https://link.springer.com	Conference,Journals, Proceedings, books
6	https://www.elsevier.com/	Journals, Proceedings, books
7	https://www.sciencedirect.com/	Journals, Proceedings, books
8	https://ieeexplore.ieee.org/	Conference, Journals, Proceedings

BACKGROUND

PA practices using modern technology to reduce agricultural inputs by site-specific applications, as it better target inputs to spatial and temporal needs of the fields, which can result in lower greenhouse gas emissions through applying adequate quantity of fertilizers and agrochemicals. There is encouraging impact of PA system on productivity of different crop and cost-saving to increase farmers' income, it's producing more or equal quantity of crops with minimum production cost than conventional practices(Balafoutis et al., 2017).

Various researches have been reported that assess the impacts of PA technologies. This approach can not only decrease costs but can also increase yields. Furthermore, accurately applying chemicals and fertilizers only where needed reduces the potential for ground and surface water pollution,PAwill not only help but also has considerable environmental benefits (Krishnan, Foster, Strosser, Glancey, &SUN, 2006),

The aim of this work is to clarify the important role of PAin developing the agricultural sector and increasing crop productivity to provide adequate enough food for humanity. This work exploring the different technologies used in PA and practical application for PA in various branches, also explain the importance of adoption to this system in developing countries to increase food production and provide enough food for the local community, at the end of this chapter, there is explaining for some threats that face PA applications.

Why We Need Precision Agriculture?

Under climate change conditions in the current era, there are different problems face the agricultural sector and affect crop productivity, like drought and lack of water management (Montzka, Dlugokencky, & Butler, 2011), salinity, diseases, pests, agrochemicals management, shortage of storage management, and weed management, all these problems can be controlled by adoption of PA methods.

Due to the expected increase in world population which will reach about 9.6 billion by 2050, and to 12 billion by 2100 (Food and Agriculture Organization [FAO] 2017), so, humanity need to produce more crops to feed every mouth, so, the food production must significantly double from current levels for two or three folds(Foote, 2015), therefore, using advanced technology of precision farming and using varieties with high yield crops, could produce enough food quantity to feed more people from the same arable land, without more hazards for the environment, also, the farmers could be used tools of PA to increase the efficiency of their business to increase their profitability (Gregory & George, 2011).

PRECISION AGRICULTURE

The term PA or precision farming appeared in the 1980s in the United States, it's management strategy that collecting data, developing wireless networks, minimizing of sensors to monitor, evaluate, and control agricultural practices (Navulur, Sastry, &Giri Prasad, 2017). There are various definitions of PA, while the definition of PA according to (Sulecki, 2018) "Precision agriculture is a management strategy that uses electronic information and other technologies to gather, process, and analyze spatial and temporal data for the purpose of guiding targeted actions that improve efficiency, productivity, and sustainability of agricultural operations." will be used as a definition of PA in this work.

PA is a management system based on information and technology for determining the variability intra-field on crop productivity to optimizing profitability, sustainability, and environmental protection (McBride &Daberkow, 2003).PA aims to use minimum inputs and optimize outputs while sustaining different natural resources (Whelan, McBratney, & Stein,2003).In particular, it is related to the site-specific crop management (SSCM) with a wide array of pre- and post-production aspects of agriculture, ranging from horticultural crops to field crops (Bhunia, 2019).

The PA is not a new concept, there is a different application used by agriculturalists in both agriculture and livestock like seed planter, rice transplanter machines, milking machines, also, farmers, agriculturalists and Livestock growers collecting data by different methods to making proper decisions to reduce costs of various inputs and increase profitability.

Historically there are two-stage for PA technology, the first one starts with satellite and aerial imagery, applications for variable fertilizing rate, and weather prediction, and followed by collecting adequate data for machinery for precise planting, topographical mapping, and soil sensors.

PA integrates the utilization of machines, satellites, and computers to enhance the various inputs utilization, minimizing the costs, and reducing different environmental hazards, also, PA system using adequate nutrients, seeds, and agrochemicals in the proper site at the right time to achieve the maximum productivity and increase profitability, also, the minimization of inputs results in the least negative effects on the environment. (Marino &Alvino, 2014).

PA is a key component of the third wave of modern agricultural revolutions, the basic principles of PA in managing soil and crop variability within a field is certainly not new, the management practice of Site-specific crop has been developed by the advent of GPS and Global navigation satellite systems (GNSS) [Reichardt&Jürgens (2009), and (Li, et al. 2019)]. There are numerous technologies used in PA such as soil mapping, yield monitoring, and automatic guidance to enhancing input efficiency and collect data to facilitate future production decisions, PA permits farmers to use the optimum rate for different inputs includes fertilizers, seed, pesticides, and irrigation water in the specified place, at the right time to maximize yield production, save manpower, and harvest costs (Keskin, Sekerli, Say, &Topcueri, 2017).

While in livestock farming PA system aims to use advanced technologies to optimize meat and milk production, reduce costs, and improve animal production, it allows farmers and ranchers to consistently collect information at the animal level and improve animal production (MacDonald & McBride, 2009).

Importance of PA

PA system allows using accurate application of crop and livestock management inputs such as fertilizer, seeds, pesticides, and Livestock feeding [Takacs-Gyorgy, 2009, pp. 217–223), also, it relies on advanced technology to provide a proper application for management inputs in both agricultural and livestock to reduce costs and develop production, resulting in lower costs and enhances outputs (crop yields and meat & milk production), and protecting the environment (Banhazi, Babinszky, Halas, &Tscharke, 2012).

a) PA offers a significant contribution to producing more crops to enhance food security.
b) Providing new solutions for enhancing food safety.
c) PA support sustainable impact use of different resources in the agricultural sector.
d) PA has influenced on work practices and living conditions on farmers' community, and increase new agribusiness models.

The Scope of PA

The scope of PA is minimizing inputs (Seeds, water, agrochemicals, energy) and increasing the output in both terms of quantity and quality of various crop and livestock products, to achieve the sustainability of crops, farm management, and economics.

PA uses different technology to determine soil conditions, reporting about crop health, pest attacks and predicting the final output of various crops, which assist farmers to adapt to changed conditions to improve the final outputs and increase farmers' incomes (Yadav, Yadav, Kumar, Yadav, &Verma, 2017).

Precision Agriculture Provides Different Aspects Like

1. Improve the operation system of the farms and field goals.
2. Environmental protection goals.
3. Social license to farm at risk.

Actually accurate information about soil fertility, crop productivity, water supply, climate change, and spreading diseases and pests is an important modeling input, which is helpful to farmers for decision making to establish the right policies. Using PA technologies increases management practices efficiency and improves precision in different agriculture stages from seeding to harvesting (Pierpaoli, Carlia, Pignattia, &Canavaria, 2013).

Grisso et al. (2009) reported that the farmers get various advantages through using PA-technologies including:

1. Increase the accuracy of field works,
2. Higher operation speeds,
3. Easy operating,
4. Working 24 h (day and night).
5. Less affecting by unbalanced weather,
6. Reduced operator fatigue,
7. less setup time,
8. Decrease overlapping,
9. Reduced production costs (fuel, fertilizer, pesticides, seeds, etc)

The PA practice has efficiently extended into some developed countries which have a good level of adoption of PA technologies like USA, Canada, Australia (Leonard, 2014), and some EU countries like Germany, Finland, Denmark, England, and Sweden (Bligaard, 2013, pp.12), in Japan (Liao, 2017), and in Russia (Dokin et al., 2016).

Fountas et al. (2005) reported that there are about 90% of the yield monitors worldwide were operating in the US due to there are many innovative technologies. Currently, the PA technologies in Sub-Sahara countries like Rwanda, Ethiopia, Nigeria, and Kenya at the nascent stage due to various reasons like majority of farms are family-owned, small scale farm (less than 1hectar), and shortage of local technical expertise (Getinet&Getachew, 2019).

PA provides various benefits for farmers include:

- Adding the exact quantity of materials at the correct time and in the precise location in the field.
- Preserve enough information as a record of their farm
- Improving traceability for all farm components
- Enhancing various products marketing
- Improve the lease relationship with landowners.
- Develop the quality of different products of the farm.

PA TECHNOLOGY

There is a wide range of technologies used in PA (Figure1) includes Guidance Technology,Global Navigation Satellite System (GNSS), Recording Technology, soil mapping, variable rate application,automatic steering, Reacting Technology, yield monitoring mapping, and automatic guidance, these technologies are allowing farmers to better target inputs and take the proper decisions to adapt to local conditions, these technologies are used for various objects like identification, geo-referencing, measurement of specific parameters, GNSS, connectivity, data storage and analysis, advisory systems, robotics and autonomous navigation, modern technology has created new opportunities to provide proper information for the farmers to improve agricultural production(Shafi, et al., 2019).

Figure 1. Schematic for Precision Agriculture Technology

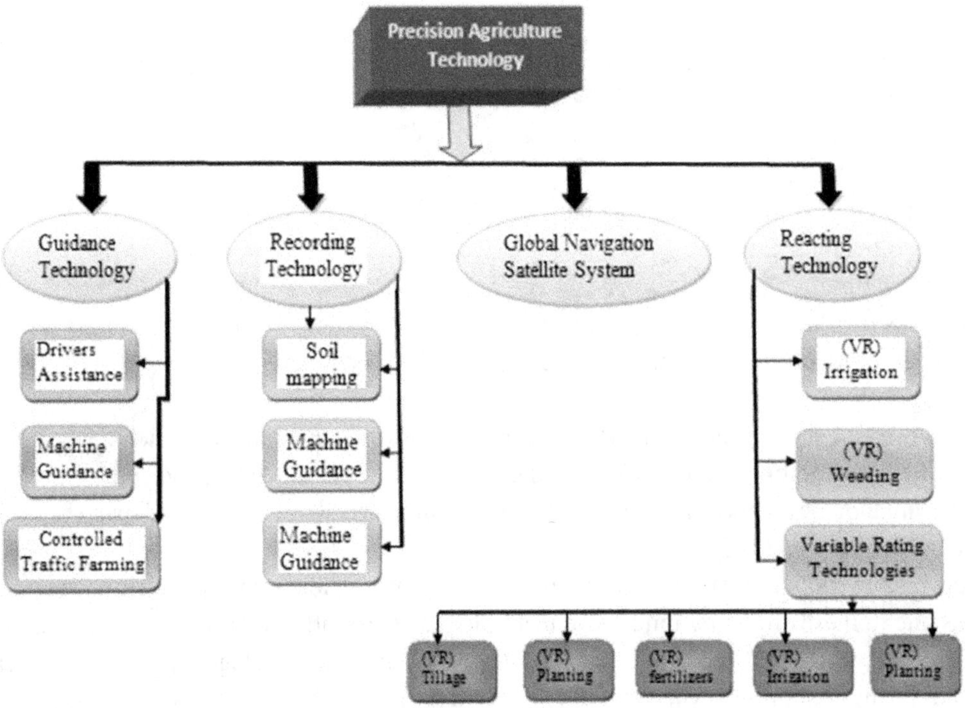

There are different technologies used in PA include:

- Portable soil analysis labs
- Satellite and drone photographs.
- New weather prediction models
- Also, the spread of the use of mobile phones facilitates collection and delivery information in a very short time.

The technology of PA provides novel solutions for producing more agricultural products with fewer inputs and helping in sustainability issues, for example, soil monitoring systems offer farmers with enough information and alarms on the status of crops and enhanced crop forecasting.

PA plays a major role in livestock husbandry, for instance, in milk production the milking robots leading to automatic milking like in Sweden and Finland whereas about 90% of the dairy sector depending on robotic milking, also, there are currently more trending in Netherlands, Germany, and France to shift towards automatic milking (Steeneveld, Tauer, Hogeveen, &Oude Lansink, 2012).

The yield monitoring technology and variable rate technology were more dominant earlier in both developed and developing countries, while the auto-guidance systems had more popularity in the last decade.Currently, there is an increasing trend in developed and some developing countries for increasing the adoption rate of PAtechnologies to maximizing the crop productivity and increase profitability by minimizing inputs, decreasing operation costs, also, through using fields mapping and sensors, farmers could recognize their crops during various growth stages (Keskin, et al. 2017).

There are three main elements represent the backbone of precision farming:

1. Data and Information
2. Technology
3. Management (decision support)

The combining of these elements in the PA system reducing inputs, increase productivity, enhance product quality, and protecting the environment (Finger, Swinton, El Benni, & Walter, 2019).

While there are main steps required to promote PAat farmer's level include:

1. Identify the proper sites for the promotion of crop-specific precision farming
2. Formation of working groups involving agricultural scientists in different fields, engineers, manufacturers and economists to study all objectives of PA.
3. Start with a small pilot of PAand providing full technical support for the farmer
4. The model or pilot must be in the field inside the community of farmers to show the positive effect of PAtechnologies.
5. Increase farmers' awareness about the hazards of continuously applying different doses of various inputs like fertilizers, irrigation, and pesticides.

All these components are correlated and responsible for developing PAat the farmer's levels.

Remote Sensing

There is a different application of Remote Sensing in PA system (Figure 2) include Soil maps, drainage maps, observe plant health, water shortage, weed control, detection of Pests and disease, nutrient deficiency, and yield forecasting (Thessler, Kooistra, Teye, Huitu, &Bregt, 2011) & (Lakhiar et al 2018).

Figure 2. Schematic of different applications of remote sensing in Precision Agriculture

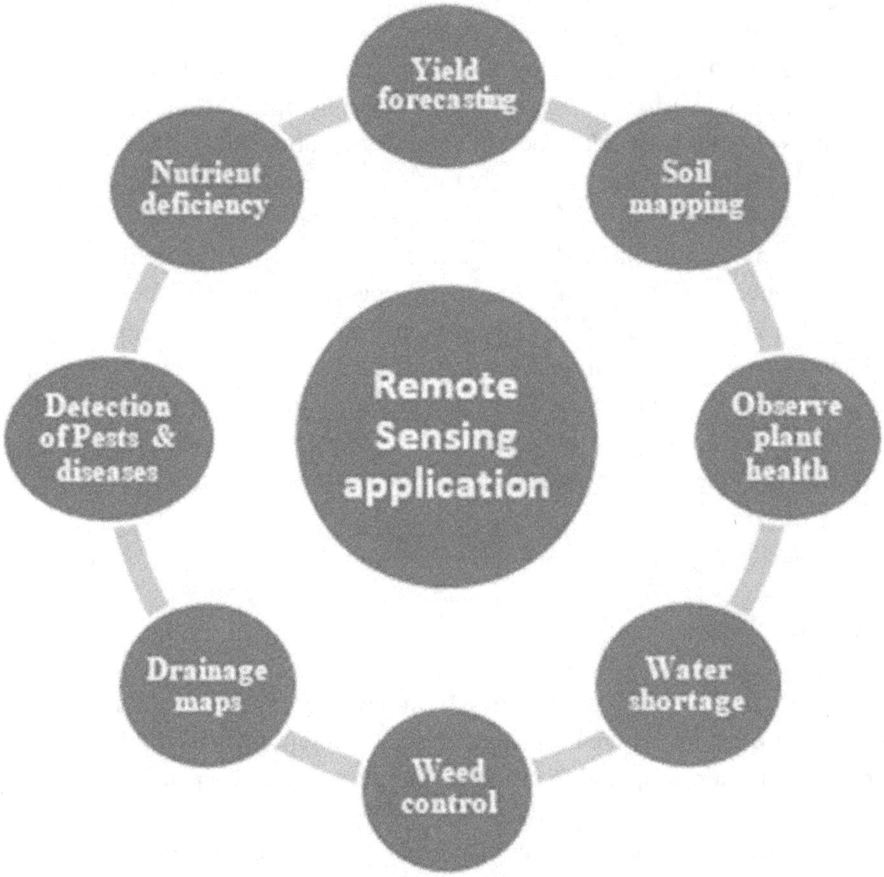

Variable Rating application

Variable-rate application considered an important factor in the adoption of PA technologies, variable rate application used in tillage, seeding, irrigation, fertilizing, and pesticide, it'sallow the farmers using different inputs by different rates to reach the precise requirements of management areas on the farm like measuring canopy volume of citrus trees by laser scanner to estimate fruit yield and using a varying fertilizing rate (Zaman, Schumann, & Miller, 2005), using GPS to determine field position, read monitoring map and/or explain data from sensors, anduse soil mapping to prepare adequate fertilizing program for olive orchards (Fountas et al., 2011).

Yield Monitoring

Applications in mechanically harvested for fruits like in pecan and apple orchards, and vegetables such as tomato and potato, yield monitors are effective in estimating the yield of each plant and collected fruits are weighed in the site directly for each plant or block in the field [Rains et al., (2002);Pooja, Uday, Nagesh, &Talekar, (2017);&Hofstee&Molema (2002)].

Fertilizers management

PAsystem permitting use the adequate fertilizers at the right time with proper methods to maximizing crop productivity (Patil, Nadagouda, & Al-Gaadi, 2012), the variable-rate application of fertilizers one of the most important elements in adoption to PA system, it's allowed farmers to use various rates of fertilizers depend on crop requirements, and improve the efficient use of different nutrients, decreasing production cost, reducing leaching of fertilizers and protect the environmental (Hornung et al. (2006); &Bramley et al. (2019).

Precision Agriculture Tools

PA uses different machines that have auto-guidance systems like tractors, combines, sprayers and planters (Figure 3), the small unit that uses GIS of auto-guidance in all machines considered the brain of the PA age (Yousefi&Razdari, 2014)through this brain all equipment connected with the data systems and other systems (Schmitz & Moss, 2015).

The main tools in PA including:

1. Geographical information system (GIS).
2. Variable-rate technology (VRT).
3. Global positioning system (GPS).
4. Grid sampling.
5. Remote sensors (Satellites & Drones).

PA uses remote sensing data of high spatial resolution which are georeferenced with the use of GPS measurements, these data used to identify potential problems at specific locations of the field such as nutrient deficiency, so the right decision takes to treat with the problem in the specific location, also, collecting data of (remote sensing data, soil mapping, and weather forecasting) use to accurate estimating yield crop(Toscano, et al. 2019).

In addition to above components, PA uses a vast array of technologies tools including hardware, software, and various tools include:

1. Mobile Devices
2. Robotics
3. Internet Of Things (IOT)
4. Weather Modeling
5. Irrigation systems
6. Fertilizing Modeling

Figure 3. Schematic of Precision Agriculture tools

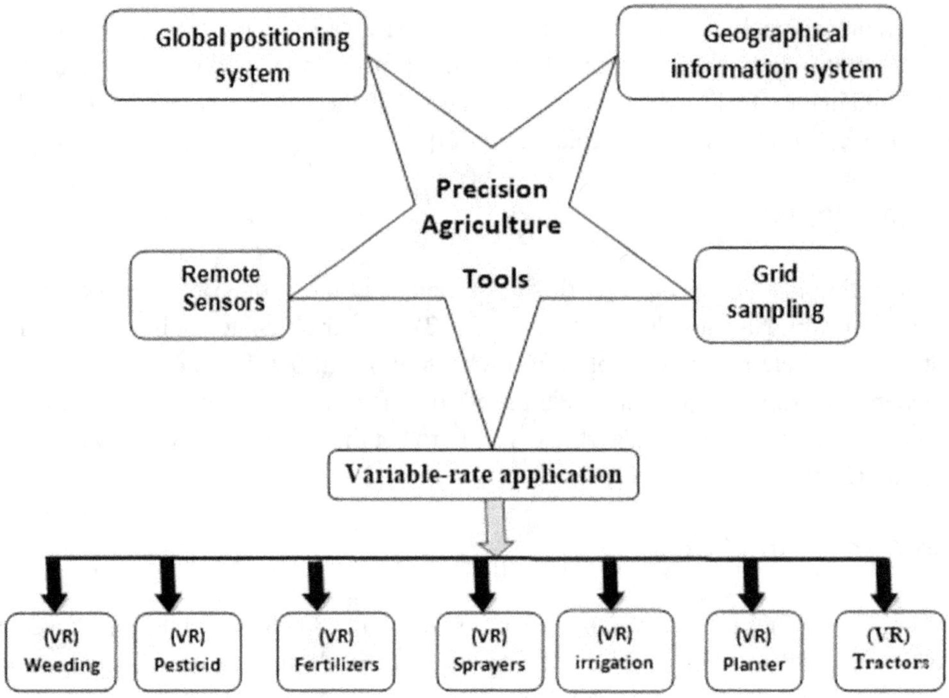

7. Yield Maps
8. Standardization.

Advantages of Precision Agriculture

Adoption to PA system providing different benefits for farmers, and allow predicting the proper time for various agricultural practice (Abdulwaheed, 2019, pp. 19-34), the main advantages of PA system include:

1. Enhancing agricultural productivity.
2. Prevent soil degradation in cultivable land.
3. Sustained agricultural development.
4. Reducing excessive agrochemical usage in crop production.
5. Increase water resources use efficiency.
6. Soil mapping.
7. Divide non-uniform farms to small plots based on their requirements.

CLIMATE AND PRECISION AGRICULTURE

PA technology reducing inputs like fertilizers, pesticides, herbicide and avoiding excessive agrochemicals which help in reducing greenhouse gas emissions, also, PA technologies like variable-rate seeding, fertilizing, spraying, weeding and irrigation which has great potential to alleviate greenhouse gas emissions (Kumari, Singh, & Prasad, 2019).

The Impact of Precision Agricultural Measures on Climate

It's well documented that about 10-15% of all gas emissions produced from agricultural activities, there are two gas nitrous oxide and carbon dioxide emitting from cultivated lands. Nitrous oxide considered more dangerous gas and has negative effects more than carbon dioxide about 300 times. PA techniques have a positive contribution to reducing the emission of greenhouse gases, through variable-rate applications that can reduce fertilizer use by 10-30%, so, reducing the emission of nitrous oxide without affecting yield productivity, therefore, transform agriculture to PA system could significantly decrease the emission of greenhouse gas [Balafoutis et al., 2017). Increasing the adoption of PA techniques will have a great impact on mitigating climate change by reducing greenhouse gas emissions (De Ollas et al., 2019).

NANOTECHNOLOGY AND PRECISION AGRICULTURE

Precision agricultural technology used to increase crop productivity and reducing pollution of soil and waterby excessive usage of agrochemicals, in addition, to enhance nutrients long-term incorporation by soil microorganisms (Abobatta, 2018, pp. 99-102).

There are different applications of nanotechnology in the PA system includes:

1. Precise management and control of inputs particularly agrochemicals.
2. Nanotools like Nanobiosensors.
3. Support the progress of high-tech agricultural farms.

Nanotechnology improving crop productivity, through enhancing plant health, increase fertilizer efficiency, reducing pesticides usage, detecting the pathogens (Anjum&Pradhan, 2018), in addition, there is a wide range of using nanosensors in precision farming for monitors and estimation crops productivity, soil mapping, spreading diseases, and, it's allowed monitoring of penetration of agrochemicals and environmental pollution.From another side, nanotools enhanced the human control of soil and plant health, increase quality control and safety assurance contributing to sustaining the agriculture resources and environmental systems (Shang et al., 2019).

PRECISION AGRICULTURE AND HORTICULTURE

There is a variation in fruit yield of horticultural crops in the same site even in small areas, therefore, horticultural production represents a challenge for PA technology (Gemtos et al 2013).There are different aspects in horticultural sector considered a promising field for adoption for PA system like using yield mapping technology for estimating fruit yield before harvesting particularly in perishable fruits like plum and apricot, PA technology could play important role in estimation orange fruit quality(Whitney, Ling, Miller, & Wheaton, 2001), also, supported other agricultural practice like thinning, reduce impacts of climate change like heat waves and frost on fruit quality, pruning, and harvest represent a wide range for PA system (Aggelopoulou et al., 2010).

Application of PA technology in horticultural sector aims to increase profitability, maximizing fruit production and reduce inputs particularly fertilizers and pesticides, improving fruit quality which increase farmers profitability(Tamirat& Pedersen, 2019), sustain different natural resources like water and soil, protect environment and reducing water pollutions (Kathner&Zude-Sasse, 2015).Under PA system, there is a control in using of different agrochemicals like nutrients, herbicides, and pesticides in fruit orchards, each tree receive determined amount of the agrochemicals which calculated automatically by using sensors, computers and GPS tools, also, PA system avoiding spray spaces between trees or missing trees to reduce inputs, saving costs, and protect environment (Zude-Sasse, Fountas, Gemtos, & Abu-Khalaf, 2016).

Some Application of PA in Horticultural Sector:

1. Estimate fruit yield by measuring tree canopy in citrus orchards (Colaco, Trevisan, Karp, &Molin, 2015).
2. In Plum orchards calculating fruit quantity per each tree before manual harvesting(Kathner&Zude-Sasse,2015).
3. Determine the weight of yield per each block in watermelon fields (Sandri, Pereira, & Vargas, 2014).
4. Measuring stem diameter of trees (Fernandez-Pacheco, Molina-Martinez, Jimenez, Pagan, & Ruiz-Canales, 2014).
5. Exploratory fruit and vegetable optical properties(Cubeddu et al., 2001).

PA TECHNOLOGY AND DEVELOPING COUNTRIES

In developing countries like India, Serbia, Romania, Greece, Egypt, and Ethiopia per-capita, agricultural output is very low due to the majority of landowners have small areas less than one hectare approximately, old technology and cultivation practices, which represent a challenge for feeding the local community, so, the solution could be by exploring the opportunity and investment in the agricultural sector and livestock through using technology and transform into automation agriculture [Tagarakis et al., (2018); Gemtos et al., (2013); Getinet and Getachew (2019);& Abobatta (2020).

In developing countries, most PAtechnologies were beyond the reach of most farmers, also, there was a difficulty in using remote sensing to obtain proper data about soil and weather forecasting and other information (Ehsani, 2011, pp. 11-12), but, during the last two decades the adoption levels of PA technology increasing yearly in different developing countries, while, the adoption of new techniques should start with a basic, affordable, and effective mix of technologies and practices, like using portable soil analysis labs and widespread of mobile phones use facilitate collection data and supply adequate information to the farmers at the proper time to increase the efficiency of various inputs and enhancing crop and livestock production, (Silva et al., 2007), The adoption levels of PA technology increasing yearly in different developing countries like India, Brazil, Argentina, Turkey, Kazakhstan Egypt (Abobatta, 2020), South Africa, and Ethiopia, the variable-rate technology was the earliest technologies used then followed by the auto-guidance systems (Getinet&Getachew, 2019). therefore, the adoption of PA in developing countries considered a key solution to increase crop production to feed the population from the available lands and sustain various resources like soil, water, and energy (Das, Pathak, Meena, &Mallikarjun, 2018), also, PA could be an effective investment when choosing the right crops like in Brazil, whereas cultivation of soybean and maize was profitable under PA system, so, PA system considered a powerful tool when applied properly based on local farmers conditions (Borghi, Avanzi, Bortolon, Junior, &Bortolon, 2016).

There are various constraints in developing countries limiting the adoption of PA system include:

1. Small size of the majority of farms.
2. Low socio-economic condition of farmers.
3. Wide gaps in Knowledge and technology.
4. Large variation of cropping system
5. Shortage of market performance.
6. Scarcity of local technical expertise.
7. Lack of data collection and information availability.

Magruder (2018) reported that, there are various factors affecting the cost of adoption of PA include:

a. Costs of machines.
b. Navigation technique.
c. Costs of services.
d. Farm size.

CHALLENGES FOR PRECISION AGRICULTURE

There are different factors affecting PAincludes Awareness, characteristics of the farms, personality and family structure of the farmer, features of the equipment, characteristics of the technology, legal affairs, social interaction, etc. (McBratney, Whelan, &Ancev, 2005).

One of the most important factors in favor of the adoption of the PA technologies is farm size, therefore, the countries with wide farms like the US, Australia, Canada, Brazil, and Argentina tend to adopt these technologies in bigger areas (Say, Keskin, Sehri, &Sekerli2017), while, in developing countries, there are various challenge for adoption of PA(Mondal&Basu, (2009), some of this challenges are:

1. The majority of farms are small in size (less than 2 hectare).
2. Difficulty of collecting adequate data to extract the required knowledge.
3. The lack of a professional workforce.

The lack of a workforce considered the main challenge for adoption to PA system in developing countries, experienced manpower that has technology literate, creative, innovative, fully trained in their discipline, able to utilize and interpret information gained from information-age technologies to make smart management decisions and has the ability to convert locally collected information into practical solutions, therefore, under developing countries conditions with small farm size, farmers must form large entities to be able to using PA partially by providing customized information and services that increase productivity and take advantage of the precision farming applicationslike profitability, protect the environment, sustainability, also, it is preferable to introduce PAto high-value crops to encourage farmers to use this technology (FAO, 2019).

From another side, different developing countries like India, China, Kenya, Egypt, Ethiopia, and Bangladesh, are starting from the last decade preparing to follow the experience of the developed world in PAand are starting to investigate the new technology (Katke, 2019, pp.863-869).

Some reason for slow adoption to AP in developing countries:

1. The lack of experience of using these technologies by the majority of farms in developing countries (Say et al., 2017).
2. High capital costs may discourage farmers to not adopt this method of farming (Kanter, Bell, &McDermid,2019).
3. There are various PA techniques are still under developing and requires expert advice before actual implementation (Bagheri&Marzieh, 2014).
4. PAneedsmany years to gathering the enough data to fully implement the system (Yirga and Hassan, 2010).
5. Adoption to PA system requires collecting and analyzing big data which represent the main challenges for adoption to PA, especially in developing countries (Wiseman&Sanderson, 2018).
6. Shortage of high technology companies in developing countries (Getinet&Getachew, 2019).

The Potential Threats to Precision Agriculture

Due to PA connected online 24/7 and considered a mechanical intensive industry, so, there are various threats faces PA systems, while, until now there does not fully understand for potential threats to PA, or maybe not treated seriously (Window, 2019), some potential threats to PA include:

1. Data theft and stealing resources which lead to reputation loss.
2. Destruction of equipment.
3. Obtaining information that harms competitors 'financial interests.

There are different vectors of threats through using USB thumb drives, spear-phishing, and other malicious cyber-attacks which need more careful treating to avoid the threat and stopped any attack quickly (West, 2018), the threats for PA not only cyber-attacks, its include any vector that affects negatively on agricultural production like natural disasters which affect crop and livestock productivity, terrorist attacks, equipment breakdown, or insider threats, threats can attack any part of PA system, like attack a vulnerability part of machine and disrupt significant amounts of equipment, which has negative impact on reputation loss of producer machines company or software companies ("Public-Private Analytic Exchange Program" 2018).

PA is unique, however, because it took a highly mechanical labor-intensive industry and connected it online, dramatically increasing the attack space available to threat actors. Due to this, otherwise common threats may have unique and far-reaching consequences on the agricultural industry (Jawad, Nordin, Gharghan, Jawad, & Ismail, 2019).

A consequence of this rapidly advancing digital revolution is the increased exposure to cyber and other vulnerabilities to the agricultural sector.

PA system allows employs a variety of embedded and connected technologies that rely on remote sensing, global positioning systems, and communication systems to generate big data, data analytics, and machine learning. These technologies allow for more precise application of agricultural and livestock management inputs such as fertilizer, seeds, and pesticides, resulting in lower costs and improved yields.

A consequence of this rapidly advancing digital revolution is the increased exposure to cyber and other vulnerabilities to the agricultural sector. PA system attacked by different threats ("Public-Private Analytic Exchange Program" 2018), some of the potential threat scenarios are explained below:

Potential Threats

1. **Threats to Confidentiality:**
   In the PA system, data privacy is a top concern issue, all information like crop varieties, yield data, land prices, and herd health need full protective all the time, because there are serious effects for misuse of this information or losing data(Jones et al., 2017).
2. **Threats to Integrity:**
   PA strongly moved to smart farming in both crop and livestock sectors, as a result, to introduce and build of huge sensors network. PA increasingly adopts robotics, machine learning, and automation equipment, also, it relies on information collection and utilization in proper time to take the right decision about crop and livestock farming (Devitt, 2018). Therefore, there are various threats to data integrity are manifesting attacked by nontraditional ways in the agriculture sector.
3. **Threats to Availability**:
   PA system is reliant on heavy use of equipment in all crop and livestock sectors, complex embedded tools, and a sophisticated suite of communication and guidance systems (Yost, Sudduth, Walthall, & Kitchen, 2019), consequently, threats to the machines and equipment availability are results for both cyber-related attacks and natural disasters.
4. **Crowding of the spectrum signal:**
   Due to crowding of the spectrum with different signals, loss of guidance signal considered risk faces farmers routinely which has negative impacts in PAsystem, also, access to the guidance systems could be suffered because some conflict or during various crises, which reduces the ability to fully using AP equipment.
5. **Failing of Smart livestock facility:**
   New house buildings or smart houses of livestock which connected through the satellite and internet create new threats and open the PA system to attack by malicious threat, also, mankind errors increase the complexity of management and affect the health and productivity of animals (Banhazi et al., 2012).

There is various security control tools used in securing PA technologies includes:

1. Implement Email and Web Browser Protections
2. Use a critical control for Network, Protocols, and minimize unsecured ports.
3. Survey and control all devices for avoiding unauthorized ones.
4. Control software assets and regular updating for authorized software.
5. Determine authorized users to control the AP system.
6. Install and separate a virtual local area network (VLAN) from business networks to increase the safety of operational machines.

The main threats to PA related to information and data security, therefore, farmers need to use some security keys (Figure 4) which employed in other industries to save the data from malicious threats (Suciu, Istrate, &Dițu, 2019), protect various machines, and their users like:

1. Ability to recovery required data.
2. Protection privacy information and data
3. Guarantee ownership of data and information

Figure 4. Schematic of main data security keys

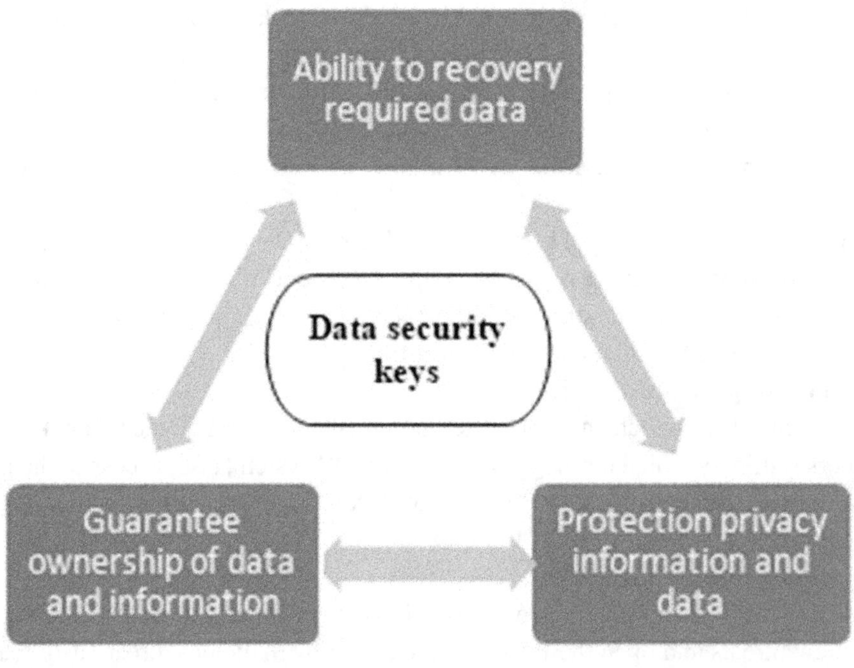

CONCLUSION

PA is an advanced management system that aims to decrease inputs, optimizing productivity, increase farmer profitability, sustain natural resources, and protect the environment, by using information, technology like field mapping, variable rate control, yield monitoring. Under harsh climate change conditions, adoption for PA could be a novel key to providing enough food production, there are different factors affect adoption to PA system include personality characters of farmers, farm size, features of the machines, characteristics of the technology, legal affairs, and social interaction. There is a wide range of technologies used in PA like soil mapping, yield monitoring mapping, and automatic guidance, these technologies are used to provide numerous advantages for farmers such as using the adequate quantity of seeds and agrochemicals at the accurate time and in the precise location in the field.

Adoption to PA system in developing countries could be a key solution to provide enough food for the local community and could be an effective investment when choosing the right crops, in the same time there are different factors slowly adoption to PA like shortage of data collection and information availability, the small scale of the majority of farms, also, it's considered a family-owned business.

The main threats to PA associated with information and data security, therefore, using of security key is necessary to protect the data from malicious threats and provide the ability to recovery required data at the proper time.

ACKNOWLEDGMENT

This research received no specific grant from any funding agency in the public, commercial, or not-for-profit sectors.

REFERENCES

Abdulwaheed, A. (2019). Benefits Of Precision Agriculture in Nigeria. London Journal of Research in Science: Natural and Formal. 19(2). *Compilation*, *1*(0), 19–34.

Abobatta, W. F. (2018). Nanotechnology Application in Agriculture. *Acta Scientific Agriculture*, *2*(6), 99–102.

Abobatta, W. F. (2020) Precision Agriculture Age. *Open Acc J Agri Res.*

Aggelopoulou, K. D., Wulfsohn, D., Fountas, S., Gemtos, T. A., Nanos, G. D., & Blackmore, S. (2010). Spatial variation in yield and quality in a small apple orchard. *Precision Agriculture*, *11*(5), 538–556. doi:10.100711119-009-9146-9

Anjum, M., & Pradhan, S. N. (2018). Application of nanotechnology in precision farming: A review. *International Journal of Chemical Studies*, *6*(5), 755–760.

Bagheri, N., & Bordbar, M. (2014). Solutions for fast development of precision agriculture in Iran. *Agricultural Engineering International*, *16*(3), 119–123.

Balafoutis, A., Beck, B., Fountas, S., Vangeyte, J., van der Wal, T., Soto, I., ... Eory, V. (2017). Precision Agriculture Technologies Positively Contributing to GHG Emissions Mitigation, Farm Productivity and Economics. *Sustainability*, *9*(8), 1339. doi:10.3390u9081339

Banhazi, T. M., Babinszky, L., Halas, V., & Tscharke, M. (2012). Precision Livestock Farming: Precision feeding technologies and sustainable animal production. *International Journal of Agricultural and Biological Engineering*, *5*(4), 1–7.

Bhunia, G. S. (2019). Horticulture Development in India: Issues and Scenario of Space Technology. *World J Agri& Soil Sci.*, *3*(1), 1–5. doi:10.33552/WJASS.2019.03.000555

Bligaard, J. (2013). *Identified user requirements for precision farming in Germany, Finland and Denmark*. Project Report. Available online at https://www.landbrugsinfo.dk/Planteavl/Praecisionsjordbrug-og-GIS/Filer/pl_13_1370_b1.pdf

Borghi, E., Avanzi, J. C., Bortolon, L., Junior, A. L., & Bortolon, E. S. O. (2016). Adoption and use of precision agriculture in Brazil: Perception of growers and service dealership. *The Journal of Agricultural Science*, *8*(11), 89–104. doi:10.5539/jas.v8n11p89

Bramley, R. G. V., Ouzman, J., & Gobbett, D. L. (2019). Regional scale application of the precision agriculture thought process to promote improved fertilizer management in the Australian sugar industry. *Precision Agriculture*, *20*(2), 362–378. doi:10.100711119-018-9571-8

Colaco, A. F., Trevisan, R. G., Karp, F. H. S., & Molin, J. P. (2015). Yield mapping methods for manually harvested crops. *Precision Agriculture*, *15*, 225–232. doi:10.3920/978-90-8686- 814-8_27

Cubeddu, R., D'Andrea, C., Pifferi, A., Taroni, P., Torricelli, A., Valentini, G., Dover, C., Johnson, D., Ruiz-Altisent, M., & Valero, C. (2001). Nondestructive quantification of chemical and physical properties of fruits by time-resolved reflectance spectroscopy in the wavelength range 650–1000 nm. *Applied Optics*, *40*(4), 538–543. doi:10.1364/AO.40.000538 PMID:18357029

Das, U., Pathak, P., Meena, M. K., & Mallikarjun, N. (2018). Precision Farming a Promising Technology in Horticulture: A Review. *Int. J. Pure App. Biosci.*, *6*(1), 1596–1606. doi:10.18782/2320-7051.3088

De Ollas, C., Morillon, R., Fotopoulos, V., Puertolas, J., Ollitrault, P., Gomez-Cadenas, A., & Arbona, V. (2019). Facing Climate Change: Biotechnology of Iconic Mediterranean Woody Crops. *Front. Plant Sci*, *10*, 427. doi:10.3389/fpls.2019.00427 PMID:31057569

Devitt, K. S. (2018). Cognitive factors that affect the adoption of autonomous agriculture. *Farm Policy Journal*, *15*(2), 49–60.

Dokin, B. D., Aletdinova, A. A., & Kravchenko, M. S. (2016). Prospects and features of robotics in russian crop farming. *IOP Conf. Series: Journal of Physics: Conf. Series, 803*(2017). https://iopscience.iop.org/article/10.1088/1742-6596/803/1/012032/

Ehsani, R. (2011). Precision Agriculture for Small Growers. *Resource Magazine*, *18*, 11–12.

Ess, D., & Morgan, M. (2003). *The Precision-Farming Guide for Agriculturists*. Deere & Company.

Fernandez-Pacheco, D. G., Molina-Martinez, J. M., Jimenez, M., Pagan, F. J., & Ruiz-Canales, A. (2014). SCADA Platform for regulated deficit irrigation management of almond trees. *Journal of Irrigation and Drainage Engineering*, *140*(5), 04014008. doi:10.1061/(ASCE)IR.1943-4774.0000718

Finger, R., Swinton, S. M., El Benni, N., & Walter, A. (2019). Precision Farming at the Nexus of Agricultural Production and the Environment. *Annual Review of Resource Economics*, *11*(1), 313–335. doi:10.1146/annurev-resource-100518-093929

Food and Agriculture Organization. (2017). *The future of food and agriculture – Trends and challenges*. Available online at http://www.fao.org/3/a-i6583e.pdf

Food and Agriculture Organization. (2019). *Digital technologies in agriculture and rural areas – Status report*. Available online at http://www.fao.org/3/ca4985en/ca4985en.pdf

Foote, W. (2015). *To Feed the World in 2050, We Need to View Small-Scale Farming as a Business*. Skoll World Forum. Available online at https://www.forbes.com/sites/skollworldforum/2015/04/06/to-feed-the-world-in-2050-we-need-to-view-small-scale-farming-as-a-business/

Fountas, S., Aggelopoulou, K., Bouloulis, C., Nanos, G. D., Wulfsohn, D., Gemtos, T. A., Paraskevopoulos, A., & Galanis, M. (2011). Site-specific management in an olive tree plantation. *Precision Agriculture*, *12*(2), 179–195. doi:10.100711119-010-9167-4

Fountas, S., Van der Wal, T., Eory, V., & Gómez-Barbero, M. (2019). *The contribution of precision agriculture technologies to farm productivity and the mitigation of greenhouse gas emissions in the EU*. Available online at: https://ec.europa.eu/jrc/en/publication/contribution-precision-agriculture-technologies-farm-productivity-and-mitigation-greenhouse-gas

Gemtos, T., Fountas, S., Tagarakis, A., & Liakos, V. (2013). Precision Agriculture Application in Fruit Crops: Experience in Handpicked Fruits. *Procedia Technology*, *8*, 324–332. doi:10.1016/j.protcy.2013.11.043

Getinet, A., & Getachew, A. (2019). Precision Agriculture and the Need to Introduce in Ethiopia. *Ethiop. J. Agric. Sci*, *29*(3), 139–158.

Gregory, P. J., & George, T. S. (2011). Feeding nine billion: The challenge to sustainable crop production. *Journal of Experimental Botany*, *62*(15), 5233–5239. doi:10.1093/jxb/err232 PMID:21841178

Grisso, R., Alley, M., & Groover, G. (2009). *Precision Farming Tools: GPS Navigation*. Virginia Cooperative Extension. Publication No 442-501.

Hofstee, J. W., & Molema, G. J. (2002). *Machine vision based yield mapping of potatoes*. Paper number 021200, 2002 ASAE Annual Meeting. Available online at www.asabe.org

Hornung, A., Khosla, R., Reich, R., Inman, D., & Westfall, D. G. (2006). Comparison of Site-Specific Management Zones: Soil-Color-Based and Yield-Based. *Agronomy Journal*, *98*(2), 407–415. doi:10.2134/agronj2005.0240

Jawad, H. M., Nordin, R., Gharghan, S. K., Jawad, A. M., & Ismail, M. (2017). Energy-Efficient Wireless Sensor Networks for Precision Agriculture: A Review. *Sensors (Basel)*, *17*(8), 1–45. doi:10.339017081781 PMID:28771214

Jones, J. W., Antle, J. M., Basso, B., Boote, K. B., Conant, R. T., Ian Foster, I., ... Wheeler, T. R. (2017). Toward a new generation of agricultural system data, models, and knowledge products: State of agricultural systems science. *Agricultural Systems*, *155*, 269–288. doi:10.1016/j.agsy.2016.09.021 PMID:28701818

Kanter, D. R., Bell, A. R., & McDermid, S. S. (2019). Precision Agriculture for Smallholder Nitrogen Management. *One Earth*, *1*(3), 281–284. doi:10.1016/j.oneear.2019.10.015

Kathner, J., & Zude-Sasse, M. (2015). Interaction of 3D soil electrical conductivity and generative growth in Prunusdomestica. *European Journal of Horticultural Science*, *80*(5), 231–239. doi:10.17660/eJHS.2015/80.5.5

Katke, K. (2019). Precision Agriculture Adoption: Challenges of Indian Agriculture. *International Journal of Research and Analytical Reviews*, *6*(1), 863–869.

Keskin, M., Sekerli, Y. E., Say, S. M., & Topcueri, M. (2017). Farmers' satisfaction level with GNSS-based tractor auto guidance systems in Adana province of Turkey. *13th Int Congress on Mechanization and Energy in Agriculture & Int. Workshop on Precision Agriculture*. Available online at https://www.researchgate.net/publication/320024304_Farmers'_Satisfaction_Level_with_GNSS-Based_Tractor_Auto_Guidance_Systems_in_Adana_Province_of_Turkey_Adana_Ilinde_GNSS_Esasli_Traktor_Otomatik_Dumenleme_Sistemi_Kullanan_Ciftcilerin_Memnuniyet_Duzeyi

Krishnan, M., Foster, Ch. A., Strosser, R. P., Glancey, J. L., & Sun, J.-Q. (2006). Adaptive modeling and control of a manure spreader for precision agriculture. *Computers and Electronics in Agriculture*, *52*(1–2), 1–10. doi:10.1016/j.compag.2005.11.005

Kumari, S., Singh, T. P., & Prasad, S. (2019). Climate Smart Agriculture and Climate Change. *International Journal of Current Microbiology and Applied Sciences*, *8*(3), 1112–1137. doi:10.20546/ijcmas.2019.803.134

Kutter, T., Tiemann, S., Siebert, R., & Fountas, S. (2011). The role of communication and cooperation in the adoption of precision farming. *Precision Agriculture*, *12*(1), 2–17. doi:10.100711119-009-9150-0

Lakhiar, I. A., Jianmin, G., Syed, T. N., Chandio, F. A., Buttar, N. A., & Qureshi, W. A. (2018). Monitoring and Control Systems in Agriculture Using Intelligent Sensor Techniques: A Review of the Aeroponic System. Journal of Sensors Volume 2018. *Article ID*, *8672769*, 18. doi:10.1155/2018/8672769

Leonard, E. (2014). *Precision Ag Down Under*. Available online at https://www.precisionag.com/in-field-technologies/guidance/precision-ag-down-under/

Li, Z., Taylor, J., Frewer, L., Zhao, C., Yang, G., Liu, Z., &Yu, C. (2019). A comparative review on the state and advancement of site-specific crop management in the UK and China. *Front. Agric. Sci. Eng.*, 1–21. doi:10.15302/J-FASE-2018240

Liao, M. (2017). *XairCraft Launched in Japan Targeting Global Precision Farming*. Available online at https://globenewswire.com

MacDonald, J. M., & McBride, W. D. (2009). The Transformation of U.S. Livestock Agriculture: Scale, Efficiency, and Risks. *Economic Information Bulletin, 43*. Available online at https://www.ers.usda.gov/webdocs/publications/44292/10992_eib43.pdf?v=0

Magruder, J. R. (2018) An Assessment of Experimental Evidence on Agricultural Technology Adoption in Developing Countries. *Annu. Rev. Resour. Econ., 10*, 299–316. doi:10.1146/annurev-resource100517-023202

Marino, S., & Alvino, A. (2014). Proximal sensing and vegetation indices for site-specific evaluation on an irrigated crop tomato. *European Journal of Remote Sensing*, *47*(1), 271–283. doi:10.5721/EuJRS20144717

McBratney, A., Whelan, B., Ancev, T., & Bouma, J. (2005). Future Directions of Precision Agriculture. *Precision Agriculture*, *6*(1), 7–23. doi:10.100711119-005-0681-8

McBride, W. D., & Daberkow, S. D. (2003). Information and the Adoption of Precision Farming Technologies. *Journal of Agribusiness*, *21*(1), 21–38.

Shang, Y., KamrulHasan, M., Ahammed, G. J., Li, M., Yin, H., & Zhou, J. (2019). Applications of Nanotechnology in Plant Growth and Crop Protection: A Review. *Molecules (Basel, Switzerland)*, *24*, 2558. doi:10.3390/molecules24142558

Mondal, P., & Basu, M. (2009). Adoption of precision agriculture technologies in India and in some developing countries: Scope, present status and strategies. *Progress in Natural Science*, *19*(6), 659–666. doi:10.1016/j.pnsc.2008.07.020

Montzka, S. A., Dlugokencky, E. J., & Butler, J. H. (2011). Non-CARBON DIOXIDE greenhouse gases and climate change. *Nature*, *476*(7358), 43–50. doi:10.1038/nature10322 PMID:21814274

Navulur, S., Sastry, A. S. C. S., & Giri Prasad, M. N. (2017). Agricultural Management through Wireless Sensors and Internet of Things. *Iranian Journal of Electrical and Computer Engineering*, *7*(6), 3492–3499. from. doi:10.11591/ijece.v7i6.pp3492-3499

Patil, V. C., Nadagouda, B. T., & Al-Gaadi, K. A. (2012). Spatial Variability and Precision Nutrient Management in Sugarcane. *Photonirvachak (Dehra Dun)*, *41*(1), 183–189. doi:10.100712524-011-0181-4

Paustian, M., & Theuvsen, L. (2017). Adoption of precision agriculture technologies by German crop farmers. *Precision Agriculture*, *18*(5), 701–716. doi:10.100711119-016-9482-5

Pierpaoli, E., Carlia, G., Pignattia, E., & Canavaria, M. (2013). Drivers of Precision Agriculture Technologies Adoption: A Literature Review. *Procedia Technology*, *8*, 61–69. doi:10.1016/j.protcy.2013.11.010

Pooja, S., Uday, D. V., Nagesh, U. B., & Talekar, S. G. (2017). Application of MQTT protocol for real time weather monitoring and precision farming. *2017 International Conference on Electrical, Electronics, Communication, Computer, and Optimization Techniques (ICEECCOT)*, 1-6. 10.1109/ICEECCOT.2017.8284616

Public-Private Analytic Exchange Program. (2018). *Threats to Precision Agriculture (2018).* Available online at https://www.dhs.gov/sites/default/files/publications/2018%20AEP_Threats_to_Precision_Agriculture.pdf

Rains, C. R., & Thomas, D. L. (2009). Precision farming: An introduction. The University of Georgia Bulletin, 1186, 12.

Reichardt, M., & Jürgens, C. (2009). Adoption and future perspective of precision farming in Germany: Results of several surveys among different agricultural target groups. *Precision Agriculture*, *10*(1), 73–94. doi:10.100711119-008-9101-1

Sandri, D., Pereira, J. A., & Vargas, R. B. (2014). Production costs and profitability of watermelon under different water depths and irrigation systems. Irriga. *Botucatu*, *19*(3), 414–429. doi:10.15809/irriga.2014v19n3p414

Say, S. M., Keskin, M., Sehri, M., & Sekerli, Y. E. (2017) *Adoption of precision agriculture technologies in developed and developing countries.* International Science and Technology Conference, Berlin, Germany. Available online at http://tojsat.net/journals/tojsat/articles/v08i01/v08i01-02.pdf

Schmitz, A., & Moss, C. B. (2015). Mechanized Agriculture: Machine Adoption, Farm Size, and Labor Displacement. *AgBioForum*, *18*(3), 278–296.

Shafi, U., Mumtaz, R., García-Nieto, J., Hassan, S. A., Zaidi, S. A. R., & Iqbal, N. (2019). Precision Agriculture Techniques and Practices: From Considerations to Applications. *Sensors (Basel)*, *19*(17), 3796. doi:10.339019173796 PMID:31480709

Silva, C. B., do Vale, S. M. L. R., Pinto, F. A., Müller, C. A., & Moura, A. D. (2007). The economic feasibility of precision agriculture in MatoGrossodoSul State, Brazil: A case study. *Precision Agriculture*, *8*(6), 255–265. doi:10.100711119-007-9040-2

Steeneveld, W., Tauer, L. W., Hogeveen, H., & Oude Lansink, A. G. J. M. (2012). Comparing technical efficiency of farms with an automatic milking system and a conventional milking system. *Journal of Dairy Science*, *95*(12), 7391–7398. doi:10.3168/jds.2012-5482 PMID:23021754

Suciu, G., Istrate, C., & Dițu, M. (2019). *Secure smart agriculture monitoring technique through isolation.* Available online at https://ieeexplore.ieee.org/document/8766433/authors#authors

Sulecki, J. (2018). *Association seeks definitive definition of "precision agriculture"—What's your vote?* https://www.precisionag.com/events/association-seeks-definitive-definition-of-precision-agriculture-whats-your-vote

Tagarakis, A. C., van Evert, F., Kempenaar, C., Ljubicic, N., Milic, D., Crnojevic-Bengin, V., & Crnojevic, V. (2018). Opportunities for precision agriculture in Serbia. *Proceedings of the 14th International Conference on Precision Agriculture.* Available online at https://ispag.org/proceedings/?action=download&item=5042

Takacs-Gyorgy, K. (2009). Importance of precision farming in improving the environment. *Žemes Ukio Mokslai, 16*(3–4), 217–223. Available online at http://mokslozurnalai.lmaleidykla.lt/publ/1392-0200/2009/3-4/217-223.pdf

Tamirat, T. W., & Pedersen, S. M. (2019). Precision irrigation and harvest management in orchards: an economic assessment. *Journal of Central European Agriculture, 20*(3),1009-1022. DOI: /. doi:10.5513/JCEA01/20.3.2160

Tey, Y. S., & Brindal, M. (2012). Factors influencing the adoption of precision agricultural technologies: A review for policy implications. *Precision Agriculture*, *13*(6), 713–730. doi:10.100711119-012-9273-6

Thessler, S., Kooistra, L., Teye, F., Huitu, H., & Bregt, A. K. (2011). Geosensors to Support Crop Production: Current Applications and User Requirements. *Sensors (Basel)*, *11*(7), 6656–6684. doi:10.3390110706656 PMID:22163978

Toscano, P., Castrignano, A., Di Gennaro, S. F., Vonella, A. V., Ventrella, D., & Matese, A. (2019). A Precision Agriculture Approach for Durum Wheat Yield Assessment Using Remote Sensing Data and Yield Mapping. *Agronomy (Basel)*, *9*(8), 437. doi:10.3390/agronomy9080437

West, J. (2018). A Prediction Model Framework for Cyber-Attacks to Precision Agriculture Technologies. *Journal of Agricultural & Food Information*, 1-24. DOI: doi:10.1080/10496505.2017.1417859

Whelan, B. M., McBratney, A. B., & Stein, A. (2003). On-farm field experiments for precision agriculture. *Precision Agriculture, Proceedings of the 4th European Conference on Precision Agriculture*, 731–737.

Whitney, J. D., Ling, Q., Miller, W. M., & Wheaton, T. A. (2001). A DGPS yield monitoring system for Florida citrus. *Applied Engineering in Agriculture*, *17*(2), 115–119. doi:10.13031/2013.5451

Window, M. (2019). *Security in Precision Agriculture Vulnerabilities and risks of agricultural systems* (M Sc. Thesis). Lulea University of Technology Department of Computer Science, Electrical and Space Engineering. Available online at http://ltu.diva-portal.org/smash/get/diva2:1322203/FULLTEXT02.pdf

Wiseman, L., Sanderson, J., & Robb, L. (2018). Rethinking Ag data ownership. *Farm Policy Journal*, *15*(1), 71–77.

Yadav, V. K., Yadav, M. R., Kumar, R., Yadav, B., & Verma, A. K. (2017). Potentials and Prospects of Precision Farming in Indian Agriculture. *Marumegh*, *2*(2), 47-53. Available online at www.marumegh.com

Yirga, C., & Hassan, R. M. (2010). Social costs and incentives for optimal control of soil nutrient depletion in the central highlands of Ethiopia. *Agricultural Systems*, *103*(3), 153–160. doi:10.1016/j.agsy.2009.12.002

Yost, M. A., Sudduth, K. A., Walthall, C. L., & Kitchen, N. R. (2019, December 22). Public–private collaboration toward research, education and innovation opportunities in precision agriculture. *Precision Agriculture*, *20*(1), 4–18. doi:10.100711119-018-9583-4

Yousefi, M. R., & Razdari, A. M. (2014). Application of GIS and GPS in Precision Agriculture (A Review). *International Journal of Advanced Biological and Biomedical Research*, *2*(4), 473–476.

Zaman, Q. U., Schumann, A. W., & Miller, W. M. (2005). Variable rate nitrogen application in Florida citrus based on ultrasonically-sensed tree size. *Applied Engineering in Agriculture*, *21*(3), 331–335. doi:10.13031/2013.18448

Zude-Sasse, M., Fountas, S., Gemtos, T. A., & Abu-Khalaf, N. (2016). Applications of precision agriculture in horticultural crops. *European Journal of Horticultural Science*, *81*(2), 78–90. doi:10.17660/eJHS.2016/81.2.2

Section 2
Enabling Technologies

This section talks about the usage of the engineering technologies, such as electronics, sensing, information, and communication, in PA systems to collect field data; convey it; record, process, and analyze the collected data; extract decisions; and take automatic or manual actions.

Chapter 3
An Overview of Internet of Things Technology Applied on Precision Agriculture Concept

Reinaldo Padilha França
State University of Campinas (UNICAMP), Brazil

Rangel Arthur
Faculty of Technology, State University of Campinas (UNICAMP), Brazil

Ana Carolina Borges Monteiro
State University of Campinas (UNICAMP), Brazil

Yuzo Iano
State University of Campinas (UNICAMP), Brazil

ABSTRACT

The internet of things (IoT) is characterized by devices that communicate without human interference, sending and receiving data online, to which they have shaped the way of connecting household appliances, machines, and equipment, cars, among other things, also arriving at the field through characterized by the communication between devices, sensors, drones, and machines. They have great potential to improve production processes, making agriculture increasingly digital, creating solutions, connectivity, and training for specialist labor. As well as irrigation systems and other intelligent machines with the ability to talk to each other enabling management in the use of energy, resources, and inputs making the production process more efficient. Precision agriculture encompasses a series of components and factors from which the best procedures can be chosen that are appropriate in a given agricultural operation that effectively meets your needs, also related to the application of inputs at the right time and in the right place, following the growth and productivity over the entire length of a plantation by controlling pests, among other technologies, providing a reduction in production costs and spending on inputs, reducing the pollution of nature by the pesticides used, making it possible to reduce operating costs, increasing precision in obtaining results in the same way as less variability in production. Therefore, this chapter aims to provide an updated overview and review of the use of the internet of things in the precision agriculture system showing and approaching its success relation, with a concise bibliographic background, categorizing and synthesizing the potential of both technologies.

DOI: 10.4018/978-1-7998-5000-7.ch003

INTRODUCTION

The Internet of Things (IoT) has the ability to transform the world we live in; enabling better management of electricity use, creating more efficient industries, connected cars, and smarter cities are all components of the IoT equation, controlling vehicle traffic in large cities, creating the concept of smart cities, which inspired by smart farms, where management and production processes are integrated into them, where it is seen that their application of technology such as the IoT in agriculture can have the greatest impact, with sensors installed on agricultural machines, it is possible to obtain a soil information series, which can guide the actions of acidity correction, irrigation, and planting (Gilchrist, 2016).

Advances in precision agriculture, with sensors installed on equipment and networked, indicate that future agriculture should be increasingly supported by scientific knowledge, since the global population has grown over the years, where to feed in this population, the agricultural industry must embrace the IoT. Since technology can help in extreme weather conditions and increasing climate change, lessening the environmental impact resulting from intensive farming practices, generating a demand for more accurate food (Ahmad & Mahdi, 2018).

Intelligent agriculture, based on IoT technologies, will enable producers and farmers to reduce waste and increase productivity, from the amount of fertilizer used to the number of trips the farm vehicles produced. It can be considered as something that makes agricultural practice more controlled and precise when it comes to livestock and crop cultivation, wherein this approach to farm management, a key component is the use of IT and various items such as sensors, farming systems. control, robotics, autonomous vehicles, automated hardware, variable rate technology, and so on (Siddique, 2019).

The manufacturer's adoption of high-speed Internet access, reliable, low-cost mobile devices, and satellites used for imaging and positioning are key technologies that characterize the trend toward precision farming. Coupled with smart Internet-based agriculture of things, a system is built to monitor the field of cultivation with the help of sensors, capturing signals such as light, air humidity, temperature, soil moisture, among others, and can thus automate the system. Irrigation Giving farmers the ability to monitor field conditions from anywhere. IoT-based smart agriculture is highly efficient compared to the conventional approach (Jeschke, 2017).

IoT-based smart farming applications are not only intended for large conventional agricultural operations, but may also be new levers to lift other common or growing agricultural trends such as organic farming, family farming, ranging from complex or small spaces, private livestock, and livestock/or crops, even preserving particular or high-quality varieties, and improving highly transparent agriculture (Prathibha, 2017).

Making agriculture increasingly digital is a goal that depends on many factors ranging from creating solutions, connectivity to empowering the workforce. Since applications for more assertive decision-making by the farmer involve collecting thousands of data by sensors and robots or automated machines, high information and image processing, and analysis.

In terms of environmental issues, IoT-based smart agriculture can offer major benefits, including more efficient water use or optimization of inputs and treatments. Where drones are being used in agriculture to improve various agricultural practices as long as it has land and aerial applications are being used, assessing crop health, irrigation, crop monitoring, crop spraying, planting and soil, and field analysis. With key benefits from using drones from crop health imagery, integrated GIS mapping, ease of use, the potential to increase yields, and save time (Prathibha, 2017; Suma, 2017).

With strategy and planning based on real-time data collection and processing, drone technology will deliver a high-tech transformation for the agricultural industry. Where large farm owners can use wireless IoT applications to collect data on the location, welfare, and health of their livestock, this information enables them to identify animals that are sick and separate them from the herd, thereby preventing the spread of the disease. Similarly, reducing labor costs as ranchers can locate their livestock with the help of IoT-based sensors (Wolfert 2017).

Greenhouse agriculture is another technological methodology that helps to increase the yield of vegetables, fruits, crops as long as they control environmental parameters through a proportional control mechanism. Sincemanual intervention is a less effective method, which results in lost production, energy loss, and labor cost. So, a smart greenhouse can be designed with the help of the IoT; monitoring and controlling the climate of the environment, eliminating the need for manual intervention (Hanan 2017).

The IoTcan be applied to any tool or function, enabling the collection of data that will be sent to a communication center via wireless sensors, located on the ground or on tractors, which together with data analysis software will allow a more accurate field mapping; sensors that can measure the water level in the soil and collect data indicating moisture or water balance, leading to the need for irrigation; sensors that analyze soil properties, enabling intelligent seed planting and optimized application of inputs and pesticides; through plant images captured by cameras, drones, and satellites that can assist in pest detection, capture crop information and map the productivity of each part of the land; or even equipment embedded in agricultural machinery that indicates the need for maintenance; or equipment installed in silos that may indicate storage conditions; reaching even sensors with more technological content, being intracorporeal in animals, assisting in the monitoring of animal health and welfare (Tzounis 2017).

Thus, these tools contribute together with the IoTon the farm enabling the collection of data (information) through equipment (drones, cameras, sensors, satellite), sending this data, through a communication system, so that it be stored and processed, being analyzed, allowing the consequent emission of a diagnosis, generating monitoring or some information from the analysis made.Today there is a huge variety of computational and mechanized solutions, such as pesticide application and more rational irrigation, planting and harvesting according to local weather conditions, however, one of the main challenges for institutions and professionals working in the sector is to interpret this amount. data to extract relevant information and integrate it into solutions that ensure the quality of agricultural production (Kamilaris, 2016).

Yet another horizon remains that still some large landowners and small farmers need to understand the potential of the IoT market for agriculture by installing smart technologies to increase competitiveness and sustainability in their productions. Affecting a demand for an exponentially growing population that can be successfully met if farmers and small farmers successfully implement IoT agricultural solutions (Mekala& Viswanathan 2017).

Finally, agriculture has gone through several revolutions, but the difference is that this digital transformation is able to converge several areas, such as instrumentation, genomic editing tools, techniques of genetic improvement, and bioinformatics., soil and plant integrating technologies that overcome the challenges of the growing demand for food without land expansion and in a context of natural resource constraints. So, the IoT being applied to capture drone data, satellite imagery, sensors, and automated machinery uses all of this information, improving from planting to storage and logistics throughout the entire production chain (Imran,2018).

Therefore, this chapter aims to provide an updated overview and review of the use of the IoT in Precision Agriculture system, precision farming, site-specific crop management, or precision agronomy, showing and approaching its success relation, with a concise bibliographic background, categorizing and synthesizing the potential of both technologies.

METHODOLOGY

This study was based on the research of scientific articles and books that address the theme of the present work, exploring mainly a historical review and applicability of techniques related to the **IoT technology applied to Precision Agriculture**. These papers were analyzed based on the publication date of fewer than **5** years, with emphasis onpublications and indexing in renowned databases, such as IEEE and Scholar Google

PRECISION AGRICULTURE CONCEPTS

The conventional model of agriculture is one where the planting fields are understood as homogeneous, and often large regions can have similar characteristics, such as soil characteristics, they can vary drastically from one small area to another, considering that each small portion of land needs a specific amount of irrigation, fertilizers, and pesticides. In this sense, treating the regions homogeneously using average inputs and with the same formulation for the whole area, ignoring such differences, is what occurs in the traditional model, since some regions receive excessive amounts of these inputs, resulting in waste, and other regions which receive few inputs, with insufficiency. In practice, a crop does not have uniform productivity, as none of its areas is able to maximize production, either due to excess or lack of inputs (Plumecocq 2018).

In this sense, with the rapid transformations that modern agriculture has undergone in the last decades, it has become a highly competitive activity, demanding from agribusiness and rural producers a high degree of specialization in order to increase the managerial capacity of rural companies, since technology in the field is increasingly inserted in the farmer's life, and it is necessary to understand the main changes and innovations that result in greater production, with better quality and more profits (Plumecocq 2018).

Precision agriculture is a system of administration of field information using technology, which allows more effective control of agricultural production, since field information is collected and analyzed by sensors, enabling automatic decision making and application aiming at a higher quality and productivity of production, that is, it allows the control of information for better management of rural production. Associated with this administrative capacity is the producer's ability to collect data and information related to his productive area, intending to adapt his reality with the new technologies, being essential for the modern rural producer to be efficient in the application of the available resources ensuring success in its activity, relying on obtaining information on the factors that interact in the crop, seeking to positively maximize its effects (Shannon 2018).

Precision agriculture can be characterized as the use of technology and information to take more precise actions in the field activities, it is a production system adopted by farmers in advanced technology countries, which perform information management from advances in referencing and positioning technology, such as GPS (Global Positioning System) and remote sensing technologies. In this context, it is important to invest in new technologies with low computational consumption and low level of abstraction, which facilitate data transmission, without any system crashes (Shannon 2018; França 2020; França 2018).

An agricultural management system based on the spatial and temporal variability of the productive unit allows a more rational exploration of the production systems, leading to increased profitability and sustainability, to the optimization of the use of inputs, and the minimization of environmental impacts. Precision agriculture has the idea of collecting data on what occurs in the crop in a given time, studying these data, and acting in an appropriate and controlled way so that it is more productive and sustainable (Kladivko&Timmenga, 2019).

Traditional agriculture focuses on large areas and sees them as homogeneous, leading to the concept of the average need for the application of inputs such as water, fertilizers, pesticides, among others, making the same amount and formulation of an input to be used for the whole area, meeting only the average needs and not considering the specific needs of each part of the field, resulting in a crop with non-uniform productivity (Scoones, 2017).

On the other hand, precision agriculture provides for the reversal of this type of scenario, allowing the application of agricultural inputs in the correct places and the required quantities, with a philosophy of agricultural management that starts from exact, precise information and with exact decisions, managing a productive field. from the meter to meter, because each piece has different properties, focusing on the main concept of the application of inputs at the right time, the right place, the quantities of inputs needed for agricultural production, for increasingly smaller and more homogeneous areas, as much as the technology and costs involved, and through the technology of sensors to detect information about what is happening in the soil, still considering the possibility of using artificial intelligence, where this information is processed generating decision making. This consolidation of technologies as tools available to the producer allows visualization of the spatial and temporal variability of edaphoclimatic factors, that is, through environmental factors such as climate, relief, lithology, temperature, air humidity, radiation, the type of soil, wind, atmospheric composition, and rainfall, of each agricultural area, considering the peculiarities of each part of the area at the time of management, instead of handling it as if it was uniform (Mikula,2019).

With the precision agriculture production model, it relies on the use of technologies, machines, and farmer control to monitor the information on the planted area, treating each small area as unique, recognizing the differences of each region, bringing benefits and advantages such as each location have its productivity maximized, concerning the inputs they are not wasted, there is less damage to the soil and the environment, generating an increase in productivity in a more sustainable way than traditional methods. Since production and productivity estimates have always been considered essential and extremely important, guiding agriculture such as seed treatment, crop choice, the number of fertilizers applied, the option for agrochemicals; all these decisions being directed to the goal of the highest possible productivity (Duhan, 2017).

In precision agriculture, the organization of the collected information is carried out by integrating the data with the agricultural GPS to the autonomous sprayer, for example, preventing the machine from passing twice in the same place, thus the proper management of an agricultural property uses technology in favor of the producer, obtaining data and mapping the area in the smallest details characterizing the productivity maps, which consider the production with the total quantity produced and the whole productivity, taking into account the quantity produced in a unit, either per plant, per plot, per hectare or bushel (Mitchell, 2018).

Since it is important to be aware of some characteristics of the plant that affect productive capacity, evaluating its productivity as the genotype, considering grain weight, the number of grains per pod, number of pods per plant; the number of plants per meter; climatic condition; soil fertility; relief; the incidence of pests and diseases; variables that when considered concerning the areas of a property generate productivity maps indicating factors to be improved with management. The idea of precision agriculture is the knowledge of the type of soil and production characteristics pertinent to this soil, without considering the meteorological factors, which cause a different product for each part of the field and, from that, to improve the inputs of inputs within portions small in the field, as needed and that they are more economical for production (Molnar &Kinnucan, 2019).

Still considering that it is essential to have good agricultural software, feeding it with the appropriate information, is a tool that helps in decision making and allows the detailed management of the factors that influence the productive potential of the crop, which are software aimed at improving the crops. results of the rural producer, assisting in the production control and agricultural management, making the producer have the information of his property enabling him to make more assertive decisions. In this context, the application of systems related to deep learning and big data is indicated. The performance and viability of agribusiness currently depend on a set of factors, ranging from good management of operations to the use of adequate inputs, since this type of system helps to control production, identifying possible problems and extracting the best from machines and harvest (Minango 2019; Lohr, 2012).

The management of large areas with a focus on generating better results requires a lot of attention from the producer, a broad and detailed vision is needed that accurately defines the right moment to apply pesticides to crops, planting or irrigating, provided that by using software that assists the process of obtaining, analyzing and storing information, managing the activities carried out in the field, controlling and managing the inputs of the property, linked to the importance of management in agriculture with the idea of making the information all in the same place and very clear for the producer, controlling the entire process of the property, the farmer generating possibilities of the forecast until the productivity of a plot, that is, the minimum unit of cultivation of a property that is built based on relief and mechanization planning, making the planning of every harvest can be done (Reed, 2011; Moncada, 2017; Salic &Zelic; Salic &Zelic, 2018).

The agricultural management software facilitates the day-to-day tasks of agribusiness management, optimizing the efficiency of the processes, in addition to providing advantages such as increasing the productivity of the plots, reducing production costs, the greater financial security of the harvest and higher quality agricultural production. They are considered part of precision agriculture, since, with these resources, it is possible to understand everything that is happening on a property, assisting from agricultural management to production control, making the producer have "information about his property in hand "managing to make decisions more assertively and easily (Reed, 2011; Moncada, 2017; Salic &Zelic; Salic &Zelic, 2018).

PRODUCTIVITY MAPS

The use of mapping and sensor technologies to collect information about the soil and to be sure of what each plot needs, applying only what is necessary for each land space, in this sense the utility of productivity maps is the reason for so much variation in productivity from one area to another within a property, concerning the choice of the type of culture and the same cultural treatments in all plots, since an ecosystem is influenced by all biotic and abiotic factors in the same region (Piper, 2017; Huuskonen& Oksanen, 2018; Brevik, 2016).

In this context, the small details are crucial in the way a plant responds to the stimuli submitted, even taking into account that the simple characterization of the environment in which a plant is located already allows projecting variations in productivity, such as areas close to forests since these regions have higher moisture levels, favoring the appearance of fungal diseases and, consequently, a drop in productivity, with a higher incidence of pests and diseases due to the simple fact that the forests are the natural habitat of many insects annual crop pests; Valley bottoms, which have a greater thermal amplitude than in flat locations and tend to have lower night temperatures, especially in regions at risk of frost, since it is necessary to evaluate the risks of implanting cultivars that are not resistant to cold; Tractor maneuver location with the edges of plots and close to roads show planting failures and difficulty in controlling pests and diseases due to the imprecision of machinery in carrying out activities where access is difficult, making it difficult for pesticides to adhere to the leaves., such as excess dust. This is resolved with the aid of technology, such as spray drones; Regions of rugged relief where this type of slope hinders the operation of machinery requiring conservationist practices such as level planting and the adoption of terraces that mitigate the loss of nutrients in the topsoil, which in contrast compared to flat regions, the loss will be greater, negatively influencing the development of the crop (Piper, 2017; Huuskonen& Oksanen, 2018; Brevik, 2016).

In this sense, the productivity maps are obtained, basically, with the information of the grain weight and its moisture content in a certain area, through the installation of sensors in the harvester detecting the grain mass flow and its moisture content in a given area and time unit, and a type of spatial variation can be made every n square meters per collection, or n points per hectare (McEntee, 2020; Huuskonen& Oksanen, 2018).

Through precision farming software, the data obtained are interpreted by interpolating the points and determining colors for each productivity range pre-defined in the software, together with the constant calibration of the devices, which is essential to obtain reliable maps, since that the change of field, related to the difference in cultivation and the planting time are factors that affect the precision of the process, requiring the calibration of the devices, just as the sensors must show the tonnage of grains in values close to those obtained in the balance (Ristorto,2017; Vidoni, 2017).

Precision agriculture transcends the idea of only technological machines, this term means a lot about the control of data by the producer, allowing for better decision making that is smarter and more efficient, increasing productivity in each field, eliminating the need for empirical experience. rural producers to know how different portions of land can have different productivity rates, since this is related to factors such as climate variation; variation in soil temperature; variation in humidity; variation in soil fertility, since from this type of scenario it is possible to treat the different regions that have diversified characteristics with customized approaches to each of them (Hakkim, 2016).

In this way, when mapping the characteristics of the soil of a field with an understanding of climate, humidity, temperature, and other information, it is possible to treat each piece of land in the best way, making the most of its crops,taking advantage of technology making it possible to collect information from the crop every n periods of time employing satellite images, analyzing this information, acquiring the necessary knowledge for the ideal combination of fertilizers and their application rates for plantations, from planting to the development of cultivars, since the productivity maps are a way of showing the results of all the handling carried out on the property, and from a precision farming system, it optimizes all agricultural practices with preservation of the environment and substantial savings to the producer (Huuskonen& Oksanen, 2018; Brevik, 2016; Khana, 2017).

Through the productivity maps, it is possible to dismember the property, allowing to identify specific points of low or high production as well as the reasons that can contribute to the occurrence of these results, rugged relief with a higher rate of nutritional loss, as possible nutritional deficiencies, errors in the application of agrochemicals, the occurrence of nematodes and other pests or diseases. Likewise, it is necessary to use and supply the database correctly, requiring constant calibration of the equipment and the choice of productivity intervals with criteria so that there is no overestimation or underestimation of the data obtained, as well as the constant study of the variables, is essential (Brevik, 2016; Khana, 2017;Maes,2019).

In the same way that treating plots as if they have the same characteristics affects the amount produced in the field, a large amount of inputs is wasted, considering pesticides and seeds some of the inputs that can be thrown in the trash unnecessarily, so when there is knowledge about exactly the amount of inputs a plot needs, greater precision in its application can be obtained generating economies of scale for the rural property, so precision agriculture helps the producer by providing complete fertilizer schedules according to factors such as plant growth stages, through field data, crop characteristics and productivity goals (Ahmad & Mahdi, 2018; Siddique, 2019).

TYPES OF THE PRECISION MAPS

Through the **Mapping of the variability of the land**, through maps of variability or condition, the analysis of each part of the land/crop is carried out where all the management stages are mechanized, so that the management can be individualized for each part of the land. terrain, allowing to know the variability of soil and crop conditions in each region. Considering that through the varied rate of application of inputs, the distributor automatically changes the dosage of inputs; with equipped machines, during their travel, there is the possibility of recognizing the need for inputs at each point of the field. As well as concerning the varied application rate of seeds and fertilizers, they generate savings for the producer, since in the areas with less productive potential, reflecting that these application rates are reduced. And considering that in the areas with the highest productivity, more dense sowing is carried out with a higher dose of fertilizers(Borges et al., 2020; Sciortino et al., 2020).

Maps of soil fertility observe crop productivity, as areas of low fertility can determine low crop productivity, even though fertility is not the only factor influencing productivity. Through sampling and chemical analysis of the soil, it is possible to generate specific maps that demonstrate indications of its fertility making maps with the indexes of macronutrients such as calcium, magnesium, sulfur, phosphorus, potassium. Reflecting on the **Micronutrient Maps**regarding copper, zinc, manganese, boron, molybdenum, iron, related to soil pH; a sum or base saturation; among other aspects(Chen et al., 2020; Nganga et al., 2020; Paris et al., 2020).

Maps of the physical characteristics of the soil are instruments to assist in decision making regarding a certain management system, such as the irrigation system concerning the texture of the soil and its water retention are factors that greatly determine the need for irrigation. Presenting specific aspects of the soil, such as compaction, texture, water holding capacity(Paul et al., 2020; Rasaei et al., 2020).

Topography/Elevation Map contains information used to analyze soil conservation practices, such as level planting. This type of map represents the slope of each cultivated area, that is, the slope of the land surface. It also indicates the productive potential of each area according to the slope of the soil, with the face of the field exposure to the sun, among other factors(Peng et al., 2020; Bian et al., 2020).

Variability Map or remote sensing is related to the distribution of correctives and fertilizers at a variable rate, considering the application rate determined based on this type of map that will be applied according to the need of each area of the field(Chattopadhyay et al., 2020; Shekhar, 2020).

Sowing Recommendation Map refers to the variable sowing rate, since the seeders have automatic control of each planting line, it is possible to control the areas with the greatest productive potential receiving a higher sowing density. The Recommendation Map must be associated with the Harvest Map(da Cunha et al., 2020; Hartono et al., 2020).

Obtaining **Nutrient Maps** and creating the Recommendation Map is only possible after collecting georeferenced soil samples and generating Fertility Maps, with the aid of Productivity Maps. Relating the knowledge of where greater amounts of nutrients are extracted, it is possible to replace these nutrients in a localized manner(Luo et al., 2020; Hamlin et al., 2020).

With the **Recommendation Map** and the information generated in real-time by the sensors, it is possible to correlate the varied rate of application of pesticides, according to the application in the necessary dose and the correct place, avoiding overlaps or absence of an application, in addition to an alignment between passes(Luo et al., 2020; Hartono et al., 2020).

Productivity Map is made during the harvesting operation, employing productivity sensors installed in the harvesters, since the harvester is equipped with GPS sensors and receivers, elaborating this map simultaneously with the harvest, on top of the grain elevator. for deposit. With countless data that are extremely useful for management recommendations for the next crop, they are usually presented from an analysis of several crops; however, they can also be generated from a single crop, as a representative productivity data layer(González-González et al., 2020; Soltani et al., 2020).

Productivity Maps or Harvest Maps are vital for understanding the variability of our fields, as they have a large amount of data.

Spatial Production Trends Map is one that shows zones based on the percentage concerning the average normalized productivity. Also assessing that the **Production Stability Map** uses the Coefficient of Variation (CV) to identify stable and unstable land production zones(Liu et al., 2020; Chiang et al., 2020; Shelbani et al., 2020; Labuza et al., 2020; Fry et al., 2020).

Regarding the productivity zone, the **Units Map** combines information from the Spatial Production Trends map and the Production Stability map to identify possible management zones(Al-Hamdani et al., 2020; Bian et al., 2020).

Profitability Maps are obtained through tabulated production costs, helping to manage the most profitable parts of the crop(Massey et al., 2020; Sulkava et al., 2020).

Thus, the collected data and productivity mapping show the important evaluations of the productive areas, reflecting on the technological potential through the use of sensors and data processing. Impacting on the decisions made on maps of precision agriculture, which come from these analyzes, bringing reliable information and not simply "numbers".

SENSING TECHNOLOGIES AND DRONES

Another impacting factor on precision agriculture is concerning the forecast that in the coming decades the population of the planet will continue to increase dramatically, and agriculture is one of the most important business areas that help to solve the issue of food security and the problem of food. world hunger; with this focus, the generation of greater productivity in the same land space, in addition to reducing the use of agricultural inputs, is one of the greatest contributions to sustainable agriculture, since the use of precision agriculture technologies contributes to making the business more productive and sustainable (Ahmad & Mahdi, 2018; Shalloo,2018).

With the help of the technologies already mentioned, rural producers are able to access information in real-time or with a skillful periodicity to make faster decisions, not having to wait for the end of the harvest to be able to make adjustments or apply inputs in order to correct some aspects of plant development, since time is related to the continuous improvement of work with increases in the effectiveness of fertilizer application and even identify pests or disease outbreaks in less time for a solution more quickly, in the same way as identifying and alerting the farmer about nutrient imbalance, salinity risks, fertilizer incompatibility, water quality problems and current and future use of fertilizer, with the aim of ways to obtain greater productivity in each harvest and mission that knowledge brings possibilities for farming no longer considered as an opportunity, but an obligation (Shakhatreh, 2019).

The first step towards precision farming in practice is to understand what is currently going on in the soil, performing analysis of field fractions in the laboratory collecting information about what each land space needs, in the same way as the use of unmanned drones also gets information about the crop. Since the use of this technology allows the use of strategies to solve the problems of unevenness in crops, making it possible to make available a large amount of crop-specific data, which can subsidize decision-making and reduce business uncertainty, and can also be used in precision livestock, fruit growing, and precision irrigation (Wolf, 2017).

One of the most famous applications of drones in agriculture is for spraying, capable of mapping the topography of the soil, using a laser and ultrasound system, mapping the property, applying pesticides with greater precision, economy and agility, being faster in comparison application with conventional machinery. In general, these drones for aerial mapping are equipped with the NDVI system (Normalized Difference Vegetation Index), a technology that uses the reflected light of each plant to generate a mapping of the planting and the location of weeds, being able to count plants and identify flaws in the planting lines; performing management of pesticides; identifying the occurrence of pests, diseases and nutritional deficiency in the plantation; and generating variable rate application maps for the application of inputs and fertilizers with precision (Shamshiri, 2018).

In the same way that drones for irrigation are able to identify with great precision, and with just one flyover, the areas of the crop with greater water demand, relieving through hyperspectral, multispectral or thermal sensors, are more used and common in regime regions. irregular rainfall, since in these regions its use is feasible, since, in addition to the costs with the implementation and maintenance of sprinklers and central pivots, the correct calculation of the amount of water to be used is usually a difficulty for producers, where the application of this type of technology for irrigation becomes very efficient (Bhandari, 2017).

In the case of Brazil, georeferencing is an instrument required by INCRA (National Institute of Colonization and Agrarian Reform), and its terms vary according to the size of the property, and the use of drones is one of the options of georeferencing, making its use for this purpose, the certification can come out faster and speed up the legalization of the property (Machado, 2017).

Although the use of drones in agriculture, autonomous equipment, and the IoT continue to revolutionize the agribusiness sector, and not only for spraying, aerial mapping or irrigation, drones with the simple presence of a good resolution camera allow to check deforested areas; fire outbreaks; the presence of river springs and watery eyes; exploration of areas of difficult access; searching for lost animals; checking areas for opening roads; surveillance; as well as the utility of comparative images in experimental areas (demo-plots) (Saha, 2018).

The use of telemetry systems allows the agricultural producer to control the quality and quantity of the inputs applied, so when installing on sprayers, farmers control the application of fertilizers and pesticides at a distance according to the needs of each area, making management of this entire system in the field, the use of agricultural software is necessary, helping in the organization of production information and more assertive decision making by the producer. In the same way that fertilizers, tractors, harvesters, and remote-controlled sprayers already exist in the market, considered a practical part of precision agriculture, where this autonomous equipment applies fertilizers and pesticides according to the needs of each area (Playán, 2018).

Still considering the UAVs (Unmanned Aerial Vehicles) that can identify pests and diseases, in addition to finding nutritional deficiencies in specific parts of the crop, since from this information, it is possible to control the distribution of inputs, still taking into account spraying, through property maps, it is possible to spray assertively plots depending on your need.With the use of precision agriculture, crops are better explored, since it is one of the pillars of growth with the tendencies of agribusiness, it is the combination of the ancient practice of vegetable cultivation with technology, generating a more controlled and efficient production, based on an integrated system capturing information from the soil, analyzing what is happening in the region and managing the inputs for application in the correct place and in the appropriate quantity, accumulating advantages of this system, allowing a better knowledge of the production field and allowing the making decisions based on the flexibility of the distribution of inputs in the places and in the most necessary time, minimizing production costs; sustaining a high productivity contributing to the preservation of the environment, by the less use of pesticides, also reducing the risk of agricultural activity; creating a uniformity in productivity achieved by correcting the factors that contribute to its variability, obtaining an overall increase in productivity (Mogili, 2018; Wolf, 2017).

Precision agriculture can also be understood as a process that involves the precise spatial knowledge of the agricultural activity, whose foundation is often based on the use of data obtained with the aid of satellites, also being considered a philosophy of farm management in which producers are able to identify variability within a field, and then manage that variability to increase productivity, that is, it encompasses the idea of commitment to land use, about future generations, enabling sustainable management implying more than just maintaining productivity indexes (Vidoni, 2017).

THE USE OF IOT TECHNOLOGY IN CONTEXT OF AGRICULTURE

IoT is what the use of technology to improve society, the quality of life of people in cities and the countryside, technically is the connection of devices, such as sensors and actuators, in a network such as anInternet, which these devices will be controlled by a software system, which analyzes the data received and relates them to other sources of information, with business knowledge, to make automatic and/or manual decisions, to increase efficiency and decreasing the costs of a business process. In this way, IoT is used to refer to the interconnection of devices and everyday objects capable of exchanging data with each other via the Internet, expressing the connection capacity, encompassing technologies capable of connecting all types of digital devices to databases, networks, and the Internet, so that it has the capacity to "talk" to each other or can be accessed remotely (Tzounis, 2017).

However, given by data it is not interesting and technology by technology does not solve a problem, in this sense, it is necessary to correlate the information, wherein an agricultural scenario concerning irrigation systems that can be scheduled, if the producer receives information that on the day rain will follow, thus enabling the deactivation of irrigation; which likewise some data require the analysis of the agronomist or the agricultural engineer, since the level of elaboration of the information depends on the problem and size of the property. Thus, using intelligent sensors and appropriate software, the objects connect and become part of the large agricultural communication network (Kamilaris, 2016).

The IoT is one of the technologies that make precision agriculture more viable, applying the concept of remote sensing, it is possible to monitor all the necessary variables and to operate, precisely, in each part of a crop, still considering the meteorological information with meters of irradiation of ultraviolet rays and environmental conditions, it is possible to list input management, seed management, knowing which and how many pieces of machinery are available for possible changes, amount of pesticide, facilitating the decision when it is better to carry out the planting (Mekala, 2017).

With IoT, it is possible to improve efficiency by contributing to increased production and reduced costs, in addition to benefiting those living in the countryside and the countryside. These technologies do not bring the same communication infrastructure that people are used to, such as a broadband network in their homes or offices, however, they deliver what the connected "things" need, enabling the implementation of IoT in these areas (Elijah, 2018).

IoT is already present on farms and signals a promising future, in agriculture, IoT devices perform analysis of information collected by these IoT sensors, informing farmers what they need to know about soil, humidity, water levels, and other metrics, in real-time.However, the IoT finds barriers to be faced, especially in the field concerning connectivity, since there are remote points, which often make connections impossible (Muangprathub,2019).

The drone consists of a camera that captures the image, an application processor, a graphics processor, sensors, GPS, and connectivity, so 86=what the camera captures can be presented on a mobile device such as a tablet or the farmer's smartphone, and so on. it is possible to monitor, in real-time, everything that occurs in their lands. In the same way that from on-board systems developed for drones, their mission is to collect, process, analyze and transmit information on crops, in real-time, for both farmers and environmental agents.Thus, intelligence data is used to accurately detect crop deficiencies, pest occurrence, water scarcity, nutrient deficit, and environmental damage, allowing farmers to take precise measures avoiding overuse of crop protection products, over-fertilization, as well as enabling the irrigation of dry fields, in order to reduce the environmental impact and increase productivity (Krishna, 2018).

Regions of the property where water is indispensable, being as essential as understanding the best time to supply it and the most appropriate volume, sensors strategically positioned on the ground inform the command center, activating an irrigation system, for example, in a determined area of the property, considering that only that area has demand for irrigation, optimizing work, energy, and resources. Likewise, upon reaching satisfactory humidity levels, the sensors inform the irrigation system that it can now be turned off (Zhou, J., & Zhang, 2019).

The use of drones in agricultural properties brought a multitude of practical solutions, considering their integration via the Internet with sensors, GPS and other equipment, facilitated the obtaining of information and the taking of actions by a management system as well as by the producer, making a diagnosis the health of a crop, for example, has become a less laborious operation, without the need to travel kilometers within plots and blocks, since the drone does this with precision, and great time savings (Krishna, 2018).

Also, provided with GPS and sensors for image generation, drones connected to appropriate systems allow the analysis of plant height; definition of the stand related to the number of plants; the identification of health conditions; analysis of weed competition; identification of formation of attacked plants concerning diseases and pests and measurement of nutrient content in the plant.The digitization of agriculture is a worldwide trend providing systems capable of assisting the farmer at various times when he has to decide on the field, with technological tools that provide information in real-time, which is essential for the result of the harvest, where through resources such as satellite images, high-tech sensors, mobile or GPS applications, variable application algorithms, this decision making becomes clearer, smarter and simpler (Voronin, 2019).

Greenhouses are protected environments whose conditions of light, humidity, and temperature need to be controlled, whether for intensive agriculture or for the production of seedlings, which generally this control tends to be manual and mechanical, making use of own coverings, sprinkling, and other management measures. And with the implementation of the IoT, sensors for each of the environmental conditions "talk" with the respective systems, thus, a ventilation system can be activated and the light intensity can be regulated. Likewise, while the irrigation system starts the water supply, the opening of screens clears and airs the environment, both of which are integrated actions transmitted via the Internet to management software, which records and analyzes the various interventions carried out (Hanan,2017).

Anticipating the damage that a pest infestation can cause in the crop is a great operational and economic advantage, in that sense the control of insects to the crops can be carried out with the installation of traps using attractive pheromones, from which sensors identify the presence at a certain level of infestation and warn the agricultural management system, which maps the occurrence points. Based on this, it is possible to plan localized control actions without the need to work the entire cultivated area, increasing the economy of inputs, time, and labor, as well as a significant reduction in the amount of dispersed product, reducing the environmental impact. IoT sensors provide real-time information about the health of plantations and show the existence of pests, in the same way, that other devices can collect data on patterns of behavior of these pests (Hanan,2017; Gilchrist, 2016).

IoT assists in animal monitoring, pest management, reducing water use, and other use cases, with numerous possibilities in which technology gives producers a 360-degree view, identifying insights and being able to adapt operations to obtain results favorably.However, IoT devices have different network needs than smartphones, computers, or tablets. IoT devices often only need to send small data packets at regular intervals, yet considering the typical requirement for IoT environments is the need to connect to areas far from traditional infrastructure and an available power source (Harris, 2019).

FIELD DEVICE CONNECTIVITY

To solve this problem, one of the most reliable and cost-effective paths is the LoRa network (Long Range Network), which uses radiofrequency (unlicensed spectrum) to transmit data in a very optimized way over distances of even more 15 kilometers between the connected points, which in urban environments, this range drops to about 4 kilometers, still considering that the battery of the LoRa device is quite long-lasting; concerning technology, it has the ability to enter a type of deep sleep when it is not transmitting messages (Augustin, 2016; Lavric, 2017).

In terms of security, there are two levels, one for the network, the native security of the LoRa network ensuring authentication of the node on the network, and the other for the application, with the security level of the application, the network still has, AES standard encryption (Advanced Encryption Standard), is a cryptographic primitive designed to compose symmetric encryption and decryption systems, ie the same key for encrypting and decrypting, based on a block cipher, operates in blocks of a fixed size of 128 bits or 16 bytes (Lavric, 2017).

With the LoRa infrastructure, 5 to 10 times fewer base stations are used than in a 3G / 4G network, facilitating the monitoring and control of the environments, highlighting that this standard has a low cost of application. A barrier to the spread of Precision Agriculture is the fact that a large part of rural properties is not covered by mobile cellular signal and Internet networks, lacking the necessary infrastructure to connect the IoT devices scattered throughout the field. LoRa tags can be placed both on vehicles and on animals that spread across the vastness of the rural property, and through the adoption of LoRa networks, the field can become a connected space, with visibility and control fully aligned with the best practices of Precision Agriculture (Wixted, 2019; Davcev, 2018).

The data points, being those where part of the processing takes place, for analysis tend to be considered distant from the location of the operation, and the fact that the LoRa is a Low Power Wide Area (LPWAN) network, that is, a standard directed to the connection of millions of "things", with better coverage networks, lower connection costs, and lower battery consumption, facilitates the implementation of IoT in the field. Since the quality of service is good because the messages generated by the IoT devices spread across the rural company are not huge amounts of data, something that requires immense bandwidth (Wixted, 2019).

In this way, with LoRa tags it is possible to track a herd that moves along a vast field, ensuring the delivery of data of the most varied types indicating whether an animal is sick, lost, bogged down, or dead. Accelerometer sensors are able to distinguish up to nine different diseases of a given animal, the temperature measured by the IoT device and transmitted by the LoRa tag can indicate a dead animal that, if not detected, could spread diseases to others, in the same sense as the temperature of a live animal, however static, from the point of view of GPS, can be stuck or injured (Kamilaris, 2016; Wixted, 2019; Davcev, 2018).

The LoRa network is effective in monitoring heavy vehicles such as harvesters, trucks, tractors, among others, since they already have intelligence and sensors, which helps to equate with the connectivity between these vehicles and the processing platforms in the cloud, making it possible to receive sensor signals with information in the form of text and, from there, send this data to the platforms that digest them in an engine and return them in dashboards.Thus, IoT makes the equipment connect and perform key functions in the field, being responsible for the technological revolution that the rural environment is experiencing and, increasingly, will automate and connect the plantation, from sowing to harvest, with the implementation of integrated systems based on sensors, satellites, data collection, and analysis, making it possible to apply Big Data in this context, combined with the use of drones. In the rural environment, IoT can lead to higher agricultural yields, less food waste along the supply chain, and less input consumption per hectare (Davcev, 2018).

DISCUSSION

Precision agriculture relies on technology that reduces and/or eliminates the use of pesticides applied to properties, breeds several breeds of insects, such as bees, flies, spiders, among other types of natural predators that eat harmful insects, helping farmers to export their products to developed countries that have more restrictions on the grain trade; just as robotic milking systems are making milking cows more efficient, which clean, milk and perform the necessary procedures after milking the cow, working without the need for humans, also related to the reduced cost of the process, as well as increasing the volume of milk production. With the addition of technology in agriculture, systems based on sustainability in agriculture already exist, allowing the rural producer to place robotic water pipes in the soil, making it possible to determine the ideal temperature for a given field, enabling cases in which if the soil is very hot, the sensors can cool the water and vice versa, increasing crop yields (Shalloo,2018; Hostiou, 2017).

Precision Agriculture has numerous advantages over the conventional management system, since in the conventional modality fertilizers and inputs are applied equally throughout the area, based on an average sampling for the plots or even for the entire farm, which in On the other hand, in Precision Agriculture, several pieces of information help in decision making, allowing the application of this quantity only as necessary in each piece of property. Considering that the area is not uniform, along with the tools of technology, such as insect infestation maps for spraying; soil maps for fertilizer application; irrigation maps; variable rate seeding; automation of machines and processes, in the same way as the creation of differentiated management units.

With the concept of non-uniformity of ownership, there are several advantages such as savings in inputs and resources; financial savings; a detailed view of the property; greater control of the property through the improvement of agricultural activities; since the inputs are applied according to what each part of the crop needs, based on soil sampling, harvest maps, geolocation of information, through the use of satellites, drones, machines and other tools that are adapted from other areas for use in agriculture as a variable rate seeder; automatic pilot; distributors of fertilizers, corrective and pesticides at variable rates; harvest monitors; drones for collecting images and sensors.

Through these tools it is understood the variations of the property, making application of inputs where they are really necessary, avoiding unnecessary expenses and less environmental pollution, considering that the productivity maps are good tools for the discovery of the variability present in the plots of the property, basically pointing files collected and according to the speed of travel of the machine, the width of the platform and the time of data acquisition.

Regarding the harvesting machinery, as the grains are thrown against it, the values are recorded, and with this system, there is a humidity sensor in order to estimate the productivity in the form of dry grains, provided that all this information they are stored and georeferenced, enabling the subsequent processing and preparation of harvest maps for these cultures.

Many factors can explain places with high or low productivity, ranging from physical factors such as soil texture, to chemical attributes, such as lack of fertilizers or inadequate handling of inputs applied in the area, so the gross productivity maps must pass by prior filtering, which is related in terms of environmental issues, IoT-based agriculture can offer great benefits, including more efficient use of water or optimization of inputs and treatments.

In the same sense that drones are good tools for investigating the causes and effects of crop management, mapping and collecting information, and this information, in general, is related to vegetation health; planialtimetric surveys; pest attack; creation of MDT (Digital Terrain Model) which is a numerical model detailing only the terrain, and MDS (Digital Surface Model) that details all objects above the ground, under the influence of vegetation and buildings; contour lines; as well as georeferencing properties, among others.

Through the use of wireless IoT applications to collect data about the location, owners of large properties visualize the welfare and health of their cattle or other specific animal production, where this information helps to identify animals that are sick, for that can be separated from the herd, thus preventing the spread of the disease, reducing labor costs, still considering the help of sensors based on IoT facilitating the location of animals.

Among the challenges of adopting IoT in precision agriculture is the discovery of IoT solutions that overcome the problems of connectivity and network coverage in the field, considering that this is the main challenge of technology. Since the costs involved in making optimal connectivity, at distant points, are high and, in general, make it impossible to return on investments, which may limit the application technology scenario with the use of IoT.

Even so, there are some disadvantages and difficulties when practicing Precision Agriculture, which are related to the lack of qualified people for processing and analyzing information; lack of technical knowledge to transform information into useful knowledge for producers, as well as lack of dissemination of knowledge about the technologies employed.

Thus, the agricultural applications of the IoT enable the collection of significant data, which through the installation of intelligent technologies that increase competitiveness and sustainability in agricultural production, makes the demand can be successfully met by both large farmers and small farmers. in a prosperous way. Since precision agriculture is not only related to the use of sophisticated machines and technologies, this principle goes further, constituting a system of actions that lead to the more efficient management of the production factors associated with the diverse conditions of an agricultural area.

TRENDS

Machines that communicate without human interference, exchanging data over the network, are already a reality in the field, and in the future agricultural equipment will have more and more sensors connected to the Internet, together with artificial intelligence applied to agriculture, which can improve production processes and support the taking decision-making by the farmer, bringing more income and reducing costs. Still considering the advance in the area of precision agriculture, with sensors installed in the equipment and connected in a network, indicating that the agriculture of the future should be increasingly supported by scientific knowledge.

Precision agriculture is considered the basis of intelligent agriculture, bringing cutting edge technology and innovation to the field, enabling the management of the entire production system in more detail, from the identification of the most productive points in the soil to the increase in the capacity of forecasting climatic conditions, including the production of strategic data on the application of inputs, pesticides, and fertilizers.

Among the advantages is the ability to promote the mapping of land and crops, favoring the planning of agricultural production, increasing productivity, as well as reducing costs and impacts on the environment, which in intelligent agriculture based on IoT, a system is built to monitor the cultivation field with the help of sensors acquiring signals such as air humidity, temperature, light, soil moisture, among others, automating the irrigation system. Related to environmental issues, smart IoT-based agriculture offers the most efficient use of water and the optimization of inputs.

In this way, the adoption of management practices that take into account the diversity of edaphoclimatic conditions of an agricultural area, leading the cultures to the possibility of greater expression of the genetic potential in the entire cultivated area, and not only in a part of the crop, where conditions are more favorable. The applications of intelligent IoT-based agriculture are not only intended for large conventional agricultural operations, but can also be new levers to raise other common or growing trends in agriculture such as organic farming, family farming, considering small or complex spaces, private animal production, and cultures, preserving the particular variety.

Fog Computing is an extension of Cloud Computing since it was conceived with a focus on devices that are on the edge of the network, optimizing and improving data management and processing, allowing connections closer to the user and IoT devices. Considered a disruptive and decentralized tendency, technology has allowed greater agility and flexibility in several activities, allowing a greater approximation and connectivity with any IoT device, referring to the data processing carried out on the rural property, in equipment that aggregates the collected data. on the sensors. Still considering its potential for greater automation of agricultural processes, capable of integrating different technologies, such as IoT, with Big Data analysis and cloud computing, resulting in an implementation of applications in the intelligent management of rural properties, increasing their productivity and decreasing costs production costs.

CONCLUSION

The IoT can radically change the world that people live in, abstracting from advanced industries, connected vehicles and smarter cities are all components of the IoT equation, however, applying technology like IoT to the agricultural industry has had a positive impact, regarding that the global population is expected to reach close to 10 billion by the year 2050, which sums up to feed this amount of population, the agricultural industry must adopt technologies such as IoT, facing challenges such as the increase in climate change as well as extreme weather conditions in certain regions, and the environmental impact resulting from intensive agricultural practices, correlating the demand for more food has to be met.

Precision agriculture is one of the most famous applications of IoT in the agricultural sector and can be considered as something that makes agricultural practice more controlled and accurate when it comes to livestock and crop cultivation, about the farm management approach, a key component is the use of information technology (IT) and various items such as control systems, robotics, sensors, autonomous vehicles, automated hardware, variable rate technology, among others, together with the adoption of high-speed Internet access, reliable and low-cost mobile devices and satellites for image acquisition and positioning, featuring the key technologies that are the trend in precision agriculture.

Precision Agriculture is a concept of differentiated management of crops, considering that crops are not uniform, that is, they consider that each piece of a property is different whether in relief, temperature, soil, plants, among other factors already explicit. It is a management system that seeks to optimize and make better use of each portion of the area, meaning working with the system's sustainability in mind.

In IoT-based agriculture, a system is built to monitor the crop field with the help of sensors that capture signals such as light, humidity, temperature, soil moisture, and automates the irrigation system, allowing farmers to monitor the conditions of the field from anywhere, being highly efficient when compared to the conventional approach, that is, the role of Precision Agriculture is to assist producers in the acquisition and interpretation of the data collected in order to optimize the handling done on the farm.

REFERENCES

Ahmad, L., & Mahdi, S. S. (2018). Recent Advances in Precision Agriculture. In *Satellite Farming* (pp. 129–138). Springer. doi:10.1007/978-3-030-03448-1_10

Al-Hamdani, A. S., & Al-Shimmary, H. M. (2020). Characterization and Classification of Soil Map Units by Using Remote Sensing and GIS in Bahar Al-Najaf, Iraq. In *Environmental Remote Sensing and GIS in Iraq* (pp. 61–92). Springer. doi:10.1007/978-3-030-21344-2_4

Augustin, A., Yi, J., Clausen, T., & Townsley, W. M. (2016). A study of LoRa: Long range & low power networks for the internet of things. *Sensors (Basel), 16*(9), 1466. doi:10.339016091466 PMID:27618064

Bhandari, S., Raheja, A., Green, R. L., & Do, D. (2017, May). Towards collaboration between unmanned aerial and ground vehicles for precision agriculture. In *Autonomous Air and Ground Sensing Systems for Agricultural Optimization and Phenotyping II*. International Society for OpticsandPhotonics.

Bian, Z., Guo, X., Wang, S., Zhuang, Q., Jin, X., Wang, Q., & Jia, S. (2020). Applying statistical methods to map soil organic carbon of agricultural lands in northeastern coastal areas of China. *Archives of Agronomy and Soil Science, 66*(4), 532–544. doi:10.1080/03650340.2019.1626983

Borges, L. F., Bazzi, C. L., Souza, E. G. D., Magalhães, P. S. G., & Michelon, G. K. (2020). Web software to create thematic maps for precision agriculture. *Pesquisa Agropecuária Brasileira, 55*, 55. doi:10.15901678-3921.pab2020.v55.00735

Brevik, E. C., Calzolari, C., Miller, B. A., Pereira, P., Kabala, C., Baumgarten, A., & Jordán, A. (2016). Soil mapping, classification, and pedologic modeling: History and future directions. *Geoderma, 264*, 256–274. doi:10.1016/j.geoderma.2015.05.017

Chattopadhyay, A., Singh, A. P., Singh, S. K., Barman, A., Patra, A., Mondal, B. P., & Banerjee, K. (2020). Spatial variability of arsenic in Indo-Gangetic basin of Varanasi and its cancer risk assessment. *Chemosphere, 238*, 124623. doi:10.1016/j.chemosphere.2019.124623 PMID:31545212

Chen, S., Lin, B., Li, Y., & Zhou, S. (2020). Spatial and temporal changes of soil properties and soil fertility evaluation in a large grain-production area of subtropical plain, China. *Geoderma, 357*, 113937. doi:10.1016/j.geoderma.2019.113937

Chiang, Y. Y., Duan, W., Leyk, S., Uhl, J. H., & Knoblock, C. A. (2020). Creating Structured, Linked Geographic Data from Historical Maps: Challenges and Trends. In *Using Historical Maps in Scientific Studies* (pp. 37–63). Springer. doi:10.1007/978-3-319-66908-3_3

da Cunha, J. A., Rolim, P. M., & da Silva Chaves, K. S. F. (2020). From seed to flour: Sowing sustainability in the use of cantaloupe melon residue (Cucumis melo L. var. reticulatus). *PLoS One, 15*(1), e0219229. doi:10.1371/journal.pone.0219229 PMID:31895921

Davcev, D., Mitreski, K., Trajkovic, S., Nikolovski, V., & Koteli, N. (2018, June). IoT agriculture system based on LoRaWAN. In *2018 14th IEEE International Workshop on Factory Communication Systems (WFCS)* (pp. 1-4). IEEE. 10.1109/WFCS.2018.8402368

Duhan, J. S., Kumar, R., Kumar, N., Kaur, P., Nehra, K., & Duhan, S. (2017). Nanotechnology: The new perspective in precision agriculture. *Biotechnology Reports (Amsterdam, Netherlands)*, *15*, 11–23. doi:10.1016/j.btre.2017.03.002 PMID:28603692

Elijah, O., Rahman, T. A., Orikumhi, I., Leow, C. Y., & Hindia, M. N. (2018). An overview of Internet of Things (IoT) and data analytics in agriculture: Benefits and challenges. *IEEE Internet of Things Journal*, *5*(5), 3758–3773. doi:10.1109/JIOT.2018.2844296

França, R. P., Iano, Y., Monteiro, A. C. B., & Arthur, R. (2020). Potential proposal to improve data transmission in healthcare systems. In *Deep Learning Techniques for Biomedical and Health Informatics* (pp. 267–283). Academic Press. doi:10.1016/B978-0-12-819061-6.00011-2

França, R. P., Iano, Y., Monteiro, A. C. B., & Arthur, R. (2020). Improvement for Channels With Multipath Fading (MF) Through the Methodology CBEDE. In Fundamental and Supportive Technologies for 5G Mobile Networks (pp. 25-43). IGI Global.

França, R. P., Iano, Y., Monteiro, A. C. B., & Arthur, R. (2020). A Proposal of Improvement for Transmission Channels in Cloud Environments Using the CBEDE Methodology. In Modern Principles, Practices, and Algorithms for Cloud Security (pp. 184-202). IGI Global. doi:10.4018/978-1-7998-1082-7.ch009

França, R. P., Iano, Y., Monteiro, A. C. B., & Arthur, R. (2020). Improvement of the Transmission of Information for ICT Techniques Through CBEDE Methodology. In Utilizing Educational Data Mining Techniques for Improved Learning: Emerging Research and Opportunities (pp. 13-34). IGI Global. doi:10.4018/978-1-7998-0010-1.ch002

Fry, J. E., & Guber, A. K. (2020). Temporal stability of field-scale patterns in soil water content across topographically diverse agricultural landscapes. *Journal of Hydrology (Amsterdam)*, *580*, 124260. doi:10.1016/j.jhydrol.2019.124260

Gilchrist, A. (2016). *Industry 4.0: the industrial internet of things*. Apress. doi:10.1007/978-1-4842-2047-4

González-González, M. G., Gómez-Sanchis, J., Blasco, J., Soria-Olivas, E., & Chueca, P. (2020). CitrusYield: A Dashboard for Mapping Yield and Fruit Quality of Citrus in Precision Agriculture. *Agronomy (Basel)*, *10*(1), 128. doi:10.3390/agronomy10010128

Hakkim, V., Joseph, E., Gokul, A., & Mufeedha, K. (2016). Precision farming: The future of Indian agriculture. *Journal of Applied Biology and Biotechnology*, 68–72. doi:10.7324/JABB.2016.40609

Hamlin, Q. F., Kendall, A. D., Martin, S. L., Whitenack, H. D., Roush, J. A., Hannah, B. A., & Hyndman, D. W. (2020). Quantifying Landscape Nutrient Inputs With Spatially Explicit Nutrient Source Estimate Maps. *Journal of Geophysical Research: Biogeosciences, 125*(2).

Hanan, J. J. (2017). *Greenhouses: Advanced technology for protected horticulture*. CRC press. doi:10.1201/9780203719824

Hanan, J. J. (2017). *Greenhouses: Advanced technology for protected horticulture*. CRC press. doi:10.1201/9780203719824

Harris, T. (2019). IoT: A Business Perspective. In *Fundamentals of Internet of Things for Non-Engineers* (pp. 279–301). Auerbach Publications. doi:10.1201/9780429274992-15

Hartono, A., Barus, B., Tejo, R. K., Rostiana, O., Manohara, D., & Wahyuno, D. (2020, January). Smart farming on pepper for fertilizers recommendation in Bangka Belitung. In *IOP Conference Series: Earth and Environmental Science* (Vol. 418, No. 1, p. 012049). IOP Publishing. 10.1088/1755-1315/418/1/012049

Hostiou, N., Fagon, J., Chauvat, S., Turlot, A., Kling-Eveillard, F., Boivin, X., & Allain, C. (2017). Impact of precision livestock farming on work and human-animal interactions on dairy farms. *RE:view*.

Huuskonen, J., & Oksanen, T. (2018). Soil sampling with drones and augmented reality in precision agriculture. *Computers and Electronics in Agriculture*, *154*, 25–35. doi:10.1016/j.compag.2018.08.039

Imran, M., Rizwan, M., Kerkhi, S. A., & Ali, N. (2018). Bioinformatics for Crop Breeding and Sustainable Agriculture. *Journal*, *6*(4), 1–5. www. ijbpr. in

Jeschke, S., Brecher, C., Meisen, T., Özdemir, D., & Eschert, T. (2017). Industrial internet of things and cyber manufacturing systems. In *Industrial internet of things* (pp. 3–19). Springer. doi:10.1007/978-3-319-42559-7_1

Kamilaris, A., Gao, F., Prenafeta-Boldu, F. X., & Ali, M. I. (2016, December). Agri-IoT: A semantic framework for Internet of Things-enabled smart farming applications. In *2016 IEEE 3rd World Forum on Internet of Things (WF-IoT)* (pp. 442-447). IEEE. 10.1109/WF-IoT.2016.7845467

Khanal, S., Fulton, J., & Shearer, S. (2017). An overview of current and potential applications of thermal remote sensing in precision agriculture. *Computers and Electronics in Agriculture*, *139*, 22–32. doi:10.1016/j.compag.2017.05.001

Kladivko, E. J., & Timmenga, H. J. (2019). Earthworms and agricultural management. In *Rhizosphere dynamics* (pp. 192–216). CRC Press. doi:10.1201/9780429304798-8

Krishna, K. R. (2018). *Agricultural drones: a peaceful pursuit*. Apple Academic Press. doi:10.1201/9781315195520

Labuza, T. P., &Altunakar, B. (2020). Water activity prediction and moisture sorption isotherms. *Water activity in foods: Fundamentals and applications*, 161-205.

Lavric, A., & Popa, V. (2017, July). Internet of things and LoRa™ low-power wide-area networks: a survey. In *2017 International Symposium on Signals, Circuits and Systems (ISSCS)* (pp. 1-5). IEEE. 10.1109/ISSCS.2017.8034915

Liu, S., Zhu, Y., Yue, E., & Guo, X. (2020). Key Technology and Application Platform of Agricultural Economic Spatial Information Service—Chinese Agricultural Economic Electronic Map. In *China's e-Science Blue Book 2018* (pp. 269–289). Springer. doi:10.1007/978-981-13-9390-7_16

Lohr, S. (2012). The age of big data. *New York Times, 11*(2012).

Luo, L., Hou, X., Cai, W., & Guo, B. (2020). Incremental route inference from low-sampling GPS data: An opportunistic approach to online map matching. *Information Sciences*, *512*, 1407–1423. doi:10.1016/j.ins.2019.10.060

Machado, J., de Fátima Padilha, M. D. R., de Lira, F. P., de Oliveira, J. G., da Silva, R. S., & Caetano, M. B. C. (2017). Agricultura de Precisão e abertura de novas fronteiras no Brasill Precisionagriculture-andopening new frontiers in Brazil. *Revista Geama*, *4*(1), 49–53.

Maes, W. H., & Steppe, K. (2019). Perspectives for remote sensing with unmanned aerial vehicles in precision agriculture. *Trends in Plant Science*, *24*(2), 152–164. doi:10.1016/j.tplants.2018.11.007 PMID:30558964

Massey, R. E., Myers, D. B., Kitchen, N. R., & Sudduth, K. A. (2008). Profitability maps as an input for site-specific management decision making. *Agronomy Journal*, *100*(1), 52–59. doi:10.2134/agronj2007.0057

McEntee, P. J., Bennett, S. J., & Belford, R. K. (2020). Mapping the spatial and temporal stability of production in mixed farming systems: An index that integrates crop and pasture productivity to assist in the management of variability. *Precision Agriculture*, *21*(1), 77–106. doi:10.100711119-019-09658-6

Mekala, M. S., & Viswanathan, P. (2017, August). A Survey: Smart agriculture IoT with cloud computing. In *2017 international conference on microelectronic devices, circuits and systems (ICMDCS)* (pp. 1-7). IEEE.

Mikula, K., Izydorczyk, G., Skrzypczak, D., Mironiuk, M., Moustakas, K., Witek-Krowiak, A., & Chojnacka, K. (2019). Controlled release micronutrient fertilizers for precision agriculture–A review. *The Science of the Total Environment*. PMID:31935544

Minango, P., Iano, Y., Monteiro, A. C. B., França, R. P., & de Oliveira, G. G. (2019). *Classification of Automatic Skin Lesions from Dermoscopic Images Utilizing Deep Learning. SET International Journal of Broadcast Engineering*. doi:10.18580etijbe.2019.14

Mitchell, S., Weersink, A., & Erickson, B. (2018). Adoption of precision agriculture technologies in Ontario crop production. *Canadian Journal of Plant Science*, *98*(6), 1384–1388. doi:10.1139/cjps-2017-0342

Mogili, U. R., & Deepak, B. B. V. L. (2018). Review on application of drone systems in precision agriculture. *Procedia Computer Science*, *133*, 502–509. doi:10.1016/j.procs.2018.07.063

Molnar, J. J., & Kinnucan, H. (2019). Introduction: The biotechnology revolution. In Biotechnology and the new agricultural revolution (pp. 1-18). CRC Press.

Moncada, K. M. (2017). *Introduction to agronomy: food, crops, and environment*. Delmar Cengage Learning.

Muangprathub, J., Boonnam, N., Kajornkasirat, S., Lekbangpong, N., Wanichsombat, A., & Nillaor, P. (2019). IoT and agriculture data analysis for smart farm. *Computers and Electronics in Agriculture*, *156*, 467–474. doi:10.1016/j.compag.2018.12.011

Nganga, W. B., Ng'etich, K. O., Macharia, M. J., Kiboi, N. M., Adamtey, N., & Ngetich, K. F. (2020). Multi-influencing-factors' evaluation for organic-based soil fertility technologies out-scaling in Upper Tana Catchment in Kenya. *Scientific American*, *7*, e00231.

Padilha, R., Iano, Y., Monteiro, A. C. B., Arthur, R., & Estrela, V. V. (2018, October). Betterment Proposal to Multipath Fading Channels Potential to MIMO Systems. In *Brazilian Technology Symposium* (pp. 115-130). Springer.

Paris, J. O., Gontijo, I., Partelli, F. L., & Facco, A. G. (2020). Variability and spatial correlation of soil micronutrients and organic matter with macadamia nut production. *Revista Brasileira de Engenharia Agrícola e Ambiental, 24*(1), 31–36. doi:10.1590/1807-1929/agriambi.v24n1p31-36

Paul, S. S., Coops, N. C., Johnson, M. S., Krzic, M., Chandna, A., & Smukler, S. M. (2020). Mapping soil organic carbon and clay using remote sensing to predict soil workability for enhanced climate change adaptation. *Geoderma, 363*, 114177. doi:10.1016/j.geoderma.2020.114177

Peng, H., Jia, Y., Zhan, C., & Xu, W. (2020). Topographic controls on ecosystem evapotranspiration and net primary productivity under climate warming in the Taihang Mountains, China. *Journal of Hydrology (Amsterdam), 581*, 124394. doi:10.1016/j.jhydrol.2019.124394

Piper, C. S. (2017). *Soil and plant analysis*. Scientific Publishers.

Playán, E., Salvador, R., Bonet, L., Camacho, E., Intrigliolo, D. S., Moreno, M. A., ... Sánchez-de-Ribera, A. (2018). Assessing telemetry and remote control systems for water users associations in Spain. *Agricultural Water Management, 202*, 89–98. doi:10.1016/j.agwat.2018.02.015

Plumecocq, G., Debril, T., Duru, M., Magrini, M. B., Sarthou, J. P., & Therond, O. (2018). The plurality of values in sustainable agriculture models. *Ecology and Society, 23*(1), art21. doi:10.5751/ES-09881-230121

Prathibha, S. R., Hongal, A., & Jyothi, M. P. (2017, March). IoT based monitoring system in smart agriculture. In *2017 International Conference on Recent Advances in Electronics and Communication Technology (ICRAECT)* (pp. 81-84). IEEE. 10.1109/ICRAECT.2017.52

Rasaei, Z., Rossiter, D. G., & Farshad, A. (2020). Rescue and renewal of legacy soil resource inventories in Iran as an input to digital soil mapping. *Geoderma Regional, 21*, e00262. doi:10.1016/j.geodrs.2020.e00262

Reed, C. A. (Ed.). (2011). *Origins of agriculture*. Walter de Gruyter.

Ristorto, G., Gallo, R., Gasparetto, A., Scalera, L., Vidoni, R., &Mazzetto, F. (2017). *A mobile laboratory for orchard health status monitoring in precision farming*. Academic Press.

Saha, A. K., Saha, J., Ray, R., Sircar, S., Dutta, S., Chattopadhyay, S. P., & Saha, H. N. (2018, January). IoT-based drone for improvement of crop quality in agricultural field. In *2018 IEEE 8th Annual Computing and Communication Workshop and Conference (CCWC)* (pp. 612-615). IEEE. 10.1109/CCWC.2018.8301662

Salic, A., &Zelic, B. (2018). Introduction to environmental engineering. *Physical Sciences Reviews, 3*.

Sciortino, M., De Felice, M., De Cecco, L., & Borfecchia, F. (2020). Remote sensing for monitoring and mapping Land Productivity in Italy: A rapid assessment methodology. *Catena, 188*, 104375. doi:10.1016/j.catena.2019.104375

Scoones, I. (2017). *Free-Book: Pathways To Sustainable Agriculture*. Academic Press.

Shakhatreh, H., Sawalmeh, A. H., Al-Fuqaha, A., Dou, Z., Almaita, E., Khalil, I., Othman, N. S., Khreishah, A., & Guizani, M. (2019). Unmanned aerial vehicles (UAVs): A survey on civil applications and key research challenges. *IEEE Access: Practical Innovations, Open Solutions*, *7*, 48572–48634. doi:10.1109/ACCESS.2019.2909530

Shalloo, L., O'Donovan, M., Leso, L., Werner, J., Ruelle, É., Geoghegan, A., ... O'Leary, N. (2018, September). *Simulation of the variation in grass growth, milk production and residency time in a 40-hectare farm composed of either 6 or 40 paddocks*. Academic Press.

Shamshiri, R. R., Hameed, I. A., Balasundram, S. K., Ahmad, D., Weltzien, C., & Yamin, M. (2018). *Fundamental research on unmanned aerial vehicles to support precision agriculture in oil palm plantations*. Agricultural Robots-Fundamentals and Application.

Shannon, D. K., Clay, D. E., & Sudduth, K. A. (2018). An introduction to precision agriculture. *Precision Agriculture Basics,* 1-12.

Sheibani, S., Asgharipour, M. R., Ghanbari, A., & Abolpour, B. (2020). Modeling the yield spatial distribution based on the basic production factors under uncertainty conditions in Kamin, Iran. *Arabian Journal of Geosciences*, *13*(2), 1–11. doi:10.100712517-019-5015-7

Shekhar, C. (2020). Spatial Variability of Ground Water Quality for Irrigation of Mansa District, Punjab. *International Journal of Agriculture Sciences.*

Siddique, A., Prabhu, B., Chaskar, A., & Pathak, R. (2019). A review on intelligent agriculture service platform with lora based wireless sensor network. *Life, 100*, 7000.

Soltani, A., Alimagham, S. M., Nehbandani, A., Torabi, B., Zeinali, E., Zand, E., Ghassemi, S., Vadez, V., Sinclair, T. R., & van Ittersum, M. K. (2020). Modeling plant production at country level as affected by availability and productivity of land and water. *Agricultural Systems*, *183*, 102859. doi:10.1016/j.agsy.2020.102859

Sulkava, M., Sepponen, A. M., Yli-Heikkilä, M., & Latukka, A. (2015). Clustering of the self-organizing map reveals profiles of farm profitability and upscaling weights. *Neurocomputing*, *147*, 197–206. doi:10.1016/j.neucom.2013.09.063

Suma, N., Samson, S. R., Saranya, S., Shanmugapriya, G., &Subhashri, R. (2017). IoT based smart agriculture monitoring system. *International Journal on Recent and Innovation Trends in Computing and Communication, 5*(2), 177-181.

Tzounis, A., Katsoulas, N., Bartzanas, T., & Kittas, C. (2017). Internet of Things in agriculture, recent advances and future challenges. *Biosystems Engineering*, *164*, 31–48. doi:10.1016/j.biosystemseng.2017.09.007

Vidoni, R., Gallo, R., Ristorto, G., Carabin, G., Mazzetto, F., Scalera, L., & Gasparetto, A. (2017, November). ByeLab: an agricultural mobile robot prototype for proximal sensing and precision farming. In *ASME 2017 International Mechanical Engineering Congress and Exposition*. American Society of Mechanical Engineers Digital Collection. 10.1115/IMECE2017-71216

Voronin, B., Mitin, A., & Pichugin, O. (2019, June). The complexity of digitization of agriculture in Russia. In *International Scientific and Practical Conference "Digital agriculture-development strategy" (ISPC 2019).* Atlantis Press. 10.2991/ispc-19.2019.29

Wixted, A. J., Kinnaird, P., Larijani, H., Tait, A., Ahmadinia, A., & Strachan, N. (2016, October). Evaluation of LoRa and LoRaWAN for wireless sensor networks. In *2016 IEEE SENSORS* (pp. 1-3). IEEE.

Wolf, J., Bhandari, S., & Raheja, A. (2017). *Unmanned Aerial Vehicles for Precision Agriculture in Orchard Crops*. Academic Press.

Wolfert, S., Ge, L., Verdouw, C., & Bogaardt, M. J. (2017). Big data in smart farming–a review. *Agricultural Systems, 153*, 69–80. doi:10.1016/j.agsy.2017.01.023

Zhou, J., & Zhang, B. (Eds.). (2019). *Agricultural Robots: Fundamentals and Applications.* BoD–Books on Demand. doi:10.5772/intechopen.74631

Chapter 4
Towards Fully Automated Decision-Making Systems for Precision Agriculture:
Soil Sensing Technologies - "The Missing Link"

Marios Sophocleous
https://orcid.org/0000-0002-9669-0581
University of Cyprus, Cyprus

ABSTRACT

As the world population increases, food demands continue with it. This puts the challenge to feed the world in a continuously rising priority position, which requires to be timely and properly addressed. Precision agriculture (PA), a concept that has been around for decades with the potential to address this challenge by maximizing agricultural efficiency and effectiveness, hasn't yet seen worldwide adoption due to missing enabling technologies. Soil sensing technologies able to monitor soil fertility/quality can close the loop in maintaining optimum soil conditions for maximizing crop yield and quality through the implementation of agricultural automated decision-making systems (ADMS). This chapter provides an overview of the existing soil sensing technologies towards the implementation of ADMS and highlights the major challenges in the development of such systems. The chapter continues to give an insight on how different technologies could be combined to form sophisticated systems for agricultural applications.

INTRODUCTION

Current and future generations face an ever-increasing challenge concerning food supply. "As the population increases, there is an escalating demand for food, especially fruits and vegetables" (M. Sophocleous & Georgiou, 2017). The world population is projected by the Food and Agriculture Organization (FAO) of the United Nations to reach more than 9 billion people by 2050(FAO, 2009). At the same time, it is

DOI: 10.4018/978-1-7998-5000-7.ch004

agreed by analysts that inter-country and inter-regional inequalities within the developing world would become more pronounced (Zarco-Tejada, Hubbard, & Loudjani, 2014). Furthermore, the projected global economic growth (2.9% annually) would lead to a major reduction of absolute economic poverty in developing countries; however, this will be far from solving the problem of economic deprivation and malnutrition of large parts of the global population (FAO, 2009). These trends suggest that the market demand for food will continue to grow(Alexandratos & Bruinsma, 2012). The rise of biofuels has the potential to alter some of these trends and cause an eruptive upsurgeon food demand that will mainly depend on energy prices and governmental policies(Stafford, 2000). Nowadays, that rural labor force is decreasing and the rise of a huge upcoming bioenergy market is on the horizon, farmers are given the burden to produce more food and fiber to feed the continuously increasing population. Furthermore, agriculture is now required to adapt to new, more efficient, sustainable and environment friendly production methods. The projections show that with a world population exceeding 9 billion people in 2050, overall food production must rise by approximately 70% comparing to 2005/07 production (FAO, 2009). Especially in developing countries, the production would need to double implying major increases in key commodities production. FAO expects that 90% of the global growth in crop production would come from higher yields and increased cropping intensity. Only 10% increase will be achieved by land expansion. Additionally, irrigable land would expand by 11% and harvested irrigated land by 17% with the increase happening entirely in developing countries.Fresh water resources follow a similar trend to the land availability(FAO, 2009; Zarco-Tejada et al., 2014). Overall, the total amount of water resources is sufficient but it is distributed in certain regions. There is an increasing number of countries reaching alarming levels of water scarcity whilst other countries have enough water to provide for more people than the country's population. However, there are still plenty of opportunities to increase water use efficiency. The potential to improve crop yields, even with the existing technologies is still significant but socio-economic reasons do not encourage the farmers to invest significant amount of money to bridge the gaps.Unfortunately, the same projections show that production increases alone would not be enough to ensure food for everyone(Bongiovanni & Lowenberg-Deboer, 2004). Governments need to ensure food access even to their most vulnerable groups. These projections emphasize the importance of establishing effectivepoverty reduction strategies, safety netsand rural development programs. The problem is further accentuated by healthier diet trends and an increase in veganism. A significant increase in demand of agricultural products puts an even heavier burden on farmers to improve their crop yields (Zhang, Wang, & Wang, 2002).

In the past centuries, in many surviving writings, it is obvious that especially for larger fields the spatial and temporal differences within the same field were highly appreciated. The first seeds of a concept to adopt to these variations were planted back in the 1930s by Linsley & Bauer (Linsley & Bauer, 1929) and possibly also by Eden & Maskell(Eden & E.J., 1928). The technology that actually initiated the progress of this concept was the development of the Global Positioning System (GPS) back in the 1970s. In 1983, the concept of custom prescribed tillage was developed, paving the way to use technology and automation for increased crop production (Stafford, 2000).

Although in previous centuries, due to the smaller agricultural areas, these spatial and temporal variations were manually controlled, nowadays due to much larger production demand and agricultural land, these variations are impossible to be controlled manually. The concept of using sensing technologies and automation to respond to spatial and temporal variations within a field is now called Precision Agriculture (PA). PA is also called Satellite Farming or Site Specific Crop Management (SSCM) and is described as a key component of the third wave of modern agricultural revolutions (Marios Sophocleous, 2016). PA is currently one of the very few concepts that show the potential to provide the crop yield growth needed to overcome the upcoming challenges of food production and food supply.

PA features two main advantages compared to current agricultural practices. Firstly, it promises to decrease the operating costs of crop production by optimizing the use of natural resources such as fertilizers and water, whilst decreasing labor costs with the goal to minimize or even eliminate the need of human presence. PA technologies allow monitoring and fast tuning of crop production. Due to the quick responses to spatial and temporal variabilities within the field and the elimination of unnecessary use of natural resources, the operational cost can be significantly decreased. Additionally, farmers can perform economic analysis of crop yield of a field and within a field for a more precise risk assessment. By optimizing the amount and time of fertilizers and water input as well as minimizing the use of pesticides and insecticides, farmers can also estimate cash return for each field or hectare(Niu, Fratta, & Wang, 2015).

Secondly, the significantly lower environmental footprint of PA has attracted the attention of multiple environmental agencies. Currently, the monitoring of agricultural footprint on the environment is not a tangible quantity and environmental, governmental and European agencies are on the look for technologies that can provide those tangible controls. Several European, Australian and American agencies are in the process of establishing directives to enforce the farmers to minimize and even eliminate the use of non-environmentally friendly chemicals and the excessive use of fertilizers. Phosphorus have been recently added to the list of the critically low materials of the European Union, skyrocketing its price and sounding the alarm of its excessive agricultural use. The precise and optimized application of fertilizers, water and other chemicals in agricultural practices has the potential to minimize the agricultural footprint on the environment, whilst significantly increasing crop yield and productivity(NRCS East National Technology Support Center, 2011).

Although PA features major benefits, it has yet to see worldwide adoption due to several reasons. Firstly, the denial from the farmers to deviate from the methods they have been using for over a millennia is probably the most essential factor. Thus, even the younger farmers that are keen to use and implement agricultural technologies are discouraged by other factors. For example, most agricultural equipment that have the ability to automate a major part of the process, are very expensive. Looking at the current state-of-the-art for agricultural equipment, the capital investment needed by the farmers is so high that the cash return they will have from the implementation of the concept is not enough.

These two reasons are the most prominent factors as to why PA was not been adopted yet, even in the 21st century. In order to overcome these barriers, new technologies are needed that will advance the state-of-the-art to provide a much higher benefit to the farmers compared to the required capital investment. Hence, the price of such technologies need to decrease significantly as to match a cash return value that the farmers will be tempted by it(M. Sophocleous & Georgiou, 2017). This type of technological advancements include soil quality monitoring, plant growth and plant health monitoring. Although plant growth and plant health monitoring are the most obvious, soil quality is very important for optimized farming but also soil serves many other environmental purposes than just agriculture. Therefore, soil quality monitoring technologies have attracted a significant attention by researchers all over the world.

BACKGROUND

"Soil is the world's most vital component for food and fiber production therefore; protection of such a crucial natural resource is of life-threatening importance"(Marios Sophocleous, 2016). This was understood even back in the 1990s and highlighted by many scientists(Papendick & Parr, 1992). Soil serves multiple functions that not only affect crop yield and productivity but also is a key component for much wider cycles of nature. Soil health contributes to clean air and water, abundance of forests and crops, diverse wildlife and productive and eye-catching landscapes. There are four main functions that soil serves ensuring environmental health and optimized food production(NRCS East National Technology Support Center, 2011).

Firstly, soil is responsible for balanced nutrient cycling. It stores, controls the release and cycles of nutrients such as nitrogen, phosphorus and potassium along with other elements. These nutrients are held in the soiland lost to water or air but most importantly are transformed by soil to forms that plants can absorb through complex biogeochemical processes.Secondly, soil is the habitat of a wide variety and quantity of microorganisms. The population of those microorganisms is of vital importance to soil health as most degradation processes happening in the soil are performed by them. Altering the balance between different species can greatly affect soil's ability to degrade or transform chemicals to forms that are useful and non-harmful to plants and animals. Hence, soil biodiversity is widely accepted as one of the most important soil quality parameters. The European Commission has dedicated multiple Horizon 2020 calls to soil biodiversity for this reason. Thirdly, soil acts as a filter for water, and nutrient flow control. It has the ability to filter toxic compounds or excess nutrients in water and air, which can be degraded or guaranteed from plants and animals. Additionally, it can regulate the water and nutrient flow in the water whilst storing useful nutrients like nitrogen and maintaining their levels within the proper range. Moreover, pesticides and insecticides are also filtered by soil avoiding the pollution of underground waters. Healthy soil partitions water for groundwater recharge and for use by plants and soil animals. Finally, the last soil function is to maintain physical stability and support. Soil has a porous structure that allows water to flow through, which can withstand erosive forces and provide a nourishing environment for plants' roots (Fuglie, 2016).

Soil quality is defined as "the soil's capacity to function within natural or managed ecosystem boundaries and to sustain plant productivity while reducing soil degradation" (NRCS East National Technology Support Center, 2011).Soil must maintain its ability to properly function and provide all the previously explained properties. During the last three decades, there has been a vast number of studies where soil scientist have managed to correlate multiple soil parameters with specific soil functions. Those soil parameters have been later called soil quality indicators.

Soil Quality Indicators

"Soil Quality Indicators (SQIs) are defined as physical, chemical and biological properties of soil that is sensitive to disturbance and represents performance of ecosystem function in the soil of interest. Indicators are dynamic soil properties"(NRCS East National Technology Support Center, 2011). Soil scientists are using these SQIs to evaluate how well soil performs because soil functions cannot be measured directly. SQIs are usually separated in three main categories, where each category affects multiple soil functions (Figure 1).

Physical SQIs can provide information on soil hydrological properties such as soil permeability and permittivity, which are related to how well the soil can maintain water and/or how fast can the water penetrate the soil. This has a major impact on water availability for plants but also on how easily plant roots can penetrate the soil to grow. Some of these indicators are also related to the availability of nutrients based on the effect of rooting volume and aeration status, whilst others provide information on erosional status. An overview of the physical SQIs are shown below (Figure 2).

Figure 1. Most important SQIs and their importance

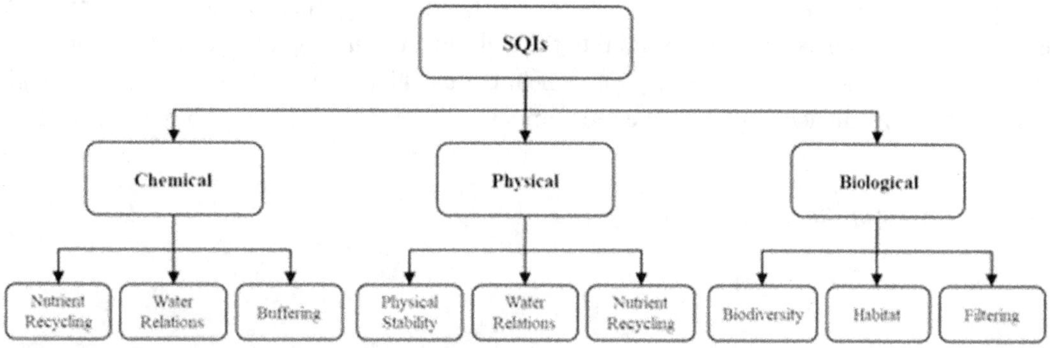

Figure 2. Overview of Physical SQIs

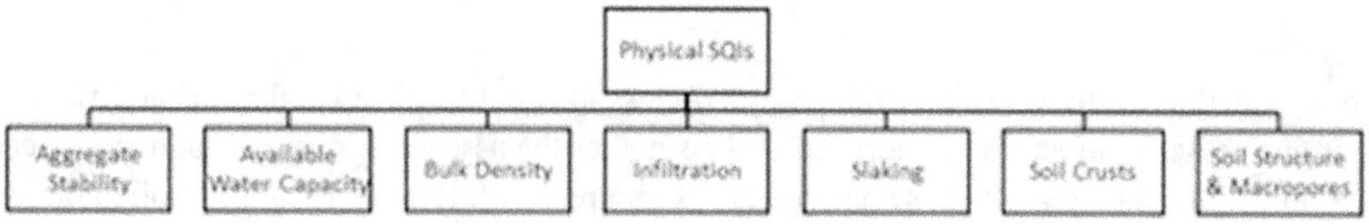

Chemical SQIs are able to provide information on nutritional characteristics of soil-plant root interaction. More specifically, they can describe the equilibrium between soil-water/nutrients and nutrient exchange sites, which is vital for plant nourishing and growth. These SQIs can further show plant health by controlling the nutritional requirements of plant roots and microorganisms, whilst define the soil contaminant concentrations that can have a negative effect on plants and animals. Some of the most important chemical SQIs are shown below (Figure 3).

Figure 3. Overview of Chemical SQIs

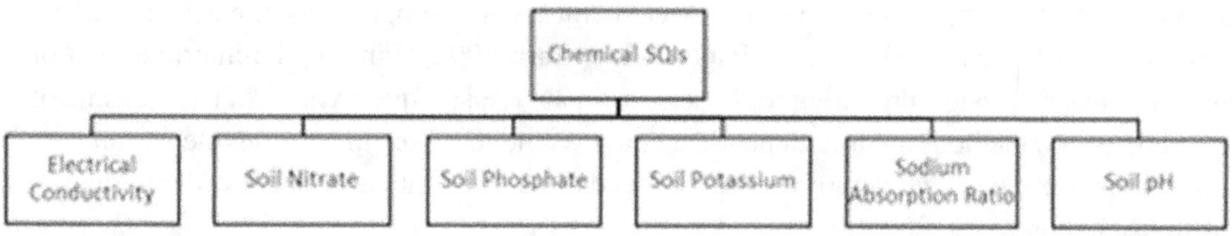

Biological indicators are used to monitor the population of microorganisms in the soil, which are responsible for decomposition of organic matter and nutrient cycling. Several studies suggested that not only the population of microorganisms is important but also the diversity and specific species are responsible for different processes making then essential to the right functioning of soil. These microorganisms are directly related to the ability of soil to regain its health after it has been disturbed in some way giving it the resistance and resilience required. The most important biological SQIs are shown below (Figure 4).

Figure 4. Overview of Biological SQIs

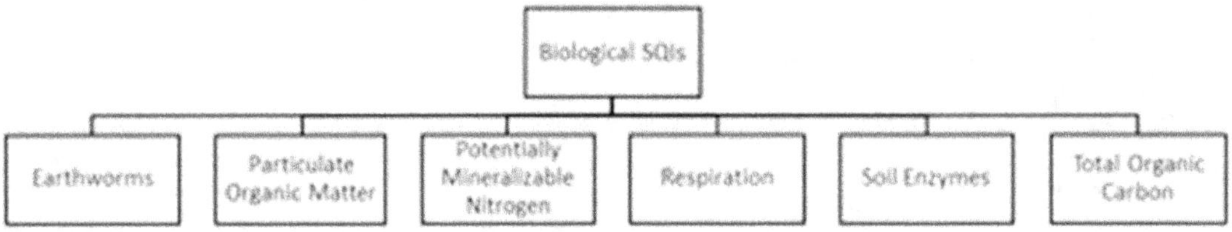

Although all these SQIs are a measurable way to assess soil quality, different parameters affect various functions and these parameters are interrelated. This makes the assessment of soil quality a much more complex procedure than treating these parameters as independent. Hence, a range of soil parameters or indicators has been suggested by soil scientists after years of research to estimate soil quality. Mathematical or statistical frameworks have been developed to generate the soil quality index, which can provide an overall rating of soil quality. Yet, it is still unclear which method is the best to use for a more reliable soil quality index. Soil quality index was developedto take into account not only productivity, which may result in soil degradation, but also environmental issues. Therefore, an appropriate soil quality index may have three component goals: environmental quality, agronomic sustainability, and socio-economic viability(Amacher, O'Neill, & Perry, 2007).

Very few of the previously mentioned parameters can be measured directly in the soil whilst most chemical and biological parameters require extensive laboratory procedures(Mukherjee & Lal, 2014).

Conventional Soil Quality Detection Methods

Conventional soil-quality detection methods involve several multistep procedures from field sampling to complex laboratory analysis ranging from wet chemistry to biological tests, particle size investigations and other compaction-related tests(Papendick & Parr, 1992). Finally, the interpretation of those results is used as a decision-providing tool use to estimate the soil quality index. The steps involved in these soil detection methods are interdependent and possible mistakes in previous steps, can greatly affect the results of the next. Therefore, all aspects need to be taken into account in order to draw reliable conclusions(Benton Jones, 2001).

The first step to aclassical soil test begins with field sampling and sample preparation before the laboratory analysis. When the sample is ready for laboratory analysis, several tests are performed to determine the SQIs' levels needed to calculate the soil quality index. It is extremely important to understand that the significance of the analysis is only as worthy as the quality of the sample analyzed, whose quality is determined by multiple factors, such as field sampling technique, transportation conditions, sample preparation techniques, sample aliquot measurement, laboratory influences and sample storage (Figure 5). Other on-site techniques exist to measure the mechanical properties of soil, such as standard penetration, cone penetration, dilatometers, pressure meters, vanes, flat and stepped blades, hydro-fracture, borehole shear and torsional probes(Mayne, 2006).

Figure 5. Standard soil sampling and testing procedure (Benton Jones, 2001)

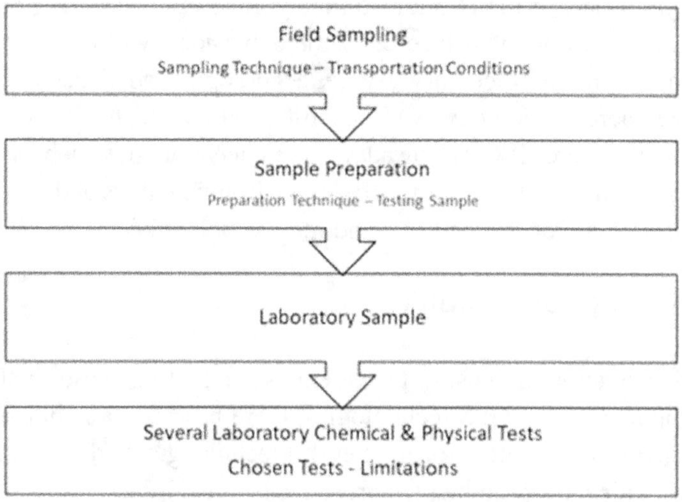

For example, accurate and consistent measurements of soil pH are not easy as there are a number of aspects that can considerably affect the measurement. Usually a salt solution is used (0.01 M calcium chloride ($CaCl_2 \cdot 2H_2O$) or 1 M potassium chloride (KCl)) to overcome the "salt-effect" on pH measurement especially in the case of sandy soils or those soils with relatively low cation exchange capacities. The level of soluble salts found in the sample can be estimated by measuring the electrical conductivity of the solution. As soluble salts concentration increase, the usual effect is decreased plant growth; therefore, soluble salt monitoring is of vital significance. Soils affected by high soluble salt levels are also difficult to manage, particularly when sodium is the major cation contributing to the high salt level. Sodium concentration does not only affect the plant growth but also affects soil aggregation through the Sodium Absorption Ratio (SAR). SAR is the ratio between sodium concentration and the multiplication of calcium and magnesium concentrations. In natural environment, soils with high soluble salt content are found in low-rainfall areas(Pichu & Alla, 2011). Soluble salt problems occur because of the use of salt-laden irrigation waters, or of improper fertilizer placement, high fertilizer application rates, or accumulation from repeated fertilizer applications. Salinity affects about 25% of the croplands in the world and is becoming an increasing problem in most irrigated croplands. Fertilizer placing very close to plant seeds can be very easily damaged. Additionally, high concentrations of soluble salt are a common problem when growing plants in pots, especially when over fertilized. Soluble salt monitoring is recommended as an important procedure for the container and hydroponic grower, indicating when containers should be leached or hydroponic solutions replenished. The cations Na^+, K^+, Ca^{2+}, Mg^{2+}, and NH^{4+}, and the anions Cl^-, NO^{3-}, HCO^{3-}, SO_4^{2-}, and CO_3^{2-}significantly contribute to the resistivity of the soil solution or irrigation water(Marios Sophocleous, 2016).

Conventional soil analysis techniques mentioned above demonstrate several advantages and disadvantages that can restrict their use in applications where fast response is an important requirement. Soil sampling is a very timely and labor-intensive procedure(Bar-on, Jog, & Shacham-Diamand, 2019). Furthermore, it is very sensitive and highly complex making it non-viable for non-experts to use. Timely procedures for soil analysis result in a slow response eliminating the opportunity farmers and soil scientists to ensure on-time responses. In addition, sampling rate is another issue since soil chemical properties vary widely not only spatially but also timely. Likewise, analyzing physical soil properties using the existing technologies one will face the same downsides.All these methods are off-grid and very expensivetechniques."In the 21stcentury, where technological advances allow for automated systems to provide remote operation and observation, a different approach is needed to provide real-time measurements of SQIs" (Stafford, 2000). Nowadays, using technologies such as the Internet of Things (IoT), all the information can be brought to the farmer, environmentalists and soil scientists for a deeper understanding of nature's complex mechanisms and more efficient PA.

Soil Quality Monitoring Technologies

Since the initial implementation of sensing technologies for custom prescribed tillage back in 1983 and the further development of the GPS technologies, there has been a major redirection of research resources in PA and environmental monitoring. Several technological advances have been reported in the last two decades with the most prominent approach to be the remote sensing of the environment and agricultural fields(Stafford, 2000). Remote sensing (RS) or satellite imaging (SI) approach has received an increasing attention from environmental engineers to the level that several scientific journals have been created with supporting scientific societies and worldwide conferences(Lingyan, Yanning, Wei, & Qilin, 2017). However, other technologies have also been reported that feature several advantages and disadvantages compared to the classical satellite imaging approach. A brief overview of these technologies is given below (Figure 6).

Figure 6. Overview of Agricultural & Environmental Monitoring Technologies

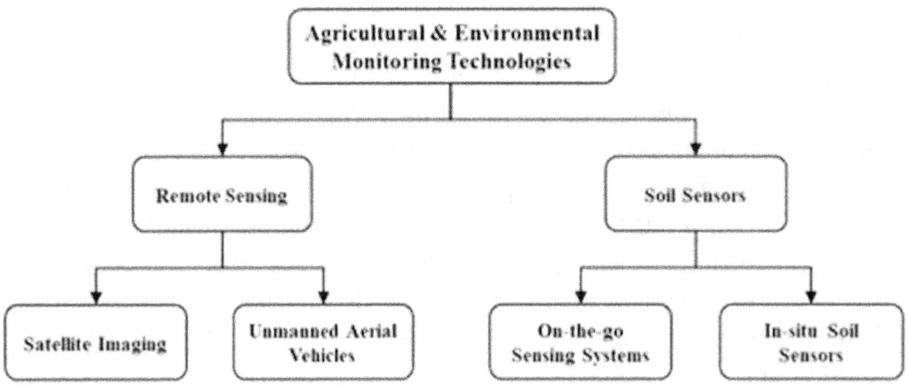

Remote Sensing–Satellite Imaging

Remote Sensing is understood as the concept of acquiring information about an object or phenomenon without having physical contact with it. This concept is nowadays mostly implemented in Earth observations either for climate changes and predictions but also for automatic mapping. It has gained a lot of attention over the years as it can provide information through several types of image processing. One of the most common uses of RS is the use of hyperspectral or multispectral images from satellites to monitor agricultural and non-easily access land(GE, Thomasson, & Sui, 2011). This application is called satellite imaging (SI) and with the available technological advances allowing almost every country to have its own satellite or even giving access openly to certain satellites, it has bloomed all over the world. Reported research has shown that certain soil parameters can be monitored through image processing of those hyper/multi spectral images. A significant amount of work has been reported over the years on building soil maps, showing how soil parameters vary within the field, and it has become one of the most common techniques to monitor soil quality remotely (Figure 7).

Figure 7. Common soil maps constructed using SI(Ceinsys, 2018)

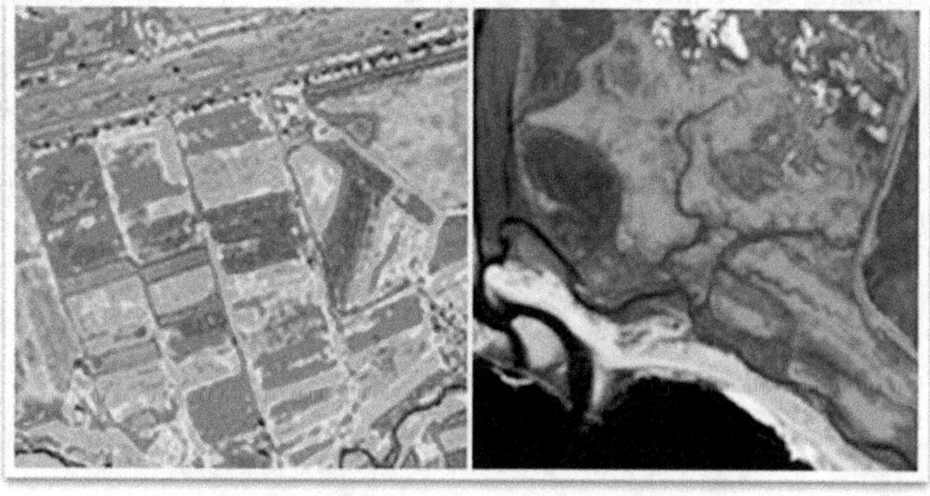

SI is a very well-know and well-used method to monitor both soil quality but also plant health. The general concept of this technique is to investigate the intensity of the images at a certain frequencies, which are correlated with certain soil properties, such as soil moisture, and generate soil maps like the ones shown in Figure 7 above. However, this technology has major trade-offs in terms of capital investment, real-time data and spatial resolution (Angelopoulou, Tziolas, Balafoutis, Zalidis, & Bochtis, 2019).

Satellite imaging and hyper/multi spectral images have been used by scientists since the 1990s to detect soil properties remotely. Soil maps have been developed from researchers all over the world for multiple soil properties. The most common soil properties that can be monitored using RS and SI specifically include but are not limited tosoil types such as clay, silt or sand; organic matter; nitrogen; and moisture (Vermeulen et al., 2017).

Using an orbiting satellite to take images of a small agricultural area is not preferable. Usually, these technologies are available to scientists with enough funding to rent the satellite for a certain amount of time. On the other hand, in the case of farmers this is not a very convenient method since the farmers are not familiar with running the image processing algorithms or understanding how to control satellites. Especially farmers of the previous generation that are already reluctant to adopt the general principle of motoring and acting accordingly, the use of satellites further intimidates them. Additionally, agricultural businesses do not have the required profits to expand to such expensive technologies, since the required capital investment is incredibly high and the return period is not attractive enough. Technically, satellite imaging does not require any human presence, does not affect agricultural practices of any kind and has the ability to obtain information for large areas very easily with a relatively good spatial resolution. One of the most important drawback of this approach is that it can only provide information about the surface soil properties (only a few centimeters of depth), which although useful, it is far from sufficient for optimized agricultural practices since plant roots are much deeper than that (Syvitski, Overeem, Brakenridge, & Hannon, 2012).

Remote Sensing–Unmanned Aerial Vehicles

Another approach to RS instead of SI is the use of Unmanned Aerial Systems (UAVs) to acquire field images using hyper and/or multispectral camera (Stehr, 2015). UAVs are defined as flying objects/vehicles without a human pilot onboard. The most common type of UAVs are drones (Figure 8) and with all the technological advancements of the GPS systems, UAVs can operate on their own and travel pre-determined paths at specific times of the day to acquire the images (Daponte et al., 2019).

Figure 8. Available drones for agricultural and environmental monitoring (DJI, 2019)

Despite their ease of use and effectiveness, UAVs have a major drawback. They need to be calibrated for specific tasks lacking the ability to provide a sufficient overview of the agricultural process and subsequently soil and plant quality. For example, UAVs have been used over the years for tasks like identifying vegetation types, water quantities and soil parameters. However, the lack of interoperability creates more work for the farmers and in a field that they are not utterly familiar. Although they can eliminate some of the tasks that are currently completed by the farmers, the fact that they require significant amount of time for image or video processing by the farmer himself, defeat their very own purpose of using them for soil quality monitoring and PA in general.

Future trends in this area are moving towards the use of cheap commercial mini or micro drones to decrease the initial capital investment but at the same time, the technical specifications and detection accuracies and reliabilities of the UAVs decrease significantly(Veroustraete, 2015).

Soil Sensors–On-the-go Sensing Systems

In comparison with the RS approach, attempts have been made by scientists and agricultural technologies companies to develop soil sensors that will measure soil parameters either by directly inserting them in the soil (in-situ soil sensors) or by attaching them on existing agricultural machinery (on-the-go sensors). The concept of on-the-go sensors (OTGS) has been introduced towards the end of 1990s by (Sudduth, Hummel, & Birrell, 1997), who were the pioneers in this approach(Adamchuk, Hummel, Morgan, & Upadhyaya, 2004). The concept was that since agricultural machinery is already used by the farmers, adding soil-sensing systems on them that could take readings while performing other compulsory agricultural practices would provide the farmer with extra information on the quality of the soil (Figure 9). In the left photo of Figure 9, the OpticMapper of Veris Technologies is shown that has a dual-wavelength optical sensor mounted within a specially configured planter row unit. As the OpticMapper moves along within the field, it maps the soil underneath for crop residue and surface soil moisture. Soil measurements are obtained through a sapphire window at the bottom of a metal part. The measurements are collected and matched to their GPS location. The OpticMapper module is usually combined with Soil Electrical Conductivity sensing on the sensing platform, so that soil texture and organic mattervariations are simultaneously mapped. In the right photo of Figure 9, the MSP3 system by Veris Technologies is shown. This system can measure the same soil parameters but it is also equipped electrodes to measure soil pH with 5-10 pH samples per acre (Veris, 2014).

Figure 9. Examples of On-the-go sensors (Veris, 2014)

This idea is much easier to be adopted by most farmers since it does notrequire farmers to change anything from the recipes they are already using and the initial capital investment would be much lower. Instead of buying new machinery or equipment that they are not familiar with, a simple addition to their existing equipment could provide very important soil quality information.

Systems have been reported over the years can measure a wide variety of soil parameters and in contradiction with RS technologies, soil properties can be detected as deep as the equipment can go giving much more valuable information that surface properties. Soil parameters that can be measured using OTGS sensing systems include but are not limited to electrical resistivity, capacitance or inductance, take images for image processing similar to RS approaches, porosity and permeability and electrochemical sensors monitoring the activity of hydrogen, potassium, nitrate, sodium ions and many others.

Although this approach seems simpler, more economically viable and easier to adopt by the farmers, its potential for productivity and yield increase is not enough for what is coming by 2050(Alexandratos & Bruinsma, 2012; FAO, 2009). The reason is that these systems cannot provide continuous measurements to allow for automated operation of the agricultural farms. The readings on soil quality will only be taken whenever the farmer decides to use the equipment in the field, meaning that human presence is required and it is very time consuming especially for larger fields. Comparing OTGS sensing systems and RS, RS can cover a much larger area with much less effort. However, it does not have the spatial resolution and measurement accuracies featured by OTGS systems. On the other hand, OTGS systems require a significant amount of labor time, which eventually defeats the purpose of automatic control. Furthermore, OTGSsystems cannot provide continuous measurements or real-time data, meaning that this approach's potential is limited and possibly inadequate to overcome future food and agricultural challenges(Adamchuk et al., 2004).

Soil Sensors–In-Situ Soil Sensors

The most recent approach to soil sensing, especially after the price reduction of electronic devices, is the use of soil sensing systems directly in the soil able to measure both physical and chemical soil parameters without the need of soil sampling. The general concept of in-situ sensors (ISS) is to revolutionize the soil sampling approach and instead of taking the soil to the laboratory, to take the laboratory to the field(Atkinson et al., 2013; Bar-on et al., 2019; M. Sophocleous & Georgiou, 2017; Marios Sophocleous, 2017; Marios Sophocleous, Garcia-Breijo, Atkinson, & Georgiou, 2019). This is a very promising approach however; it faces numerous fundamental issues and questions making this field very attractive for scientists to create the standards and the knowledge in order to implement this concept successfully. Soil ISS are classified by the soil parameter that they can detect, which is subsequently the soil properties mentioned earlier in this chapter. Therefore, the categories are very similar to those of soil properties. They are conceptually separated into physical, chemical, biological and gas sensors where each category corresponds to the physical, chemical and biological properties of soil. Despite this categorization, the sensing phenomena used to detect each of these parameters can also be used to classify the sensors into more subcategories such as electrochemical, optical, electromagnetic and thermal. Due to the complexity of those sensing systems, there are even more ways to classify them, for example depending on the sensing technique such as potentiometric, amperometric, voltammetric and impedimetric. An attempt to show the different classifications of soil sensors is given below (Figure 10).

Figure 10. Overview of soil ISS

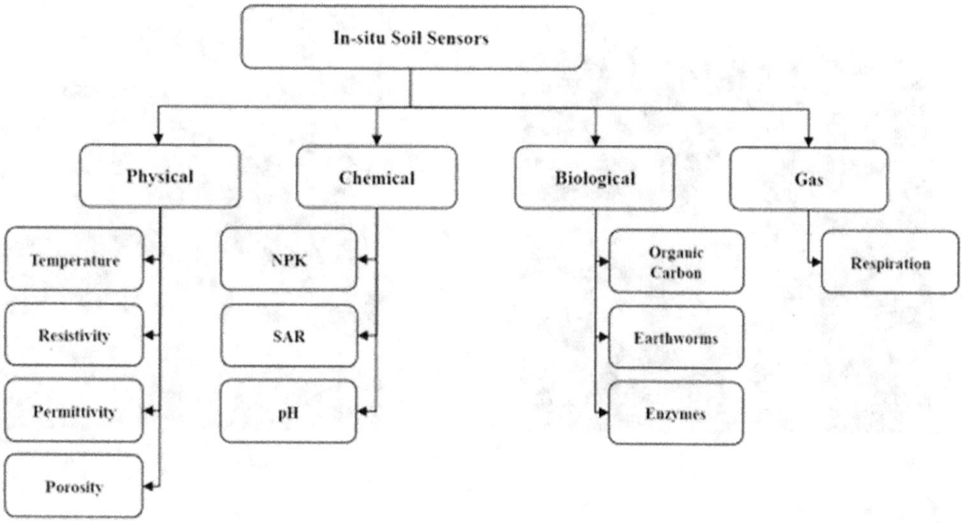

Physical sensors are the most-investigated category from all because they are the easiest to measure in soil. Parameters such as temperature do not depend on the medium, hence a huge variety of soil temperature sensors are already available in the market. The electrical resistivity of soil is directly related to a multiple of soil properties with the most dominant being the soil salinity. As the ion concentration in the soil increases, the soil electrical resistivity decreases making the soil more conductive. However, other factors can affect this parameter such as soil compaction, soil type i.e. particle size and others. Electrical resistivity sensors have been reported in the literature with some of them actually reaching the market. On the other hand, the most commonly available soil sensor is the water content sensor, which gives information on one of the most critical soil properties. Soil water content sensors have been around for several year with a wide variate of them reaching the market, such as Delta-T ML3 probe (Figure 11) (Delta-T Devices Ltd, 2017) that features a very impressive performance. This type of sensor measures the relative permittivity of the soil at a specific frequency and that parameter is related with the amount of water in the soil. This can provide real-time data on soil water content and allows the use of automated systems for irrigation control.

Water potential is also measured using a tensiometer (Charles, Anthony, & Junjun, 2019), which is a device made of a low-pressure chamber placed in the soil. As the water in the soil increases, the moisture in the soil is absorbed in the low-pressure chamber and that change is proportional to the water content in soil. Furthermore, other soil physical parameters are important, which have not yet been detected using ISS. For example, soil porosity is still measured using ground penetrating radars (Department of Agriculture, 2009) that are very expensive and require the presence of highly experienced personnel(Charles et al., 2019).

Chemical and biological soil properties are probably the most important factors to gauge soil quality but also to evaluate whether current agricultural practices are compatible with sustainable soil quality. Chemical properties such as soil pH and Nitrogen, Phosphorous and Potassium (NPK) concentrations are vital for evaluating soil fertility. Unfortunately, such chemical sensors have yet to see their implementation in soil for direct and in-situ measurements of soil. pH is probably the most commonly

Figure 11. Delta-T ML3 probes for soil water content monitoring (Delta-T Devices Ltd, 2017

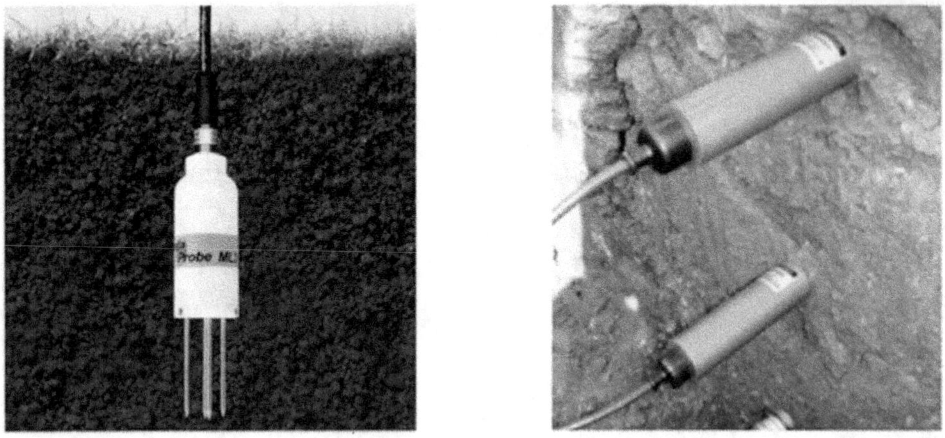

investigated type of soil sensors by scientists. Potentiometric, amperometric and voltammetric sensors have been reported over the years that have the ability to be used directly in the soil without the need of soil sampling but also to provide real-time data(Edaphic, 2019). NPK sensors have not been reported anywhere in the literature so far and nothing is yet available commercially. The main reason for that is the deployment method to be used to directly insert the sensors in the soil while being able to acquire real-time data. Usually, the most common problem is the interface between the sensor and the soil and the establishment of a comparable standard that can be used to assess soil NPK content. Biological soil properties such as earthworm population, enzymes and organic carbon cannot be monitored directly due to the lack of direct, in-situ biosensors that can provide continuous and real-time data. Although there is a plethora of biosensors that can detect almost all of the pre-mentioned properties in the laboratory, none of them has the ability to be used directly in the soil.

Gas sensors on the other hand, since the gasses to be monitored are the ones evaporating above ground, they are much easier to implement. There is a huge range of gas sensors available in the market that don't really need to be customized for soil purposes since the measurements are taking place above soil surface. Several gases escape from soil during agricultural cycles. Water vapor, carbon dioxide (CO_2), methane (CH_4) and nitrous monoxides (NO) are the main gas emissions in agriculture. The efflux of these gases is related to soil bio-content thus, there is a significant interest in observing the gas emissions in a specific agricultural exploitation. Two technologies are the most appropriate for monitoring this assortment of gases. Optical sensors is one key category for measuring gas emissions. Near and mid-infrared spectroscopies can detect the absorption and emission radiation interactions of the gases. Electrochemical sensors is the other significant category. Metal oxides, polymers and membranes have been published devoted to gas sensing by measuring the changes of the electrochemical properties of the sensing material. Infrared spectroscopy is a robust, reliable and simple technique; however, the technology needs of quite complex instrumentation. A spectrometer requires of an infrared source, a diffraction grating, lenses, an infrared detector and many other components. The average cost for a gas to be measured can be around 50 € (commercial equipment for gas sensing can be found at a price around 200 €). In the case of a facility in which CO_2, carbon monoxide (CO), water vapor, NO, nitrogen dioxide (NO_2), nitrous oxide (N_2O), ammonia (NH_3) want to be measured for each individual plant or surface area (for instance, a complete sensing system each squared meter) the cost turns out to be excessive (higher than €350/m^2). This does not satisfy the requirement of low cost sensing making the technology unsuitable for this particular application.

Summarizing, soil ISS show a huge potential for implementation in automated systems for agriculture since they offer the best features such as continuous and real-time data but also can show x-y-z variability of those properties, something that the rest of the available sensing technologies are unable to provide. Additionally, soil ISS are much more sensitive than RS and OTGS sensors; however, they have a much shorter lifetime and require high-density sensor networks, which have a labor-intensive installation. Additionally, they have many remaining issues to be overcome such as the need of long lifetimes, the ability to operate electrochemical sensors in a medium like soil that can be dry or wet, maintaining a good enough electrical/ionic conduct between the sensor surface and the soil medium and many other. There are several technologies that can be implemented to overcome these challenges such as microfluidic chambers and 3D printed matrices to withhold soil and water at a fixed ratio however, the only reported approach is by enclosing the sensors using an absorptive green foam, as described by Sophocleous in 2016 (Marios Sophocleous, 2016).

Towards Automated Decision-making Systems

Although the various soil sensing technologies explained in the previous sections of this chapter, have a strong potential to meet the future food demands as predicted by FAO(FAO, 2009), independent systems and nodes without intelligence can only reach a certain yield and productivity level. The integration of those technologies into a more autonomous and intelligent, a higher order system has the potential to offer much higher yields and productivity, whilst decreasing operational costs to the minimum. Despite the benefits of integrating these systems to a more intelligent version, the cost of the system increases significantly. This implies that the application demands integrated technologies with lower costs making their implementation more viable for the farmers. Currently, there are commercially available systems that can be controlled by the operators remotely and digitally. These technologies are currently, the state-of-the-art but they are only the first step towards fully automated decision-making systems.

Available Technologies for Digitally Controlled Farms

There are currently several commercially available technologies that can be used to digitally-control a farm or greenhouse, most commonly, and a table with a short summary of them showing their functions is given below. It is clear that environmental parameters such as air temperature, soil water moisture, soil salinity and sunlight can be monitored by almost all systems. Moreover, almost all systems have a graphical user interface, which makes the control so much easier for the non-experienced user to control. Controls include environmental temperature control for greenhouses and irrigation control for both greenhouses and open fields. These functionalities make the systems very intimidating to the farmers however; they have not been widely implemented. The initial capital investment for such systems is in the range of tenths of thousands of euros and although the most prominent farmers are willing to make the investment, it is almost impossible for smaller farms. Smaller farmers cannot afford to invest that amount of money to a system that can only provide a certain level of automatic control and a limited potential for yield and productivity increase.

Most importantly, as shown in Table 1 below, the most important farming parameters such as soil chemical parameters and plant health are still not included in these systems. These systems lack the ability to monitor these parameters in continuously and this is one of the most important limitations of the existing agricultural technologies.

Table 1. Overview of commercially available technologies for farming digital control (Hoogendoorn, 2020; HortiMax growing solutions, 2020b, 2020a; Priva, 2019a, 2019b)

	Priva-maximiser	**Hortimax synopta**	**Priva-connext**	**Hortimax multima**	**Hoogendoorn isii**
Environmental control	✓	✓	✓	✓	✓
Environmental gas concentration monitoring	✓	✓	✓	✓	✓
Visual interface	✓	✓	✓	✓	✓
Automated irrigation &fertilizer supply	✓	✓	✓	✓	✓
Access on tablet or smartphone		✓	✓	✓	✓
Customization	✓	✓	✓	✓	
Predictive environmental control	✓	✓		✓	✓
IoT platform			✓	✓	✓
Compatibility with other components (sensors etc.)			✓		
Data analytics					✓
Automated decision making					✓
Soil nutrient concentration monitoring					
Soil gases efflux monitoring					
Disease detection					
Low cost					
Low capital investment					
Self-learning based on historical data					

Some of the more intelligent systems such as Hoogendoorn iSii have IoT compatibility features and most importantly provide the software for agricultural management that uses data analytics algorithms, paving the way for the first systems with automated decision-making abilities. Vendors such as John Deere, IBM and SensorCloud currently provide solutions for crop performance monitoring that support specific sensors. However, vendor-specific solutions feature limited or no interoperability with other IoT compatible devices. Several IoT middleware platforms recently appeared, claim to have solutions for all verticals, including PA and site-specific crop management. Examples include Xively, Thing Speak Open.Sen.se, and SensorCloud and more recently, Amazon IoT and IBM IoT platforms. A significant shortcoming of those platforms relates to edge-network-protocol-termination for heterogeneous sensor technologies, which must be bridged with a local or remote server transparently to the user. Despite the fact that Hoogendoorn iSii has the automated decision-making ability, those decisions are only drawn based on the same parameters that the rest of the systems can monitor. None of the existing systems is able to monitor the most important soil and plant parameters that will provide much more reliable decisions.

In order to ensure the reliability of the decisions, more soil and plant parameters need to be taken into account. Soil parameters such as nitrogen, phosphorus and potassium concentrations in the soil together with plant growth rates, fruit ripeness and plant diseases, when combined together can form a much more complete picture in terms of the crop yield, productivity and most importantly what the type and amount of inputs the plant needs. These parameters should be monitored without the need of human intervention, such as soil sampling but they should also account for any spatial and time-based variability within the same field. Although the relationships between these parameters and the quality of the crops/fruits are well established by soil and plant scientists, they cannot be used in real-time since they currently require soil sampling and destructive plant tests. Overcoming these problems can be achieved soil ISS and possible plant imaging systems with image processing capabilities. Combining the readings of all those sensors with environmental conditions, together with intelligent optimization algorithms, much more accurate and reliable decisions can be made to maintain optimum levels of water, fertilizer and pesticide/insecticides maximizing crop yield, productivity, efficiency and minimizing any environmental pollution caused by overuse of natural resources.

Fully Automated Decision-Making Systems

The concept of fully Automated Decision-Making Systems (ADMS) being used in PA has been discussed by several researchers in the last decade, especially with the development of low-cost electronics and sensors. The implementation of such a system has been recently described by Sophocleous et al. showing the different technologies and specifications required to run such systems. ADMS are defined as systems that have the ability to operate a farm/field without the need of human presence or human input. In order to achieve this goal, ADMS must be able to monitor soil and plant parameters, assess soil and plant condition and make the right decisions as to maintain those parameters at optimum levels by controlling various inputs. As described by Sophocleous et al. the architecture of ADMS is shown below.

Figure 12. Architecture of ADMS (M. Sophocleous & Georgiou, 2017)

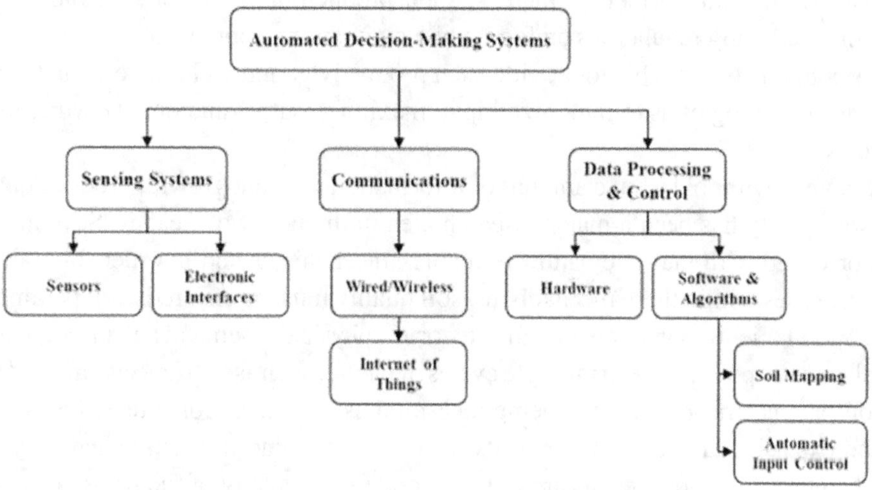

ADMS and digitally controlled farms have many things in common but they also have several differences. Their most important difference is that ADMS do not require the user to provide any input as they are self-operating using their own artificial intelligence (AI). In the systems described in the previous section, the user was able to control the farm remotely with the exception of Hoogendoorn iSii, which is the first to provide decision-making features. Agricultural systems with AI capabilitieshave much better potential to solve the future food challenges but have also the ability to create knowledge for the farmer, agricultural institutes and agencies, as well as governmental agencies and policy makers. With the exponential rise of regulations on food quality and controlling the use of Genetically Modified foods, agencies and governments are in the need of a way to monitor the growing parameters of agricultural produce to ensure high quality standards. None of the pre-mentioned systems can monitor the quality of produce at the farm during production and subsequently cannot use historical data to learn and create new knowledge. Machine Learning (ML) techniques have gained a lot of attention in the last twenty years due to their ability to learn from the obtained data and create new knowledge based on that with automatic interpretation of those data. Hence, integration of these systems together with ML techniques can have a significant upgrade to the systems in terms of autonomy and new knowledge generation.

Finally, the implementation of ADMS, combined with ML techniques and AI, has the ability to significantly improve the quality of produce, minimize the use of natural resources, considerably reduce operating costs by eliminating the need of human intervention but also to minimize any deleterious impact on the environment from fertilizer and pesticide/insecticide usage. However, the initial capital investment required by the farmers will remain unapproachable unless, customized, ultralow-power electronics i.e. Application Specific Integrated Circuits (ASICs) are developed to match the requirements of such systems together with the development of low-cost soil/plant sensors using technologies such as Thick-Film Technologies or 3D printing. The goal for such technologies will be to have a less than 5 year Return on Investment with less than €10000 initial capital investment.

CONCLUSION

Human population will exceed 9 billion people by 2050 as predicted by FAO; hence, food demand will skyrocket at that point, making the energy debate pale into insignificance. Therefore, a huge burden will be added to agriculture and farmers to increase their productivity to match demand, whilst simultaneously abide the increasing regulations on food while maintaining approachable prices for their produce. The only agricultural concept able to provide such productivity and yield increase is PA with the use of state-of-the-art technologies as to minimize input, maximize output and quality while maintaining low operating costs.

Soil is known to withhold a huge amount of information on plant growth, produce quality and yield. Therefore, soil quality has been a major research area in the last two decades. Several soil parameters have been correlated with multiple nature's complex mechanisms and in order for soil to maintain its functional properties within these mechanisms, soil quality must be ensured. Soil parameters have been categorized by soil scientists into three main categories, physical, chemical and biological. Each of these categories plays an important role in multiple cycles of nature, such as nutrient cycling and water filtering.

These soil parameters are currently being measured using conventional techniques that involve soil sampling and extensive laboratory-based tests, which are extremely sensitive and very laborious. The decisions taken based on those results are only as good as the quality of the steps taken during the process and since the steps are interdependent with each other, a minor modification in one of the steps can completely deteriorate the quality of the next.

In the 21st century and in the era of IoT, new technologies have been developed to eliminate the need for soil sampling and laborious experimental procedures. These technologies have the advantage of improving the agricultural yield and significantly increasing productivity. However, most of them require a huge capital investment and the farmers' willingness to change their classical agricultural practices and adopt these new technologies. These technologies can be separated into four different categories, satellite imaging, OTGSsensing systems, drones or other aerial systems and soil ISS. Each of these categories has its own advantages and disadvantages however, for real-time soil quality monitoring, soil ISS are the most prominent for the job.

The ability to obtain real-time data without the need of human presence, has envisioned the concept of fully automated decision-making systems for agriculture. In the second decade of the 21st century, there are commercially available technologies that can transform the classical agricultural farms to digitally controlled farms. Unfortunately, digitally controlled farms lack intelligence to run on their own. A new concept of ADMS has been recently described in the literature that uses a combination of sensing nodes talking to each other and eventually to a central controller. The central controller uses machine learning algorithms and artificial intelligence to make decisions on its own without the need of human control. This system has the advantage that due to the machine learning algorithms and the use of historical data, new optimum points can be found and a new pool of knowledge can be generated without any further human input. However, the implementation of such a system has not yet been reported even today.

Major challenges remain in the area of sensor development for in-situ and real-time soil quality monitoring, such as deploying the sensors underground for an extended period of time while being able to reliably monitor soil quality at low cost. Along with the sensor development, the electronic interfaces required to run those sensors should fulfil the strict specification requirements of the application, such as power reduction, size and cost. Communication of these sensor nodes with the central controller in high-density sensor networks without any interference, with low power and at large distances are challenges that still remain unanswered. Adding intelligence to these systems along with machine learning and neural networks will highly increase agricultural productivity, improve food quality, reduce food prices and optimize the use of natural resources.

ACKNOWLEDGMENT

The author would like to thank the Holistic Electronics Research Laboratory of the University of Cyprus for the support, Professor Julius Georgiou and Professor John K. Atkinson for the continuous guidance. Moreover, the author would like to thank the Research & Innovation Foundation of Cyprus for the funding, under the project No. POST-DOC/0718/0163.

REFERENCES

Adamchuk, V. I., Hummel, J. W., Morgan, M. T., & Upadhyaya, S. K. (2004). On-the-go soil sensors for precision agriculture. *Computers and Electronics in Agriculture*, *44*(1), 71–91. doi:10.1016/j.compag.2004.03.002

Alexandratos, N., & Bruinsma, J. (2012). *World Agriculture Towards 2030/2050*. Academic Press.

Amacher, M. C., O'Neill, K. P., & Perry, C. H. (2007). Soil vital signs: A new soil quality index (SQI) for assessing forest soil health. *USDA Forest Service - Research Paper RMRS-RP*, (65 RMRS-RP), 1–14. doi:10.2737/RMRS-RP-65

Angelopoulou, T., Tziolas, N., Balafoutis, A., Zalidis, G., & Bochtis, D. (2019). Remote sensing techniques for soil organic carbon estimation: A review. *Remote Sensing*, *11*(6), 1–18. doi:10.3390/rs11060676

Atkinson, J. K., Glanc, M., Prakorbjanya, M., Sophocleous, M., Sion, R., & Garcia Breijo, E. (2013). Thick film screen printed environmental and chemical sensor array reference electrodes suitable for subterranean and subaqueous deployments. *Microelectronics International*, *30*(2), 92–98. doi:10.1108/13565361311314485

Bar-on, L., Jog, A., & Shacham-Diamand, Y. (2019). Four Point Probe Electrical Spectroscopy Based System for Plant Monitoring. In *Proceedings of ISCAS 2019, Japan* (pp. 1–5). 10.1109/ISCAS.2019.8702623

Benton Jones, J. J. (2001). *Laboratory Guide for Conducting Soil Tests and Plant Analysis*. CRC Press. doi:10.1201/9781420025293

Bongiovanni, R., & Lowenberg-Deboer, J. (2004). Precision Agriculture and Sustainability. *Precision Agriculture*, *5*(4), 359–387. doi:10.1023/B:PRAG.0000040806.39604.aa

Ceinsys, T. L. (2018). *Applications of satellite imagery & remote sensing data*. Retrieved from https://www.ceinsys.com/blog/applications-of-satellite-imagery-remote-sensing-data/

Charles, W. W. N., Anthony, K. L., & Junjun, N. (2019). *Plant-Soil Slope Interaction*. CRC Press.

Daponte, P., De Vito, L., Glielmo, L., Iannelli, L., Liuzza, D., Picariello, F., & Silano, G. (2019). A review on the use of drones for precision agriculture. *IOP Conference Series: Earth and Environmental Science*, *275*(1). 10.1088/1755-1315/275/1/012022

Department of Agriculture. (2009). *Ground Penetrating Radar soil suitability maps*. Retrieved from https://www.nrcs.usda.gov/wps/portal/nrcs/detail/soils/survey/geo/?cid=nrcs142p2_053622

Devices Ltd. D.-T. (2017). *ML3ThetaProbe*. Retrieved from www.delta-t.co.uk

DJI. (2019). *DJI introduces P4 Multispectral for Precision agriculture and land management*. Retrieved from https://www.suasnews.com/2019/09/dji-introduces-p4-multispectral-for-precision-agriculture-and-land-management/

Edaphic, S. (2019). *Soil pH Sensor & Meter*. Retrieved from https://www.edaphic.com.au/products/soils/soil-ph-meter/

Eden, T., & Maskell, E. J. (1928). The influence of soil heterogeneity on the growth and yield of successive crops. *The Journal of Agricultural Science*, *18*(2), 163–185. doi:10.1017/S0021859600009023

FAO. (2009). How to feed the world 2050 - Global Agriculture towards 2050. Rome, Italy: FAO.

Fuglie, K. (2016). The growing role of the private sector in agricultural research and development worldwide. *Global Food Security*, *10*, 29–38. doi:10.1016/j.gfs.2016.07.005

Ge, Y., Thomasson, A., & Sui, R. (2011). Remote sensing of soil properties in precision agriculture: A review. *Frontiers of Earth Science*, *5*(3), 229–238. doi:10.100711707-011-0175-0

Hoogendoorn. (2020). *The next generation iSii process computer*. Retrieved from https://www.hoogendoorn.nl/en/product/isii-process-computer/

HortiMax growing solutions. (2020a). *HortiMaX MultiMa*. Retrieved from https://www.hortimax.co.uk/water-and-nutrition/hortimax-multima/

HortiMax growing solutions. (2020b). *HortiMax Synopta*. Retrieved from https://www.hortimax.co.uk/operation-and-analysis/hortimax-synopta/

Lingyan, R., Yanning, Z., Wei, W., & Qilin, Z. (2017). A hyperspectral image classification framework with spatial pixel pair features. *Sensors (Basel)*, *17*(10), 2421. doi:10.339017102421 PMID:29065535

Linsley, C. M., & Bauer, F. C. (1929). Illinois Agricultural Experiment Station. *Circular, 346*.

Mayne, P. W. (2006). In-situ Test Calibrations for Evaluating Soil Parameters. In T. S. Tan, K. K. Phoon, D. W. Hight, & S. Leroueil (Eds.), *Singapore Workshop* (pp. 1–56). 10.1201/NOE0415426916.ch2

Mukherjee, A., & Lal, R. (2014). Comparison of soil quality index using three methods. *PLoS One*, *9*(8), e105981. Advance online publication. doi:10.1371/journal.pone.0105981 PMID:25148036

Niu, Q., Fratta, D., & Wang, Y.-H. (2015). Precision agriculture - a worldwide overview. *Journal of Hydrology (Amsterdam)*, *522*(20150806), 475–487. Retrieved from https://linkinghub.elsevier.com/retrieve/pii/S002216941401066X

NRCS East National Technology Support Center. (2011). *Soil Quality for Environmental Health*. Retrieved from http://soilquality.org/home.html

Papendick, R., & Parr, J. (1992). Soil quality—The key to a sustainable agriculture. *American Journal of Alternative Agriculture*, *2–3*(1-2), 2–3. Advance online publication. doi:10.1017/S0889189300004343

Pichu, R., & Alla, M. (2011). Cation ratio of soil structural stability (CROSS). *Soil Research (Collingwood, Vic.)*, *49*(3), 280. doi:10.1071/SR10105

Priva, B. V. (2019a). *Maximizer|Priva*. Retrieved from http://www.priva-asia.com/en/products/maximizer/

Priva, B. V. (2019b). *Priva Connext*. Retrieved from https://www.priva.com/horticulture/products/connext

Sophocleous, M. (2016). *Thick-Film Underground Sensors*. LAP LAMPERT Academic Publishing.

Sophocleous, M. (2017). Electrical Resistivity Sensing Methods and Implications. In Electrical Resistivity and Conductivity. doi:10.5772/67748

Sophocleous, M., Garcia-Breijo, E., Atkinson, J. K., & Georgiou, J. (2019). Modeling & Performance Comparison of Screen-Printed, Impedance Spectroscopy Probes for Harsh Environments. *IEEE Sensors Journal*. Advance online publication. doi:10.1109/JSEN.2019.2951200

Sophocleous, M., & Georgiou, J. (2017). Precision agriculture: Challenges in sensors and electronics for real-time soil and plant monitoring. *2017 IEEE Biomedical Circuits and Systems Conference (BioCAS)*, 1–4. 10.1109/BIOCAS.2017.8325180

Stafford, J. V. (2000). Implementing precision agriculture in the 21st century. *Journal of Agricultural Engineering Research*, *76*(3), 267–275. doi:10.1006/jaer.2000.0577

Stehr, N. J. (2015). Drones: The Newest Technology for Precision Agriculture. *Natural Sciences Education*, *44*(1), 89–91. doi:10.4195/nse2015.04.0772

Sudduth, K. A., Hummel, J. W., & Birrell, S. J. (1997). Sensors for Site-Specific Management.

Syvitski, J. P. M., Overeem, I., Brakenridge, G. R., & Hannon, M. (2012). Floods, floodplains, delta plains - A satellite imaging approach. *Sedimentary Geology*, *267–268*, 1–14. doi:10.1016/j.sedgeo.2012.05.014

Veris. (2014). *Veris Technologies*. Retrieved from https://www.veristech.com/the-sensors/u-series

Vermeulen, P., Flémal, P., Pigeon, O., Dardenne, P., Fernández Pierna, J. A., & Baeten, V. (2017). Assessment of pesticide coating on cereal seeds by near infrared hyperspectral imaging. *Journal of Spectral Imaging*, *6*, 1–7. doi:10.1255/jsi.2017.a1

Veroustraete, F. (2015). The Rise of the Drones in Agriculture. *EC Agriculture*, *2*(2), 325–327.

Zarco-Tejada, J. P., Hubbard, N., & Loudjani, P. (2014). *Precision Agriculture: An opportunity for EU farmers - Potential Support with the CAP 2014-2020*. Academic Press.

Zhang, N., Wang, M., & Wang, N. (2002). Precision agriculture - a worldwide overview. *Computers and Electronics in Agriculture*, *36*(2-3), 113–132. doi:10.1016/S0168-1699(02)00096-0

KEY TERMS AND DEFINITIONS

Artificial Intelligence (AI): AI is defined as the ability of a digital computer or computer-controlled robot to perform tasks commonly associated with intelligent beings. The term is frequently applied to the project of developing systems endowed with the intellectual processes characteristic of humans, such as the ability to reason, discover meaning, generalize, or learn from experience.

Automated Decision-Making Systems (ADMS): Automated decision-making systems (ADMS) are systems, which have the ability to operate a farm/field without the need of human presence or human input. In this context, ADMS are able to monitor soil and plant parameters, assess soil and plant condition and make the right decisions as to maintain those parameters at optimum levels by controlling various inputs.

In-Situ Sensors (ISS) for Soil: Soil ISS are defined as sensors that are buried in the soil, which are able to monitor soil parameters directly in order to assess soil quality.

Internet of Things (IoT): IoT is defined as the interconnection via the internet of computing devices embedded in everyday objects, enabling them to send and receive data.

Machine Learning (ML): ML is the scientific study of algorithms and statistical models that computer systems use to perform a specific task without using explicit instructions, relying on patterns and inference instead.

On-the-Go Sensing Systems (OTGS): On-the-go sensing systems are defined as sensing systems that are installed on moving agricultural equipment, such as tractors that in this context are used to monitor soil parameters in order to assess soil quality.

Precision Agriculture (PA): The concept of using sensing technologies and automation to respond to special and temporal variations within a field.

Remote Sensing (RS): Remote sensing (SI) is defined as the acquisition of information about an object or phenomenon without any physical contact. This concept is mostly used for observing Earth's crust and several phenomena on Earth's atmosphere. Usually, this is achieved be processing of images acquired by satellites.

Satellite Imaging (SI): Satellite imaging (SI) is an approach that uses images from satellites to monitor environmental conditions. In this context, SI is used for monitoring soil quality through the image processing of multi-spectral images of specific fields to assess several soil parameters.

Sodium Absorption Ratio (SAR): SAR is defined as the ratio between sodium concentration and the multiplication of calcium and magnesium concentrations in soil.

Soil Quality Index: Soil quality index is calculated by mathematical or statistical frameworks that use a range of soil parameters or indicators to assess soil quality taking into account multiple goals, such as environmental quality, agronomic sustainability, and socio-economic viability.

Soil Quality Indicators (SQIs): SQIs are defined as physical, chemical, and biological properties of soil that is sensitive to disturbance and represents performance of ecosystem function in the soil of interest. Indicators are dynamic soil properties.

Chapter 5
Emerging Microwave Technologies for Agricultural and Food Processing

Kok Yeow You
https://orcid.org/0000-0001-5214-7571
Universiti Teknologi Malaysia, Malaysia

Man Seng Sim
https://orcid.org/0000-0001-7776-2239
Universiti Teknologi Malaysia, Malaysia

Suhail Najm Abdullah
https://orcid.org/0000-0003-4897-9396
Universiti Teknologi Malaysia, Malaysia

ABSTRACT

This chapter focuses on microwave research and measurements for agricultural and food processing applications. Normally, microwave devices and components are used as a moisture detector for soil, agricultural, and food products. For instance, the moisture content in fruits normally is correlated with the maturity and sweetness of the fruits. In addition, the moisture content can be used to determine the storage period and the quality of the agricultural products after processing by an industrial factory. In this chapter, several microwave applications of agri-food products are selected to be reviewed comprehensively, such as microwave heating mechanism for several agri-food products, heating/drying, or freeze-drying process in food industry to control pathogenic and spoilage microorganisms in packaged foods, moisture soil testing, fruit moisture measurement, ripeness/storage period determination, fruit sweetness detection, microwave radiation for agricultural pest control.

DOI: 10.4018/978-1-7998-5000-7.ch005

INTRODUCTION

Microwave technology is expanding rapidly and is becoming more and more common in our daily lives, such as mobile phone and microwave oven. Microwave technology application can be divided into two groups, namely communication and non-communication applications. In this chapter, microwave non-communication applications have been described and specifically focus on the use of such technologies in agriculture/food fields. This chapter reviews the microwave applications for agricultural in this era based on a detailed literature survey and the author's experience in microwave researches.

In fact, over the past 40 years, a lot of research on microwave applications in agriculture has been conducted. Most of the research related to the field has been published in journals, such as Journal of Microwave Power and Electromagnetic Energy (Formerly Journal of Microwave Power) and Transactions of the ASAE (American Society of Agricultural Engineers). However, commercial microwave products/instruments for agricultural use are rarely found in comparison with optical-based (infrared) and acoustic-based (ultrasound) technologies. This is due to the higher price of microwave electronic components, such as monolithic microwave integrated circuit (MMIC) chips, compared to others. Foremost, microwave device components require high precision machining. Also, normally, microwave raw materials used in electronic components are the expensive synthetic materials, which materials have good thermal resistance and lossless dielectric at the high operating frequency. Besides, the design of microwave components is not easy and requires experts in the field of microwave engineering, especially in the last 30 years where it is difficult to obtain workstation computers and simulation software for microwave components/electronics design.

Nowadays, advances in the mechanized industry (industry 4.0) and electronic communication technologies, such as internet of things (IoT) and fifth-generation (5G) wireless technologies, are increasing globally. Thus, precision machining, such as computer numerical control (CNC) milling machine, high-end personal computers, and engineering simulation software are becoming common. In addition, techniques and knowledge in the production of new synthetic materials and microwave components are matured, and that information can be easily obtained from books, journal or internet. Hence, recently, the microwave component prices have reached a level nearly similar to optical and acoustic electronic components. This has provided the advantages and opportunities for more microwave technologies to be applied in agricultural / food processing.

In fact, microwave technology is capable of applying in sensors for agricultural/food processing, such as grain/soil moisture measurement, fruit ripeness/storage period determination, fruit sweetness detection, control of milk of lime, monitoring of nitrogen/phosphorus content in fertilizer. The microwave also can be used for heating/drying or freeze-drying process (sterilization/pasteurization) in the food industry to control pathogenic and spoilage microorganisms in packaged foods. Microwave applications for heating and crushing normally use high microwave power which is up to megawatts. Besides, microwave energy also has been implemented for agricultural pest control.

BACKGROUND PRINCIPLE OF MICROWAVE AGRICULTURAL AND FOOD PROCESSING

Interaction Between Substances in Agri-Food Products And Electromagnetic Waves

In fact, both optical and microwave agri-foods processing use the equivalent measurement principle, namely resonant technique and non-resonant technique. Differentiations between microwave and optical (infrared) methods are the application of operating frequency (Radio waves: 3 kHz to 300 MHz; Microwaves: 300 MHz to 300 GHz; TeraHertz: 300 GHz to 10 THz; Infrared: 10 THz to 430 THz) and excited source as shown in Figure 1.

Figure 1. Electromagnetic spectrum

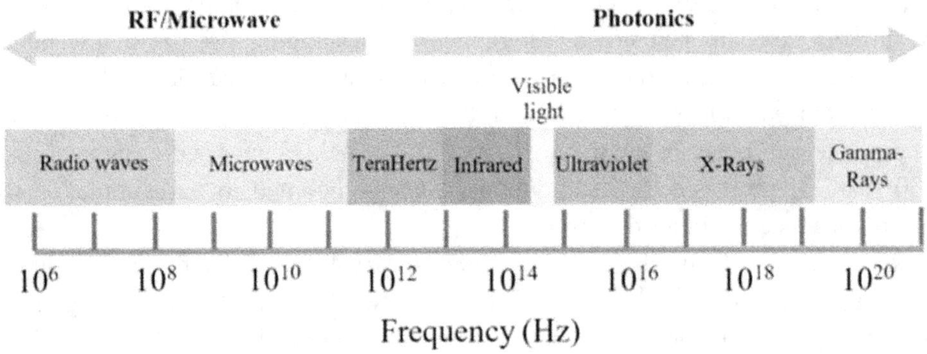

The interaction between agri-foods specimens with microwave is described by the complex relative permittivity, ε_r as:

$$\varepsilon_r = \varepsilon_r' - j\varepsilon_r'' \tag{1}$$

where the real part, ε_r¢ and imaginary part, ε_r^2 are the dielectric constant and the dielectric loss factor, respectively. The ε_r¢ influences the electric field distribution and the phase of waves traveling through the specimen. In contrast, the ε_r^2 influences the energy absorption or attenuation by the specimen. On the other hand, the interaction between agri-foods specimens with photonic (optical) is observed based on complex refractive index, n (= n¢ - jn^2) parameter. Despite this, the relationship between n and ε_r is given as $n^2 = \varepsilon_r$. Figure 2 shows the response of the relative complex permittivity, ε_r from Radio frequencies (RF) to Ultraviolet frequencies.

At lower RF (3 kHz to 300 MHz), the loss factor, ε_r^2 is mainly dominated by the ionic conductivity, σ response in the substance and the relationship between the frequency-dependence of ε_r^2 and σ is expressed as:

Figure 2. Frequency dependence of relative complex permittivity, $\varepsilon_r = \varepsilon_r¢ - j\varepsilon_r^2$

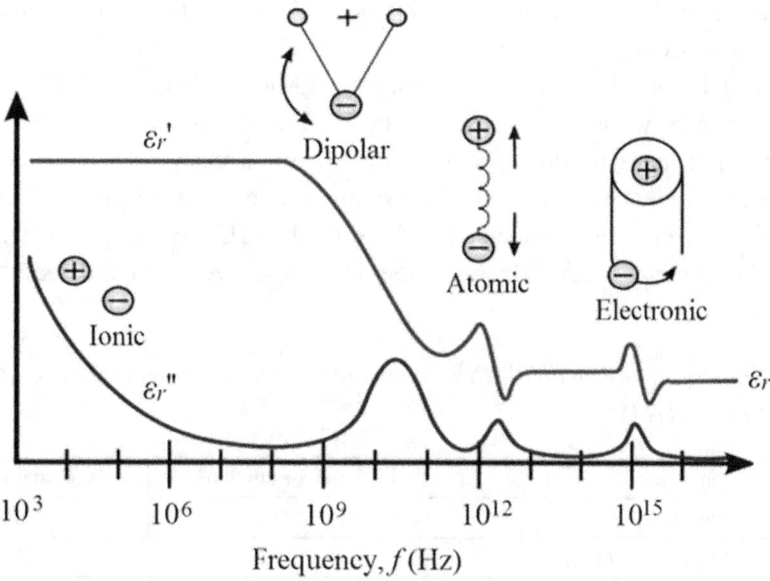

$$\varepsilon_r'' = \frac{\sigma}{2\pi f \varepsilon_o} \qquad (2)$$

where ε_o (»8.8541878×10^{-12} Fm^{-1}) is the permittivity of free-space. At microwave frequencies (300 MHz to 300 GHz), $\varepsilon_r¢$ and ε_r^2show the reaction caused by dipole relaxation phenomena of the substance. On the other hand, at infrared frequencies, $\varepsilon_r¢$ and ε_r^2exhibit a significant reaction due to the resonant occurring between the molecular vibration of the substance and the infrared frequency.

For instance, as known that the vibration frequency, *f* of water molecules is approximately 30 THz (Wavelength, $\lambda = 10\ \mu m$). When the frequency of the irradiated far-infrared rays is same as the vibration frequency range of molecular motion of water, molecules group of water becomes less and water molecules motion become fast due to the resonance absorption effect. Therefore, the irradiated far infrared sensor is normally used for moisture measurements and the characterization of substance (Different substances have different vibrational molecular frequencies).

Microwave Measurement Regulations

Microwave technology has been widely used in communication and non-communication applications. Owing to this, each application requires a specific operating frequency band to avoid any frequency interference that exists between the applications. Therefore, a portion of the RF/microwave spectrum was proposed and fixed to be mainly used for industrial, scientific and medical (ISM) applications (non-communication purposes) (Osepchuk, 1984).

The first three industrial, scientific, and medical (ISM) frequency allocations (at 13.66 MHz, 27.32 MHz, and 40.98 MHz) were designated by US Federal Communications Commission (FCC) in 1945. However, this regulation was not specifically enunciated until the FCC (United States and Canada) adopted Part 18 of its rules at the International Telecommunications Conference of the International Telecommunication Union (ITU) in Atlantic City in 1947. The rules of Part 18 in 1947 is implemented to impose some severe restrictions on out-of-band radiation and set aside three frequencies, namely 13.56 MHz, 27.12 MHz, and 40.68 MHz, use for ISM purposes (Allen and Garlan, 1962; Osepchuk, 1984). Recently, there are two microwave frequencies allocated by the FCC for widely ISM usage, namely 915 MHz and 2.45 GHz. For instance, the domestic microwave oven currently used is operating at a frequency of 2.45 GHz. Some researchers have been used operating frequency at 5.8 GHz for ISM applications. Besides that, the other current ISM frequencies defined by the ITU are 433.92 MHz, 5.8 GHz, 24.125 GHz, 27.129 GHz, 61.25 GHz, 122.5 GHz, and 245 GHz, respectively, as tabulated in Table 1.

Table 1. Industrial, scientific, and medical (ISM) operating frequency band defined by the International Telecommunication Union (ITU)

	Frequency range (f)	Bandwidth (BW)	Center frequency (f_o)
MHz	*6.765–6.795 MHz	*30 kHz	*6.780 MHz
	13.553–13.567 MHz	14 kHz	13.560 MHz
	26.957–27.283 MHz	326 kHz	27.120 MHz
	40.660–40.700 MHz	40 kHz	40.680 MHz
	**433.050–434.790 MHz	*1.74 MHz	*433.920 MHz
	***902.000–928.000 MHz	*26 MHz	*915.000 MHz
GHz	2.400–2.500 GHz	100 MHz	2.450 GHz
	5.725–5.875 GHz	150 MHz	5.800 GHz
	24.000–24.250 GHz	250 MHz	24.125 GHz
	*26.975–27.283 GHz	*308 MHz	*27.129 GHz
	*61.000–61.500 GHz	*500 MHz	*61.250 GHz
	*122.000–123.000 GHz	*1 GHz	*122.500 GHz
	*244.000–246.000 GHz	*2 GHz	*245.000 GHz

*Subject to local acceptance
**Only Europe, Africa, the Middle East/Middle West of the Persian Gulf, the former Soviet Union, and Mongolia
***Only Americas, Greenland, and some of the eastern Pacific Islands

However, during recent years, ISM bands have also been shared with license-free error-tolerant communications applications, such as wireless local area networks(WLANs). For example, domestic wireless fidelity (Wi-Fi) based on IEEE 802.11x technology is also operating at 2.45 GHzdue to it offers wide bandwidth and high wall-penetrating properties. Despite this, domestic wireless and communication circuits operating on these frequencies are required to accept any ISM interference that may be received.

In addition to frequency interference issue, the safety and health of using microwave devices should be emphasized. The health risks associated with microwave radiation have been investigated for many years, focusing on the absorption of microwave energy radiated from microwave source into human head tissues. Normally, the rate of the absorption of microwave energy by the human body is measured based on the specific absorption rate (SAR) in watts per kilogram (W/kg) and is defined as:

$$SAR = \frac{\sigma}{\rho}|E|^2 \tag{3}$$

where σ is the conductivity of the human tissues (S/m). The ρ is the mass density of the human tissues (kg/m^3). The E is the intensity of the electric field within the tissues (V/m). The measured SAR values are dependent on the volume or mass tissues under test. Generally, the mass is fixed to be 1 g or 10 g and the volume is defined in the shape of a cubic. In general, there are two radiation protection standards, which are defined by the European Committee for Electrotechnical Standardization (CENELEC) (joined by at least 33 European countries) and FCC, respectively. For the CENELEC standard, the maximum SAR for commercially sold is limited to 2 W/kg averaged over the 10 g of body tissues. On the other hand, for FCC standard, the maximum SAR limit is 1.6 W/kg averaged over the 1 g of body tissues.

Nevertheless, microwave oven heating applications have different safety and health standards than above applications. The safety and health standards for the heating have been defined by several international committees or organizations, such as International Electrotechnical Commission (IEC), the International Committee on Electromagnetic Safety (ICES) of the Institute of Electrical and Electronics Engineers (IEEE), and also the CENELEC. Those committees or organizations have set a microwave emission limit(maximum allowable leakage) of 50 mW/cm^{2}at 5 cm away distance from the external microwave oven surfaces.

Water Molecule Mechanism In RF/Microwave

Normally, agri-foods specimens containing water are the polar substance and the dielectric properties (ε_r¢, ε_r^2, and ion conductivity, σ) of agri-foods specimen can be modelled by Debye/Cole-Cole model as:

$$\varepsilon_r = \varepsilon_\infty + \frac{\varepsilon_s - \varepsilon_\infty}{1 + (j\omega\tau)^\alpha} - j\frac{\sigma}{\omega\varepsilon_o} \tag{4}$$

On the other hand, ε_s, $\varepsilon_{¥}$ are the relative permittivity at DC and extremely higher frequency stages, respectively. In additional, τ and σ are the relaxation time and ion conductivity of the agri-foods material. Symbolsω andα are the angular frequency and exponent coefficient ($0 < \alpha < 1$), respectively. Normally, the ion conductivity, σ is only dominated at lower operating frequency (< 2 GHz).

Most of the microwave measurements in agricultural and food processing are implemented to monitor the moisture content, *m.c.* in the agricultural products and foods under test. This is caused by the polarization of water molecules contained in the agri-foods products which are sensitive and showed a significant response when exposed to microwaves as shown in Figure3. The tendency of water to absorb microwave energy has made microwave device to sense the moisture content, *m.c.* changes containing in agricultural products.

The mechanism water molecules in agri-foods specimens are different when exposed to different ranges of operating frequency, f as shown in Figure 4. However, in the measurement, only three states of water contained in the agricultural products are usually analyzed and discussed, namely ionic conductivity state, bound water state, and free water state.

Figure 3. Microwave heating mechanisms: Water molecules are oriented when exposed to microwave (You et al., 2017)

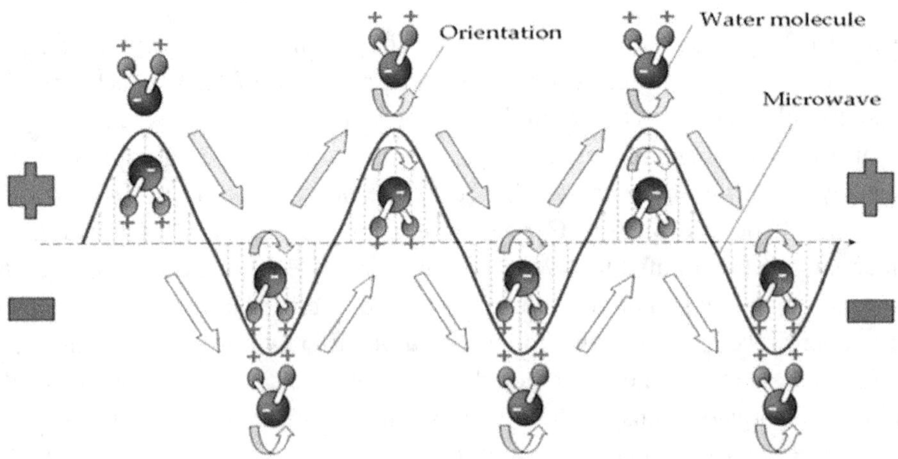

Figure 4. Mechanism of water molecules in different operating frequency ranges (You, 2015)

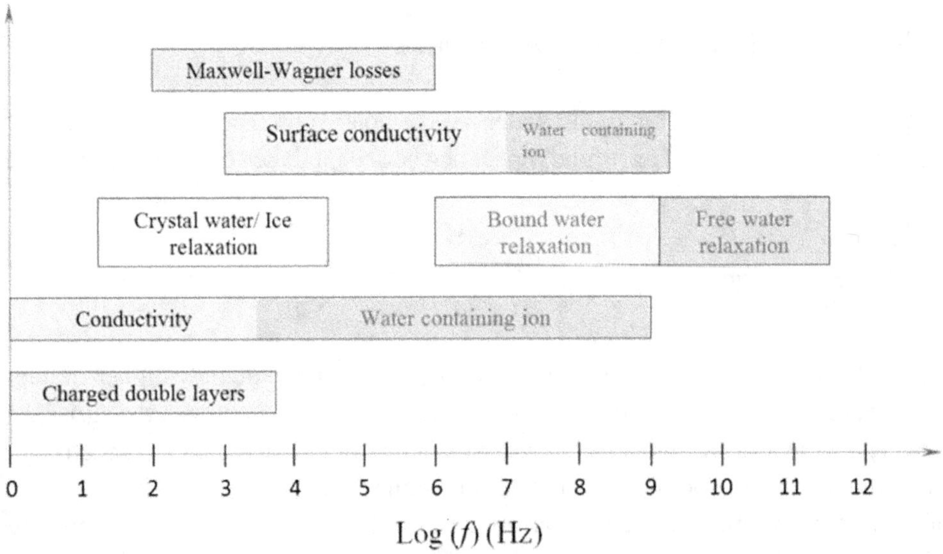

Figure 5 shows the correlation between loss factor, ε_r'', mechanism of water molecules and operating frequency, f, which have provided some measurement guidelines. For example, if ionic conductivity, σ of an agricultural product is desired to be measured, then the low operating frequency (f< 1 GHz) must be considered. Mechanism water molecules are affected by the total volume of water contained in the agricultural product. Generally, when the total volume of water in the agricultural product is less than 30%, there is a bond between water molecules and other substances in the product.

Figure 5. Relationship between molecular mechanism of water, loss factor, εr", and operating frequency, f

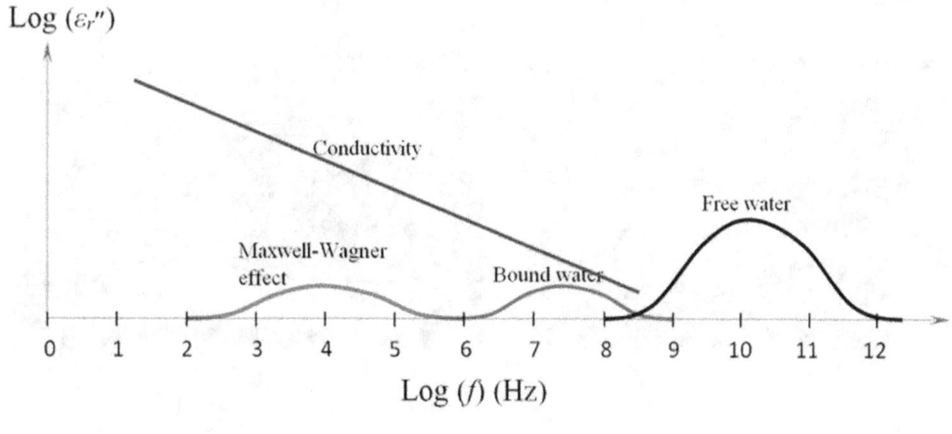

Microwave Measurement Techniques

Microwave agri-foods processing techniques can be categorized into either resonant technique (based on shifting resonant frequency or wavelength) or non-resonant technique (based on *S*-parameters or impedance, *Z* measurements). Typically, the *S*-parameters (interpreted in term of impedance) of the agri-food under test is initially measured using applicator connected with an impedance analyzer or *LCR* meter below GHz (in Figure 6) or vector network analyzer (up to GHz) (in Figure 7). Based on the raw measurements, the values of ε_r of the agri-foods can be predicted. Hence, before the values of ε_r is predicted, the raw information required are the reflection coefficient, S_{11} (or input impedance) and the transmission coefficient, S_{21} for the agri-food under test in which the measurements are performed using applicators or sensors.

Figure 6. (a) Agilent 4294A Precision impedance analyzer (40 Hz to 110 MHz). (b) Keysight (Formerly Agilent Technologies Inc) E4991B Impedance analyzer (1 MHz to 3 GHz) with terminal adapter

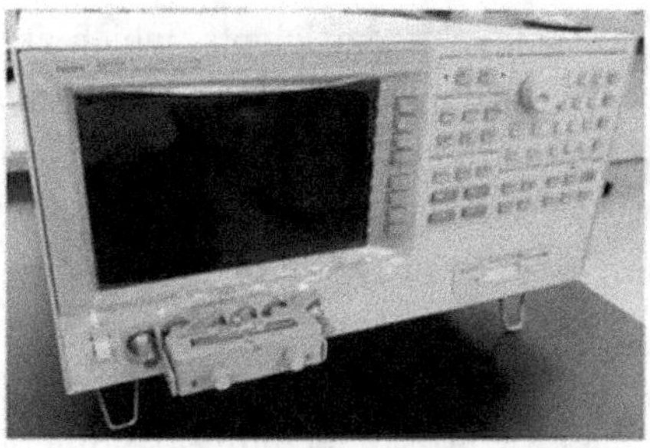

(a)

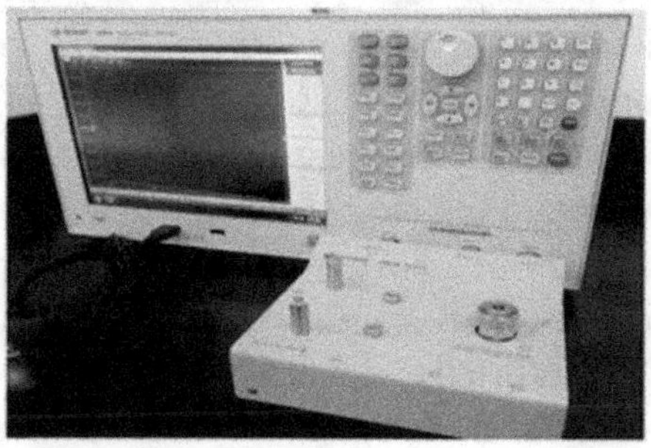

(b)

Figure 7. (a) Keysight E5071C network analyzer (300 kHz to 20 GHz). (b) Rohde & Schwarz ZNB 8 vector network analyzer (9 kHz to 8.5 GHz)

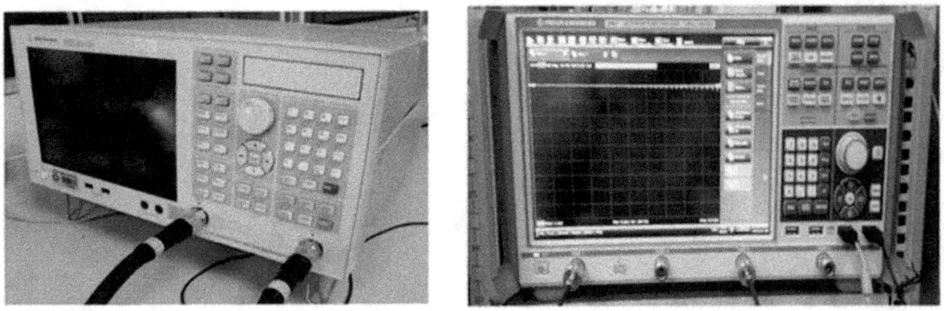

(a) (b)

Resonant Technique

For resonant methods, a resonator (applicator) is filled with a sample as illustrated in Figure 8. This produces a resonance frequency shift, Δf and also a broadening of the resonance curve compared to the resonator without filled with any sample. The particular resonance frequency, f_r for the resonator without filled with sample is depends on its shape and dimensions. The properties of the sample can be characterized from the measurements of shifting resonance frequency, Δf and quality of factor Q. The resonance techniques are suitable choices for determining low loss tangent, tan δ ($< 10^{-3}$) values for the low-loss sample, but they cannot be used for the measurement of swept frequency. Although this method provides accurate measurements, it requires high resolution and frequency sweep instruments, as well as good environmental control measure.

Free-Space Measurement Technique

Typically, the one-port measurements can only yield the ε_r value. However, the two-port measurements can give the ε_r and μ_r values simultaneously, since the measurements enable sufficient information to be obtained from the measured reflection, S_{11} and transmission, S_{21} coefficients, which are the wave propagation, Ö($\varepsilon_r \times \mu_r$) and impedance, Ö($\varepsilon_r / \mu_r$) properties of the sample. A free-space and transmission/reflection measurement techniques are grouped in the category of non-resonant methods. The free-space technique is a far-field measurement, and a horn antenna is used as the radiator as shown in Figure 9 (Tosaka*et al.*, 2015; Abbas *et al.*, 2001; Kemptner*et al.*, 2011; Bourreau*et al.*, 2006). However, this method provides a less precise measurement due to the sensing field is highly dispersed. Furthermore, the distance between the sample surface and the horn aperture is difficult to gauge precisely. The free-space technique is suitable for the broadband measurement for low-loss or medium-loss thin film sample with high temperature because horn radiators do not come into direct contact with the sample, and thus, the RF circuits of the instrument are safe from heat damage. However, it is not suitable for low tan δ measurements due to highly dispersed sensing field.

Figure 8. (a) TE_{011} parallel plate dielectric resonator. (b) Cavity resonator. (c) $TE_{01\delta}$ dielectric resonator for solid sample measurements. (d) Dielectric resonator for liquid/powder sample measurements. (e) Resonant-mode and (f) Evanescent-mode split cylinder dielectric resonators for thin solid sample measurements. (g) Split cylinder dielectric resonator. (h) Open-ended coaxial resonator. (i) One-port open resonator. (j) Two-port open resonator (Fabry-Perot resonator). (k) Open-ended rectangular resonator. (l) Cavity perturbation. (m) TE_{01n} cylindrical cavity resonator. (n) Microstrip ring resonator. (o) Microstrip line resonator.

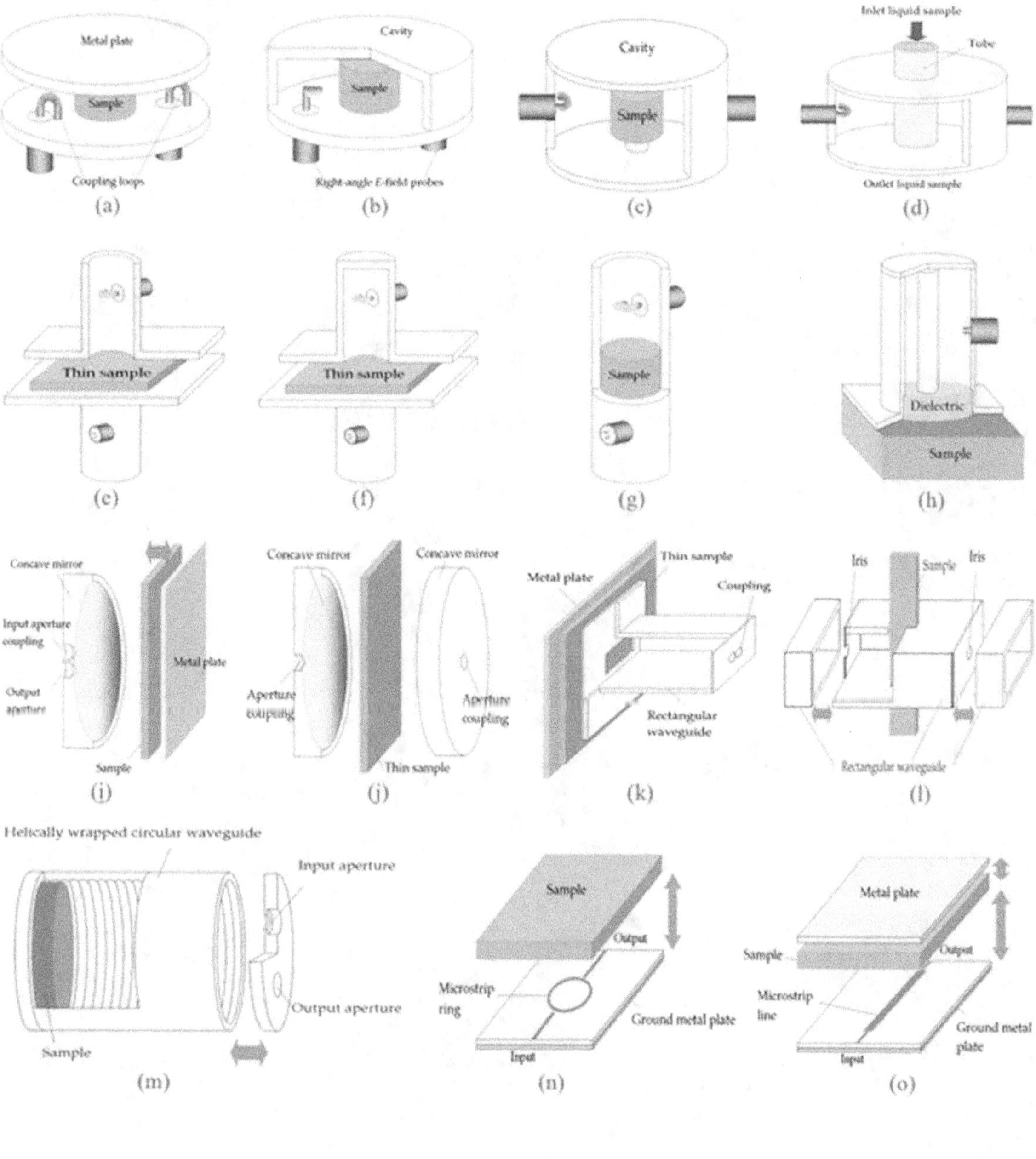

Figure 9. Free-space measurement setup for dielectric measurement of the sample. (a) Spot focusing horn set-up. (b) Bistatic radar set-up. (c)Rectangular dielectric waveguide set-up (Abbas et al., 2001). (d)Monostatic radar set-up. (e) Quasi-optical free-space measurement setup (Bourreau et al., 2006)

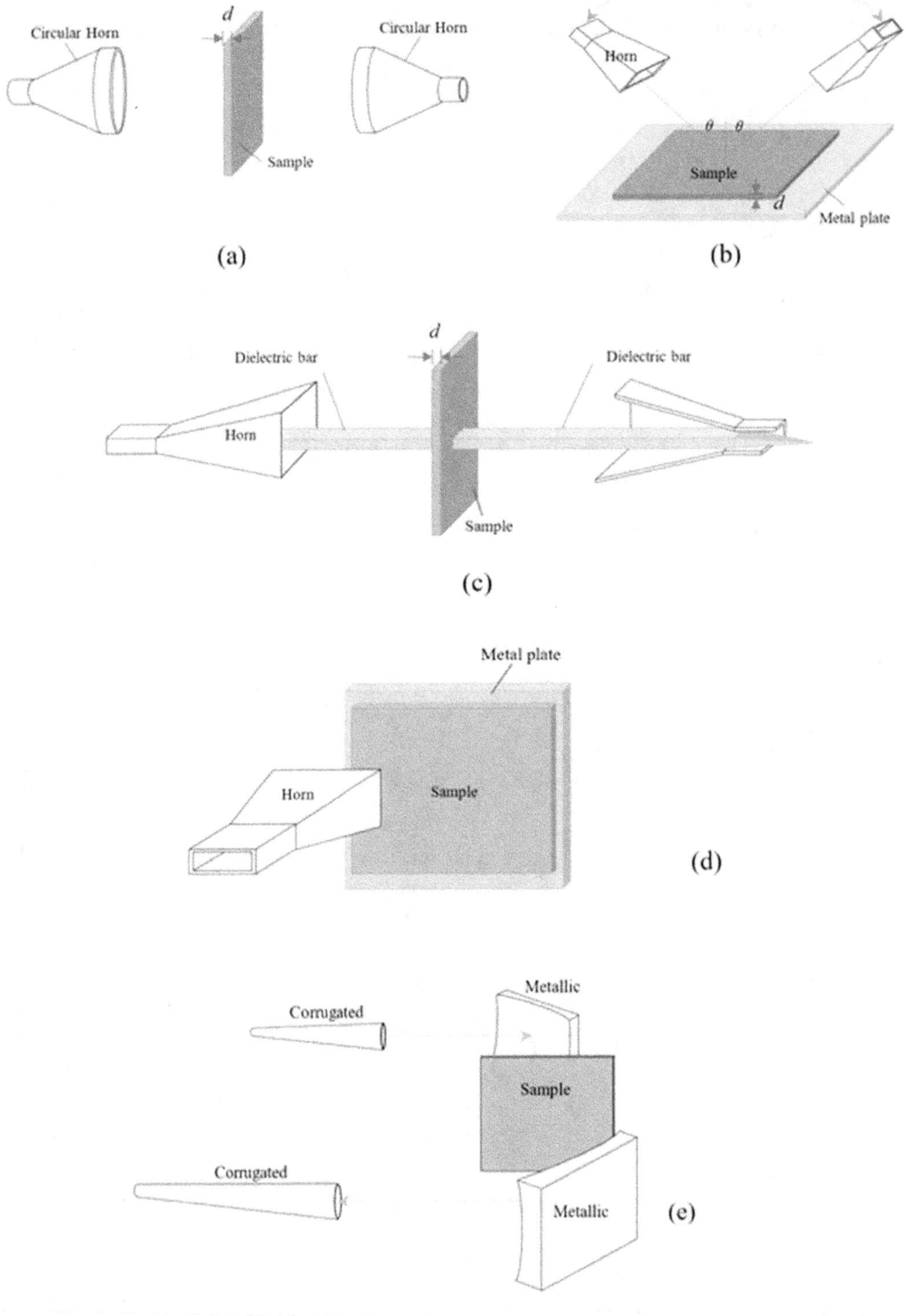

One-Port Measurement Technique

The one-port measurement is based on the principle that a reflected signal (reflection coefficient, S_{11}) through the waveguide, which end aperture is contacting firmly with the substance under test, will obtain the desired information about the material as shown in Figure 10. The main advantage of using one-port reflection techniques is that the method is the simplest, broadband, non-destructive way to measure the dielectric properties of a substance. However, one-port measurement is suitable only for measuring the relative permittivity, ε_r, of the dielectric material (non-magnetic material, $\mu_r = 1$). It is due to insufficient information to predict the permeability, μ_r, if only obtained the measured reflection coefficient, S_{11} without transmission coefficient, S_{21} as mentioned above. For Figure 10 (a) and Figure 10 (c) measurement, the sample is considered infinite, as long as the sample thickness d is greater than the radius of the outer conductor b. However, the radiation, or sensing area, for an aperture rectangular waveguide is much greater than that of a coaxial probe. For instance, the WR90 waveguide has a radiation distance up to 20 cm in the air from the aperture waveguide. Hence, the substance under test must be much thicker when the rectangular waveguide is utilized in the measurement. Besides, the coaxial probe and rectangular waveguide also capable of testing the thin film sample as shown in Figure 10(b) and Figure 10 (d). The measurements required that the thin sample is backed by a metallic plate.

Two-Port Measurement Technique

For two-port measurement, in general, the material filled in the waveguide has been conventional practice to measure the reflection coefficient and the transmission coefficient by using Nicholson-Ross-Weir (NRW) routines and convert these measurements to relative permittivity, $\varepsilon_r$and relative permeability, μ_r as shown in Figures 11 (a)-(h).

Interpretation of Substance Properties

For agri-foods heating and drying processing, dissipation/absorption and penetration depth of microwave power for the agri-food specimens are played an important role in the determination of total required energy and maximum thickness of specimens which can be heated or dried directly by microwave (Sun *et al.*, 2016). The dissipation/absorption of microwave power, P (in W/m^3)of agri-food specimen can be expressed in term of loss factor, ε_r'' as:

$$\begin{aligned} P &= 2\pi f \varepsilon_o \varepsilon_r'' |E|^2 \\ &\approx 55.63\times10^{-12} f \varepsilon_r'' |E|^2 \end{aligned} \tag{5}$$

Where f and E are the operating frequency and the supplied microwave energy (rms values of intensity of electrical field) within the agri-food specimens. While, the penetration depth, D (in unit meter) of microwave in terms of $\varepsilon_r¢$ and ε_r'' can be briefly calculated as:

Figure 10. One-port measurements using the coaxial probe and the rectangular waveguide: (a) Half-space infinite sample. (b) Thin sample backed by metal plate. (c) Sample as a part of coaxial line. (d), (e) Sample placed at the end of the coaxial line. (f) Monopole driven from coaxial line inserted into sample. (g) Similar to (a). (h) Similar to (c). (i) Similar to (b)

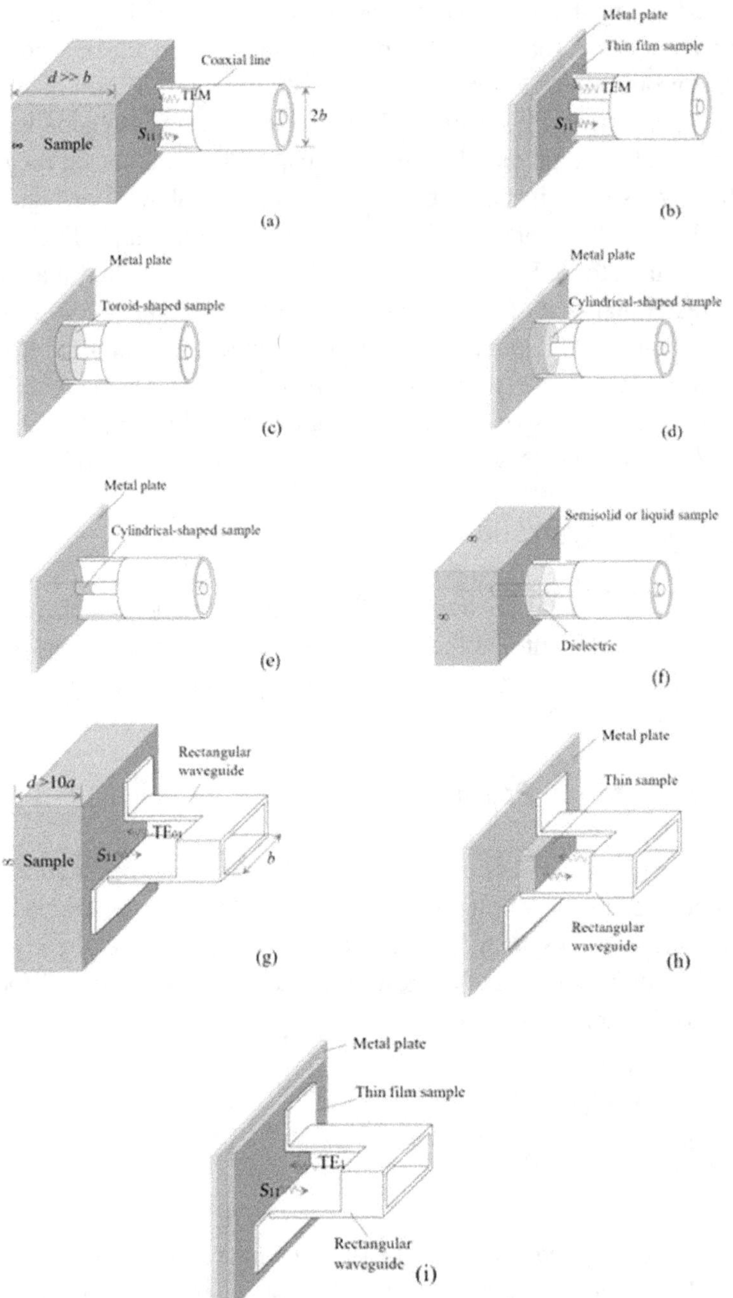

Figure 11. Sample (a) fully and (b) partly filled the cross-section of rectangular waveguide. (c) Similar to (a). (d) Thin sample placed at the center between two waveguides. (e), (f) similar to (d). (g), (h) Thin sample on the top of transmission line

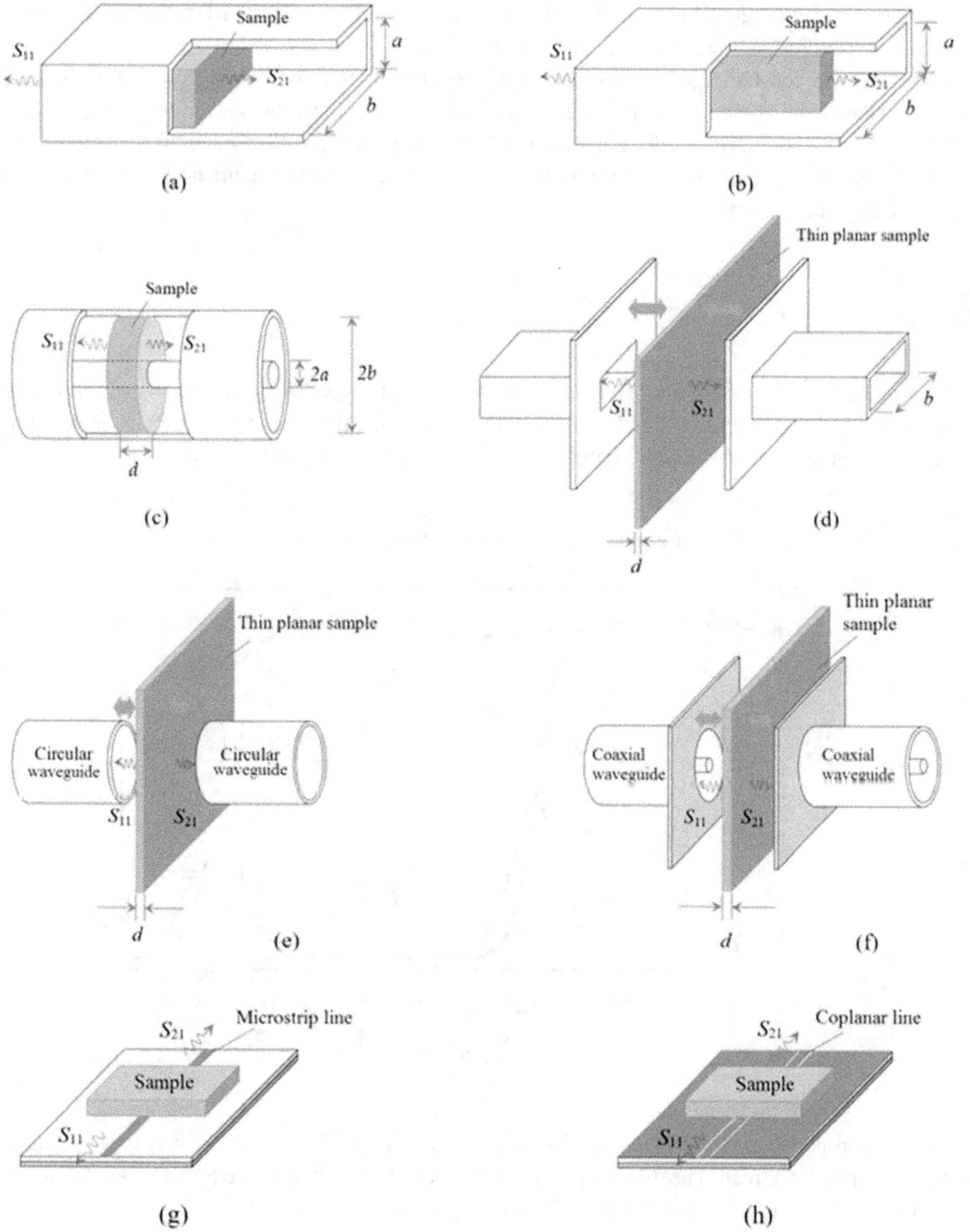

$$D \approx \frac{\lambda_o \sqrt{\varepsilon_r'}}{2\pi\varepsilon_r''} \tag{6}$$

where λ_o is the wavelength of the supplied microwave. Normally, operating frequency of 915 MHz or 2.45 GHz is used for microwave heating/drying applications.

The dielectric constant, ε_r¢ and loss factor, ε_r'' are the temperature-dependent parameters and its influences the power absorption of the heated specimen. For instance, the pure water with temperature of 0 °C and 25 °C has different values of ε_r¢ and ε_r'' as shown in Figure 12. Hence, In agri-food heating/drying applications, the rate of temperature rising, *T*/*t* of the agri-food specimen which can be defined as a function of power, *P* as:

$$\frac{\partial T}{\partial t} = \frac{P}{\rho C_p} \tag{7}$$

where ρ is the mass density of the specimen (in kg/m^3), C_ρ is the specific heat capacity of the specimen [in J/(kg °C)] and *t* is the time (s).On the contrary, the dissipation/absorption power, *P* can be determined using (7) once the rate of increase in temperature, *T*/*t*,ρ, and C_p are known.

Figure 12. Temperature dependence of relative complex permittivity, ε_r

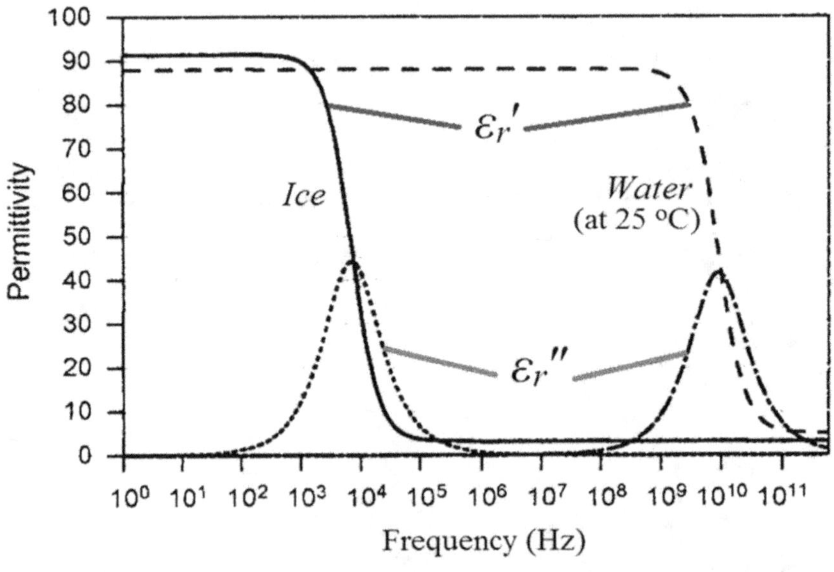

Besides, propagation microwave in agri-food specimen is also depended on the density, ρ and temperature, *T* of the specimen. The density, ρ of the specimen can be roughly predicted based on measured ε_r¢ and ε_r'' over certain operating frequency range as (Trabelsi *et al.*, 1999):

$$\frac{\varepsilon_r''}{\rho} = \kappa\left(\frac{\varepsilon_r'}{\rho}\right) - \eta \tag{8}$$

whereκ is the slope of the fitting linear line. On the other hand, η is the intercept constant of the fitting line. The value of density,ρ is required to be adjusted in order to obey the linear equation (8). Thus, in general, the density-independent moisture content, $m.c$ of specimen can be determined as (Kraszewski, 2005):

$$m.c = \sqrt{\frac{b\left(\varepsilon''/\varepsilon'\right)}{\varepsilon_r'' - a\varepsilon_r'}} \tag{9}$$

In fact, the uncertainty in the density, ρ and temperature, T of the substance will influence the accuracy of moisture, $m.c$ measurement using microwave technique. Thus, sometimes the gamma rays sensor is installed in conjunction with the microwave sensor system, in which gamma rays are used to determine the density, ρ of the substance as shown in Figure 13. Finally, the measured density, ρ values will be included in equations (8) and (9) to achieve the goal of density-dependence moisture measurement.

Figure 13. The gamma rays are used for density compensation during the moisture, m.c determination using microwave free-space measurement technique (Berthold Technologies, 2007)

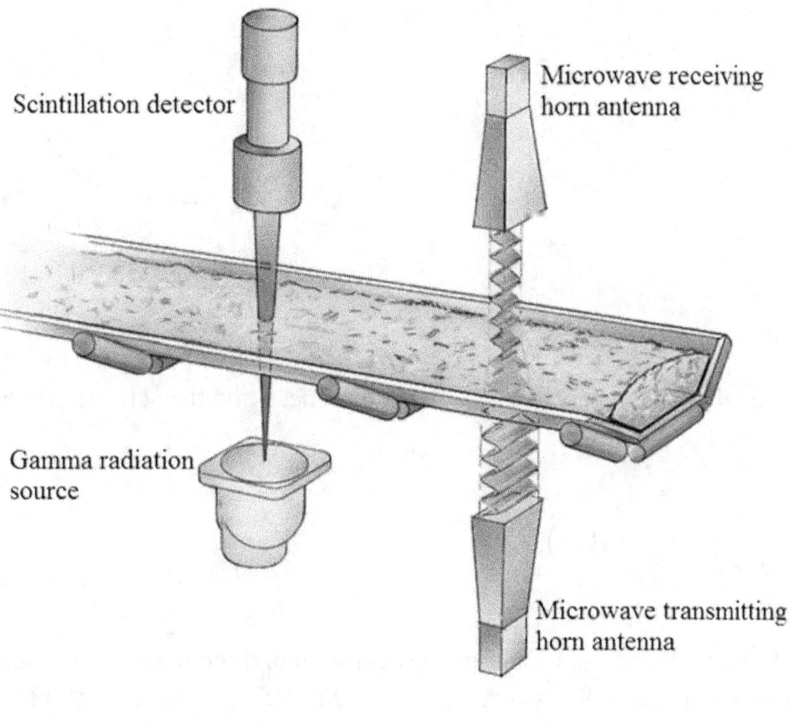

In fact, from the measurements of ε_r and σ, the activation energy, Q_e (absorption energy) and entropy change, ΔS of the agri-food speciemen under test can be interpreted. In addition, some physical phenomena, such as glass transition temperature, T_g, (frozen and boiling state transitions), non-linear properties of specimen, charge transport in specimens, timing of the chemical reaction processes, phase compositions of mixed specimens, and conductivity/resistivity of specimens, can be predicted. For instance, the relaxation time, τ (in unit second) of specimen can be predicted as (You *et al.*, 2010):

$$\tau = \frac{\varepsilon_s - \varepsilon_r'}{2\pi f \varepsilon_r''} \tag{10}$$

The crosslink network on rigidity of food's fiber chains is affecting the change in relaxation time, τ as shown in Figure 14 (a) (Ahmad, 2012). On the other hand, based on the sudden change in the value of ε_r¢ at certain temperatures, the glass transition temperature, T_g of the material under test can be determined as shown in Figure 14 (b) (Yrö H. Roos, 1995).

Figure 14. (a) The relaxation time, τ increase with the increase in amount of cross-linking reaction (Ahmad, 2012). (b) The dielectric constant, ε_r¢ increase with the increase in molecular mobility (Yrö H. Roos, 1995)

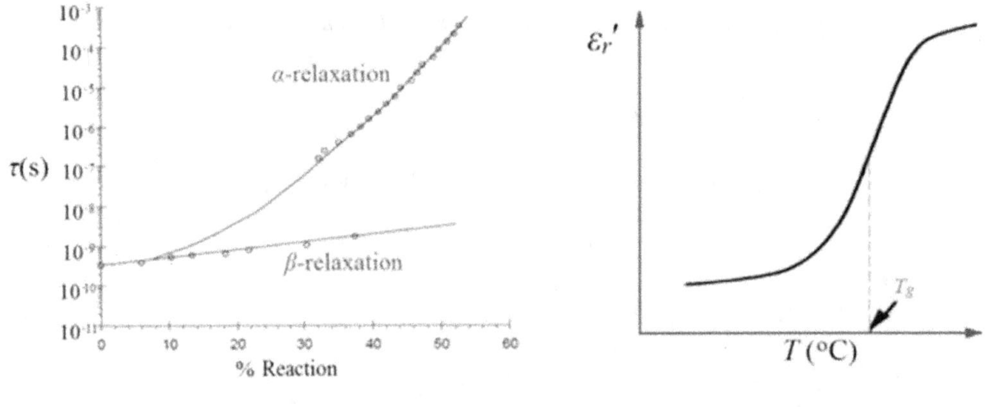

In addition, the relationship between the relaxation time, τ and the activation energy, Q_e (in unit kJ mol^{-1}) is given as (You *et al.*, 2014):

$$\ln(\tau) = Q_e\left[\frac{1}{R(T+273)}\right] + \ln(\tau_o) \tag{11}$$

where R (= 8.3143 J mol^{-1} K^{-1}) and T are the gas constant and temperature (in unit ^{o}C), respectively. On the other hand, when value of T is very high and $1/R(T+273)\approx 0$ in equation (11), the value of τ_o is equal to relaxation time, τ at the particular temperature. In fact, the τ_o can be written as (Hasted, 1973):

$$\tau_o = \frac{h}{k(T+273)}\exp(-\Delta S/R) \tag{12}$$

where k (= 1.38066 × 10^{-23} J K^{-1}) and h (= 6.62608 × 10^{-34} J s^{-1}) are the Boltzman constant and the Phanck constant, respectively. Indirectly, the entropy change, ΔS (in unit J (K mol)$^{-1}$) can be determined from the relaxation time, τ.

APPLICATION OF MICROWAVE IN AGRICULTURE AND FOOD PROCESSING

Soil Moisture Monitoring

Soil moisture content, *m.c* is one of the most essential physical characteristics in agriculture, civil engineering, landscaping, irrigation engineering and hydrology, since the consistency and workability of a clayey soil strongly depend on its moisture content, *m.c*. For measuring soil moisture, direct and indirect methods are being implied. The standard method of measuring soil moisture content is gravimetric technique (so-called oven drying technique), which is a direct method. In this method, a known volume of soil is dried up using oven and then by calculating the weight loss, one can determine the soil moisture content, *m.c*. However, the direct method consumes a lot of time. Thus, recently, a lot of end-users are opting for indirect methods which are fast in terms of response. Several indirect methods are currently available in the market, namely tensiometer, resistance blocks (two electrodes method), time-domain reflectometry (TDR), frequency-domain reflectometry (FDR), as well as neutron probes. Normally, TDR and FDR methods operating at microwave range frequency are considered as a microwave soil measurement techniques.

In indirect methods, the soil moisture content, *m.c.* is evaluated by calibrating a relationship with other measured physical variable. For instance, the change in electrical properties that can be directly correlated with a change in the actual moisture content, *m.c.* of the soil obtained from oven drying method (direct method). Besides, the pH levels in the soil can be determined based on the measured ion conductivity, σ of the soil. The variations in relative dielectric constant, ε_r¢ and ionic conductivity, σ of the soil with volumetric moisture content, $m.c_v$ are plotted in Figures15 (a) and (b) including a solid fitting line for measurement data (You *et al.*, 2013). The dielectric of dry soil is much lower than that of wet soil.

Generally, the water contained in the soil can be in bound or free water conditions. The fraction of bound water is different for different types of soil, depending on the texture of the soil. Compared to other soils, loam is a soil with a high water holding capacity because it contains a lot of organic matter. Organic matter in loam can hold more water than mineral matter. Therefore, it is possible to increase the water retention capacity and the maximum fraction of bound water in the soil by adding organic matter. The measured reflected voltage, V_r (in volts) using monopole sensor for various soil types in the range from 0 $m.c_v$ to 0.45 $m.c_v$ is shown in Figure 17. The small monopole driver from coaxial line is buried in the soil under test, and its measured reflected voltage, V_r is correlated with corresponding soil moisture content, $m.c_v$ which obtained by the oven dying method.

Figure 15. Variations in (a) relative dielectric constant, εr¢ and (b) conductivity, σ of soil with volumetric moisture content, m.cv at (25 ± 1) °C (You et al., 2013)

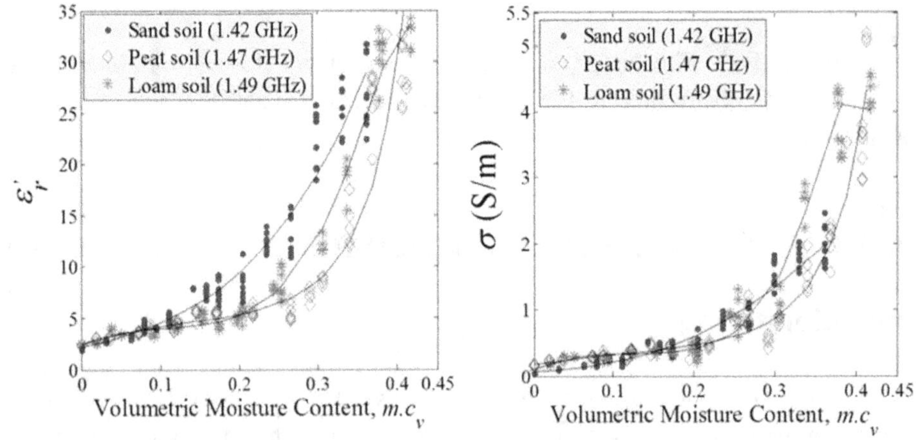

Figure 16. (a) A simple soil microwave reflectometer connected to (b) monopole sensor via SMA connector

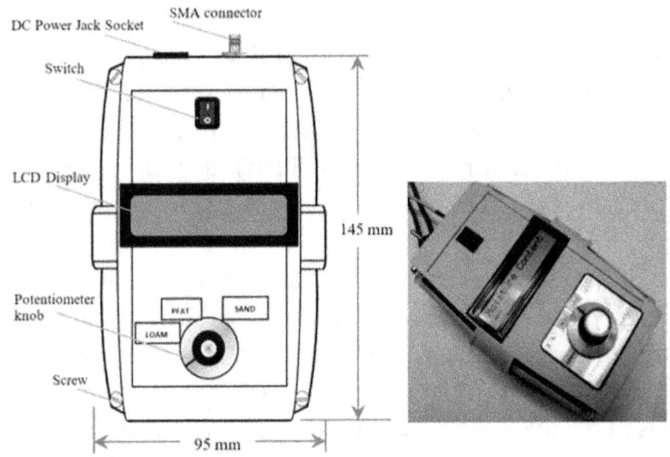

(a)

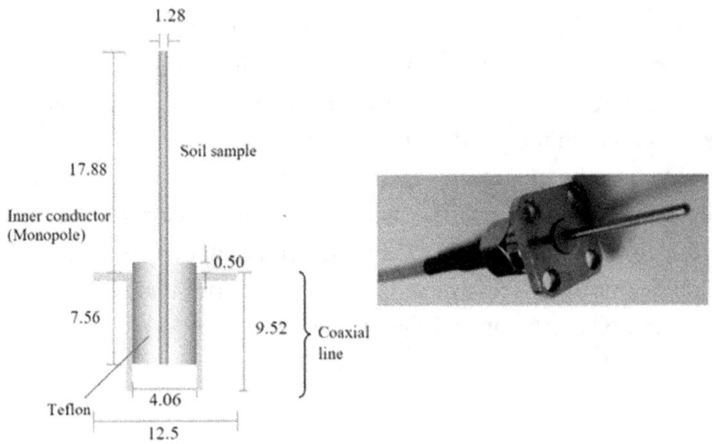

(b)

From Figure 17, peat soil is a soil with a higher water holding capacity of 0.45 $m.c_v$. In fact, it contains a lot of organic matter (0.643%) compared to loam (0.335%) and sandy soil (0.058%). As mentioned earlier, the organic matter in peat soil can hold more water than mineral substances. On the other hand, the sand soil is saturated with water only 0.3 $m.c_v$.

Figure 17. (a) Monopole sensor. (b) Variations in reflected voltage, Vr. with volumetric moisture content, m.cv of three kinds of soils at (25 ± 1) °C (You et al., 2013)

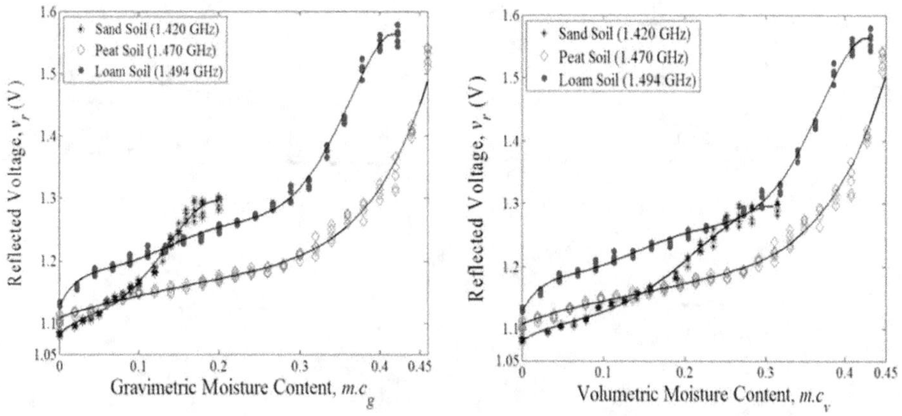

Rubber Latex Quality Measurements

Natural rubber latex is complex biological product containing 50% to 80% of water, 15% to 45% of rubber hydrocarbon and approximately 2% to 4% of non-rubber constituents (Chow, 1979). In manufacturing standard, the latex product is required to have dry rubber content (DRC) of 60% and approximate 39% of moisture content, *m.c*. Thus, the dry rubber content (DRC) in the latex can be indirectly estimated from moisture content measurement, since there are only two major components in the product.

Figures 18 (a) and (b) show the temperature, *T* dependence of ε_r¢ and ε_r^2 of heve a latex at 0.6 GHz and 10 GHz, respectively. The rubber latex specimens are diluted fresh latex (*m.c.* = 89%), fresh latex (*m.c.* = 56%), latex concentrate (*m.c.* = 38%), and water, respectively. Clearly, there are three different states exist in the temperature range, namely the frozen/solid state (-30 °C to -3 °C), the transition state (-3 °C to 3 °C), and the liquid state (> 3 °C). Above 3 °C and at 0.6 GHz, the increase in temperature, *T* raises the mobility of ions in the latex solution causing in an increase in ε_r^2(ionic conductivity effect). On the other hand, at higher frequency of 10 GHz,ε_r^2 decreases as temperature, *T* increases, which is similar to the trend of water (polar effect) (Kaida *et al*., 1996). The variations in relative dielectric constant, ε_r¢ and loss factor, ε_r^2 of the rubber latex specimens with frequency, *f* for various moisture content, *m.c* values are plotted in Figures19 (a) and (b).

Ansarudin *et al.* (2012) was used an insulated monopole radiator [Figure 20(a)] to monitor the moisture content in the rubber latex. The insulated monopole with 15 mm of length is fully immersed in the rubber sample and the moisture content, *m.c* in the sample is predicted based on the change of measured phase shift, ϕ. The variation in moisture content, *m.c* with the corresponding measured phase shift, ϕ (in rad) is shown in Figure 20 (b). Clearly, the phase shift, ϕ is linearly correlated with the moisture content, *m.c* in the rubber latex.

Figure 18. Relative complex permittivity, ε_r versus temperature, T of hevearubber latex at (a) 0.6 GHz and (b) 10 GHz, respectively (Khalid et al., 1996)

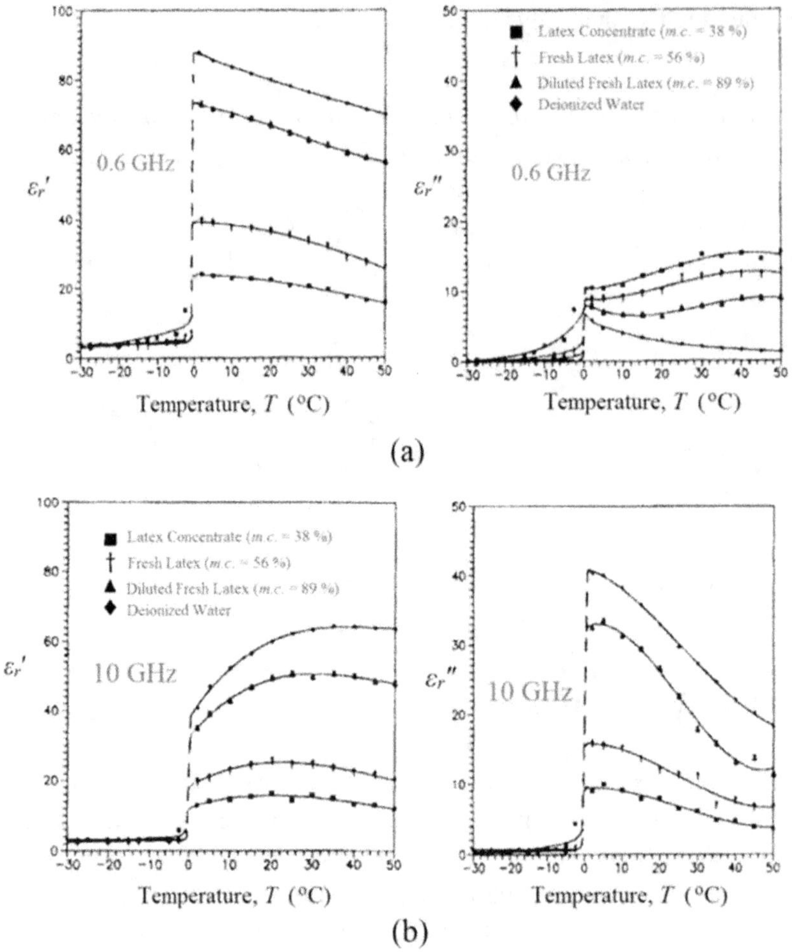

Figure 19. Variation in (a) dielectric constant, εr¢ and (b) loss factor, εr2 with frequency at room temperature (25 ± 1) oC for various water content, m.c. in rubber latex (Ansarudin et al., 2012)

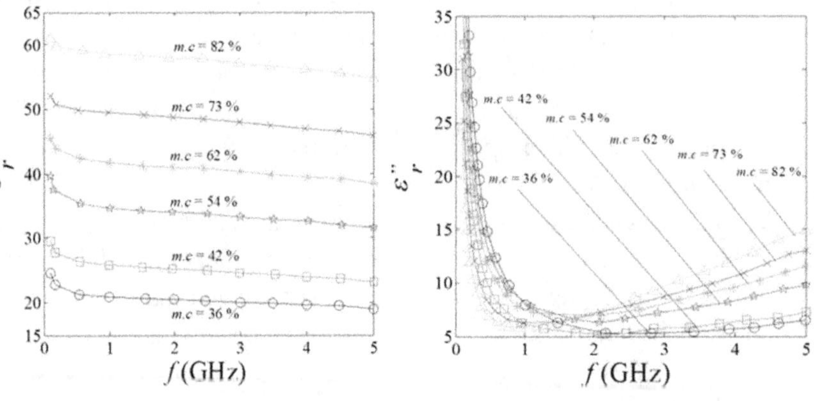

Figure 20. (a) Dimensions of the insulated monopole radiator. (b) The moisture content, m.c. in the rubber latex versus measured phase shift, ϕ using insulated monopole (Ansarudin et al., 2012)

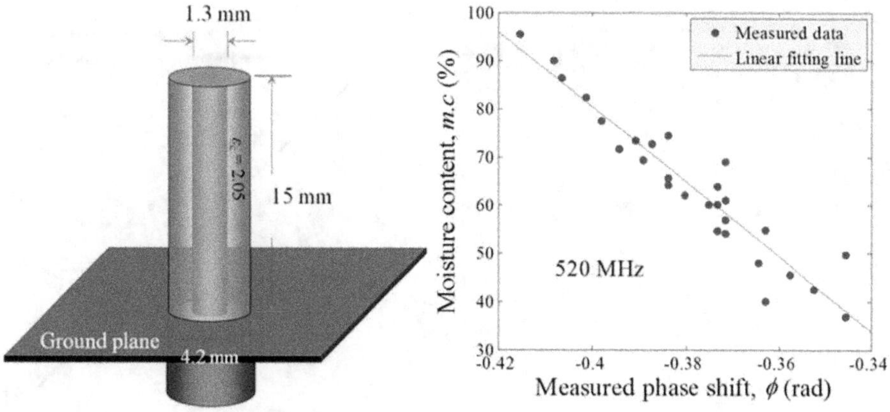

Oil Palm Fruits Ripeness Detector

Palm oil is obtained from the mesocarp of the oil palm fruits. It usually takes nearly three years for a palm tree to produce the first bunch of fruit. In Ariffin *et al.* (1990), it is found that the amount of moisture content, *m.c.* is higher at the early stage of fruit development. The excess water in the mesocarp decreases during fruit ripening, which coincides with an oil accumulation time of about 14 to 15 weeks after anthesis. The *m.c.* in the fresh mesocarp is about 80%. At 20 to 24 weeks after flowering, the *m.c.* in the ripe fruits drops rapidly to about 30-40%.As shown in Figure 21(a), the decrease in water content from 15 to 20 weeks of *m.c.* is associated with an increase in oil accumulation in the mesocarp (Khalid and Abbas, 1992). When the oil content reaches the maximum, the fruit becomes loose and falls to the ground. This is the optimum time for oil palm harvesting. In addition, Hartley, (1977) proposed that the relationship between oil and moisture content, *m.c* in a mesocarp may be expressed as:

$$Oil = 87.38 - 1.08\ Water \tag{13}$$

where *'Oil'* and *'Water'* are the percentage of oil per mesocarp and percentage of water per mesocarp, respectively. Therefore, the close relationship between the moisture and oil contents in mesocarp gives a possibility of using a percentage of moisture content, *m.c* as a parameter to gauge the mesocarp ripeness. The relationship between the oil content and moisture content, *m.c.* in oil palm mesocarp is shown in Figure 21 (b).

Until now, different microwave sensors were utilised to gauge the ripeness of oil palm fruits, for example microstrip sensor (Khalid and Abbas, 1992), coplanar sensor (Khalid and Teoh, 1998), rectangular dielectric waveguide (Abbas *et al.*, 2007), monopole probe (Lee *et al.*, 2011), rectangular waveguide (Ali, 2006), open-ended coaxial probe (Zakaria, 1998; Abbas *et al.*, 2005; You *et al.*, 2010), and microstrip ring resonator (Ahmad *et al.*, 2019) as illustrated in Figure 22. The moisture content, *m.c.* in the oil palm mesocarp under test is monitored using most of those ripeness measurement methods. It is due to the oil palm mesocarp contains polarization of water molecules that is sensitive and demonstrated a

Figure 21. (a) Variation in moisture content, m.c and oil content with weeks after Anthesis. (b) The relationship between the oil content and moisture content, m.c in oil palm mesocarp (You, 2015)

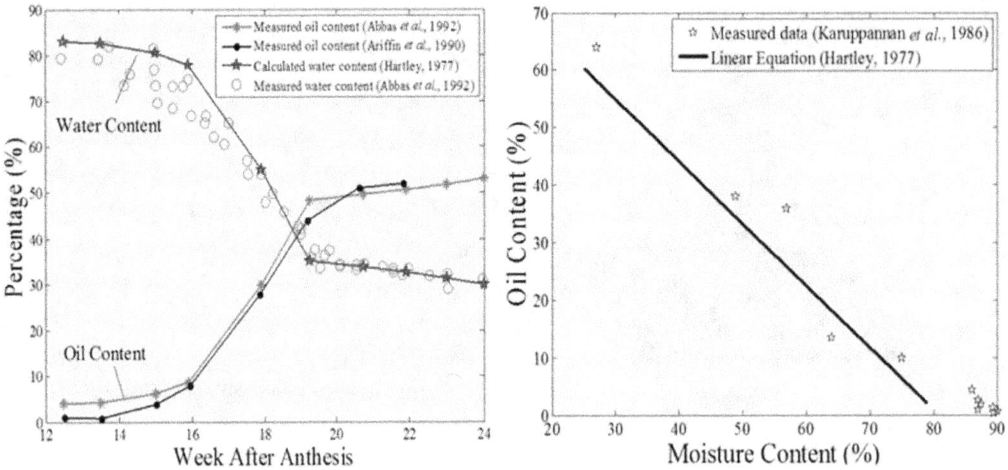

crucial response when it is exposed to microwaves. The inclination of water to absorb microwave energy, enabling the microwave device to be utilized as a measuring method to observe the moisture content, *m.c* in agri-foods products comprising water.

Microstrip and coplanar sensors (in Figures22 (a), (b)) the amount of moisture content, *m.c.* in the oil palm fruits has been successfully estimated according to the attenuation measurement. The characteristics and microstrip and coplanar line discontinuities when the line was filled by oil palm mesocarp. The microstrip and coplanar line sensor measure the ripeness of the oil palm mesocarp, in accord with the correlation between the measured attenuation and the *m.c.* inside the mesocarp fruit. For rectangular dielectric waveguide measurements, the oil palm mesocarp is placed in the sample holder between the two dielectric waveguides, as illustrated in Figure 22 (c). Nevertheless, laborious sample preparation is required by the sensors since the fresh mesocarp of the oil palm fruit has to be detached from the nut and crumbled into a form of semi-solid sample. It is undeniably time-consuming in order to prepare the fruit sample.

Figure 23 shows the moisture measurement of oil palm fruits using open-ended coaxial probe. Part of the fresh mesocarp of each fruit was sliced in the longitudinal direction to ensure good contact between the surface of the mesocarp and the open-ended coaxial probe.

As mentioned earlier, both ε_r¢ and ε_r^2 are highly correlated with moisture content since at microwave frequencies, the electromagnetic energy is mainly absorbed by water (Hasted, 1977) and the volume of moisture in the total volume of material most heavily influences the relative permittivity of the material. This is due to the relative permittivity of water ($\varepsilon_r = 80$ at dc frequencies) normally being much greater than that of the other constituents in the oil palm fruits (fiber: 2.2, oil: 2.5). Thus, when the amount of moisture changes in the oil palm mesocarp, the sensor will measure a change in reflection /transmission coefficient or resonant frequency (from the change in dielectric permittivity) that can be directly correlated with a change in *m.c.* of the mesocarp which was obtained from oven drying technique.

Figure 22. (a) Microstripwaveguide (MWG). (b) Multilayer conductor-backed coplanar waveguide (CB-CPW). (c) Rectangular dielectric waveguide (RDWG). (d) Rectangular waveguide (RWG). (e) Monopole. (f)Open-ended coaxial waveguide (OECWG). (g) Microstrip ring resonator

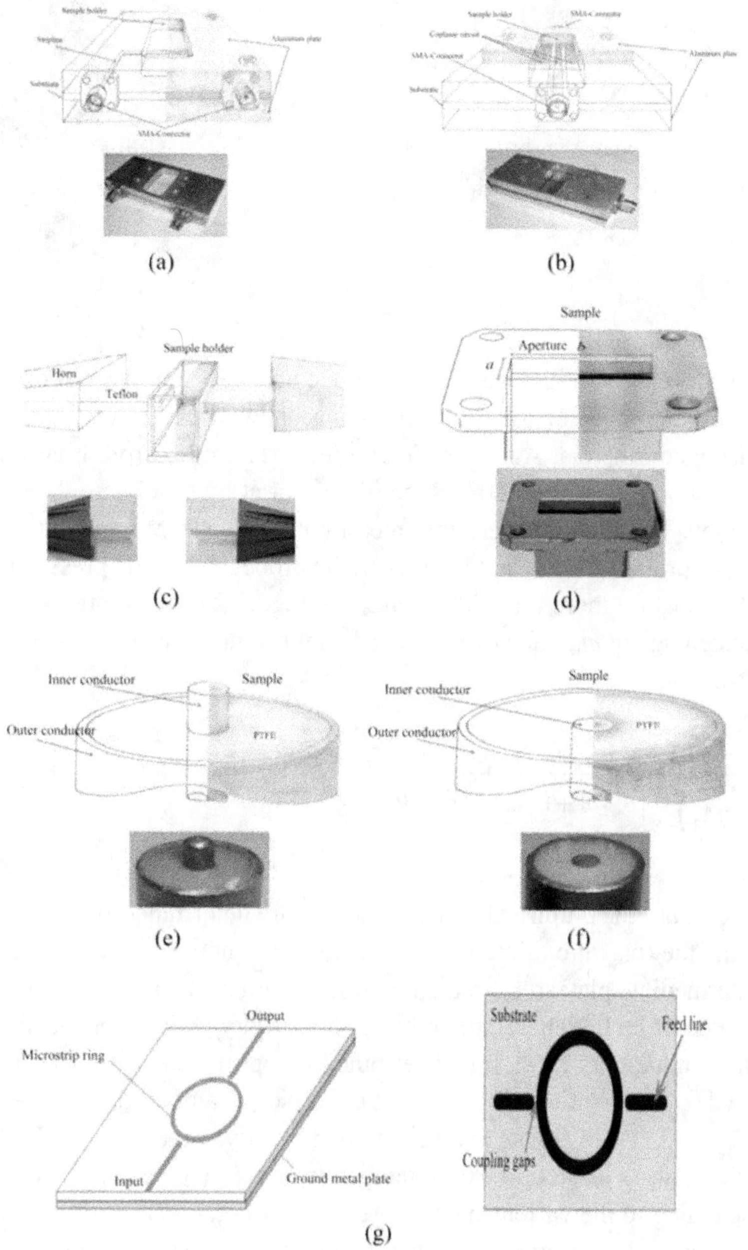

Figure 23. Moisture measurement of oil palm fruit using coaxial radiator

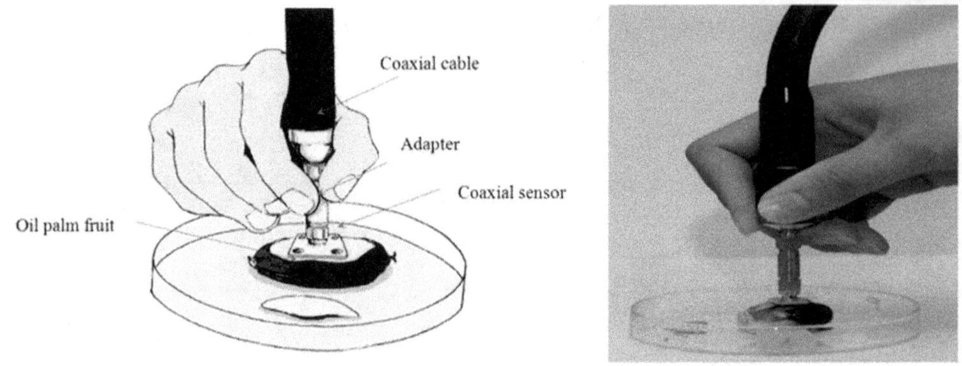

The relationship between both dielectric constant, ε_r¢ and loss factor, ε_r^2 of oil palm mesocarp and frequency for various percentage of water content, *m.c*is shown in Figure 24. The dielectric constant, ε_r¢ are measured using commercial Agilent 85070E dielectric probe. From Figure 24, dielectric properties are quite linear at *m.c.* less than 40%, suggesting that at low moisture, there is higher oil-water emulsion concentration. The higher uncertainty measurements appeared due to the water accumulated on the surface between probe aperture and fruit when the probe aperture is pressed firmly to the sample fruits that contained moisture more than 70%. Based on Figure 25, the relationship between dielectric constant, ε_r¢ and percentage of *m.c.* in mesocarp can be represented by two-relaxation Cole-Cole model as (You *et al.*, 2010).

$$\varepsilon_r = \varepsilon_\infty + \frac{\varepsilon_s - \varepsilon_1}{\left[1+\left(j\omega a_1\tau_1\right)^{a_2}\right]^{a_3}} + \frac{\varepsilon_1 - \varepsilon_\infty}{1+\left(j\omega\tau_2\right)^{a_4}} - \frac{j\sigma}{\omega\varepsilon_o} \tag{14}$$

Where a_1, a_2, a_3, and a_4 are the empirical constant, which are determined by optimized with measured data. The τ_1 and τ_2 are the relaxation time for low and high-frequency processes, respectively. The ε_1 is the height of the intermediate plateau from the baseline as well as the intersection between the line and ε_r¢-axis gives the value of ε_s. In general, the optical permittivity, $\varepsilon_{¥}$ is independent of frequency, *f* and temperature, *T*. The conductivity term, σ is determined by optimized with measured loss factor, ε_r^2 at low frequency (≤5 GHz). All parameters in (14) are expressed as a function of moisture content, *m.c* (in %) as tabulated in Table 2. The estimated values of conductivity, σ are listed in Table 3.

However, equation (14) is limited for room temperature, *T* (25 ± 1 °C). The estimated values of ε_s, ε_1, τ_1 and τ_2 corresponding to the various percentage of water content, *m.c* are revealed in Table 2. The value of the relative complex permittivity of the oil palm mesocarp is mainly contributed by the water content, *m.c.*, since the permittivity of water is far greater than other substance in the mixtures (such as oil and fiber) and it is sensitive to the microwave frequencies, especially at 2.45 GHz and 10 GHz (Nyfors and Vainikainen, 1989). The decrease in relative complex permittivity of the oil palm mesocarp is caused by the displacement of water by the oil and fibers contents.

Figure 24. Variation in dielectric constant, ε_r¢ *and loss factor,* ε_r^2 *with moisture content, m.c of 25 °C (Abbas et al., 2005)*

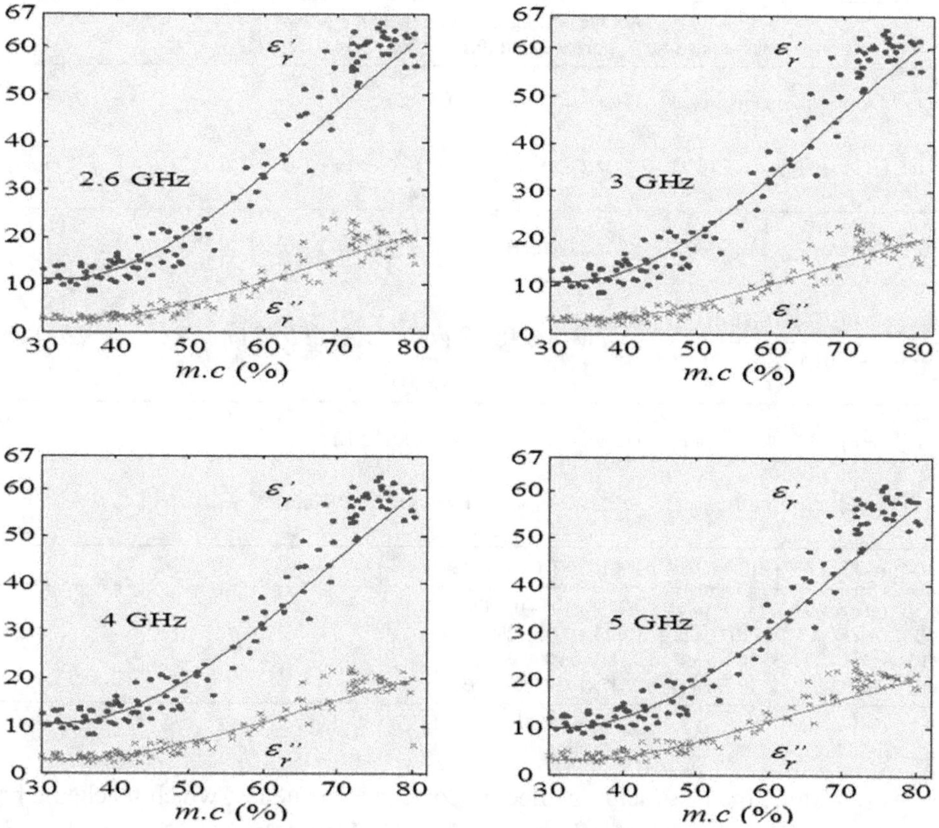

Figure 25. Variation in (a) dielectric constant, εr¢ and (b) loss factor, εr2 with frequency at room temperature (25 ± 1)oC for various water content, m.c in oil palm fruits(Abbas et al., 2005)

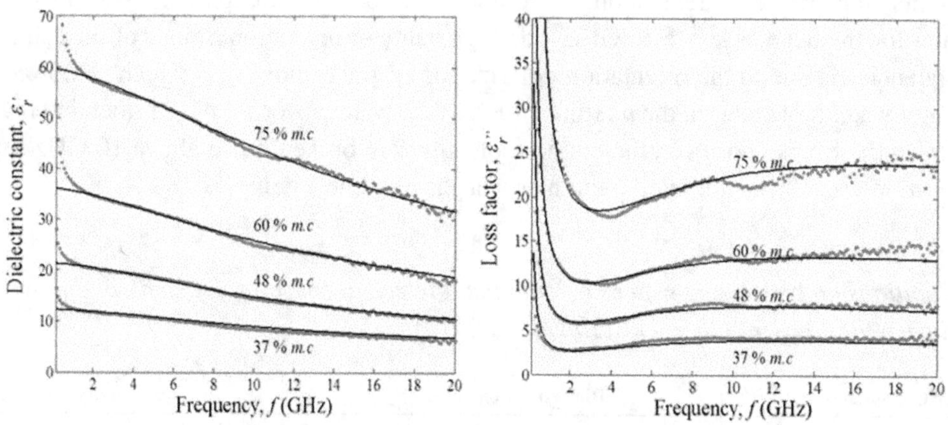

Table 2. The Debye and Cole-Cole parameters of oil palm mesocarp as a function of percentage water content, m.c (%) at (25 ± 1)°C (You et al., 2010)

Cole-Cole parameters of oil palm mesocarp **(30% £ *m.c* £ 80% ; pure water) ; (0.13 GHz £ *f* £ 20 GHz) ; (25 ± 1) °C**
$\varepsilon_s = -3.9899 \times 10^{-4}(m.c)^3 + 8.0046 \times 10^{-2}(m.c)^2 - 3.8988(m.c) + 67.258 \pm 0.9792$ $\varepsilon_1 = -1.8394 \times 10^{-5}(m.c)^4 + 3.8076 \times 10^{-3}(m.c)^3 - 0.27371(m.c)^2 + 8.7394(m.c) - 98.783 \pm 0.5093$
$\varepsilon_\infty = v_{water}\varepsilon_\infty^{water} + (0.84 - v_{water})\varepsilon_\infty^{oil} + (0.16)\varepsilon_\infty^{fiber}$, *where* $\varepsilon_\infty^{water} = 4.9,\ \varepsilon_\infty^{oil} \approx 2.5,\ \varepsilon_\infty^{fiber} \approx 2.2$ $v_{water} = \frac{0.01m.c(0.16\rho_{fiber} - 0.16\rho_{oil} + \rho_{oil})}{\rho_{water} - 0.01\rho_{water} + 0.01\rho_{oil}}$, $\rho w_{ater=}$*0.9959 gml-1,* $\rho oi_{l=0}$*.9073 gml-1,* $\rho fib_{er=0.}$*9200 gml-1*
$\tau_1 = -1.1394 \times 10^{-17}(m.c)^3 + 1.3309 \times 10^{-15}(m.c)^2 - 1.0209 \times 10^{-13}(m.c) + 1.6660 \times 10^{-11} \pm 1.119ps$ $\tau_2 = -5.5307 \times 10^{-17}(m.c)^3 + 8.7071 \times 10^{-15}(m.c)^2 - 4.6620 \times 10^{-13}(m.c) + 1.6004 \times 10^{-11} \pm 0.478ps$
σ = 6.2861×10-^{4} (m.c)2 - 3.6521×10-^{2} (m.c) + 6.3931×10-^{1} ±0.02482 Ω-^{1m}-^{1}
$a_1 = -2.2890\times10^{-4}(m.c)^2 + 3.6771\times10^{-2}(m.c) + 1.4734\times10^{-1} \pm 0.0008234$ $a_2 = -3.5634\times10^{-6}(m.c)^2 - 4.5790\times10^{-4}(m.c) + 1.0124 \pm 0.00002904$ $a_3 = 4.0888\times10^{-5}(m.c)^2 - 8.7443\times10^{-3}(m.c) + 1.1543 \pm 0.0008381$ $a_4 = 3.5374\times10^{-5}(m.c)^2 - 2.5052\times10^{-3}(m.c) + 8.7535\times10^{-1} \pm 0.004263$

In all measured fruits, the loss factor, ε_r^2 decreased at low frequency, which reached a broad minimum between 1 and 3 GHz, then increased again as shown in Figure 25 (b). This behavior is influenced by ionic conductivity, σ at lower frequencies, since dielectric properties in agriculture are believed to be primarily depending on water activity and ionic conductivity of fluids contained in their cellular structures. The oil palm mesocarp has higher relaxation time, τ below frequency of 10 GHz for different water contents, *m.c.*, while the relaxation, τ of oil palm mesocarp is approximately 8.377 ps relaxation of free water for frequencies above 10 GHz. The high value of relaxation time, τ of oil palm mesocarp at low frequencies is due to the orientation polarization of polar molecules which is slowed down by the binding of water molecules in the mixtures, such as the water absorption by a fiber and the capillary water between the fibers. However, the bound molecules are broken down above 10 GHz. In fact, the overall effect on water structure in oil palm mesocarp is extremely complex.

Table 3. Relationship between % water content, m.c, dielectric constant, ε_r¢ and conductivity, σ of oil palm mesocarp for 3 GHzat (25 ± 1)°C(You et al., 2010)

Water Content, *m.c* (%)	**Dielectric Constant, ε_r¢**	**Conductivity, σ ($\Omega^{-1}m^{-1}$)**
30 – 40	10 – 20	0.11 – 0.17
40 – 50	20 – 25	0.17 – 0.38
50 – 60	25 – 35	0.38 – 0.7
60 – 70	35 – 55	0.7 – 1.1
70 – 80	55 – 65	1.1 – 1.7

Palm Oil Quality Monitoring

The oil palm is a unique plant, which yields two different vegetable oils, namely, the red crude palm oil (CPO) and palm kernel oil (CPKO). The CPO is extracted from the mesocarp of the palm fruit while the yellow CPKO is obtained from the kernel as illustrated in Figure 26. Both palm oils are varied in their chemical compositions and physical properties as well as their industrial applications.

Figure 26. The red crude palm oil (CPO) is derived from mesocarp of the palm fruit while the palm kernel oil (CPKO) is obtained from the kernel of the palm fruit (You et al., 2014)

The CPO with low free fatty acid (FFA) is less susceptible to oxidation, good bleachability, low trace metals, low insoluble impurities and moisture content, *m.c* is labeled as good quality oil. In general, the standard quality CPO is defined at 5% FFA (max), 0.2% moisture (max) and 0.05% impurities (max). Hence, it is known that the water content in the oil influences the rate of FFA formation. This is due to the formation of FFA involve the hydrolysis reaction which can only be carried out in the water or at oil-water interface. When the water content in oil is greater than the solubility, the water will exist in droplets or emulsion state that influences the rate of hydrolysis reaction. However, the solubility of water in palm oil is very low, as in neutralized palm oil is about 0.15% (max) at room temperature, 25 °C. For palm oil containing 2.43% and 6.44% of FFA, the water solubility will increase to 0.2% and 0.25%, respectively. However, the total moisture content within the palm oil over the 26 days is about 0.60%. In general, the rate-controlling factor is referring to the water content must be kept below the solubility of water in palm oil, in order to reduce the hydrolysis of the oil. Therefore, the water content in the oil is preferably to be kept as low as possible to maintain its stability during storage.

CPKO is obtained from the kernel inside the nut of the palm fruit and it is lauric-rich oil which contains no carotene. It is saturated oil and more resistant to oxidation. The mechanisms of chemical hydrolysis and micro-organism activity of CPKO are mainly influenced by the water content as in CPO. The rate of FFA formation in CPKO is slightly higher due to the shorter hydrocarbon chains in CPKO. In addition, the water content below 7% inhibits the microbial activity in CPKO. Moreover, by keeping water content below 7%, the formation of FFA will be kept below 1.5% after 6 months storage of CPKO. Generally, the standard quality of CPKO is defined at 5% FFA (max), 7% moisture (max) and 6% impurities (max). The CPO and CPKO quotes requirements are listed in Table 4 and Table 5, respectively.

Table 4. Standard quality CPO based on Malaysian Oil Palm Growers' Council (MOPGC), Malaysian Standard (MS) and Palm Oil Research Institute of Malaysia (PORIM)

Factors	Standard quality of crude palm oil (CPO)		
	Malaysian Oil Palm Growers' Council (MOPGC)	**Malaysian Standard (MS)**	**Palm Oil Research Institute of Malaysia (PORIM, 97/98)**
FFA (max)	5%	2.5%	3.2%
Moisture (max)	0.25%	0.2%	0.18%
Impurities (max)	0.25%	0.05%	0.03%

Table 5. Standard quality CPKO based on Malaysian Oil Palm Growers' Council (MOPGC), Malaysian Standard (MS) and Standard Institute of Malaysia (SIM)

Factors	Standard quality of crude kernel palm oil (CPKO)			
	Malaysian Oil Palm Growers' Council (MOPGC)	**Malaysian Standard (MS)**	**Standard Institute of Malaysia (SIM)**	
			Grade 1	**Grade 2**
FFA (max)	5%	5%	3%	6%
Moisture (max)	7%	7%	0.35%	0.5%
Impurities (max)	6%	6% (min)	0.35%	0.5%

In order to monitor the moisture content in the palm oil (Elaeisguinensis), a rapid and nondestructive dielectric measurement using commercial coaxial probe and dielectric mixture modeling are analyzed macroscopically in this study. The measured dielectric constant, $\varepsilon¢_{r\_oil}$ and loss factor, $\varepsilon^{2}_{r\_oil}$ of CPO and CPKO versus operation frequency, *f* are respectively plotted in Figures27 (a) and (b). Obviously, the CPKO naturally has higher values of relative permittivity compare to the CPO for entire operation frequencies.

Figure 27. Variation in (a) dielectric constant, ε¢r_oiland (b) loss factor, ε2r_oilwith frequency at (25 ± 1)oC for pure crude palm oil (CPO) and crude palm kernel oil (CPKO) (You et al., 2014)

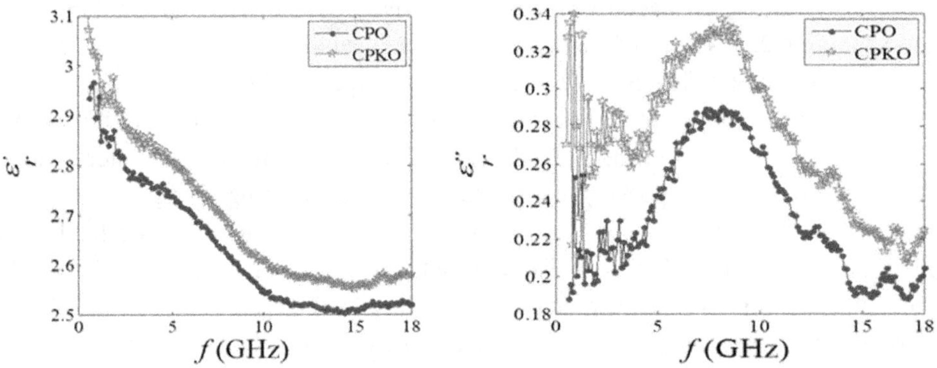

Obviously, the CPKO naturally has higher values of relative permittivity compare to the CPO for entire operation frequencies. In Figure 28, we have plotted the variation in effective dielectric constant, $\varepsilon¢_{eff}$ with frequency for various volumes of water, v_{water} in CPO. As expected, the values of the effective dielectric constant, $\varepsilon¢_{eff}$ is increased with increasing volume of water in CPO over the frequency range.

Figure 28. Variation in effective dielectric constant, ε¢eff with frequency, f at room temperature (25 ± 1)oC for various water volume in (a) CPO and (b) CPKO (You et al., 2014)

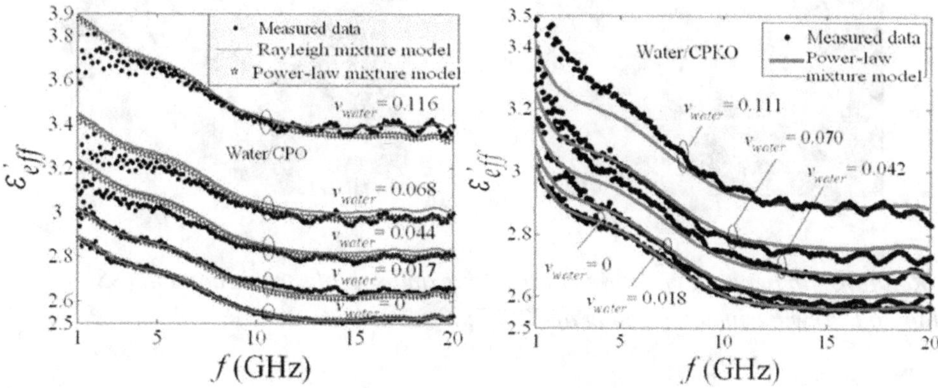

The minor differences between the results obtained from the measurement and calculation may due to the following factors:

1) The direction of polarized electric field is not fully normal to the layer.
2) The distribution of water inclusion in the mixture is not uniform because it tends to sink in oil environment.
3) The non-uniform distribution of water inclusion in oil environment around the sensing area.
4) The overlapping water inclusion can form the connected sets thus the size of water inclusion distributed in oil environment will not be uniformed.

Fruits Sweetness Detector

The correlation between the dielectric properties and fruit sweetness (or fruit maturity) using commercial open-ended coaxial probe, has been intensively studied by Nelson *et al.* (1994; 1995; 2007) and Guo *et al.* (2007; 2011) as shown in Figure 29.

The major component of soluble solids content (SSC) in fruit is the sugar content. Moreover, the SSC is linearly correlated to the moisture content, *m.c* in the honeydew melon and watermelon as shown in Figure 30, thus, the sugar content of the honeydew melon and watermelon can be indirectly determined by measuring the moisture content, *m.c* of the fruit. From Figure30, it appears that most of the watermelons are sweeter than the honeydew melons, since the average of the SSC for watermelon is almost 12%, while the average of the SSC for honeydew melon is approximately 8%. The relationship between the loss tangent, tan δ and the SSC for the watermelon at 1.8 GHz is given in Figure31.

Figure 29. Dielectric measurements of the (a) honeydew melon and (b) watermelon surfaces using open-ended coaxial probe (Guo et al., 2007; Nelson et al., 2007)

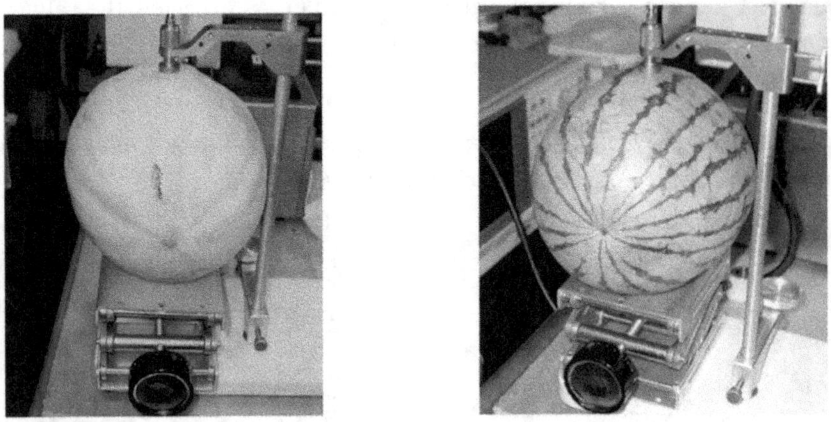

Figure 30. Relationship between moisture content, m.c and soluble solids content (SSC) for the (a) honeydew melon and (b) watermelon (Guo et al., 2007; Nelson et al., 2007)

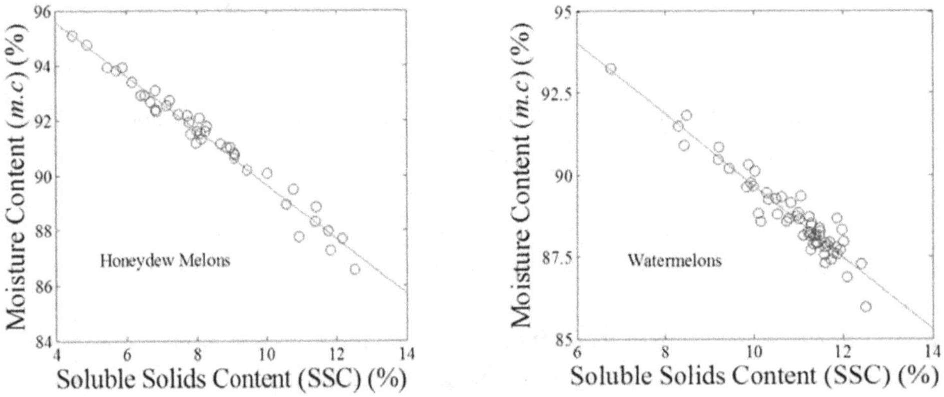

Nelson have been found that sweetness level, *SSC* of fruit can be linearly predicted based on measuring relative complex permittivity (ε_r¢andε_r^2) as:

$$SSC = \frac{\varepsilon_r'' - \kappa\varepsilon_r'}{\eta} \quad (15)$$

Symbols κ and η are fitting coefficients and its values are constant for broadband measurement. In addition, the relatively high correlations between *SCC* and relative complex permittivity (ε_r¢ and ε_r^2) can be achieved using (15). From (15), the ε_r^2/*SSC* versus ε_r¢/*SSC* at 1.8 GHz can be plotted as shown in Figure 32 (*Nelson et al., 2007*).

Figure 31. The linear relationship between loss tangent, tan δ and soluble solids content (SCC) of the watermelons (Nelson et al., 2007)

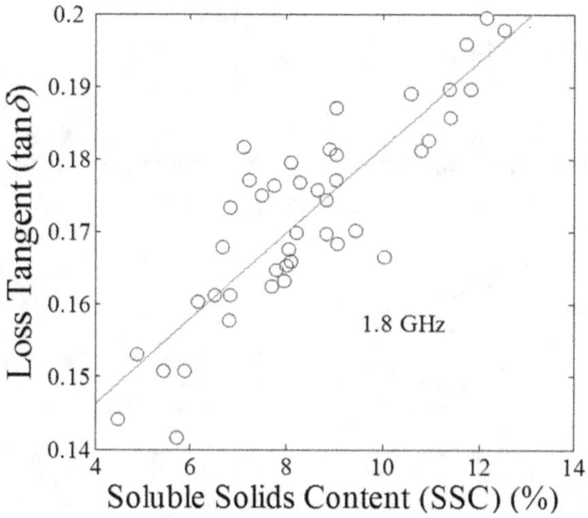

Figure 32. The linear relationship between ratio εr2/SCC and ratio εr¢/SCC of the watermelons (Nelson et al., 2007)

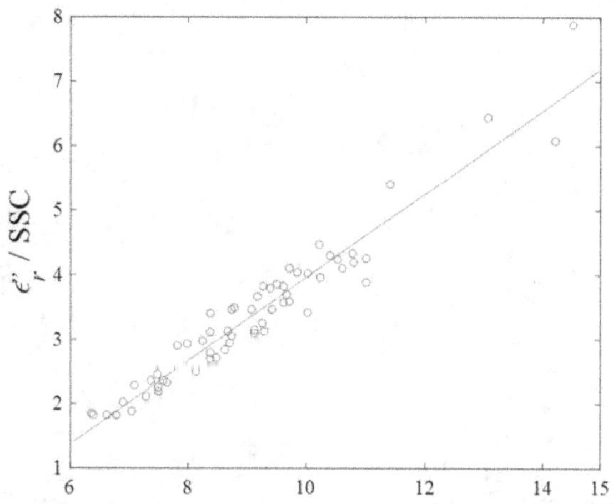

In this subtopic, another example of fruit sweetness measurement would be described, namely oranges (Rosman*et al.*, 2018). In the study, an orange sample is placed and contacted at the end of the monopole driven from coaxial line, which is connected to microwave vector reflectometer as shown in Figure 33.

When the amount of brix, °Bx changes in the orange, the monopole sensor will measure a change in reflection coefficient, $|S_{11}|$ (from the change in ε_r) that can be directly correlated with a change in brix level, °Bx of the orange, which is obtained using commercial refractometer. Figure 34 (a)shows the Brix level, °Bx of orange change is respectively caused the change of ε_r' value from 45 to 15 and ε_r'' is changing within 2 to 18 at 2.2 GHz. On the other hand, the relationship between measured reflection coefficient, $|S_{11}|$ of the orange and its sweetness level, °Bxat 2.2 GHzas shown in Figure 34 (b).The pH values for the oranges in Figure 34 (b) are measured by pH tester into the orange juice. Clearly, the significant variation of ε_r', ε_r'', and $|S_{11}|$ of the orange sample is most probable due to presence of soluble solid content (SSC), since the moisture content, *m.c.* for overall orange sample are approximately constant (~ 88% *m.c.*) as shown in Figure 34 (c).

Figure 33. The finite end of the monopole terminated by orange sample

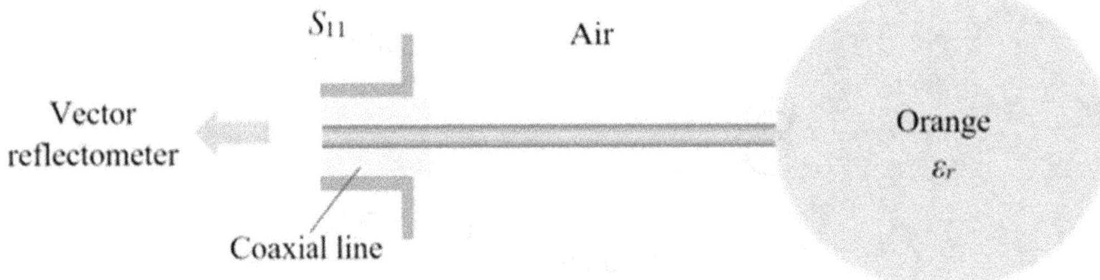

Figure 34. (a) Relative complex permittivity (εr¢ andεr2) of oranges versus Brix level (o). (b) Measured |S11| and pH versus Brix level (o). (c) Measured moisture content, m.c. versus Brix level (o).

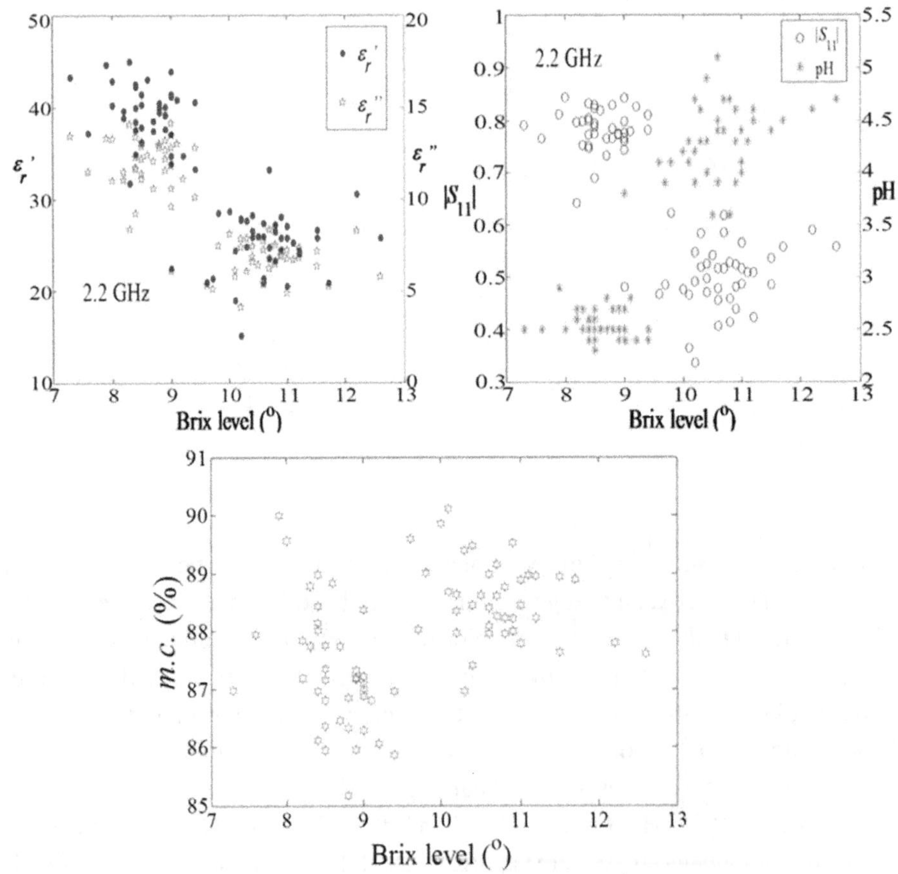

Rice Drying and Sterilization Processing

Gain moisture content, *m.c.*is the most important criterion for quality assessment and process control. The *m.c.* inside the rice grain is a key parameter for grain processing such as harvesting, milling, storage, transporting and quality control. For instance, rice grain is usually harvested between moisture content, *m.c.* 19% to 25% for maximum grain yield and needed to be dried to ~14% (wet basis) or less depending on the season and the weather for safe storage. Besides, the ideal *m.c.* for milling is 14% in order to maximize the head rice and minimize the broken rate (BR). If the *m.c.* in rice grain is too low, the milled grain will become fragile. Thus, the precise determination of the rice grain *m.c.* is important. In general, the hot air drying oven is implemented in conventional rice grain drying process. However, this method is time-consuming and causes energy loss. Recently, most of microwave equipment and measurement techniques are used to dry or monitor the moisture content, *m.c.* inside the agriculture products and foods under test. The reason is due to the tendency of water to absorb microwave energy and generate the heat within the agri-foods under test. When the moist rice grain is exposed to the microwave, the water molecules in the rice grain will be induced to rotate and produce heat as shown in Figure 3. Thus, the rate of water removal is higher than hot-air drying method. Besides, the microwave heating is capable of maintaining original texture structure and color of the rice gain compared to conventional oven drying techniques. For instance, the manufactured microwave heating system for agri-food products is shown in Figure 35.

Figure 35. Microwave heating equipment

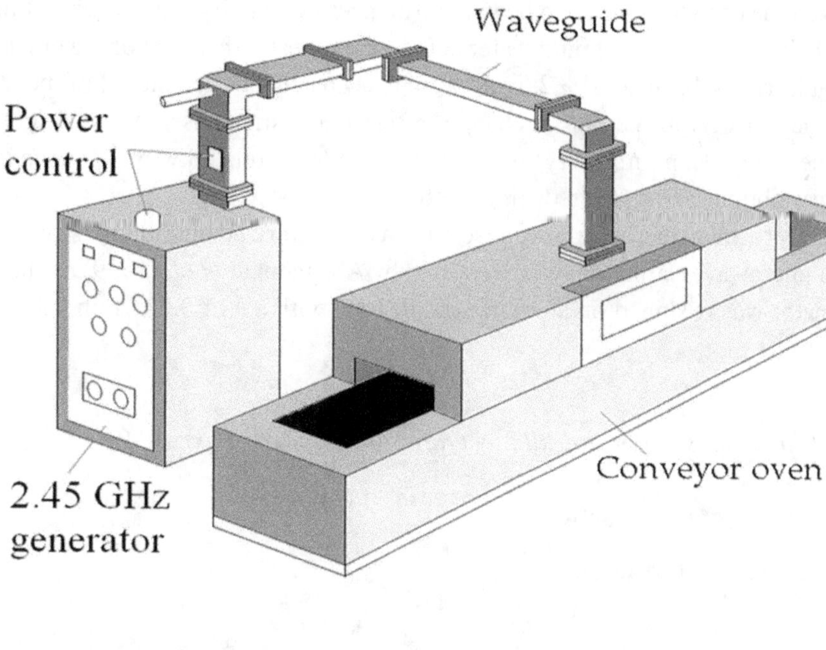

Rice Moisture Sensing

Microwave measurement techniques have been an increased interest in the determination of moisture content, *m.c.* (or dielectric properties) of bulk rice grain at microwave frequencies range (You and Nelson, 1988; Kim *et al.*, 2002; Yagiharaet *al.*, 2007). This is because the percentage of *m.c.* (range from 9% *m.c.* to 26% *m.c.*) in the rice grain has played an important role in the aspects of marketing and storage. In marketing, the price of rice is dependent on the weight of the rice bulk, more water content in the rice grain will increase the price of rice. For storage, it is the highest moisture content (*m.c.*$< 14\%$) that is present in the grain mass determines the storage duration of the rice grains, in which *m.c.*$< 13\%$ indicates a storage duration of more than 60 days (Thakur *et al.*, 2001; You *et al.*, 2011).

There are two methods of determining moisture content, *m.c.* of rice grain, which are the direct method and the indirect method. The direct method determines the *m.c.* by removing the *m.c.* of rice grain bulk using oven drying method (130° C with 24 hours) or chemical solvents. The direct method is the most accurate method to determine the grain *m.c.*, but it is time-consuming. For the direct method, the water content in the rice grain is removed totally and *m.c.* is calculated based on gravimetric method. The heating temperature must exceed the temperature of boiling water due to water molecules in rice grain was in a bind with molecules of rice substance.

In contrast, the indirect method requires the measurement of the electrical property of the rice grain using a fabricated instrument, so-called moisture meter. The change in electrical properties that can be directly correlated with a change in the actual moisture content, *m.c.* of the rice grain obtained from oven drying method (direct method). Recently, the indirect methods become more popular than the direct method due to rapid test, high sensitivity and user-friendly features. In this Section, only indirect methods that use high operating frequency for grain moisture determination are described.

As mentioned above, the polarization of water molecules contained in the rice grain is sensitive and showed a significant response when exposed to microwaves and this will allow the microwave sensor to be used as a measuring technique to sense the moisture content, *m.c.* in the moist rice grain. The volume of water in the total volume of moist rice grain heavily influences the relative permittivity of the moist rice grain due to the relative permittivity of pure water being much greater than that of the other constituents in the rice grain bulk (rice substance: ε_r » 2.5, air: $\varepsilon_r = 1$). Thus, when the amount of moisture changes in the rice grain, the sensor will measure a change in reflection /transmission coefficient or resonant frequency (from the change in permittivity) that can be directly correlated with a change in *m.c.* of the rice grain which was obtained from oven drying method.

For instance, a microwave microstrip ring resonator as miniaturized and non-destructive sensor for single wheat grain microwave estimation were reported by Abegaonkar *et al.* (1999) as shown in Figure 36. The ring resonator was designed on an alumina substrate with a h of 0.635 mm and $\varepsilon_{r,sub}$ of 9.98to resonate at 10 GHz.

Figure 36. Microstrip ring resonator overlaid by grain (Abegaonkar et al., 1999)

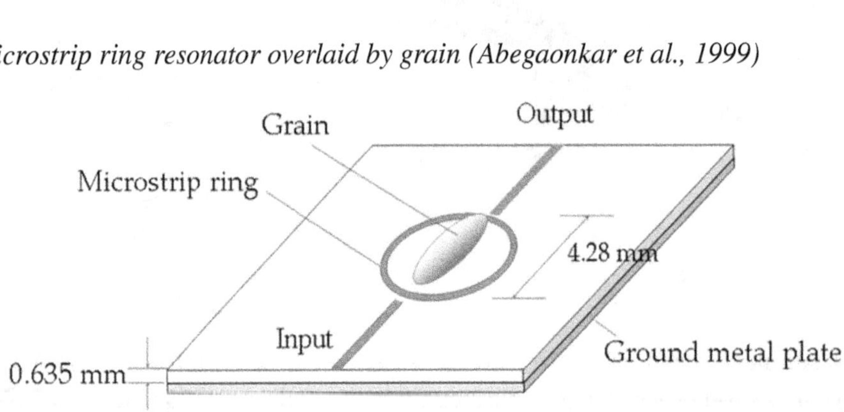

In that measurement, a wheat grain was overlaid on the ring resonator, and its corresponding measured resonant frequency f_r, bandwidth, *BW* and quality factor *Q* were obtained from a scalar network analyzer. Then, the calibration equations relate the moisture, *m.c.* with f_r, *BW* and *Q* respectively were developed. Validation test showed that the average error for *m.c.* prediction was 2.12%. The major drawback of this ring resonator is the grain orientation within the sensor significantly affects the accuracy and sensitivity of the measurement.

Kim *et al.* (2002) proposed a prototype microwave transceiver grain moisture meter based on free-space transmission method for rice grain *m.c.* prediction. The prototype grain moisture meter is mainly composed of an oscillator, isolator, horn antenna, detector and rectangular grain holder. The schematic diagram of the prototype grain moisture meter is depicted in Figure 37.A generated microwave signal at 10.5 GHz from an oscillator transmitted to the rice grain sample through an isolator and a transmitting horn antenna. The attenuated signal was then received by a receiving horn antenna and detected by a detector that will convert the signal to voltage, *V*. A calibration equation relates the moisture content, *m.c.* with moisture density, ρ, voltage, *V* and temperature, *T*, was developed and validated in that study. Validation result showed that the moisture meter can predict the *m.c.* with average error of 0.52%. However, the moisture meter is relatively large in size and operates at high frequency which will increase its cost. In addition, the output voltage, *V* of the meter needs to be substituted into the calibration equation manually for *m.c.* calculation.

Figure 37. The prototype microwave transceiver grain moisture meter (You et al., 2017)

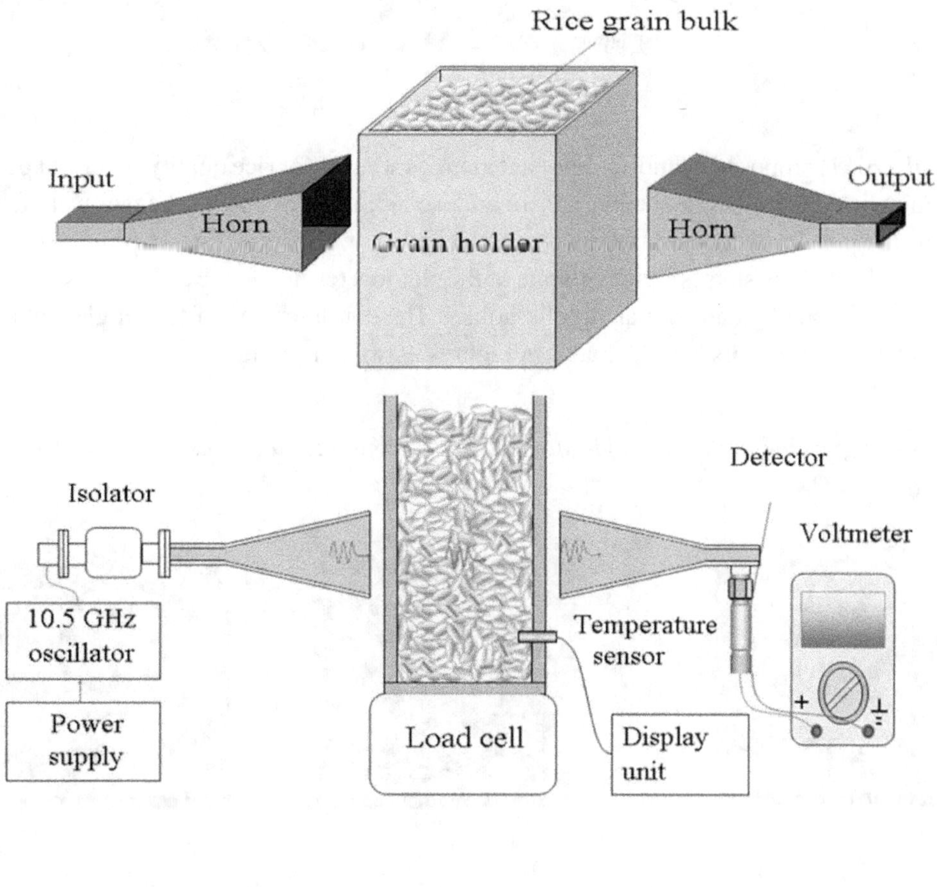

Besides, a multi-layer microstrip moisture sensor was developed by Jafari *et al.* (2010) for measuring the *m.c.* of rice grain. The sensor was designed on a laminated RT/Duroid substrate with a *h* of 1.575 mm and $\varepsilon_{r,sub}$ of 2.2 to operate at 9 GHz. The sensor consists of two main parts: the stripline section which is covered by aluminium plate, and the semi-infinite layer microstrip line which is the sensing region as shown in Figure 38. The attenuation as a function of complex permittivity of three dielectric layers (substrate, protective and grain layers) and effective dielectric constant was derived from complex propagation constant. Besides that, the relationship between the complex permittivity of rice grain and *m.c.* was also established based on mixture theory. To predict the *m.c.* of rice grain, the rice grain was loaded on the sensor and the corresponding attenuation signal was obtained from the vector network analyzer. Then, the *m.c.* was calculated from the measured attenuation through both derived attenuation and mixture equations based on numerical method. Thus, this method is complex as compared to the aforementioned studies (Abegaonkar *et al.*, 1999; Kim *et al.*, 2002).

Figure 38. (a) Multi-layer microstrip moisture sensor (b) Cross-section of microstrip sensing region (Jafari et al., 2010)

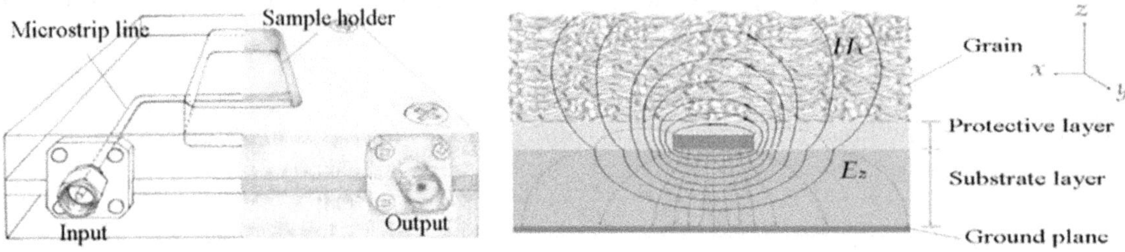

You *et al.* (2011) proposed cylindrical slot antennas as sensors for rice quality (*m.c.* and percentage of BR) determination. A single slot antenna sensor and a coupling slots antenna sensor were fabricated by using SMA stub panel with a radius of 0.65 mm and length, *l* of 1.436 cm. For coupling slots antenna, the separation distance of the slots is 1 cm with one of the slot was terminated. Both sensors were positioned on an aluminum ground plane with an acrylic holder. The configuration of the single slot antenna and coupling slot antenna sensors are illustrated in Figures39 (a) and (b), respectively.

Figure 39. Configuration of the (a) single slot sensor, (b) coupling slots sensor, and (c) Experiment-set-up (You et al., 2011)

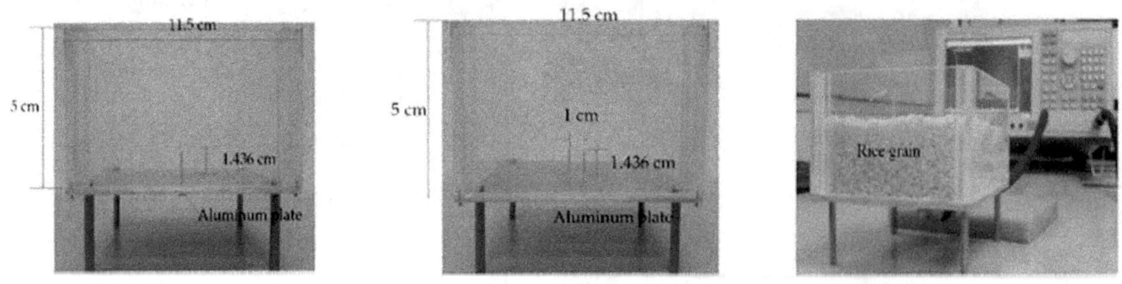

Both sensors were designed to operate at 1 GHz for *m.c.* measurement based on the measured magnitude of reflection coefficient from a vector network analyzer. Calibration equations based on the relationship between the measured magnitude of reflection coefficient and *m.c.* were also developed for both sensors. The coupling slots sensor was more sensitive to the *m.c.* of the rice grain than single slot sensor. The coupling effects between the two slots will increase the density of the sensing fields for the sensor. Recently, microstrip wide-ring, microstrip coupled-line and small coaxial probe, as shown in Figure 40, have been used for rice moisture measurements. By using wide-ring and coupled-line sensors, the bulk rice grain samples with different level of *m.c.* were placed into the sample holder of the sensor as shown in Figure 41. On the other hand, the small and slim coaxial probe has a small sensing area which can cover the size of single rice grain and provides a single grain moisture measurement. Moreover, the single grain measurement does not depend on the bulk density of the rice grains, thus the uncertainty of bulk density in the rice measurement (due to a different rate of broken rice in the bulk grain) can be ignored.

Figure 40. (a) Microstrip wide-ring sensor (Mun et al., 2015); (b) Microstrip coupled-line sensor (Mun et al., 2015); (c) Customized small coaxial probe with outer conductor radius, b = 0.33 mm and inner conductor radius, a = 0.1 mm (You et al., 2013)

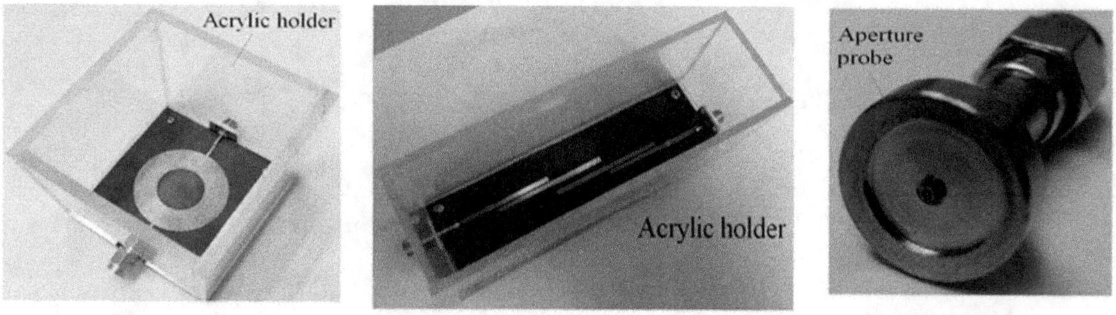

Figure 41. Measurement set-up for (a) bulk rice grain moisture measurement (Mun et al., 2015) and (b) single rice grain measurement (You et al., 2013)

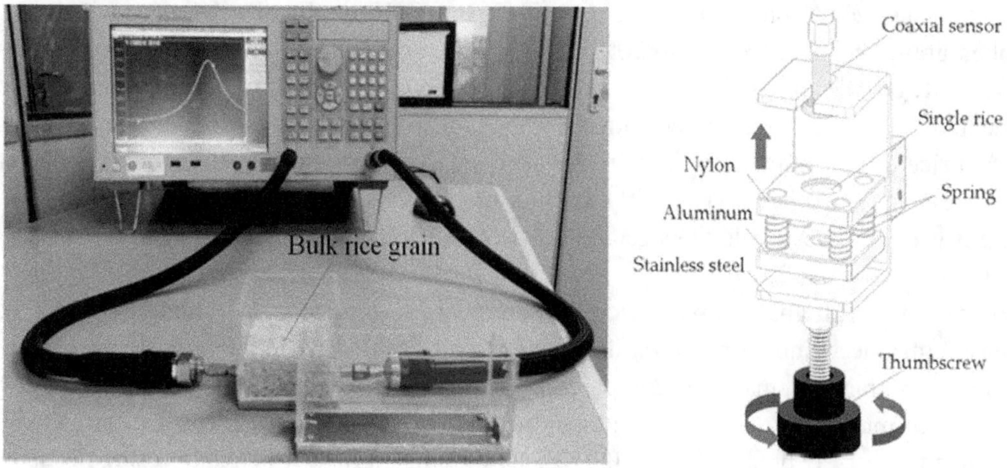

The comparison of the measured ε_r' and ε_r'' for rice grain with various *m.c.* at 2.45 GHz by using the wide-ring sensor, coupled-line sensor, literature data, and small coaxial probe are shown in Figure42 (Mun *et al.*, 2015; You *et al.*, 2013). Obviously, the value of ε_r¢ for single rice grain measurement is much higher than the bulk rice measurements. This is due to the bulk rice is composed of a mixture of air and rice grains. The measured ε_r¢ will affect by the volume of air gap in the total volume of the bulk rice grain since the value of ε_r¢ for the air is approximately unity which is being much lower than the constituents in the rice grain.

Figure 42. The variations in ε_r¢andε_r^2 of single rice grain and rice grain bulk with its moisture content, m.c. at 2.45 GHz (Mun et al., 2015; You et al., 2013)

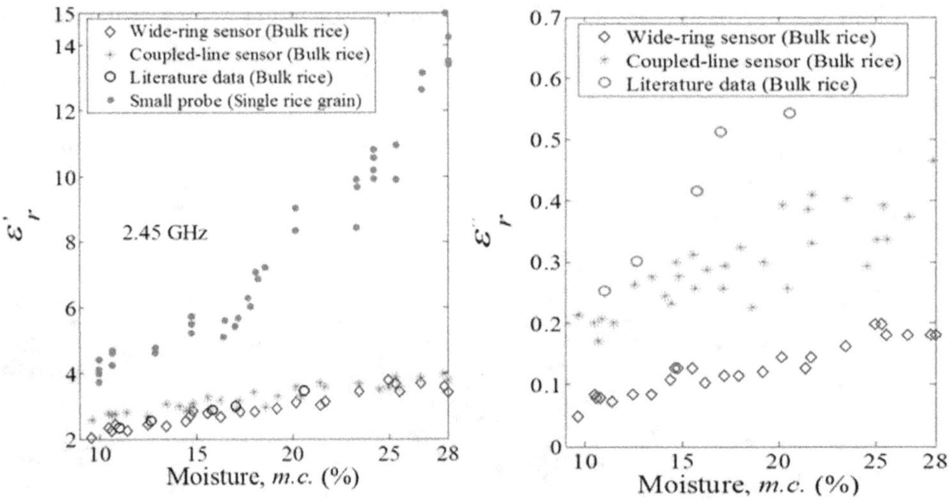

Broken Rice Detection

Besides moisture content, *m.c*, the grade and price of the rice grain are also depending on the percentage of broken grain. The rice grain has a length less than ¾, but more than ¼ of the average head grain length are categorized as broken rice as shown in Figure43. Normally, the bulk rice which is containing broken grain for 0% to 5%, 5% to 20%, 20% to 35%, and 35% to 50%, is classified as a premium grade, grade 1, grade 2 and grade 3, respectively. In conventional broken rice sorting machine, there are implemented the mechanical filter techniques. Since broken rice has a size smaller than head rice, thus, the broken rice can be filtered by the wire net with small size of mesh in the machine. Recently, the image processing techniques are popularly used for grading and classification of rice, which the optical scanning is used to differentiate the size and color of the rice grain (Yadav and Jindal, 2001; Lloyd *et al.*, 2001; Courtois *et al.*, 2010).

In this section, an alternative broken rice measurement using microwave techniques were introduced. The measurement technique is based on the change of air gap density in the bulk rice since the air gap between broken grains is smaller as compared to the normal rice grains. The electromagnetic wave is generated from microwave sensor and radiate into the bulk grain sample. The air gap density in the grain sample is inversely proportional to the density of radiated field which is covering inside the grain

Figure 43. The rice grain with (a) 0%, (b) 50% and (c) 100% broken rate

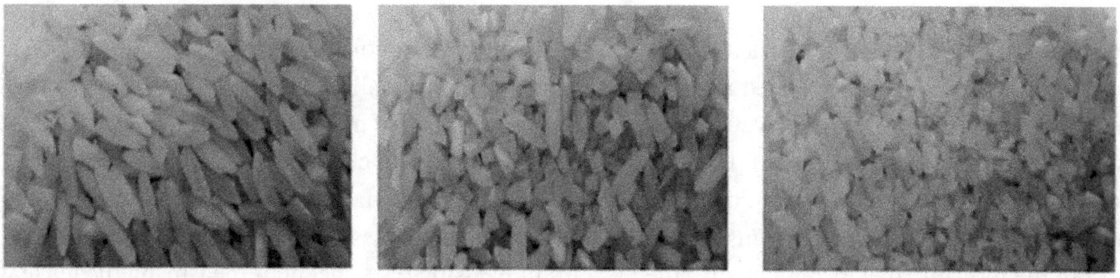

sample. The single and coupling slots radiators with one input signal are introduced in You *et al.* (2010) to measure the percentage of the broken rice. The slot radiators are constructed using commercial SMA stub panel with radius of slot, $a = 0.65$ mm and length of slot, $l = 16$ mm. The slots were driven from aluminum ground and covered by acrylic as shown in Figure 39. For coupling slots radiator, the separate distance between the coupling slots is 9 mm and one of the slot was terminated. The density of sensing field was strengthened using coupling slots radiator.

In measurements, the measured reflection coefficient, $|\Gamma|$ will vary with changes in the density of rice grains in the sample. The electromagnetic wave is generated from coaxial line and radiate into the grain sample. The measurement technique is based on the change of air gap density in the bulk rice, since the air gap between broken grains is smaller as compared to the normal rice grains, the measurement technique is based on the change of air gap density in the bulk rice. Thus, the density of rice grains in the mixed sample is high. The short wavelength signal (high operating frequency) is required to enhance the air gap sensitivity for broken rice measurement and reduces the penetration of energy into the rice grains. In fact, the air gap density in the grain sample is inversely proportional to the density of radiated field which is covering inside the grain sample. Figure 44 shows the measured reflection coefficient, $|\Gamma|$ of the rice grain vary with the percentage of broken rice in the samples at 13.5 GHz.

Figure 44. Variation in magnitude reflection coefficient, $|\Gamma|$ with percentage of broken grain in the rice sample at 13.5 GHz at room temperature 25°C using the (a) single slots sensor and (b) coupling slots sensor, respectively (You et al., 2011)

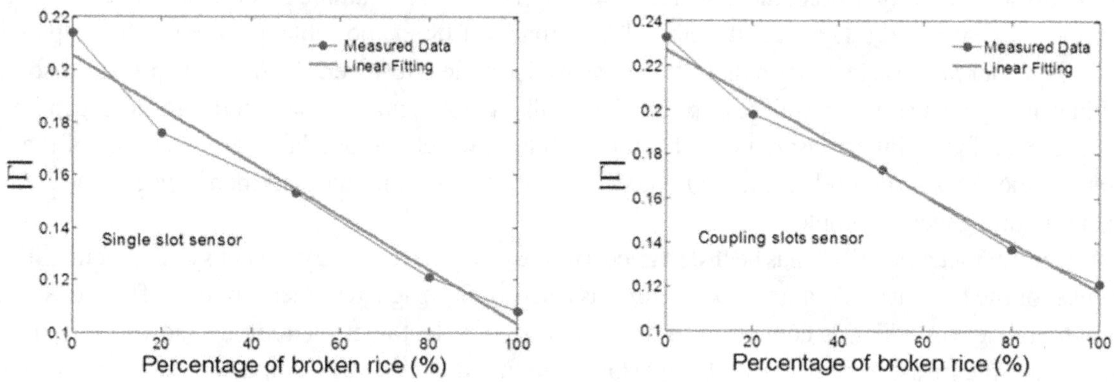

Tea Leaf Moisture Measurements

Generally, the production line of tea is mainly dependent on the moisture content, *m.c.* to decide how long the tea leaves should be dried in every drying step. Traditionally, the *m.c.* of tea leaves was estimated by the skilled workers. However, the significant percentage of measurement errors may be caused by the workers (Chen *et al.*, 2014). In fact, there are some accurate chemical methods (water extracted from tea sample) that were used to determine the moisture of tea (Hall *et al.*, 1988), but these kinds of methods are time consuming and unsuitable for large-scale tea processing in manufacturing. In addition, most of the moisture meters only involve the DC circuit by using the concept of measuring the tea leaves resistance, conductivity or capacitance (Hazarika *et al.*, 2006; Javanmard *et al.*, 2009; Li *et al.*, 2012; Okamura and Tomita, 1994; Rao and Han, 2014; Mizukami *et al.*, 2006). The ionic conductivity, σ, in the tea leaves interacts with the metal surface of the sensor (sample holder) to create a small amount of voltage in the meter. However, this measurement concept will cause less sensitivity to the dry tea leaves ($< 30\%$) measurement, due to the lack of free ionic in the dry tea leaf. Furthermore, these economical meters can only measure one type of tea leaf for a narrow measurement range of *m.c.* (normally only 3% to 18% *m.c.*). Besides, the sensitivity of measurement towards the water contained in the tea leaves requires extensive interaction between the sensor surface and tea bulk samples.

Recently, most of the study measurements (Jones *et al.*, 2017; Memon and Lim, 2018; Zarifi *et al.*, 2018; Peters *et al.*, 2018) are based on the resonant technique as the measurement principle is simple, in which the properties of the sample under test are characterized from the measurements of shifting resonance frequency and quality of factor Q. Although this method provides accurate measurements, it requires high resolution and frequency sweep instruments, as well as good environmental control measure. Therefore, this method is less suitable for large-scale product processing industries. In the past until now, there were two types of microwave sensors which have been reported and used in the monitoring of the tea drying process. The first type of sensor is one pair of horn antennas and its moisture measurement is based on the free-space transmission techniques (Zhang and Okamura, 2000; Tein *et al.*, 2017). The second type is a microstrip line sensor which is used for near field measurement (Okamura *et al.*, 2007; Zhang and Okamura, 2006). Both methods have advantages and disadvantages, respectively. The first method is suitable for high temperature tea leaves bulk measurement due to the horn antennas sensor which does not come into direct contact with the tea leaves sample; thus, the RF circuit is safe from heat damage. Nevertheless, this method provides a less precise measurement due to the sensing field being highly dispersed. Furthermore, the distance between the tea leaves sample and the horn aperture is difficult to gauge precisely. The second method has a small sensor and portable feature which is capable to measure wider *m.c.* range containing in the tea leaves sample. However, the microstrip line sensor has a thin sensing area above the microstrip line. Normally, the sensing area = (width of microstrip line)^2. Thus, the *m.c.* is mainly measured only for the tea leaves which are overlaid on the surface of the microstrip line. This situation leads to high uncertainty in the moisture measurement since the tea leaves are an inhomogeneous sample.

Recently, You *et al.* (2018) has been designed a microwave reflectometry-sensor system for measuring the *m.c.* of the tea leaves. A large coaxial probe as a sensor is designed to measure the reflected voltage, V_r of the tea leaves which is connected to the customized reflectometer circuit as shown in Figure 45 (a). During the measurement, the probe aperture is required to be firmly contacted with the tea leaves bulk as shown in Figure 45 (b). Then, the measured reflected voltage, V_r data is extracted by personal computer (PC). The corresponding moisture content, *m.c.* of the tea leaves under test are determined using oven drying method (wet basis).The variations in reflected voltage, V_r of tea leaves bulk with the percentage of *m.c.* at various operating frequencies, respectively, at (25 ± 1) °C are plotted in Figures 46 (a) to (e).As expected, the reflected voltage, V_r of tea leaves increases exponentially with *m.c.*, since relative permittivity, ε_r of agricultural product normally increases exponentially with its *m.c.* at higher operating frequencies(Khaled *et al.*, 2015).

Figure 45. (a) Microwave reflectometry-sensor system. (b)Coaxial probe for tea leaves bulk measurements(You et al., 2018)

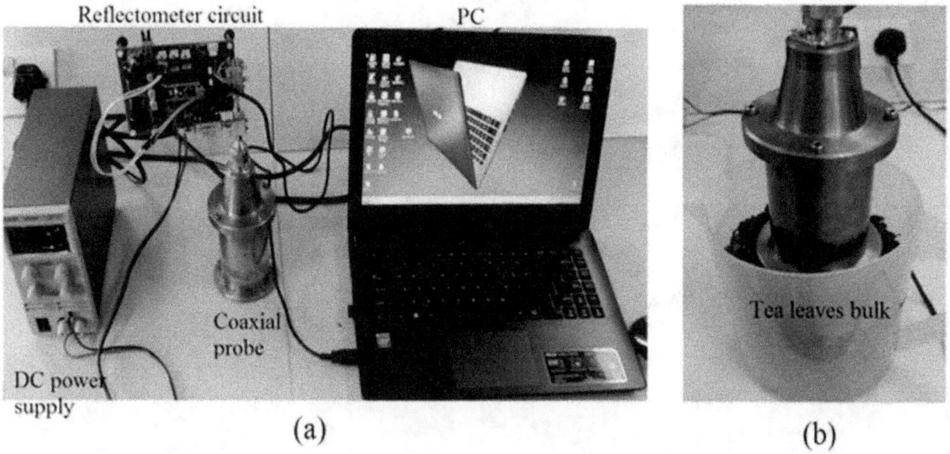

Insect Detection and Control

Insect infestation is considered as a major reason of post-harvested grain losses during storage (Kumar and Kalita, 2017). The statistical numbers reveal the great losses both in the grain products and costs. The world annual losses records an estimation of 30% in Africa and Asia while 10% in North America (Davis, 1974). A study showed the economic losses due to microorganisms and insects to require one billion dollar annually only in the United States (Brader *et al.*, 2002). The infestation of raw grain by insects causes quality degradation that decreases the nutritional value such grain (Vadivambal *et al.*, 2005). The grain crops are either stored on-farm or in the commercial handling facilities. The safety of the storage determines the quality criteria of the stored grain. On condition that the stored gain is found to be infested then the insects has to be eliminated and controlled. The common treatment of infestation is by using chemical insecticides. Such substances are contributing to the environmental hazards. Several devoted studies have been point out the possibilities of using an alternative techniques aiming the elimination of potential hazards caused by the usage of chemical substances (Nelson, 1974). Grain disinfection using microwave fields exposure has been utilised for since 1920's. Dielectric heating is the enabling feature for controlling the insects by microwave or lower frequency RF energy. The basic concept of controlling the insects is by raising the temperature to specific level that is considered as tolerant to the hosted product and lethal to the insects. Dielectric heating produces a novel solution when the water content of the insects are higher than the infested products which increases the absorption rate, hence, rapid rising to a lethal temperature and better sustain to the infested product (Nelson, 1974).

Figure 46. Variations in reflected voltage, V_r with moisture content, m.c. (%) in (a) Boh tea, (b) Pu Erh tea, (c) Green tea, (d) Sabah red tea, and (e) Oolong tea, respectively (You et al., 2018)

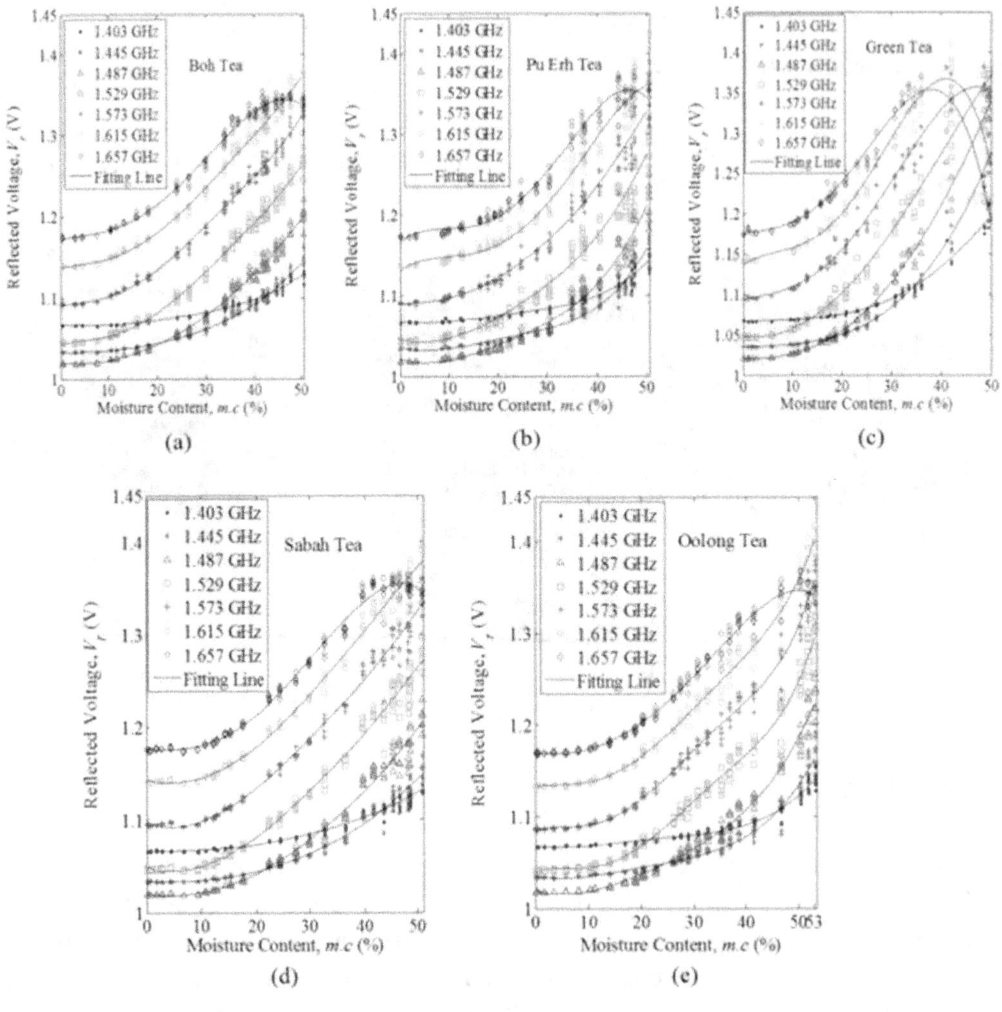

Basically, microwave treatment controls the insect lethality by the mean of either differential or localized heating of the insects' body tissues. The power caused by *E*-field of EM wave is objected to the dissipation per unit volume of the dielectric media, as expressed in equation (5). Such power dissipation increases the rate of temperature, *T* rising which defined as a function of power as given in (7). The heating rate is proportionally influenced by the operating frequency, *f* and intensity of electric field, *E*. There are several factors that affect the choice of the applied power and frequency, such as the type of the infested material which determines the effective heating threshold, the insect species where each species varies in the *m.c.* and the relative dielectric properties of body tissues. According to (7), successful insect-control depends on the dielectric properties of the host materials which cause variations in the lethal temperature to be in the range 40 °C to 90 °C. It also depends on the species and the development stage of the insect, the distribution in the host and the frequency and power of operation, where the adult insects produce high susceptibility to being controlled by the electromagnetic field as compared with embryos. The 13.6 MHz and 39 MHz are the two frequencies are examined by Nelson and Stetson, (1974) for proper insect control, at the same time without damaging the quality of germination or baking and milling.

Hence, the microwave method can be used to control and eliminate the rice weevils (in Figure 47 (a)) without affecting the texture of the rice grain. It is due to each agri-food, or biological specimens have unique energy absorption properties, which is mainly influenced by its loss factor, ε_r^2. The value of ε_r^2 for the material is frequency-dependent. The typical relationship between the loss factor, ε_r'', mechanism of water molecules and operating frequency is shown in Figure 5. For instance, the values of ε_r¢ and ε_r^2 of the rice weevils versus frequency, *f* is shown in Figure 47 (b) (Nelson, 2015). By comparing Figures 5 and 47 (b), it was found that the water content in the rice weevils is in a state of bonding at a few MHz. This means that the internal response of the rice weevils is sensitive to the external wave at few MHz. Thus, Nelson (2015) found that the complete mortality of the rice weevils can be obtained with 40 °C of grain temperature when infested grain was exposed to microwave at 39 MHz than when it was treated at 2.45 GHz (ISM band).

Figure 47. (a) Rice weevils (Sitophilus oryzae); (b) Variational in ε_r¢ and ε_r^2 of the rice weevils with frequency, f at 24 °C (Nelson, 2015)

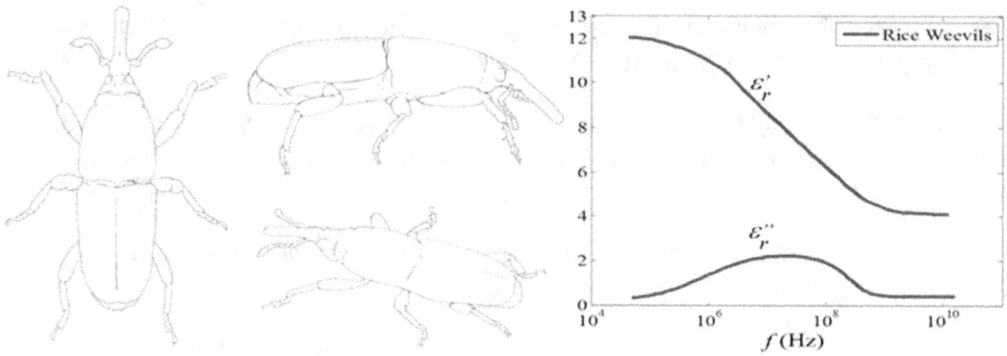

CONCLUSION

This chapter introduces selected ten sub-themes of microwave applications in the fields of agriculture and food processing, especially in microwave moisture sensing and heating applications. The descriptions of Maxwell's equations and field theory foundations in this chapter is ignored, so the content of this book can be concentrated on the description of more advanced microwave applications in this field. Nevertheless, the application of microwave in agri-foods processing is very broad and most of these technologies are new and still in the research phase. Hence, this chapter still does not cover all microwave applications, such as moisture measurement of butter/sugar, parboiling rice using microwave energy, microwave pasteurization and sterilization of foods, and microwave freeze drying of fruits/vegetables. Today most of the existing microwave devices and systems are bulky, expensive, and only used in food processing factories. However, the authors believe that in the future, several microwave devices and systems, such as microwave insect control and microwave freeze drying devices, will be scaled down and modified to be used domestically, like domestic microwave oven.

REFERENCES

Abbas, Z., You, K. Y., Shaari, A. H., Khalid, K., Hassan, J., & Saion, E. (2005). Complex permittivity and moisture measurements of oil palm fruits using an open-ended coaxial sensor. *IEEE Sensors Journal*, *5*(6), 1281–1287. doi:10.1109/JSEN.2005.859249

Abegaonkar, M. P., Karekar, R. N., & Aiyer, R. C. (1999). A microwave microstrip ring resonator as a moisture sensor for biomaterials: Application to wheat grains. *Measurement Science & Technology*, *10*(3), 195–200. doi:10.1088/0957-0233/10/3/014

Ahmad, A. F., Abbas, Z., Abrass, H. A., & You, K. Y. (2019). Optimum design of a microstrip ring resonator sensor to determine the moisture content in oil palm fruits and seeds. *BioResources*, *14*(1), 1819–1837.

Ahmed, J., Ramaswamy, H. S., & Raghavan, V. G. S. (2007). Dielectric properties of Indian Basmati rice flour slurry. *Journal of Food Engineering*, *80*(4), 1125–1133. doi:10.1016/j.jfoodeng.2006.09.004

Allen, E. W. Jr, & Garlan, H. (1962). Evolution of regulatory standards of interference. *Proceedings of the IRE*, *50*(5), 1306–1311. doi:10.1109/JRPROC.1962.288087

Ansarudin, F., Abbas, Z., Hassan, J., Yahaya, N. Z., & Ismail, M. A. (2012). A simple insulated monopole sensor technique for determination of moisture content in Hevea rubber latex. *Measurement Science Review*, *12*(6), 249–254. doi:10.2478/v10048-012-0034-5

Ariffin, A., Mat, S. R., Banjari, M., & Wan, O. W. E. (1990). Morphological changes of the cellular component of the developing palm fruit (Tenera: Elaeisguineensis). *Palm Oil Research Institute of Malaysia Bulletin*, *21*, 30–34.

Berthold Technologies. (2007). *Contactless Measurement of Water Content and Moisture-Micro-Moist LB 456*. Berthold Technologies.

Brader, B., Lee, R. C., Plarre, R., Burkholder, W., Kitto, G. B., Kao, C., Polston, L., Dorneanu, E., Szabo, I., Mead, B., Rouse, B., Sullins, D., & Denning, R. (2002). A comparison of screening methods for insect contamination in wheat. *Journal of Stored Products Research*, *38*(1), 75–86. doi:10.1016/S0022-474X(01)00006-6

Chen, A., Chen, H. Y., & Chen, C. C. (2014). Use of temperature and humidity sensors to determine moisture content of Oolong tea. *Sensors (Basel)*, *14*(8), 15593–15609. doi:10.3390140815593 PMID:25153142

Chow, H. C. (1979). *Methods for measuring the dry rubber content of field latex.RRIM Training Manual on Analytical Chemistry*. Rubber Research Institute of Malaysia.

Davis, R. (1974). Pests of stored products. *Bulletin of the Entomological Society of America*, *20*(4), 335–335. doi:10.1093/besa/20.4.335a

Deepa, G., Singh, V., & Naidu, K. A. (2008). Nutrient compostion and physiochemical properties of Indian medicinal rice –Njavara. *Food Chemistry*, *106*(1), 165–171. doi:10.1016/j.foodchem.2007.05.062

Guo, W. C., Nelson, S. O., Trabelsi, S., Kays, S. J. (2007). Dielectric properties of honeydew melons and correlation with quality. *Journal of Microwave & Electromagnetic Energy*, *41*(2), 41-2-44 – 41-2-54.

Guo, W. C., Zhu, X. H., Yue, R., Liu, H., & Liu, Y. (2011). Dielectric properties of Fuji apples from 10 to 4500 MHz during storage. *Journal of Processing and Preservation*, *35*(6), 884–890. doi:10.1111/j.1745-4549.2011.00541.x

Hall, M. N., Robertson, A., & Scotter, C. N. G. (1988). Near-infrared reflectance prediction of quality, theaflavin content and moisture content of black tea. *Food Chemistry*, *27*(1), 61–75. doi:10.1016/0308-8146(88)90036-2

Hartley, C. W. S. (1977). *The Oil Palm*. Longman Group Limited.

Hasted, J. B. (1973). *Aqueous Dielectric*. Chapman and Hall.

Hazarika, D., Laskar, S., Sarma, A., & Sarmah, P. K. (2006). PC-based instrumentation system for the detection of moisture content of tea leaves at its final stage. *IEEE Transactions on Instrumentation and Measurement*, *55*(5), 1641–1647. doi:10.1109/TIM.2006.881031

Jafari, F., Khalid, K., Yusoff, W. M. D. W., & Hassan, J. (2010). The analysis and design of multi-layer microstrip moisture sensor for rice grain. *Biosystems Engineering*, *106*(3), 324–331. doi:10.1016/j.biosystemseng.2010.04.005

Javanmard, M., Abbas, K. A., & Arvin, F. (2009). A microcontroller-based monitoring system for batch tea dryer. *The Journal of Agricultural Science*, *1*(2), 101–106.

Jones, T. R., Zarifi, M. H., & Mojgan, D. (2017). Miniaturized quarter-mode substrate integrated cavity resonators for humidity sensing. *IEEE Microwave and Wireless Components Letters*, *27*(7), 612–614. doi:10.1109/LMWC.2017.2711527

Kennedy, G., & Burlingame, B. (2003). Analysis of food composition data on rice from a plant generic resources perspective. *Food Chemistry*, *80*(4), 589–596. doi:10.1016/S0308-8146(02)00507-1

Khaled, D. E., Novas, N., Gazquez, J. A., Garcia, R. M., & Agugliaro, F. M. (2015). Fruit and vegetable quality assessment via dielectric sensing. *Sensors (Basel)*, *15*(7), 15363–15397. doi:10.3390150715363 PMID:26131680

Khalid, K., & Abbas, Z. (1992). A microstrip sensor for determination of harvesting time for oil palm fruits (Tenera: ElaeisGuineesis). *The Journal of Microwave Power and Electromagnetic Energy*, *27*(1), 3–10. doi:10.1080/08327823.1992.11688165

Khalid, K., Hassan, J., & Daud, W. M. (1996). The effect of ionic conductivity and dipole orientation on the dielectric loss of the hevea rubber latex. *Proceedings of IEEE MTTS International Microwave Symposium*, 23–26.

Kim, K. B., Kim, J. H., Lee, S. S., & Noh, S. H. (2002). Measurement of grain moisture content using microwave attenuation at 10.5 GHz and moisture density. *IEEE Transactions on Instrumentation and Measurement*, *51*(1), 72–77. doi:10.1109/19.989904

Kraszewski, A. W. (2005). Recent Developments in Electromagnetic Aquametry. In K. Kupfer (Ed.), *Electromagnetic Aquametry* (pp. 1–12). Springer. doi:10.1007/3-540-26491-4_1

Kumar, D., & Kalita, P. (2017). Reducing postharvest losses during storage of grain crops to strengthen food security in developing countries. *Foods*, *6*(1), 1–22. doi:10.3390/foods6010008 PMID:28231087

Li, X. L., Xie, C. Q., He, Y., Qiu, Z. J., & Zhang, Y. C. (2012). Characterizing the moisture content of tea with diffuse reflectance spectroscopy using wavelet transform and multivariate analysis. *Sensors (Basel)*, *12*(7), 9847–9861. doi:10.3390120709847 PMID:23012574

Memon, M. U., & Lim, S. J. (2018). Microfluidic high-Q circular substrate-integrated waveguide (SIW) cavity for radio frequency (RF) chemical liquid sensing. *Sensors (Basel)*, *18*(1), 1–13. PMID:29316646

Mizukami, Y., Sawai, Y., & Yamaguchi, Y. (2006). Moisture content measurement of tea leaves by electrical impedance and capacitance. *Biosystems Engineering*, *93*(3), 293–299. doi:10.1016/j.biosystemseng.2005.12.009

Mohammadi, A., Khalid, K., Bafti, P. S., & Homaiee, M. (2012). Development of microcontroller-based microwave system to measure solid content of Hevea rubber latex. *Measurement*, *45*(3), 571–579. doi:10.1016/j.measurement.2011.10.007

Mun, H. K., You, K. Y., & Dimon, M. N. (2013). Broken rice detection based on microwave measurement technique using microstrip wide-ring sensor and microstrip coupled-line sensor. *Australian Journal of Crop Science*, *7*(13), 2079–2090.

Nelson, S. O. (1991). Dielectric properties of agricultural products – measurements and applications. *IEEE Transactions on Electrical Insulation*, *26*(5), 845–869. doi:10.1109/14.99097

Nelson, S. O. (2015). *Dielectric Properties of Agricultural Materials and Their Applications*. Elsevier.

Nelson, S. O., Bartley, P. G., & Lawrence, K. C. (1997). Measuring RF and microwave permittivities of adult rice weevils. *IEEE Transactions on Instrumentation and Measurement*, *46*(4), 941–946. doi:10.1109/19.650805

Nelson, S. O., Forbus, W. R. Jr, & Lawrence, K. C. (1994). Permittivities of fresh fruits and vegetables at 0.2 to 20 GHz. *Journal of Microwave & Electromagnetic Energy*, *29*(2), 81–93. doi:10.1080/08327823.1994.11688235 PMID:8083786

Nelson, S. O., Forbus, W. R. Jr, & Lawrence, K. C. (1995). Assessment of microwave permittivity for sensing peach maturity. *Transactions of the ASAE. American Society of Agricultural Engineers*, *38*(2), 579–585. doi:10.13031/2013.27869

Nelson, S. O., Guo, W. C., Trabelsi, S., & Kays, S. J. (2007). Dielectric spectroscopy ofwatermelons for quality sensing. *Measurement Science & Technology*, *18*(7), 1887–1892. doi:10.1088/0957-0233/18/7/014

Nelson, S. O., & Stetson, L. (1974). Possibilities for controlling inserts with microwave and lower frequency RF energy. *IEEE Transactions on Microwave Theory and Techniques*, *22*(12), 1303–1305. doi:10.1109/TMTT.1974.1128482

Nyfors, E., & Vainikainen, P. (1989). *Industrial Microwave Sensors*. Artech House, Inc.

Okamura, S., & Tomita, F. (1994). Microwave moisture sensing system in drying process for green tea production. *Proceedings of Instrumentation and Measurement Technology Conference*, 1253-1256. 10.1109/IMTC.1994.351828

Okamura, S., Zhang, Y., & Tsukamoto, N. (2007). A new microstripline-type moisture sensor for heavily wet tea leaves. *Measurement Science & Technology*, *18*(4), 1022–1028. doi:10.1088/0957-0233/18/4/009

Peters, J., Taute, W., Döscher, C., Meier, R., Höft, M., Knöchel, R., & Breitkreutz, J. (2018). From laboratory-to pilot-scale: Moisture monitoring in fluidized bed granulation by a novel microwave sensor using multivariate calibration approaches. *Drug Development and Industrial Pharmacy*, *44*(6), 961–968. doi:10.1080/03639045.2018.1425427 PMID:29308682

Rao, N., & Han, A. T. (2014). Research on non-destructive measuring moisture content for packaged tea based on the LC resonant sensor. *Proceedings of the 26th Chinese Control and Decision Conference*, 2476-2480.

Rosman, R., Dimon, M. N., & You, K. Y. (2018). Correlation between reflection coefficient, dielectric properties and brix level of Malaysian oranges at microwave frequencies. *Indonesian Journal of Electrical Engineering and Computer Science*, *10*(3), 853–858. doi:10.11591/ijeecs.v10.i3.pp853-858

Sinija, V. R., & Mishra, H. N. (2011). FTNIR spectroscopic method for determination of moisture content in green tea granules. *Food and Bioprocess Technology*, *4*(1), 136–141. doi:10.100711947-008-0149-8

Tein, S. Y., Then, Y. L., & You, K. Y. (2017). Tea leaves moisture measurement and prediction using RF waveguide antenna. *Proceedings of 2017 Asia Pacific Microwave Conference*, 670-673. 10.1109/APMC.2017.8251535

Thakur, S. K. P., & Kolmes, W. S. (2001). Permittivity of rice grain from electromagnetic scattering. *4th International Conference on Electromagnetic Wave Interaction with Water and Moist Substance*, 203-210.

Trabelsi, S., Kraszewski, A. W., & Nelson, S. O. (1999). Unified calibration method for nondestructive dielectric sensing of moisture content in granular materials. *Electronics Letters*, *35*(16), 1346–1347. doi:10.1049/el:19990890

Wang, S., & Tang, J. (2001). Radio frequency and microwave alternative treatments for insect control in nuts: A review. *Asian Agricultural Engineering*, *10*(3 & 4), 105–120.

Yagihara, S., Oyama, M., Inoue, A., Asano, M., Sudo, S., & Shinyashiki, N. (2007). Dielectric relaxation measurement and analysis of restricted water structure in rice kernels. *Measurement Science & Technology*, *18*(4), 983–990. doi:10.1088/0957-0233/18/4/004

You, K. Y. (2015). *RF Coaxial Slot Radiators: Modeling, Measurements, and Applications*. Artech House.

You, K. Y. (2017). Materials characterization using microwave waveguide system. In S. K. Goudos (Ed.), *Microwave Systems and Applications* (pp. 341–358). InTech. doi:10.5772/66230

You, K. Y., Lee, C. Y., Chan, K. S., Lee, K. Y., Cheng, E. M., & Lee, Y. S. (2018). Monitoring moisture content for various kind of tea leaves in drying processes using RF reflectometer-sensor system. *Instruments*, *2*(18), 1–8. doi:10.3390/instruments2030018

You, K. Y., Lee, C. Y., Then, Y. L., Chen Chong, S. H., You, L. L., Abbas, Z., & Cheng, E. M. (2013). Precise moisture monitoring for various soil types using handheld microwave-sensor meter. *IEEE Sensors Journal, 13*(7), 2563–2570. doi:10.1109/JSEN.2013.2253318

You, K. Y., Mun, H. K., You, L. L., Jamaliah, S., & Abbas, Z. (2013). Small and slim coaxial probe for single rice grain moisture sensing. *Sensors (Basel), 13*(3), 3652–3663. doi:10.3390130303652 PMID:23493127

You, K. Y., Salleh, J., Abbas, Z., & You, L. L. (2011). Cylindrical slot antennas for monitoring the quality of milled rice. *Progress in Electromagnetics Research Symposium Proceedings*, 1370-1373.

You, K. Y., You, L. L., Yue, C. S., Mun, H. K., & Lee, C. Y. (2017). Physical And Chemical Characterization Of Rice Using Microwave And Laboratory Methods. In S. F. Amanullah (Ed.), *Rice* (pp. 81–99). InTech.

You, K. Y., Zulkifly, A., & Khalid, K. (2010). Application of microwave moisture sensor for determination of oil palm fruit ripeness. *Measurement Science Review, 10*(1), 7–14.

You, K. Y., Zulkifly, A., Khalid, K., & Abdul Rahman, M. Z. (2010). Improved dielectric model for polyvinyl alcohol-water hydrogel at microwave frequencies. *American Journal of Applied Sciences, 7*(2), 270–276. doi:10.3844/ajassp.2010.270.276

You, K. Y., Zulkifly, A., You, L. L., Lee, K. Y., & Cheng, E. M. (2014). Palm oil moisture based on dielectric properties at microwave frequencies. *International Journal of Microwave and Optical Technology, 9*(14), 294–302.

You, T. S., & Nelson, S. O. (1988). Microwave dielectric properties of rice kernels. *The Journal of Microwave Power and Electromagnetic Energy, 23*(3), 150–158. doi:10.1080/08327823.1988.11688050

Yu, X. R., Su, Y., Wang, Y., Zheng, T. S., Zhao, S. M., & Huang, S. Z. (1998). Study on microwave drying of grain. *Proceedings of the 7th International Working Conference on Stored-product Protection*, 1096-1101.

Zarifi, M. H., Gholidoust, A., Abdolrazzaghi, M., Shariaty, P., Hashisho, Z., & Daneshmand, M. (2018). Sensitivity enhancement in planar microwave active-resonator using metal organic framework for CO_2 detection. *Sensors and Actuators. B, Chemical, 255*, 1561–1568. doi:10.1016/j.snb.2017.08.169

Zhang, Y. J., & Okamura, S. A. (2000). Moisture content measurement for green tea using phase shifts at two microwave frequencies. *Subsurface Sensing Technologies and Applications, 1*(2), 489–496. doi:10.1023/A:1026571800893

Zhang, Y. J., & Okamura, S. A. (2006). Density independent method for high moisture content measurement using microstrip transmission line. *Journal of Microwave & Electromagnetic Energy, 40*(2), 110–118. doi:10.1080/08327823.2005.11688527 PMID:17278794

ADDITIONAL READING

Behari, J. (2005). *Microwave Dielectric Behavior of Wet Soils*. Anamaya Publishers and Springer. doi:10.1007/1-4020-3288-9

Brodie, G., Jacob, M. V., & Farrell, P. (Eds.). (2015). *Microwave and Radio-Frequency Technologies in Agriculture-An Introduction for Agriculturalists and Engineers*. De Gruyter Open Ltd.

Datta, A. K., & Anantheswaran, R. C. (Eds.). (2001). *Handbook of Microwave Technology for Food Applications*. Marcel Dekker, Inc. doi:10.1201/9781482270778

James, S., & James, C. (2014). Minimal processing of ready meals. In D. W. Sun (Ed.), *Emerging Technologies for Food Processing* (2nd ed., pp. 599–612). Elsevier and Academic Press. doi:10.1016/B978-0-12-411479-1.00032-2

Kent, M. (1978). The use of strip-line configuration in microwave moisture measurements. *The Journal of Microwave Power*, *8*(2), 189–194. doi:10.1080/00222739.1973.11688876

Kraszewski, A. W., & Nelson, S. O. (1996). Moisture content determination in single kernels and seeds with microwave resonant sensors. In D. W. Sun (Ed.), *Emerging Technologies for Food Processing* (2nd ed., pp. 599–612). Elsevier and Academic Press.

Ma, Z., & Okamura, S. (1996). Density dependence of permittivity of green tea at 9.4 GHz and moisture content determination. *Journal of Microwave & Electromagnetic Energy*, *33*(3), 178–183. doi:10.1080/08327823.1998.11688374

Metaxas, A. C., & Meredith, R. J. (2008). Industrial Microwave Heating. UK: The Institution of Engineering and Technology (IET).

Meyer, W., & Schilz, W. (1981). Feasibility study of density-independent moisture measurement with microwaves. *IEEE Transactions on Microwave Theory and Techniques*, *29*(7), 732–739. doi:10.1109/TMTT.1981.1130436

Miyai, Y. (1978). A new microwave moisture meter for grains. *The Journal of Microwave Power*, *13*(2), 163–166. doi:10.1080/16070658.1978.11689090

Nelson, S. O. (2004). Agricultural applications of dielectric spectroscopy. *Journal of Microwave & Electromagnetic Energy*, *39*(2), 75–85. doi:10.1080/08327823.2004.11688510 PMID:15719907

Okamura, S. (2000). Microwave technology for moisture measurement. *Subsurface Sensing Technologies and Applications*, *1*(2), 205–227. doi:10.1023/A:1010120826266

Orsat, V. VijayaRaghavan, G. S. (2014). Radio-frequency processing. In Sun, D. W. (Ed.), Emerging Technologies for Food Processing 2nd Ed (pp. 385-398). US: Elsevier and Academic Press.

Osepchuk, J. M. (1984). A history of microwave heating applications. *IEEE Transactions on Microwave Theory and Techniques*, *32*(9), 1200–1224. doi:10.1109/TMTT.1984.1132831

Ozkoc, S. O., Sumnu, G., & Sahin, S. (2014). Recent developments in microwave heating. In D. W. Sun (Ed.), *Emerging Technologies for Food Processing* (2nd ed., pp. 361–383). Elsevier and Academic Press. doi:10.1016/B978-0-12-411479-1.00020-6

Scaman, C. H., Durance, T. D., Drummond, L., & Sun, D. W. (2014). Combined microwave vacuum drying. In D. W. Sun (Ed.), *Emerging Technologies for Food Processing* (2nd ed., pp. 427–445). Elsevier and Academic Press. doi:10.1016/B978-0-12-411479-1.00023-1

Schubert, H. Regier, Marc. (Ed.). (2005). The Microwave Processing of Foods. Cambridge, England: Woodhead Publishing Limited and CRC Press.

Trujillo, F. J., & Geveke, D. J. (2014). Nonthermal processing by radio frequency electric fields. In D. W. Sun (Ed.), *Emerging Technologies for Food Processing* (2nd ed., pp. 259–269). Elsevier and Academic Press. doi:10.1016/B978-0-12-411479-1.00014-0

Volgyi, F. (1999). Microwave measurement for specifying the fresh of eggs. In *The Third Workshop on Electromagnetic wave interaction with water and moist substances*, Athens, Georgia, 52-56.

Walter, M., & Isengard, H. D. Microwave drying-A rapid method to determine the true water content in dairy products.In *The Third Workshop on Electromagnetic wave interaction with water and moist substances*, Athens, Georgia, 62-66.

KEY TERMS AND DEFINITIONS

Bound Water: An amount of water in a substance which are bound to other substance constituents and create a thin layer of water surrounding the constituent surfaces.

Cole-Cole Model: A relaxation model for describing dielectric relaxation of a substance.

Dry Basis: A measure of the water in a substance, which is expressed as the percentage of the ratio of the weight of water to the weight of the completely dry substance.

Fifth Generation Wireless Technology (5G): Digital cellular mobile communication networks that began wide deployment in 2019.

Free Fatty Acid (FFA): Fatty acid produced by the hydrolysis of oils and fats.

Free Water: Water that is not bound to an inorganic surface and can flow freely.

Frequency-Domain Reflectometry (FDR): A nonintrusive technique used to feed a frequency sweep signal into an electrical line. The receiver measures the interference pattern generated when the output signal of the swept RF source is added to or subtracted from the reflected signal of the fault and other reflection characteristics in the tested electrical lines, then determine the characteristics of electrical lines by observing the vector addition of the signals creates a ripple pattern versus frequency.

Internet of Things (IoT): A system of interrelated computing devices, mechanical, and digital machines provided with unique identifiers (UIDs) and the ability to transfer data over a network without requiring human-to-human or human-to-computer interaction.

Microwave: A form of electromagnetic radiation with wavelengths ranging from 1 m to 1 mm, which is corresponding to operating frequencies ranging from 300 MHz to 300 GHz.

Microwave Sensor: A type of sensor device that operates at microwave frequencies (within 300 MHz to 300 GHz).

Monolithic Microwave Integrated Circuit (MMIC): A type of integrated circuit (IC) device that operates at microwave frequencies.

Moisture Content (*m.c.*)**:** Quantity of water that exists in a substance.

Pasteurization: A process in which water and certain packaged and non-packaged foods are treated with mild heat, usually to less than 100 °C, to eliminate pathogens and extend shelf life.

Permittivity: A substance property that affects the Coulomb force between two-point charges in the substance.

Reflection Coefficient (S_{11})**:** A parameter that describes how much electromagnetic waves are reflected by impedance discontinuities in the transmission medium.

Relative Permittivity (ε_r)**:** A permittivity of substance expressed as a ratio relative to the vacuum permittivity.

Soluble Solids Content (SSC): A solution is determined by the index of refraction.

Time-Domain Reflectometry (TDR): A noninvasive technique used to measure the time between release and return of the low voltage pulse from any reflections and determine the characteristics of electrical lines by observing reflected waveforms.

Transmission Coefficient (S_{21})**:** A parameter that describes how much electromagnetic waves pass through the transmission medium.

Wet Basis: A measure of the water in a substance, which is expressed as the percentage of the ratio of the weight of water to the total weight of the substance (water + dry substance).

APPENDIX

List of Abbreviations

ε_r Relative Complex Permittivity
ε_o Permittivity of Vacuum (8.8541878×10^{-12} F/m)
ε_r¢ Real Part of Relative Permittivity /Dielectric Constant
ε_r^2 Imaginary Part of Relative Permittivity / Loss Factor
$\varepsilon_¥$ Optical Permittivity
ε_s Static Permittivity
n Complex Refractive Index
n¢ Real Part of Refractive Index
n^2 Imaginary Part of Refractive Index
ε¢$_{eff}$ Effective Dielectric Constant
ε_{water}, ε_{fiber}, ε_{oil} Relative Permittivity of Water, Fiber, and Oil, Respectively
ε¢$_{r\_oil}$ Dielectric Constant of CPO or CPKO
$\varepsilon^2_{r\_oil}$ Loss Factor of CPO or CPKO
$\varepsilon_{r,sub}$ Dielectric Constant of PCB Substrate
tanδ Loss Tangent
j Square Root of -1
μ_o Free Space Permeability ($4\pi\times10^{-7}$H/m)
μ_r Relative Permeability
σ Conductivity (S/m)
f Operating Frequency (Hz)
f_r Resonant Frequency (Hz)
t Time (s)
T Temperature (^{o}C)
T_g Glass Transition Temperature (^{o}C)
ω Angular Frequency (rad/s)
τ Relaxation Time (s)
τ_o Proportionality Factor of Arrhenius Equation (s)
Q_e Activation Energy
c Velocity of Light (299792458m/s)
λ Wavelength (m)
λ_o Free Space Wavelength (m)
a *and* b Dimensions of Waveguide (m)
d Sample Thickness (m)
D Penetration Depth (m)
l Length of Monopole (m)
P Dissipation of Microwave Power(W/m^3)
k Boltzman Constant (1.38066×10^{-23} J K^{-1})
ΔS Entropy Change(J (K mol)$^{-1}$)
Δf Resonance Frequency Shift (Hz)

h Phanck Constant(6.62608 × 10^{-34} J s^{-1})
$_R$ Gas Constant (8.3143 J/mol·K)
C_p Specific Heat Capacity of the Specimen (J/(kg °C))
ρ Density of the Specimen (kg/m^3)
E Supplied Microwave Energy or Electrical Field (V/m)
V_r Reflected Voltage (V)
$_{\Gamma or S11}$ Complex Reflection Coefficient
S_{21} Complex Transmission Coefficient
Z Impedance (Ω)
α Empirical Constant of Cole-Cole Model ($0 < \alpha < 1$)
ϕ Phase Shift of Reflection Coefficient (rad)
a_1, a_2, a_3, and a_4 Empirical Constant of Cole-Cole Model for Oil Palm Mesocarp
$\kappa$and η Linear Fitting Coefficients
v_{water} Volume Fraction of Water
ρ_{water}, ρ_{fiber}, and ρ_{oil} Density of Water, Fiber, and Oil, Respectively (g/ml or g/cm^3)
m Molar Mass (g/mole)
$m.c.$ Moisture or Water Content (%)
$m.c_v$ Volumetric Moisture Content (%)
ASAE American Society of Agricultural Engineers
BR Broken Rate
BW Bandwidth
CBCPW Conductor-Backed Coplanar Waveguide
CENELEC European Committee for Electrotechnical Standardization
CPO Crude Palm Oil
CPKO Crude Palm Kernel Oil
DRC Dry Rubber Content
EM Electromagnetic
FCC US Federal Communications Commission
FDR Frequency-Domain Reflectometry
FFA Free Fatty Acid
ICES International Committee on Electromagnetic Safety
IEEE Institute of Electrical and Electronic Engineers
IoT Internet of Things
ISM Industrial, Scientific, and Medical
ITU International Telecommunication Union
MMIC Monolithic Microwave Integrated Circuit
MOPGC Malaysian Oil Palm Growers' Council
MS Malaysian Standard
MWG Microstrip Waveguide
NRW Nicholson-Ross-Weir
OECWG Open-ended Coaxial Waveguide
PC Personal Computer
PORIM Palm Oil Research Institute of Malaysia

PTFE Polytetrafluoroethylene
RDWG Rectangular Dielectric Waveguide
RF Radio Frequencies
RWG Rectangular Waveguide
SAR Specific Absorption Rate
SMA Sub-Miniature A
SSC Soluble Solids Content
TDR Time-Domain Reflectometry
Wi-Fi Wireless Fidelity
WLANs Wireless Local Area Networks

Section 3
Applications and Solutions

This section presents specific applications and proposed solutions for automated irrigation systems, case studies for involving the use of artificial neural networks for harvested grain condition monitoring and crop disease identification, framework of precision management practices for legal cultivation of cannabis particularly under greenhouse situations, a web platform for access and use of information related to the variables of interest to farmers, and a research for increasing the accuracy of a prediction model for the power requirements of the chisel plow.

Chapter 6
Precision Agriculture:
Automated Irrigation Management Platform Using Wireless Sensor Networks

Amine Dahane
Research Laboratory in Industrial Computing and Networks, University of Oran 1, Algeria

Bouabdellah Kechar
https://orcid.org/0000-0001-8635-4667
Research Laboratory in Industrial Computing and Networks, University of Oran 1, Algeria

Abou El Hassan Benyamina
LAPECI Laboratory, University of Oran 1, Algeria

Rabaie Benameur
Research Laboratory in Industrial Computing and Networks, University of Oran 1, Algeria

ABSTRACT

Agricultural areas are getting reduced due to human negligence in providing modern irrigation methods. As a result, reflections aiming to save water use in the water network management context have been developed over the years. In this chapter, the authors propose a platform for precision agriculture which allows collecting fundamental physical phenomena (soil moisture, air temperature, humidity, water level, water flow, luminous intensity) required for the precision agriculture, which will be carried out to calculate the water quantity needed for optimal irrigation. The platform consists of a sensor/actuator node, a desktop application, and a gateway switches relay that controls the water pump according to the requirement. It uses an algorithm developed with threshold values of temperature and soil moisture (Rawls and Turq formulas) to control water quantity. The proposal is a good starting point for a smart, sustainable, and low-cost irrigation solution.

DOI: 10.4018/978-1-7998-5000-7.ch006

INTRODUCTION

Agriculture uses 85% of worldwide available freshwater resources, and this percentage will continue being dominant in water consumption because of population growth and increased food demand. There is an urgent need to create strategies based on science and technology (Gutiérrez, 2014) . The use of wireless sensor networks (WSNs) needin agriculture applications becomes a reality nowadays after the success of theoretical research contributions in the previous decade (Dahane, 2015). WSNs have recently motivated the information technology solutions adoption in crop fields within precision agriculture approaches, which contribute to the rational use of water, including technical, agronomic, managerial,... etc. Research studies on irrigation systems to achieve water savings in diverse crops from basic ones to more technologically advanced ones. To achieve water saving, irrigation system frameworks have been proposed, based on various techniques, e.g., thermal imaging, Crop Water Stress Index (CWSI), direct soil water measurements, etc.

Due to the recent advances in sensors to implement irrigation systems for agriculture, the evolution of WSN and the Internet of Things (IoT) technologies, that can be applied in the development of these systems,IoT takes a major role in the revolution of digital transformations. Today, we can, therefore, discuss the IoT applications in agriculture (See Fig. 1).

Rapid climate change affects not only agriculture, but also the agricultural system. In this crisis, IoT applications play a key role in improving agriculture. Sensors are installed outside and inside the agricultural domain. In the Intelligent Greenhouse System, IoT applications adjust the state of the climate according to the particular predefined instructions set. Above all, these sensors collect real-time data that helps control the automatic irrigation system in an intelligent greenhouse. In the livestock monitoring and management system, IoT applications have a major impact. Here, the location tracker is implemented in the livestock. They are tracked at grazing time. More importantly, the health conditions of all animals are recorded at the same time.

The agricultural drone is a crucial invention in the IoT field in agriculture. There are two types: Ground Drone and Aerial Drone.Agriculture drone benefits to the use ease, time-saving, crop health imaging, integrated GIS mapping and the ability to increase yields.

Collecting data from sensors helps defining the exact irrigation time. Although this is expected in many situations, IoT applications have facilitated timely water management. As a result, the water misuse is reducing day by day.Thus,IoT implements digitization in all possible sectors, especially in agriculture.

Within the farming activities that use water inputs, too known as watered farming, there are different conducts to convey the water. The different alternatives show different efficiency and, in a few cases, a particular way ought to be used for a particular corp. The particular irrigation practice has several forms that can be represented into categories (See Fig. 2): attending to the way of water is dispersed able to consider: (i) flood irrigation, (ii) shower water system, (iii) dribble water system, and (iv) nebulizer water system. Forthe existence of detecting frameworks ready to have: (i) water system without any thought, when the watersum isn't calculated or assessed, (ii) scheduled irrigation when the water is provided agreeing to the assessed needs in a period of the year, (iii) Ad hoc water system when the watersum is calculated based on the sensors measurements or prediction using artificial intelligence techniques. The tremendous larger part of the precision agriculture research area proposes usingpumps and valvesin order to convey the water in conjunction with sensors to degree natural parameters in arrange to calculate the water needs.

Figure 1. IoT applications on precision agriculture

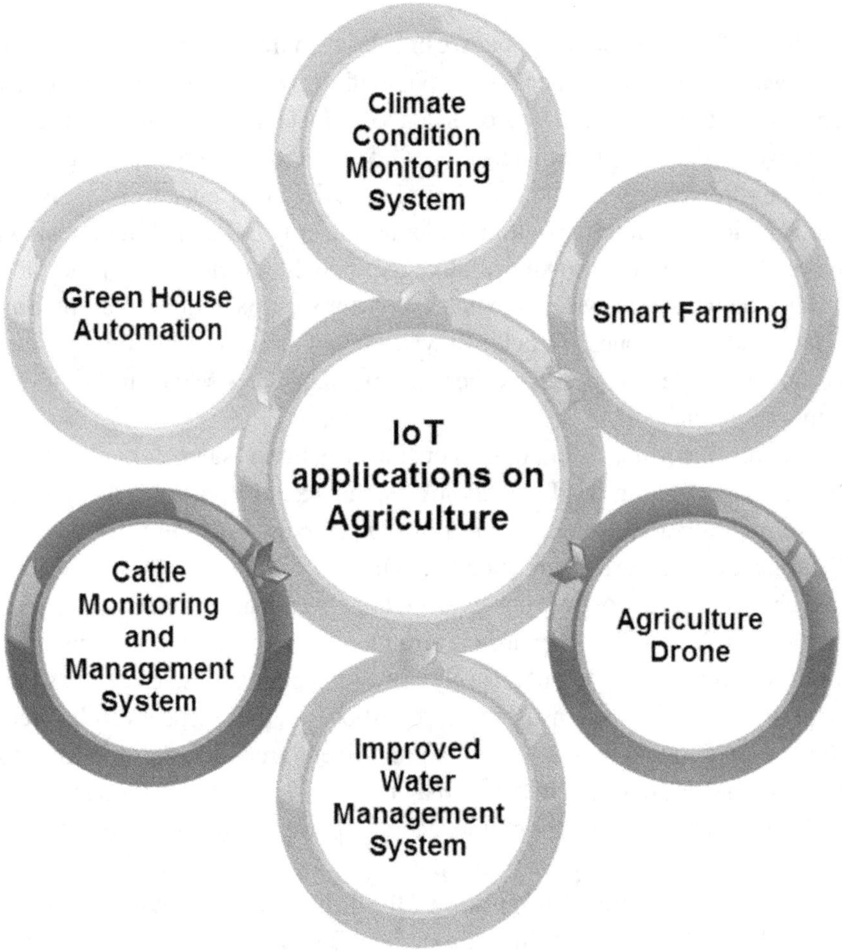

Figure 2. Water management in precision agriculture

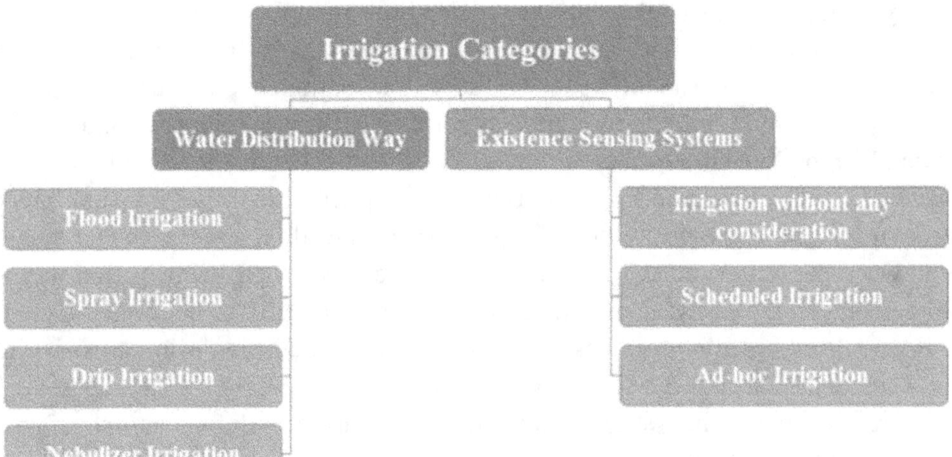

In a previous work plant, water status was monitored and irrigation scheduled based on canopy temperature distribution of the plant, which was acquired with thermal imaging (Gutiérrez, 2014; Wang, 2010). Irrigation systems can also be automated through information on the volumetric water content of the soil, using dielectric moisture sensors to control actuators and save water, instead of a pre-determined irrigation schedule at a particular time of the day and with a specific duration. An irrigation controller is used to open a solenoid valve and apply watering to bedding plants (impatiens, petunia, salvia, and vinca) when the volumetric water content of the substrate drops below a set point (Gutiérrez, 2014; Nemali, 2006). Authors suggested evapotranspiration (ET) based approach, which is an important parameter to decide crop irrigation needs which are influenced by climate parameters, e.g., solar radiation, relative humidity, temperature, wind velocity, and crop features such as the crop growth phase, assortment and plant density, soil properties, nuisance, and disease control. ET-based frameworks can save water up to 42% over time-based water irrigation scheduling (Goap, 2018; Wang, 2011; Prathibha, 2017; Rahman, 2016). An electromagnetic sensor to measure soil moisture was the basis for developing an irrigation system at a savings of 53% of water compared with irrigation by sprinklers in an area of 1000 m^2 of pasture (Gutiérrez, 2014; Goap, 2018). From the 89 evaluated papers by (García, 2020), 83 include clear information of the proposed irrigation system, the other six only mention that they include actuators for irrigation, see Figure 3. Those 83 include different levels of detail, there are 49 papers that only indicate that there are motor/pumps in their system (40 papers) or valves (nine papers) without more detail.

From those papers which or more details, 19 of them include sprinklers (the most used system), eight use drip irrigation, two propose the utilization of sprayers (Nemali, 2006) and the rest use a very specific irrigation systems (robots, pivot, rain gun or it can be applied to multiple systems).

Figure 3. Different Irrigation Systems in precision agriculture

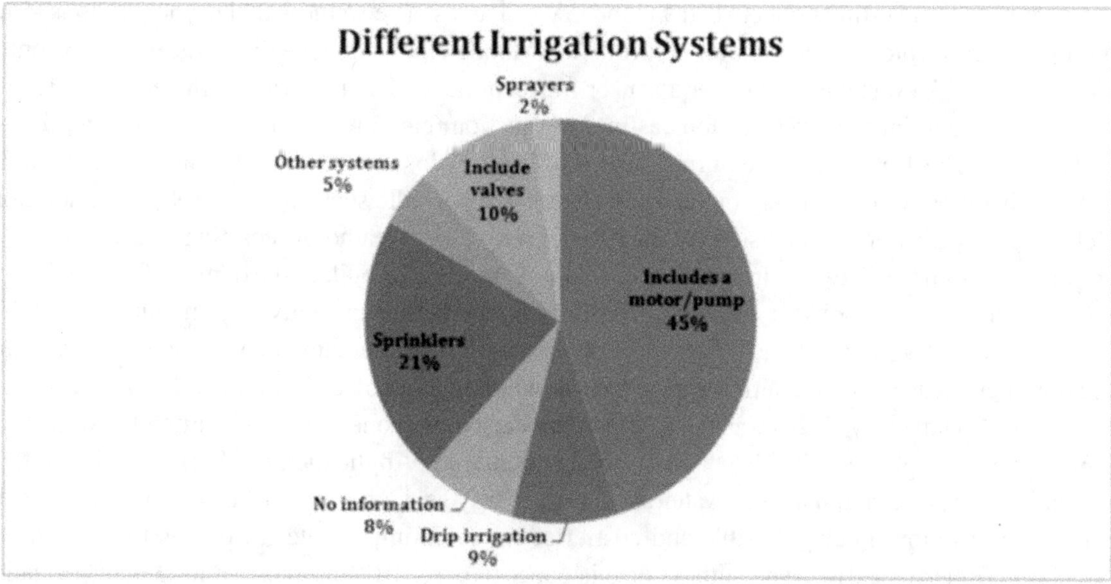

In conjunction with the principal irrigation system, three papers propose the use of a fogging system and two papers propose the use of fertigation in their systems.

A reduction in water use under scheduled systems also have been achieved, using soil sensor and an evaporimeter, which allowed for the irrigation adjustment to the daily fluctuations in weather or volumetric substrate moisture content. (Wang, 2010) developed an irrigation sensor based on a smartphone. For sensing soil moisture, the digital smartphone camerais used to process RGB to gray for theratio estimation between wet and dry soil areas. The wetness and dryness ratio is transmitted via a gateway to the water motor controller. A Mobile Application (APP) is developed to control sensor activity (like wakeup) and to set a sensor in sleep mode. The majority of the earlier irrigation systems do not consider the weather forecasting information (e.g., precipitation) while making irrigation decisions. It leads to a freshwater waste, energyand loss of crop growth (due to excess water) when rain is followed immediately by the crop watering. To handle such cases, WSNs based solutions can provide better decision support for irrigation by using weather forecasting information (e.g., precipitation) from the Network. In this chapter, we propose an automated irrigation system based on Arduino and wireless communication at an experimental scale within rural areas are presented. The implementation aim was demonstrating that the proposed platform could help making effective irrigation decisions with optimum water use.

The rest of the chapter is organized as followed. Section **2** briefly surveys the related works. Section **3** presents our proposed platform in precision agriculture,especially irrigation. Section **4** introduces and explains the tool developed to show real time monitoring and data visualization interface. Section **5** concludes the paper and outlines directions for future works.

RELATED WORKS

In this section, we outline some approaches of the irrigation system, (Wang, 2010) developed irrigation sensor based on a smartphone. For sensing soil moisture, the digital smartphone camerais used to process RGB to gray for ratio estimationbetween wetand dry soil areas. The wetness and dryness ratio is transmitted via a gateway to the water motor controller. A Mobile Application (APP) is developedto control sensor activity (like wakeup) and to set a sensor in sleep mode. The majority of the earlier irrigation systems do not consider the weather forecasting information (e.g., precipitation) while making irrigation decisions. This leads to a waste of fresh water, energy and loss of crop growth (due to excess water) when rain is followed immediately by the crop watering. To handle such cases, WSNs based solutions whichcan provide better decision support for irrigation by using weather forecasting information (e.g., precipitation) from the Network. Irrigated agriculture represents the bulk of the demand for water, being the first sector affected by water shortage, resulting in a decreased capacity of maintainingper capita food production. Therefore, the efficient water use in agriculture remains one of the most important agricultural challenges that modern technologies are helping to resolve (Bonfante, 2019). The authors objective in (Chakraborty, 2020) was developingan understanding onsite-specific suitability of the Mid Elevation Spray Application (MESA) and Low Elevation Spray Application (LESA) sprinkler systems in irrigating corn crop and potential water as well as energy savings. The automated irrigation system put in place has proven being feasible and cost-effective to optimize water resources for agricultural production (Gutiérrez, 2014]. The authors of this paper (Goap, 2018) propose an intelligent irrigation architecture based on IoT,as well as a hybrid approach based on deep learning in order to predict soil moisture. The proposed algorithm uses sensor data from the recent past and weather forecasts to predict soilmoisture. Due to the recent advances in sensors for the implementation of irrigation systems for agriculture specially smart farming (Prathibha, 2017; Kumar,2016 and Singh, 2017) and the evolution of IoT technologies that can be applied in the development of these systems, IoT takes a major role in the revolutions of digital transformations. We anticipate the proposal of IoT-based intelligent irrigation architecture with a machine learning approach to predict soil moisture to reach more convincing conclusions (Dahane, 2020]. The disadvantage of these approaches is that they set the maximum humidity and do not take into account the need for water exactly needed. They don't plan to save water.

Automated Irrigation Platform

The automated irrigation platform architecture (Fig.1) has been proposed to collect, transmit and process the physical parameters (soil moisture, air temperature, air humidity, water level, water flow, luminous intensity, Combustible Gas (vapors) for the crops security) of farming land along with the weather forecast information to manage the irrigation efficiently. The architecture system is designed for three layers: data layer, data processing layer and application layer, our architecture systemis shown in Fig.1.The platform that we present is based on the ideas proposed by (Gutiérrez, 2014; Goap, 2018; Vuran, 2018; Dahane, 2019), with changes made for our irrigation management platform. This proposed platform runs in three layers: data collection layer, data processing layer and the application layer.

The platform presentedabove is based on the ideas proposed by (Gutiérrez, 2014, Goap, 2018, Vuran, 2018; Hwang, 2017; Mekki, 2015; Patil, 2016; Kumar,2016; Singh, 2017 and Parra, 2018),with changes made for our application. As we can observe in Fig. 4, the devices of our platform in the first layer aim to collect data and transmit these parameters using NRFL01 radio module. It can bestored in a private database, the user (farmer) can analyze data and monitor the crops in real time. To lance irrigation planning, we need to calculate the useful reserve of water UR parameter in mm. There are multiple functions for quantifying UR from soil texture data.

The linear regression equations have the advantage of being simple and have been tested on a large sample of US soils (2,500 horizons taken from 32 states of the United States), their validation offered correlation coefficients of 0.80 and 0.87 for water content estimation at -15,000 hPa and -330 hPa respectively.

The Useful reserve (UR*i*) of a farm *i* is a key metric in our contribution. Initially, we calculate the field capacity with:

- W330 water content at -330 hPa(mm/m).
- W15000 water content at -15.000 hPa(mm/m).
- Ar: Clay content(%).
- Sa: Sand content(%).
- MO: Organic matter content.
- h: the thickness of the horizon(mm).

$$W330 = 257.6 - (2 * \text{Sa}) + (3.6 * \text{Ar}) + (29.9 * \text{MO}) \quad (1)$$

$$W15000 = 26 + (5 * Ar) + (15.8 * MO) \quad (2)$$

The useful reserve (UR) in mm is calculated for each horizon by the following function:

$$UR = (W330 - W15000) * h \tag{3}$$

- Wilting point: W1500 without irreversible dieback of plants.
- Field capacity: W330 after saturation and drying for 48h.

Figure 4. The proposed platform architecture

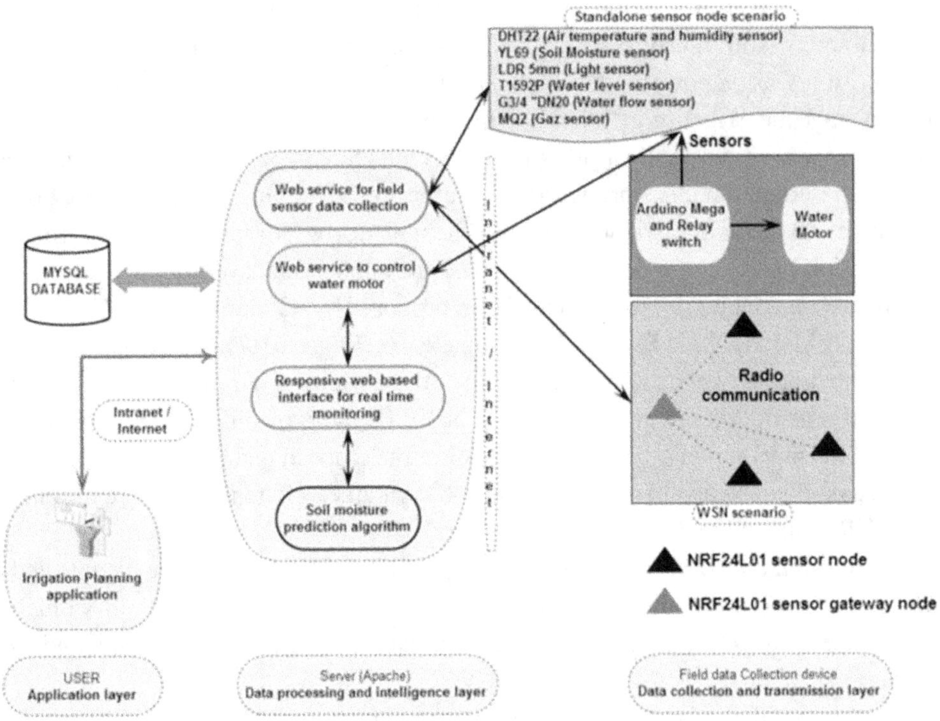

Figure 5. Field data collection device. [Legends 1: DHT22 Sensor, 2: LDR Sensor, 3: Water level Sensor, 4: Relay Switch, 5: Waterpump, 6: Power supply 9 V, 7: Waterflow sensor, 8: Soil moisture sensor, 9: LCD display, 10: MQ2 sensor, 11: LEDs, 12: Jumpers]. The mega Arduino card and NRF module are in theBox

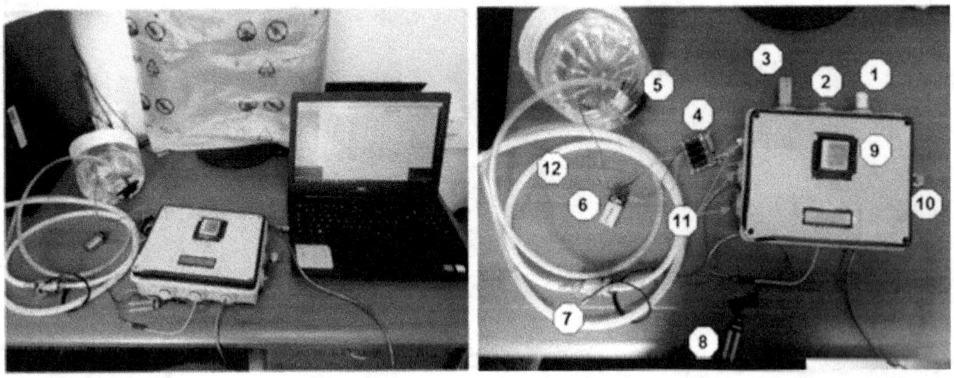

This reserve determination depends on the soil characteristics and the plant nature. Figure 13 depicts water requirement calculation for tomato crop with rooting of 30 cm. In our application,we have proposed two calculation methods, the first method calculates the RU according to the soil texture thanks to the soil structure knowledge (Clay, Limon, Sand, Organic Matter), the soil texture table that directly estimates the RU, and for the second method,we used the Rawls linear regression equations. The formula (4) is valid for relative humidity (hr)>= 50% (on the month), the formula (5) for hr<50%.

$$ETP = 0.013j(\mathrm{Rg}+50)\left(\frac{\mathrm{T}}{\mathrm{T}+15}\right) \quad (4)$$

$$ETP = 0.013j(\mathrm{Rg}+50)\left(\frac{\mathrm{T}}{\mathrm{T}+15}\right)\left(1+\frac{50-\mathrm{hr}}{70}\right) \quad (5)$$

$$B = \mathrm{ETR} - (\mathrm{Peff} + \mathrm{RFU}) \quad (6)$$

$$Peff = 0.6*\mathrm{Pmoy} - 10 \text{ if } \mathrm{Pmoy} \leq 70\mathrm{mm} \quad (7)$$

$$Peff = 0.6*\mathrm{Pmoy} - 25 \text{ if } \mathrm{Pmoy} > 70\mathrm{mm} \quad (8)$$

$$ETR = \mathrm{ETP}*\mathrm{Kc} \quad (9)$$

With:

- ETP: evapotranspiration in mm /month.
- J: number of days in the month.
- T: average temperature over the month (°C).
- hr: average relative humidity(%).
- Rg: average solar radiation (here measured) in cal / cm^2 /day.
- ETR: the actual evapotranspiration.
- RFU: it's easy reserve useful RFU.
- Peff: effective rain.
- KC: is a cultural coefficient.

Implementation and Results Analysis

This section presents the proposed platform implementation using the programming languages IDE Arduino, JAVA and the analysis of the obtained results. We update all the results,especially the prediction part of our approach according to space limitation.

The main objective of this experiment remainshow to create your own wireless network under the Arduino environment to transmit data from one or more sensors at any time of the day. Along with weather forecast information for developing an algorithm for prediction of the upcoming dayssoil moisture. The system was tested in our laboratory (RIIR) in the Oran1 university and we plan to test the platform in a real farm of the vine yard to El-Malah city, close to Oran city (Algeria). Figs. 8, 9, 10 and 11 depict the physical parameters (soil moisture, air temperature, air humidity, water level, water flow, luminous intensity, combustible Gaz (vapors), for the crops security from burning),of farming land in real time. An algorithm has been developed (see Fig. 7) to predict the soil moisture based on field sensors data and weather forecasting data, using support vector regression model and k-means clusteringalgorithm. The algorithm shows information regarding the upcoming dayssoil moisture.It also provides irrigation suggestions, based on the soil moisture defined level and predicted precipitation. In order to save water and energy, the generated information by algorithm and device is stored in MySQL Database at the server. The algorithm is discussed in detail in our future works.

Figure 6. Sequence diagram of the system

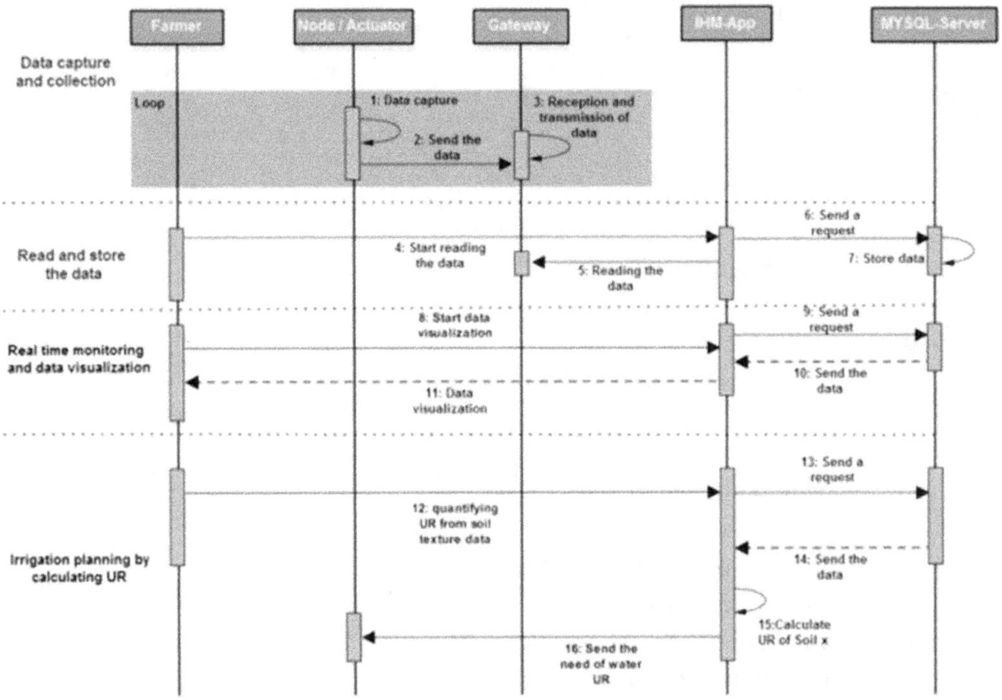

In order to understand how our platform works,we model it by representing conceptually all communications involved in its operation. To do this, we use the UML (Unified Modeling Language)modeling approach to model configuration tasks and the information exchange between different communicating objects that are involved in the test environment used in our project context. Sequence diagrams allow us modeling these exchanges.This diagram (see Figure 6) depicts the various data collection exchange, storageanddisplay messages that can take place between communicating objects: sensor/actuator, Workstation (Gateway, Java application, MySQL server) and the intermediary (the farmer).

Our goal remains the creation of an autonomous platform for environmental data measurements, such as air temperature, air humidity, solar radiation, soil humidity ...etc, in real time. In this work, a sensor/actuator node has the collecting data role and sending them to the gateway, which is connected to a serial port on the PC. A desktop application developed with Java allows us reading received data via this serial port, and storing them in a MySQL-type database, then visualizing them through graphical curves

Figure 7. The Irrigation planning flowchart

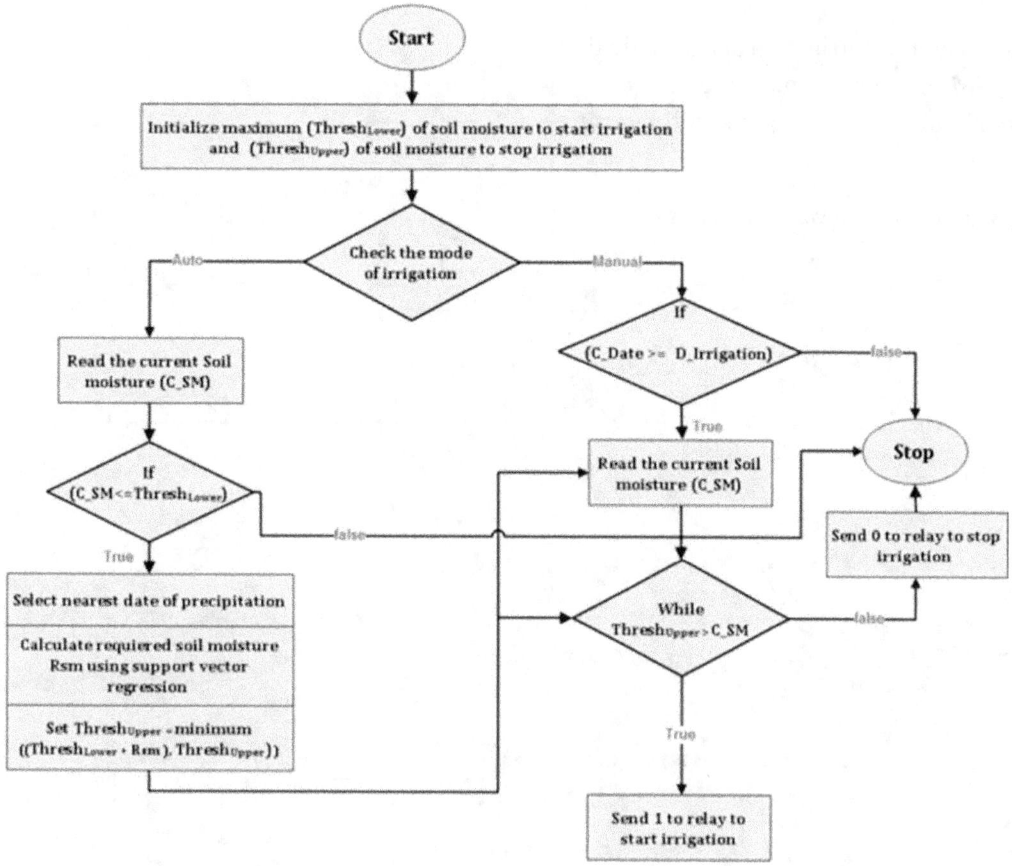

and calculating the need for soil water with the different defined methods. After that, The application sends this need to the sensor/actuator node to perform the irrigation process. The main steps related to the operation of the platform are presented in Figure 7.

These results are preliminaries because the test was made inside the laboratory, so we plan testing the platform in a real vine farm in order to demonstrate the proposed platform performance. All data were uploaded each hour to the web server for remotesupervision.

For instance, real time experimentation datais shown in Fig.11. The first graph in Fig.8 shows temperature value analysis. The vertical bars indicate automated irrigation periods triggered by temperature when soil temperature was above the threshold value (30%).

In figure 12 we evaluate the irrigation planning algorithm for real-time monitoring and decision support system, which shows the current soil moisture recorded by the sensor.

precipitation information will help user/farmer in planning/scheduling of optimum irrigation by calculating the real useful reserve (UR). If the moisture has increased to 30%, irrigation will start automatically.

The variables used in the algorithm for irrigation planning:

- **Threshupper:** selection of minimum soil- moisture required to maintain crop growth.
- **Threshlower:** selection of minimum soil- moisture required to maintain crop growth.

- **C_Date:** read precipitation information of upcoming days and select the nearest precipitation date.
- **D_Irrigation:** irrigation at a specific date.
- **C_SM:** current soil moisture.
- **Rsm:** calculate the required soil moisture.

Figure 8. Temperature value analysis (°C)

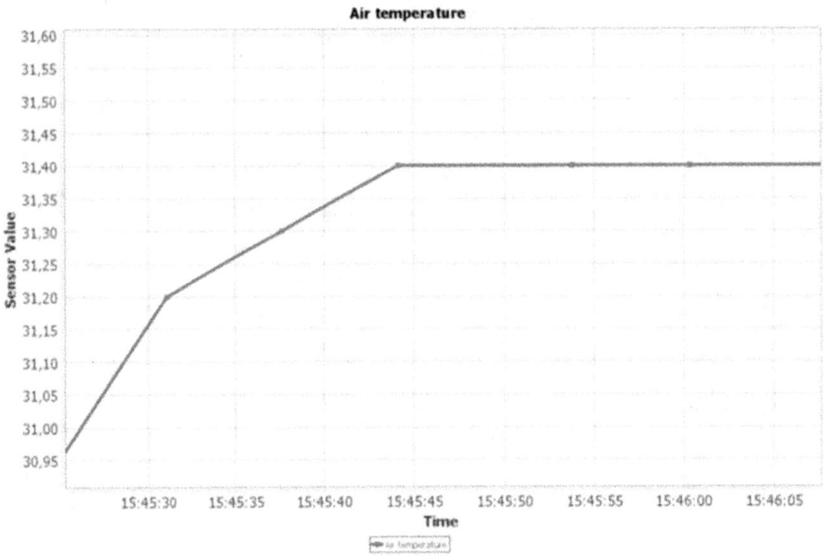

Figure 9. Humidity value analysis (%)

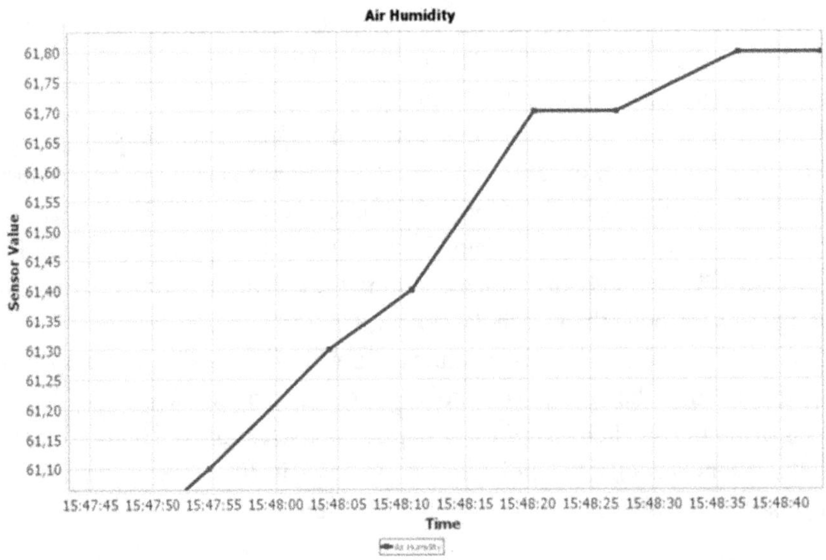

Figure 10. Moisture value analysis (%)

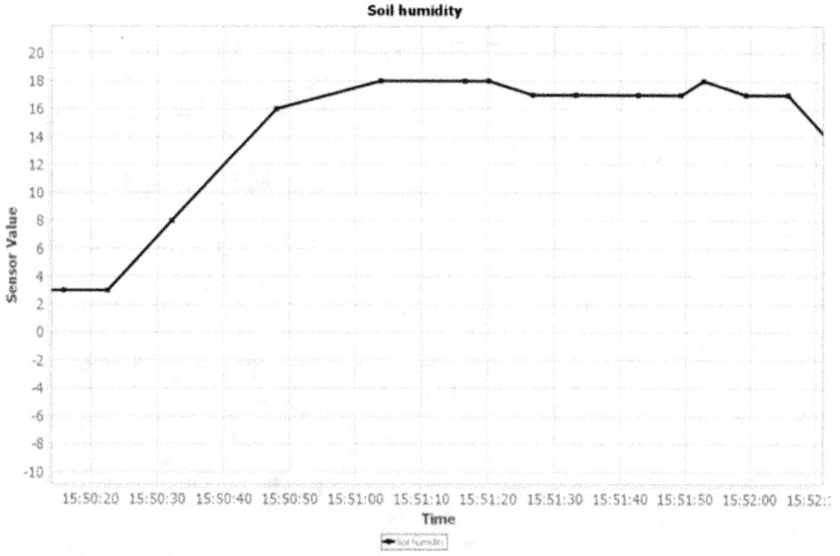

Figure 11. Real time monitoring and data visualization

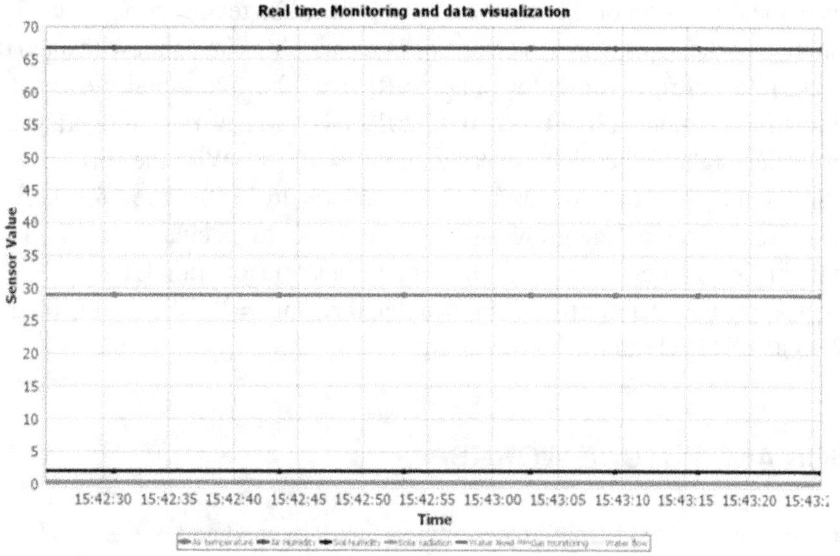

Table 1. Growth-stage-specific K_C for tomato

Crop	Clay $_{\%}$	Silt $_{\%}$	Sand $_{\%}$	Organic matter $_{\%}$	Rooting depth $_{cm}$	Crop coefficients (K_C)		
						K_c ini	K_c mid	K_c fin
Tomato	28,0	55,0	17,0	1,75	30-60	0,2	0,8-0,9	0,8- 0,7

Figure 12. Analysis of irrigation launching

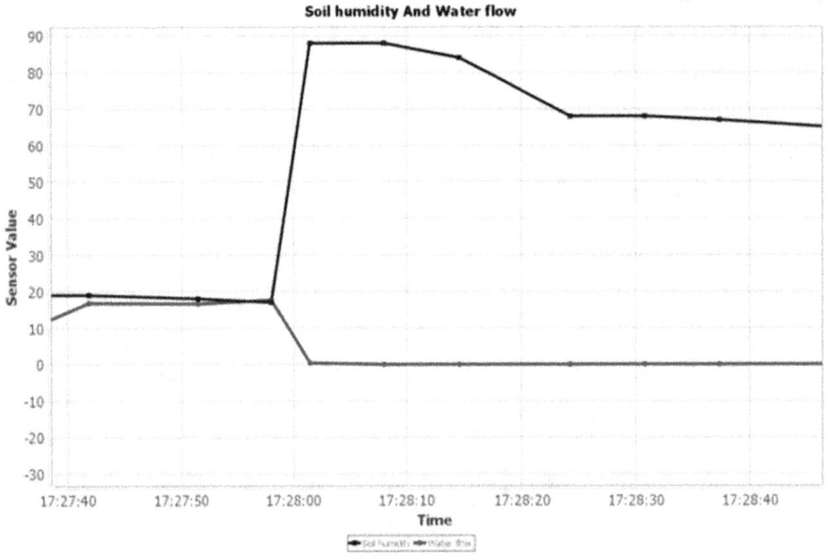

Growth-stage-specific K_C for potato and tomato used in this study are mentioned bellow the table 1. Figure 13 depicts the estimation of the reserve of a soil from the texture triangle is: RFU = 12.20, for visual representation of the dependence the texture class was: slity clay loam with a red dot. The estimation of the reserve of a soil from Rawls formulas is: RFU = 12.0. We can also estimate of actual ETR evapotranspiration for the month 2019-03-22 until 2019-04-22. T: average temperature over the month = 20 (° C). Hr: average relative humidity over the month= 46 (%). With the surface = 10 hectares and Kc ini = 0.8. So the monthly water requirement available = 76.33, the need for daily water available =2.46. We have also considered the growth-stage-specific K_C to calculate the UR, It has been found that for a growth-stage-specific K_C fin = 0.8 the actual evapotranspiration ETR = 9.83 (see Figure 13). Similarly, ETR is also calculated for the K_C ini = 0.2 and K_C mid = 0.9 the results obtained are successively ETR=2.45 and ETR=11.06.

CONCLUSION AND FUTURE WORKS

In this chapter, a new automated irrigation system was implemented and presented. It has been deemed to be feasible and cost-effective for optimizing water resources for agricultural production. This irrigation system allows cultivation in places with water scarcity thereby improving sustainability. . This system main experimentation objective is collectinga farming land physical parametersusing sensors, and usingthe collected data along with weather forecast information for developing an algorithm for predicting soil moisture of the upcoming days using machine learning and open sources technologies. Besides the monetary savings in water use, the importance of this natural resource preservationjustifies the use of this kind of irrigation system. We anticipate the proposal of IoT-based intelligent irrigation founded on new EDGE-Fog-IoT-Cloud platform with a machine learning approach, to predict soil moisturein order to reach more convincing conclusions.

Figure 13. Calculation of water needs for tomato cultivation with rooting of 10 cm

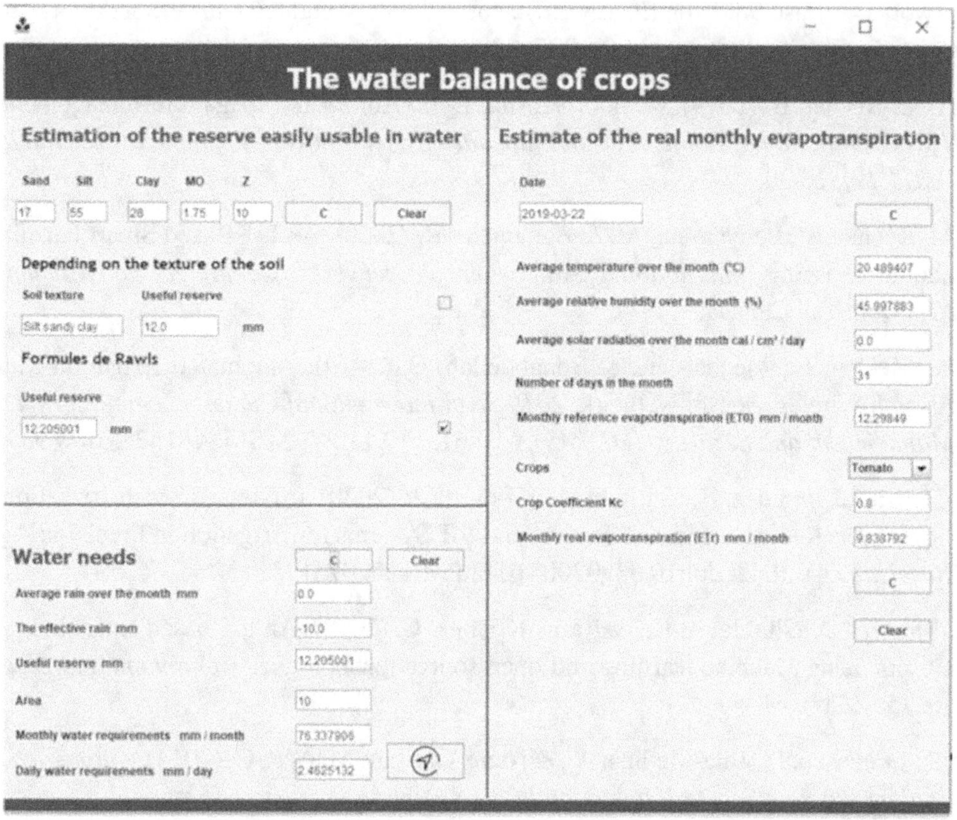

ACKNOWLEDGMENT

This work has been supported scientifically by RIIR laboratory/LAPECI laboratory. This paper is supported by the PRIMA programme under grant agreement No 1821, project WATERMED4.0. The PRIMA programme is supported by the European Union.

REFERENCES

Bonfante, A. (2019). LCIS DSS—an irrigation supporting system for water use efficiency improvement in precision agriculture: A maize case study. *Network Protocols and Algorithms.*, *176*, 102646.

Chakraborty, M., Khot, L. R., & Peters, R. T. (2020). Assessing suitability of modified center pivot irrigation systems in corn production using low altitude aerial imaging techniques. *Information Processing in Agriculture*, *7*(1), 41–49. doi:10.1016/j.inpa.2019.06.001

Dahane, A., Berrached, N. E. (2019). *Mobile wireless and sensor, A clustering algorithm for energy efficiency and safety*. Apple Academic Press.

Dahane, A., Berrached, N. E., & Loukil, A. (2015). A Virtual Laboratory to Practice Mobile Wireless Sensor Networks: A Case Study on Energy Efficient and Safe Weighted Clustering Algorithm. *Journal of Information Processing Systems*, *11*(2), 205–228.

Dahane, A., & Kechar, B. (2019). Remote Sensing Data: Automated Irrigation Management Platform Using a WSNs. *International Conference on Embedded Systems in Telecommunications and Instrumentation (ICESTI'19)*, 28-30.

Dahane, A., Kechar, B., Benyamina, A., & Benameur, R. (2020). An IoT Based Smart Farming System Using Machine Learning. *International Symposium on Networks, Computers and Communications (ISNCC'20)*, Montreal, Canada.

Dahane, A., Kechar, B., Meddah, Y., & Benabdellah, O. (2019). Automated Irrigation Management Platform using a Wireless Sensor Network. *2019 Sixth International Conference on Internet of Things: Systems, Management and Security (IOTSMS)*, 610-615. 10.1109/IOTSMS48152.2019.8939170

García, L., Parra, L., Jimenez, J. M., Lloret, J., & Lorenz, P. (2020). IoT-Based Smart Irrigation Systems: An Overview on the Recent Trends on Sensors and IoT Systems for Irrigation in Precision Agriculture. *Sensors (Basel)*, *20*(4), 1042. doi:10.339020041042 PMID:32075172

Goap, A., Sharma, D., Shukla, A. K., & Rama Krishna, C. (2018). An IoT based smart irrigation management system using Machine learning and open source technologies. *Computers and Electronics in Agriculture*, *155*(May), 41–49.

Gutiérrez, J., Francisco, J., Villa-Medina, A., & Nieto-Garibay, M. Á. P.-G, (2014). *Automated Irrigation System Based on Wireless Sensor Network and GPRS Module*. Academic Press.

Hwang, S. (2017). Monitoring and Controlling System for an IoT Based Smart Home. *Int. J. Control Autom.*, *10*(2), 339–348. doi:10.14257/ijca.2017.10.2.28

Kumar, H. S., & Kusuma, S. (2016). Automated Irrigation System Based on Wireless Sensor Network and GPRS Module. *Int. Res. J. Eng. Technol.*, *3*(4), 148–151.

Mekki, M., Abdallah, O., Amin, M. B. M., Eltayeb, M., Abdalfatah, T., & Babiker, A. (2015). Greenhouse monitoring and control system based on wireless Sensor Network. *Proc. - 2015 Int. Conf. Comput. Control. Networking, Electron. Embed. Syst. Eng. ICCNEEE*, 384–387. 10.1109/ICCNEEE.2015.7381396

Nemali, K. S., & Van Iersel, M. W. (2006). An automated system for controlling drought stress and irrigation in potted plants. *Scientia Horticulturae*.

Parra, L., Rocher, J., García, L., Lloret, J., Tomás, J., Romero, O., Rodilla, M., Falco, S., Sebastiá, M. T., Mengual, J., González, J. A., & Roig, B. (2018). Design of a WSN for smart irrigation in citrus plots with fault-tolerance and energy-saving algorithms. *Network Protocols and Algorithms.*, *10*(2), 95–115. doi:10.5296/npa.v10i2.13205

Patil, A., Beldar, M., Naik, A., & Deshpande, S. (2016). Smart farming using Arduino and data mining. *Proc. 10th INDIACom; 2016 3rd Int. Conf. Comput. Sustain. Glob. Dev. INDIACom 2016*, 1913–1917.

Prathibha, S.R., Hongal, A., & Jyothi, M.P. (2017). IOT Based Monitoring System in Smart Agriculture. *Proc. - 2017 Int. Conf. Recent Adv. Electron. Commun. Technol. ICRAECT*, 81–84.

Rahman, L. (2016). *Precision agriculture in the World and its prospect in Bangladesh Research in Agriculture, Livestock and Fisheries*. Academic Press.

Singh, P., & Saikia, S. (2017). Arduino-based smart irrigation using water flow sensor, soil moisture sensor, temperature sensor and ESP8266 WiFi module. *IEEE Reg. 10 Humanit. Technol. Conf. 2016, R10-HTC 2016 - Proc.*

Vuran, M. C., Salam, A., Wong, R., & Irmak, S. (2018). Internet of underground things in precision agriculture: Architecture and technology aspects. *Ad Hoc Networks*, *81*, 160–173. doi:10.1016/j.adhoc.2018.07.017

Wang, X., & Qi, Q. (2011). Design and realization of precision agriculture information system based on 5S. *Proc. - 19th Int. Conf. Geoinformatics, Geoinformatics*, 2–5.

Wang, X., Yang, W., Wheaton, A., Cooley, N., & Moran, B. (2010). Wheaton, Cooley, A.N., and Moran, B., Efficient registration of optical and IR images for automatic plant water stress assessment. *Computers and Electronics in Agriculture*, *74*(2), 230–237. doi:10.1016/j.compag.2010.08.004

Chapter 7
On the Use of Artificial Intelligence Techniques in Crop Monitoring and Disease Identification

Muzaffer Kanaan
Erciyes University, Turkey

Rüştü Akay
Erciyes University, Turkey

Canset Koçer Baykara
Turkish Grain Board (TMO), Turkey

ABSTRACT

The use of technology for the purpose of improving crop yields, quality and quantity of the harvest, as well as maintaining the quality of the crop against adverse environmental elements (such as rodent or insect infestation, as well as microbial disease agents) is becoming more critical for farming practice worldwide. One of the technology areas that is proving to be most promising in this area is artificial intelligence, or more specifically, machine learning techniques. This chapter aims to give the reader an overview of how machine learning techniques can help solve the problem of monitoring crop quality and disease identification. The fundamental principles are illustrated through two different case studies, one involving the use of artificial neural networks for harvested grain condition monitoring and the other concerning crop disease identification using support vector machines and k-nearest neighbor algorithm.

DOI: 10.4018/978-1-7998-5000-7.ch007

INTRODUCTION

Agricultural production and trade is a major contributor to the global economy, accounting for 1.6 trillion USD in 2016, based on data from the Food and Agriculture Organization (FAO) of the United Nations (FAO, 2018). Given this huge trade volume, it is clear that producers of agricultural products will need to offer high-quality products to the marketplace. In this context, one typically has to contend with two main problems.

The first problem is that there is typically a delay between the time agricultural products are harvested and the time they end up in the hands of consumers. This delay can be anywhere between a few days and a few months. This implies that the end products will need to be stored in special storage areas to ensure that the products will stay in peak quality prior to being shipped to end customers. Many kinds of agricultural products, such as grains, fruits and vegetables, are typically stored in this manner. While the products are being stored, it is possible that their quality will be degraded due to suboptimal storage conditions, the onset of disease-producing agents (such as bacteria and fungi), as well as insect and/or rodent infestation. It is critical that this possible degradation in product quality be monitored and detected. It is equally crucial that any products with degraded quality be immediately withdrawn from circulation, so as to avoid any danger to human health. The impact of this problem is profound, as illustrated by the fact that up to approximately 5.5% of the total weight of harvested grains, such as rice, can be lost due to pest damage, making the harvested product unfit for consumption, or even as seed(Togola et al., 2013). From a purely economic perspective, the impact is even more astounding, with some studies reporting that approximately one-third of the food produced (which amounts to some 1.3 billion tons) is lost globally during postharvest operations every year at an estimated cost of some 1 trillion USD (FOOD, 2016; Kumar & Kalita, 2017).

Another typical problem is that a disease-producing agent will attack agricultural products, prior to a harvest, making it unfit for consumption. Previous studies in the open literature indicate that approximately 25% of total harvested produce in the US alone is lost in this manner, with the loss figures in tropical areas reaching a staggering 50% (Wilson & Wisniewski, 1989),(Savary, Ficke, Aubertot, & Hollier, 2012). Again, it is critical that these agents be detected and that prompt action (such as the application of insecticides or fungicides) be taken in a prompt manner in order to avoid losing an entire harvest.

Today, both problems can be effectively addressed thanks to advances in a number of technological fields, namely sensor technology, low-power embedded systems, wireless networking, and artificial intelligence. Advances in sensor technology are enabling suboptimal storage conditions and disease-producing agents to be detected with a high-degree of precision (West, Canning, Perryman, & King, 2017), (M. Zhang, Qin, Liu, & Ustin, 2003). Advances in low-power embedded systems, particularly microcontrollers, are making it possible to integrate these sensors into powerful computing infrastructures that can effectively process the data collected by the sensors(Eisenhauer, Rosengren, & Antolin, 2010), (Atmojo, Salcic, Kevin, Wang, & Park, 2015). Advances in wireless networking, particularly wireless sensor networks and machine-to-machine (M2M) communications, are enabling storage infrastructures and agricultural fields typically spanning many square kilometers to be monitored at a distance in a cost-effective manner(Ojha, Misra, & Raghuwanshi, 2015), (Yu, Wu, Han, & Zhang, 2013). Finally, artificial intelligence techniques, particularly machine-learning algorithms, are enabling producers of agricultural products to make sense out of the vast quantities of data collected, to quickly arrive at conclusions and take corrective action where necessary(Park, Im, Jang, & Rhee, 2016).

In this chapter, we aim to illustrate how the aforementioned technologies can help with effective monitoring of harvested grain conditions, and also with disease identification on agricultural products. With this backdrop, the rest of this chapter is organized in three parts. In part 2, we illustrate how effective grain condition monitoring can be accomplished using wireless sensor networks and machine learning techniques, with a particular focus on neural networks. We do this via a case study devoted to the actual development of an actual system prototype for the Turkish Grain Board (TMO). Part 3 is devoted to addressing the problem of disease identification on crops, through image processing via machine learning techniques. We end the chapter with some conclusions and suggestions for future work in the field.

HARVESTED GRAIN CONDITION MONITORING USING WIRELESS SENSOR NETWORKS

Wirelesss Sensor Networks (WSNs) are generally defined as network of low-power sensor nodes, and control nodes. They have been, and continue to be used, in many diverse fields (such as environmental monitoring, military and public safety, and healthcare), of which agriculture is a prime example, and the focus of this chapter (Akyildiz, Su, Sankarasubramaniam, & Cayirci, 2002). Within a specific agriculture focus, the use of WSNs aims to achieve certain concrete goals. These can be summarized as: reduction in required labor, ensuring greater efficiency in the farming process (thus ensuring farmers' welfare), obtaining a harvest that is higher in both quality and quantity, as well as proliferate environment-friendly farming practices (by detection of overuse of pesticides or fertilizers, for example) (Dilay, Soy, & Bayrak, 2012), (Y. Zhang, 2011).

From the perspective of both farmers and final consumers of harvested grain products (such as corn, rice, wheat etc.) there are two main issues to be considered. The first is the effective use of the harvested product, and the second is sustaining its quality from the time it is harvested to the time it gets in the hands of consumers. In Turkey, as in many other countries, these objectives are within the purview of a public sector organization, known as the Turkish Grain Board (TMO). The TMO specifically aims to achieve these objectives by operating different channels to buy the harvested product from farmers, as well as to store the harvested product and sell it. In this manner, market stability for these products is maintained.

The TMO currently operates with a storage capability of approximately 4 million metric tons, specifically devoted to harvested grain and distributed throughout Turkey. This storage capability is mostly in the form of enclosed storage structures, such as vertical silos. However, in some regions building such storage structures has been found to be either cost-prohibitive or unfavorable regional conditions. In these cases, open-air stacks have also been deployed as an alternative. However, open-air stacks expose the harvested grain to harmful environmental elements such as rain, as well as insect and/or rodent damage. As an alternative solution, the TMO has deployed the *Silobag* system for open-air grain storage. The *Silobag* system uses large, hermetically sealed bags for storing grain in an open-air environment.

Of course, a hermetically sealed bag like the Silobag system cannot be expected to protect the stored grain from all detrimental environmental elements. Therefore, a system to monitor the status of the stored grain is critical. The currently deployed Silobag system accomplishes this function solely via measurements of CO_2 concentration (manually taken by TMO personnel onsite using handheld terminals). Increased CO_2 concentration values are associated with increased rodent, insect or microorganism activity in the stored grain pile. CO_2 measurements from a particular grain pile are sent to a web-based database system, which then classifies the pile as "safe", "risky", or "dangerous". In the course of making this determination, the monitoring system also takes into account such variables as the relative humidity, and the general physical condition of the silobag (e.g. whether the silobag is torn or damaged etc.). Figure 1 below illustrates the main features of the current grain condition monitoring system.

Figure 1. Grain condition monitoring for the currently deployed Silobag system

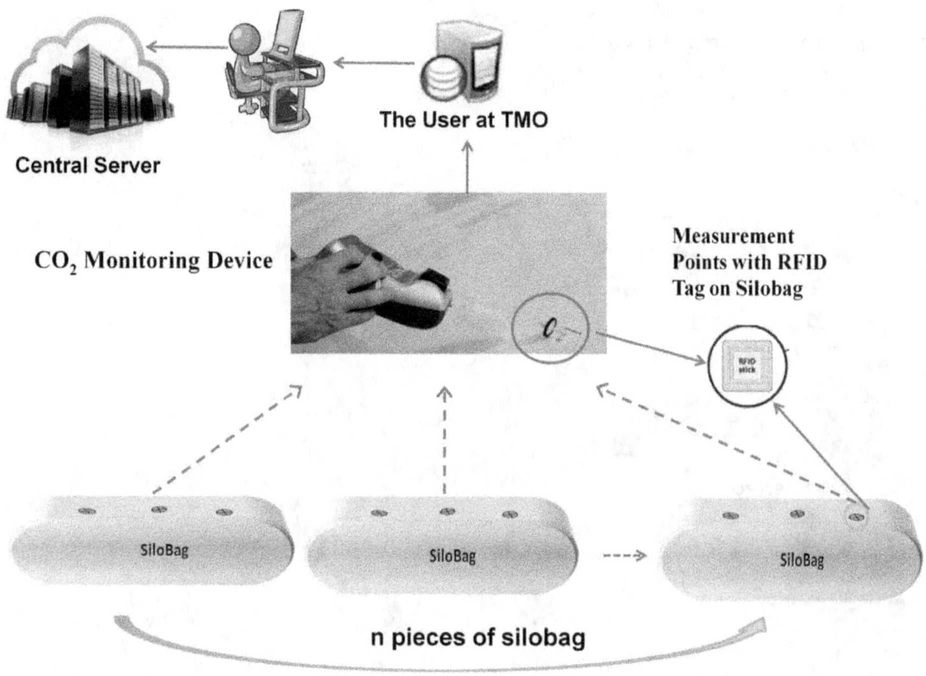

Grain condition monitoring with the current Silobag system, as explained above, has some problems, which we have attempted to rectify through our work. The most important problem we have noted is that determination of grain condition is solely on the basis of CO_2 concentration measurements. This methodology implicitly assumes that the temperature and relative humidity within a grain pile stays constant, which is far from realistic in actual field scenarios. A better method would be to take into account temperature, relative humidity, as well as CO_2 concentration values for determining the condition of stored grain. Another problem with the current monitoring system is that CO_2 measurements are manually taken by TMO personnel. This measurement method is inherently error-prone, not to mention time- as well as resource-intensive. Considering the fact that many silobags are deployed in out-of-the-way, rural areas (which will require considerable fuel consumption so personnel can get to a particular silobag), and the general trend towards environmentally friendly farming practices (such as the reduction of the carbon footprint), it is obvious that there is an opportunity for further optimization of this monitoring system.

Our solution to the aforementioned problems comes in the form of an Improved Grain Condition Monitoring System (I-GCMS). In this system, WSNs are utilized continuous monitoring of stored grain condition from a distance, thus removing the need for personnel to travel to distant locations to take measurements, and enhancing measurement accuracy quality by removing the error-prone human element from the equation. The proposed system also takes into account temperature and relative humidity within a grain pile. Machine learning algorithms, specifically artificial neural networks (ANNs), are utilized to classify the stored grain condition. The results of our experiments with the current prototype of the I-GCMS indicate that stored grain condition can be classified with approximately 99% accuracy with this system. The structure of I-GCMS is pictorially depicted in Figure 2 below.

Figure 2. The general structure of I-GCMS

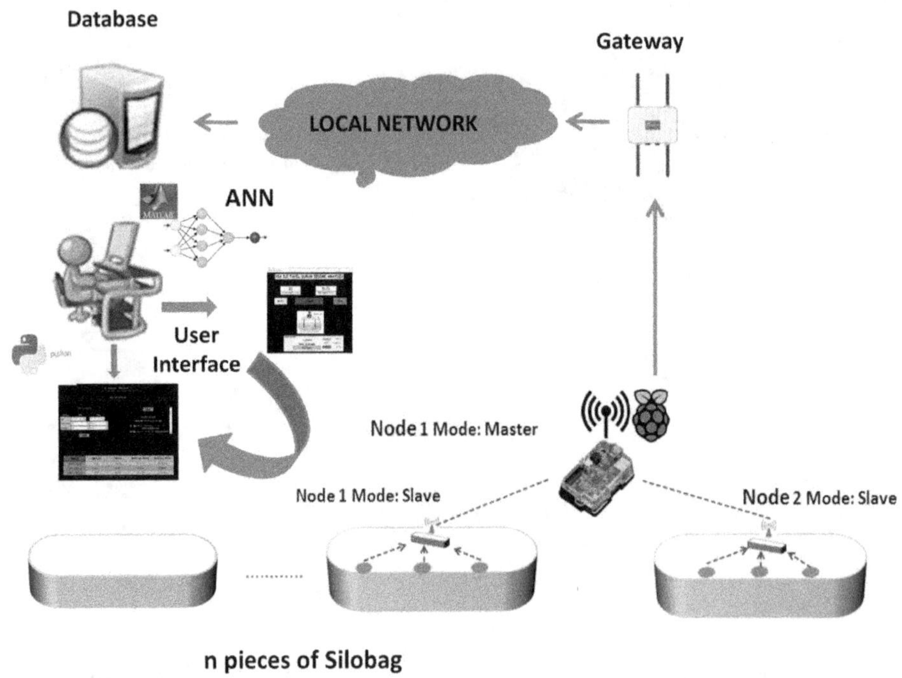

In the next few subsections, we explain the general structure of I-GCMS (including the WSN component), give some details regarding the ANN algorithms used for classifying stored grain condition, and give some performance results obtained from the prototype system.

The General Structure of I-GCMS

We are now ready to explain the general structure of the I-GCMS. Referring to Figure 2, there are N sensor modules for the N silobags, which measure temperature and relative humidity measurements within each silobag. Associated with each sensor module is a combined 3G/Wi-Fi module that relays the measurement data periodically to a gateway and, from there, to a database server via the local network. Within the database system, ANN algorithms are run in the background. The ANNs come up with a classification of the stored grain in each silobag on the basis of periodically received measurements. Classification results are presented to the user via the user interface.

We have adopted a modular, layered architecture perspective for the design of the I-GCMS, with each layer specializing in a certain function. This is illustrated in Figure 3 below. The system consists of three layers, namely the hardware layer, software layer and user layer. The hardware layer basically consists of the sensors that make temperature and relative humiditymeasurements, as well as the WSN component to relay the measurements to the central database system. The software layer then processes the measurement data collected by the hardware layer via ANNs and comes up with a classification of the grain condition. The user layer consists of a Graphical User Interface (GUI), which presents the results in a user-friendly manner.

Figure 3. Layered architecture of I-GCMS

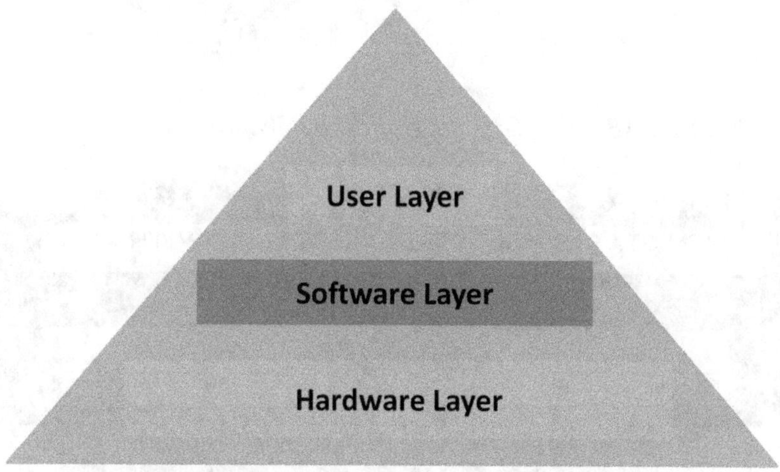

As illustrated in Figure 2, each silobag has a network node associated with it. Each of these nodes consist of the temperature and relative humidity sensors to make measurements in that grain pile, as well as the radio transceivers to relay the data. We have used the SHT-10 sensor module (see Figure 4) which could make both temperature and relative humidity measurements. The whole network node is controlled by a Raspberry Pi 3 microcontroller, with one node being designated as a master node and the others being designated as slave nodes. The master node serves as the interface to forward the data on to the server. At the server side, the database archives all the data received from the sensors. This database has been implemented through MySQL, the rest of the server side code was implemented using Ruby 2.3, running on a Windows 10 machine employing the Ruby on Rails 5.0 framework.

All that remains to be discussed is the ANN structures to accomplish the actual grain condition classification. For the purposes of our prototype, we chose the MATLAB™ environment for ease of implementation. The details of the ANN structures chosen for this study, as well as the issue of training the network will be addressed in the following subsections. We have also designed a MATLAB-based GUI to make the system more user-friendly. The GUI simply requires that the user input a temperature and relative humidity value, and returns a classification as "safe", "risky", or "dangerous", as seen in Figure 5.

Figure 4. SHT-10 temperature and moisture sensor

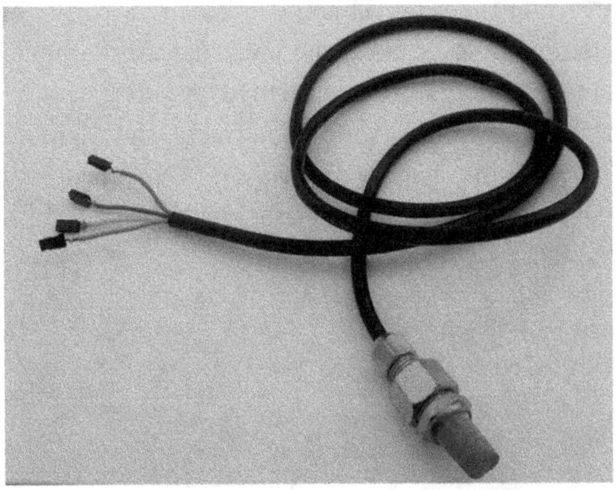

Figure 5. A view of the MATLAB-based GUI for grain classification

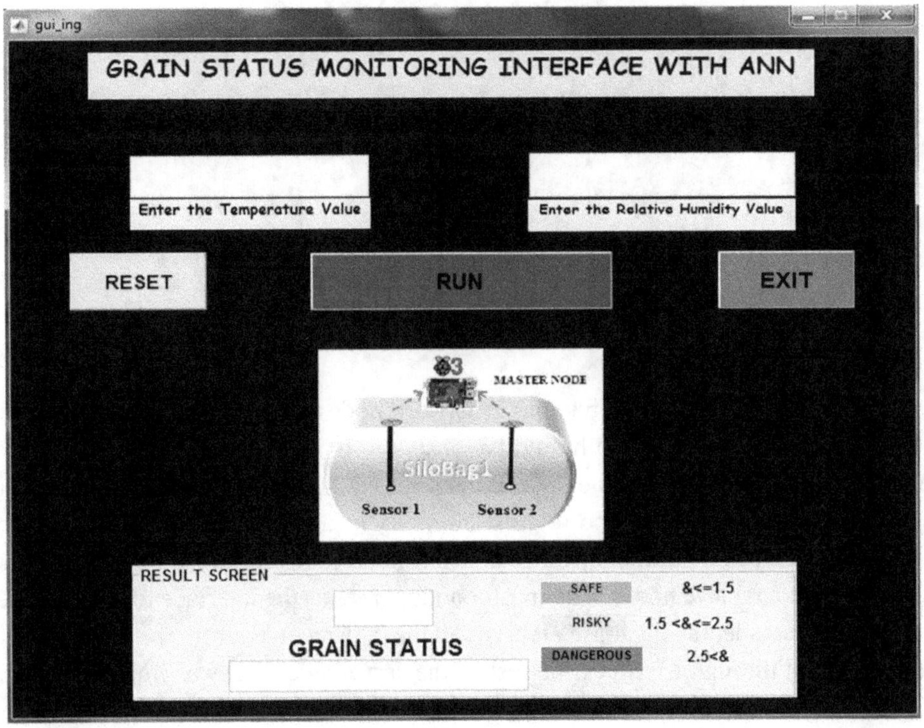

ANN-Based Classifier Structures for I-GCMS

ANNs have been widely used for classification and data analysis problems in many different applications, and the field of agriculture is no exception (Haykin, 1994), (Panda, Ames, & Panigrahi, 2010). Within an agriculture context, it is possible to find ANN applications for problems such as classification of different grain types using image processing and ANNs (Visen, Paliwal, Jayas, & White, 2002), classification of different varieties of durum wheat (Taner, Tekgüler, & Sauk, 2015), and improving classification accuracy by exploiting the color and formal characteristics of fruits(Li & Zhu, 2011), (Majumdar & Jayas, 2000).

When using ANNs to solve any classification problem, it is first necessary to train the network, i.e. update the synaptic weights between the neurons forming the network. For the I-GCMS, this entails first creating a database of representative temperature and relative humidity data for each of the grain condition classes ("safe', "risky" and "dangerous"). The training data representing each class is appropriately labeled (so the network knows which data belongs to which class). The network is then trained with this data, so it will be able to recognize a given class when presented with unfamiliar data. In ANN literature, this training methodology is known as supervised learning (Haykin, 1994).

The Training Database

The first step in the creation of the training database was to come up with three prototype silobags (one representing each class) that could hold approximately 2.5 kg of grain. Care was taken to ensure that the starting point for all three classes was identical, i.e. that the exact type of grain with exactly the same base temperature and relative humidity was used. These base temperature and relative humidity values were measured by qualified personnel using calibrated equipment.

The data representing the "safe" class was collected from one of the three piles with a nominal relative humidity value of 13.8% (generally considered acceptable for safe storage, per TMO standard procedures)(Baykara, 2018). To create the "risky" and "dangerous" datasets, it was necessary to "spoil" the grain in a controlled manner. For the "risky" dataset, this meant adding a certain amount of water to the associated grain pile, until the relative humidity reached 17% (generally considered risky for grain storage, per TMO processes). As for the creation of the "dangerous" data set, a small amount of grain from the associated pile was doused with water and then allowed to germinate; the grain would normally rot during the germination activity. The rotten grain was then added back to the pile, which would cause the rest of the pile to rot. Following this process, sensor modules were embedded in each pile, and all the bags were then hermetically sealed (just as they would be for an actual silobag in the field; see Figure 6 below). Relative humidity and temperature data were collected from each silobag over a period of approximately three months. Approximately 100.000 readings were logged from each silobag, which resulted in about 300.000 data points in total.

Figure 6. Prototype silobags and the measurement setup

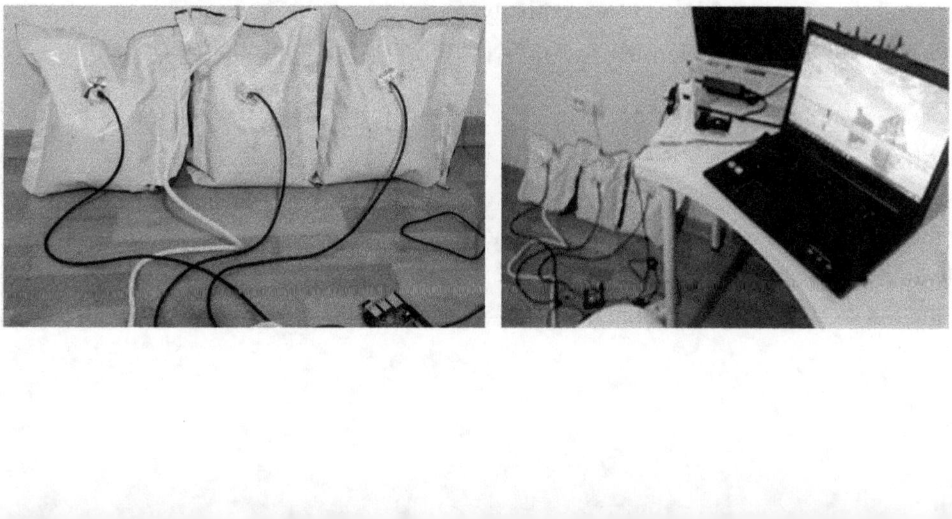

The Training Process

The process of training the network was carried out in MATLAB™ environment, and specifically the Neural Network Toolbox, as it allowed us to do rapid prototyping of the system in a flexible manner. We have considered the use of three popular training algorithms for this purpose, namely the Gradient Descent with Momentum (GDM), Levenberg-Marquardt (L-M) and Scaled Conjugate Gradient (SCG) algorithms. A common ANN topology, namely the feedforward backpropagation network, was selected. As mentioned above, a grand total of 300.000 total measurements were collected (100.000 for each class). This whole data set was further partitioned into the training data set (80% of the total data, or 240.000 data points), and test data (20% of the data, or 60.000 data points).

The choice of the feedforward backpropagation network as the ANN topology warrants some discussion. This choice has not been made arbitrarily, but is the result of a literature review, which has indicated that this network topology indeed exhibited better performance than the alternatives when applied to similar classification problems in the agriculture space. Specifically, the performance of the backpropagation network has been compared against nonparametric classifiers for the problem of classifying five different types of grain kernels from different regions of Canada; and the backpropagation network was found to exhibit superior performance(Paliwal, Visen, Jayas, & White, 2003). A more extensive study devoted to classifying four different varieties of rice compared the performance of the backpropagation network against other machine-learning approaches (support vector machines, nearest neighbor algorithm and the Bayesian classifier) and the backpropagation network was again observed to outperform the other techniques(Singh & Chaudhury, 2016). Yet another study investigated the use of the backpropagation network for classifying different types of wheat on the basis of nine different physical features (such as diameter, kernel weight etc.) (Taner et al., 2015). The performance of the backpropagation network was compared against three different types of multilayer perceptron and the backpropagation network once again exhibited superior performance. All this can be considered sufficient basis for our fundamental working hypothesis that the backpropagation network would be a good choice for the I-GCMS.

The next step in the training process was to select optimal values of the training parameters for all three algorithms. To come up with these values, we have followed a systematic, iterative process which we describe as follows. First, we assigned values to all the parameters that are common to all three algorithms, namely the adaptation learning function, performance function, number of layers, transfer function, and the number of epochs. In this process, care was taken to ensure that the values of all parameters, except for the parameter to be optimized, were kept constant. To determine the optimal value of a given parameter, the required classification results and the results produced by the network were compared, and the best value was determined based on the error rate performance (i.e. the number of misclassifications). The ANN was programmed to give an output of '1', '2', or '3' for the safe class, risky class, and the 'dangerous' class, respectively. Once this process was completed for the common parameters, algorithm-specific parameters were optimized in a similar manner. The full parameter list is given in Table 1.

Classification Results

As mentioned in the previous section, 20% of the total number of data points (60.000 in total) were used for testing the classification performance of the network. To do this, the results produced by the network, and the expected classification results based on the test data set were compared. Note that, due

Table 1. Optimal parameter values for the three training algorithms

	Training Algorithm		
	GDM	**LM**	**SCG**
Network Type	Feed Forward Back Propagation	Feed Forward Back Propagation	Feed Forward Back Propagation
Training Function	Learngdm	Learngdm	Learngdm
Activation Function	Purelin	Purelin	Purelin
Number of Nerons	12	12	12
Hidden Layer Number	3	3	3
Epoch Number	2500	2500	2500
Momentum Coefficient	0,6	not applicable	not applicable
Learning Coefficient	0,4	not applicable	not applicable
Mu value	not applicable	0,00002	not applicable
Mu_dec value	not applicable	0,0005	not applicable
Mu_inc value	not applicable	10	not applicable
Mu_max value	not applicable	10000000000	not applicable
Sigma value	not applicable	not applicable	5,00E-01
Lambda Value	not applicable	not applicable	5,00E-01

to the nonlinear nature of the ANN structure, the output would not always be exactly 1, 2 or 3; there would always be a small amount of deviation (reflecting a small amount of ambiguity characteristic of any real-life decision-making problem) as can be observed from the sample results of Table 2. For each data point, the difference between the expected classification result and the result produced by the ANN were computed, and the result was expressed as a percentage of the original expected value. Subtraction of this percentage from 100% resulted in the *success rate* with which the ANN could make an accurate classification. This success rate was adopted as the principal performance criterion in this study.

For all three algorithms, an average value of the success rate (averaged over all test data points) was calculated, and the results are given in Table 3. It is clear that all three algorithms can classify the grain condition with approximately 99% accuracy, but GDM still has the best performance among the three algorithms, although the edge it has over the other algorithms is decidedly small.

For a more graphical perspective on these results, we refer the reader toFigure 7, where the expected classification results (based on the test data set), and the results produced by the network are superimposed. Based on this figure, the ANN can classify the 'safe' data set (class 1) and the 'dangerous' data set (class 3) with essentially no uncertainty. However, for the 'risky' data set (class 2) that lies in the middle between these two extremes, the network produces results with more uncertainty (evidenced by the substantially higher deviation between the expected results and the results produced by the ANN). Part of the reason for this behavior might be that the network simply needs more data to characterize the 'risky' class.

Table 2. Sample classification results

DataRecordNo	Temperature	Moisture	ExactValue	TestResult	Difference	SuccessRate
1	22,37	47,53	1	1,00	0,00%	100,00%
2	22,24	47,53	1	1,00	0,00%	100,00%
3	22,23	47,53	1	1,00	0,00%	100,00%
4	22,24	47,50	1	1,00	0,00%	100,00%
5	22,25	47,53	1	1,00	0,00%	100,00%
20000	23,17	46,14	1	1,00	0,00%	100,00%
20001	22,52	85,84	2	2,20	9,89%	90,11%
20002	22,61	85,59	2	2,15	7,65%	92,35%
20003	22,62	85,53	2	2,14	7,19%	92,82%
20004	22,64	85,45	2	2,13	6,47%	93,54%
40000	23,30	84,57	2	1,98	1,18%	98,82%
40001	23,48	94,73	3	2,95	1,62%	98,38%
40002	23,54	94,60	3	2,95	1,69%	98,31%
40003	23,56	94,48	3	2,95	1,77%	98,23%
40004	23,57	94,46	3	2,95	1,78%	98,22%
59999	24,86	98,69	3	2,99	0,37%	99,63%

Table 3. Success rate for all three algorithms considered

Training Function	Success Rate (%)
GDM	98,9924
SCG	98,9912
LM	98,9855

Figure 7. Comparison of the expected classification results vs. results produced by the ANN-based grain classifier

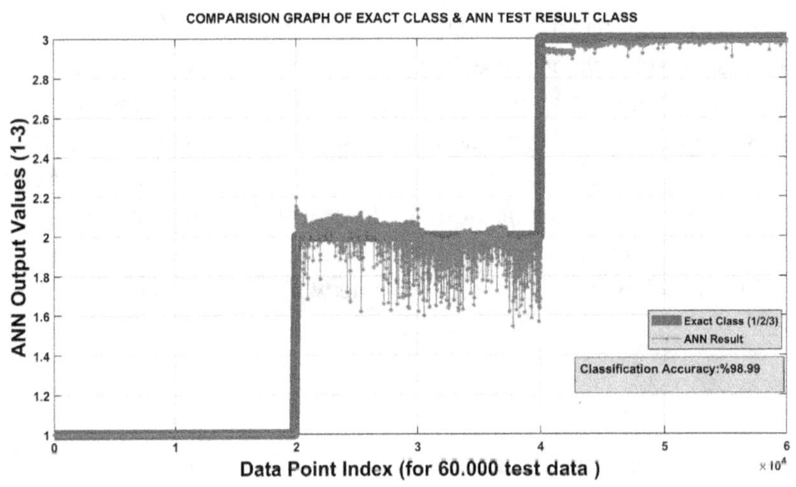

From a practical perspective, an important question arises: out of the parameters measured (temperature andrelative humidity), which one has greater impact in terms of determining the status of the stored grain? In other words, do temperature and relative humidity both carry equal significance in terms of determining the status of stored grain? An answer to this question could determine whether the sensor structure could potentially be simplified. To help address this question, we next present Figure 8 and Figure 9, which depict the temperature and relative humidityvariation for all three classes, respectively. In comparing these figures, we can clearly observe that relative humidity values exhibit a much greater variation across all three classes than temperature (42 – 48% range for the 'safe' class, 82-86% for the 'risky' class, and 92 – 99% for the 'dangerous' class). Therefore, it can be said that the relative humidity parameter has a much bigger impact, and is thus a bigger determinant of the status of stored grain.

Figure 8. Temperature variation for each grain class

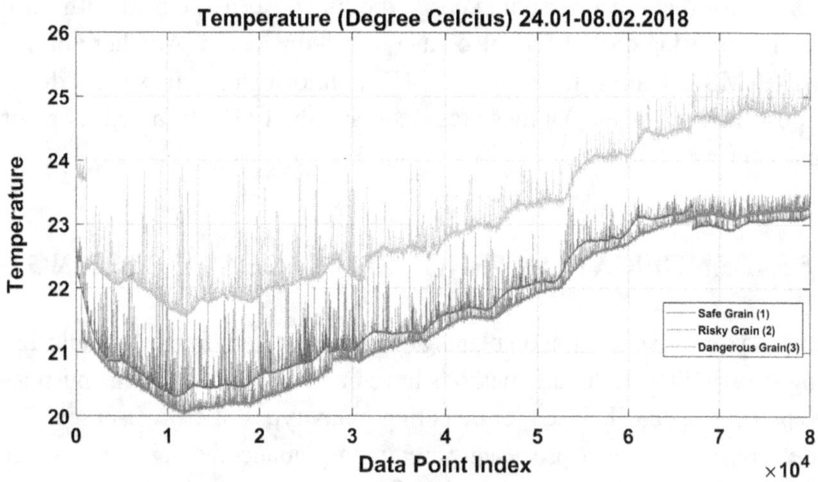

Figure 9. Relative humidity variation across all three grain classes

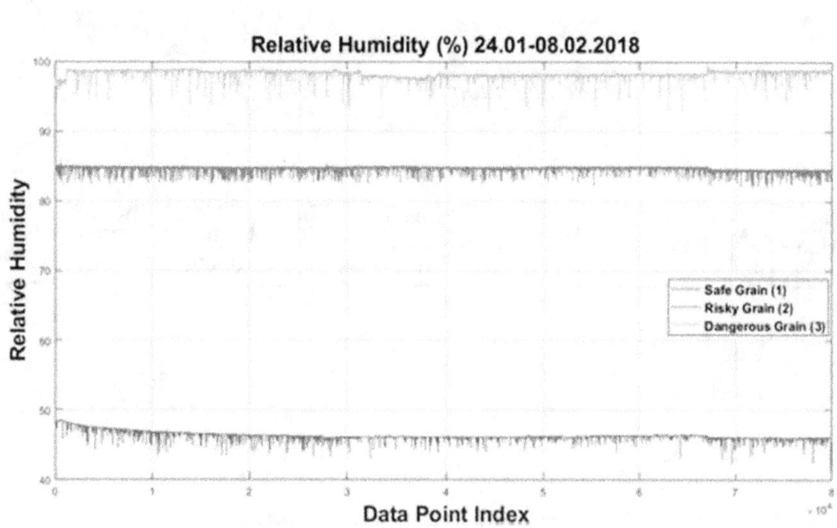

Concluding Remarks on the I-GCMS

As mentioned in the previous sections, the I-GCMS was designed to solve concrete practical problems associated with the existing generation of grain condition monitoring systems. As of the time of writing, the I-GCMS is in prototype stage and has not yet been deployed in a pilot field application. For this reason, a detailed quantitative comparison between the I-GCMS and the existing system is not yet possible. However, some general quantitative comparisons can be made on the basis of reasonable assumptions.

Within the context of the existing system, a TMO field technician examines the silobags 8 times a month, on average. A typical TMO silobag deployment consists of 8 silobags. Allowing 20 minutes' travel time for one technician to physically get to a silobag location, and approximately 10 minutes' measurement time per silobag, we can calculate the average time that one technician spends for measurement to be 660 minutes, or 11 hours / month. Assuming a regular 8-hour work day, a technician will need to spend over one day per month on a non-productive measurement job that can be automated. Please note that this analysis is just for the direct personnel costs; the fuel costs associated with travel to access the silobag, and the ever-increasing cost of fuel also have to be considered. Another noteworthy item here is the fact that the I-GCMS will allow for proactive 24/7 monitoring of grain stacks, drastically reducing the possibility of post-harvest losses. All these factors make the I-GCMS much more attractive from a business case standpoint.

CROP DISEASE IDENTIFICATION THROUGH IMAGE PROCESSING

Growth monitoring and disease detection on plants are very important for sustainable agriculture. Since manual monitoring is very difficult, many methods have been developed for this purpose. One of these techniques, image processing, can be used for detecting many types of crop diseases. The main steps in the plant disease detection using image processing are; image enhancement, image segmentation, feature extraction and classification, as depicted in Figure 10 below. In the next few subsections, we aim to give the reader an overview of all these different stages of crop disease detection. We end this section with a brief application example.

Figure 10. Main steps for plant disease detection

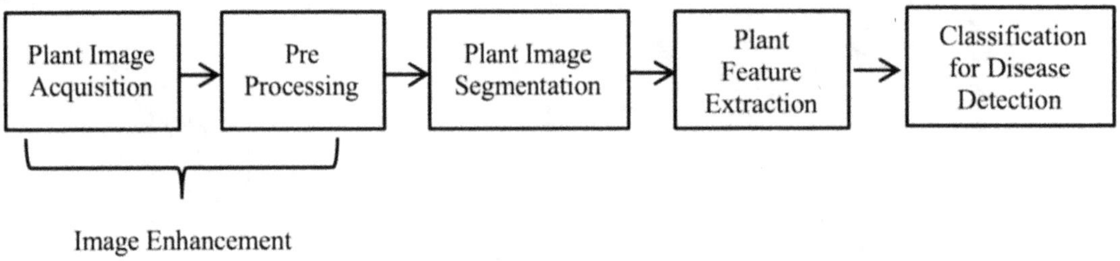

Plant Image Enhancement

Image processing begins with image acquisition where data is captured as in the form of digital images. The performance of the image processing isdramatically affected by the image quality. The development in image capturing technology made it possible to get higher quality images in larger areas. Also, employment of unmanned aerial vehicles and drones for monitoring of plant growth from the air is increasing day by day.

Once an image has been captured, various processing methods can also be applied to improve the image quality. Among them, image clipping can remove the outer parts of an image to improve the framing; image smoothing can reduce the noise and histogram equalization can improve the image contrast. Also, various filters are used to emphasize important features in the image and color space conversion can convertRGB images to grey images.

Plant Image Segmentation

Image segmentation, which extracts objects within an image, is the most important and critical stepbecause it influences the success of all the subsequent steps. Most commonly used image segmentation techniques are thresholding, edgebased techniques and region based techniques.

The thresholdingcreates binary images from gray scale images based on a threshold value which maximizes the interclass variance. The thresholding techniques can be categorized into global and local thresholding. While the global thresholding techniques use the same threshold for the whole image, local thresholding techniques choose different threshold values for every pixel in the image depending on its neighboring pixels.

The edge based techniques detect edges to extract an object of interest from the image. Sobel, Canny, Robert and Prewitt are widely used edge detection operators. Once the edges are detected, they are connected together to form the object boundaries.

The region based techniques adopt the idea that the pixels belonging to one element of the object can have similar properties and split the image into various regions with similar characteristics.

Plant Feature Extraction

Feature extraction methods capture relevant information in the object such as texture, color, shape, size, etc. The dimensionality can also be reduced by the feature extraction methods of which some examples are gray level co-occurrence matrix, gabor filter, histograms of oriented gradients. Classification accuracy is influenced by the feature extraction method used.

Classification

Classification is a supervised learning approach thatbuilds a model from the given labeled data and predicts the class of an unseen data on this model. Logistic Regression, K-Nearest Neighbors, Decision Tree, Support Vector Machines, Random Forest, Adaboost, Multilayer Perceptron and Convolution Neural Network are some classification methods.

An Application: Disease Detection on Apple Leaves

In this chapter, we give a brief example of disease detection. Here, we demonstrate a classification approach utilizing apple images with three different types of diseases (or healthy) which is available through the Plant Village project (Hudges, & Salathé, 2015). Examples of the images can be seen in Figure 11. In this study, the features are extracted from images by using the Gray Level Co-Occurrence Matrix (GLCM) method (Haralick, Shanmugam, & Dinstein, 1973), trained K-Nearest Neighbors (KNN) (Altman, 1992) and Support Vector Machine (SVM) (Cortes & Vapnik, 1995) for classification. The features are extracted from images by using the GLCM method, which generates a feature vector of seven parameters including contrast, correlation, energy, homogeneity, entropy, standard deviation and mean values. Then, the feature vector is passed to different classifiers. We trained KNN and SVM on a dataset of 3171 apple image data where 1645 of them correspond to the normal (i.e. non-diseased) case, and 1526 of them correspond to different types of tree diseases. Disease names and sample size for each disease are listed in Table 4. Classification results are presented in Table 5.

Figure 11. Examples of apple leaf images from the PlantVillage dataset, 1) Scab, 2) Black Rot, 3) Cedar Rust, and 4) Healthy Leaf

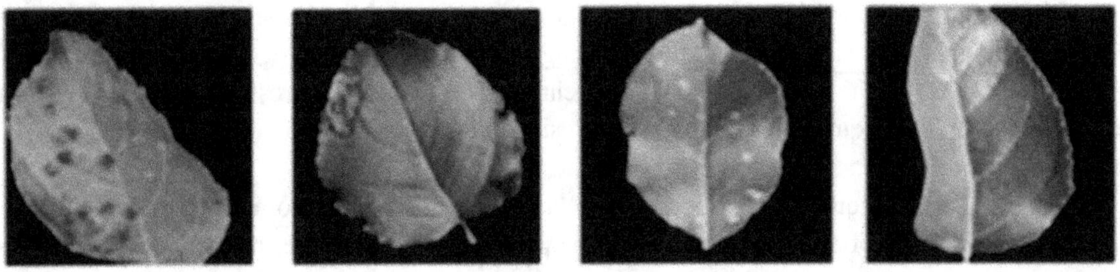

Table 4. Details of the utilized dataset for detection of plant diseases

Class ID	Disease Name	Sample Size
1	Scab	630
2	Black Rot	621
3	Cedar Apple Rust	275
4	Healthy Leaf (no disease)	1645

As shown in Table 5, the accuracy of SVM methods are higher than KNN methods; however their training time is longer than the KNN methods. The cubic SVM algorithm is the best classifier with 86.1% classification accuracy. Confusion matrix and Receiver Operating Characteristic (ROC)curve graphs of Cubic SVM are given in Figure 12 and Figure 13, respectively.

On the confusion matrix of Figure 12, the rows correspond to the predicted class (Output Class) and the columns correspond to the true class (Target Class). The diagonal cells correspond to observations that are correctly classified. The off-diagonal cells correspond to incorrectly classified observations. Both the number of observations and the percentage of the total number of observations are shown in

Table 5. Performance comparison of the GLCM-based SVM and KNN classifier

Classification methods		Accuracy %	Training time (Sec)
Support Vector Machine	Linear	80.8	8.18
	Quadratic	85.5	122.01
	Cubic	**86.1**	512.62
	Fine Gaussian	79.1	4.01
	Medium Gaussian	79.2	5.99
K-Nearest Neighbors	Fine	72.8	1.40
	Medium	77.4	1.20
	Coarse	72.6	1.60
	Cubic	77.1	1.96
	Weighted	78.1	1.44

each cell. Based on the results of Figure 12, we can say that the cubic SVM can correctly classify the four classes considered (scab, black rot, cedar apple rust and healthy leaf) with 83%, 84%, 65%, and 92% probability, respectively.

The ROC curve depicted on Figure 13 is actually another metric commonly used to check the quality of a machine-learning classifier. For a general classification problem consisting of *N* classes, two parameters are calculated: the True Positive Rate (TPR) and the False Positive Rate (FPR). A good classifier will have a high TPR and low FPR. The ROC curve depicted in Figure 13 is for the fourth class (healthy leaf with no disease) and the results clearly indicate that this class can be distinguished from the other classes with 92% probability. The area under the ROC curve (AuC) is another figure of merit for the overall system. The larger this value, the higher the reliability of the system.

CONCLUSION

It should be clear to the reader at this point that agricultural activity in the 21st century is expected to be increasingly technology-driven. The overarching vision is that this technology push is expected to drive improvements in production quality and quantity, as well as to reduce costs and post-harvest losses to the greatest extent possible. In this chapter, we have attempted to illustrate how the convergence of sensor technology, wireless sensor networks, and artificial intelligence techniques are helping to realize this vision by improving monitoring of crop quality. We have elaborated on fundamental concepts and attempted to illustrate them with two case studies, one focused on proactive, 24/7 monitoring of stored grain condition using a wireless sensor network and ANNs, and another focused on crop disease monitoring via image-processing techniques. Our results clearly illustrate the promise of these technologies. With regards to stored grain condition monitoring, the results obtained indicate that it is possible to classify the condition of stored grain using temperature and humidity sensors with an accuracy level approaching 99%. On the issue of crop disease monitoring, we have showed that it is possible to distinguish diseased produce through image processing with an accuracy of over 86%.

Figure 12. Confusion matrix for GLCM-based Cubic SVM

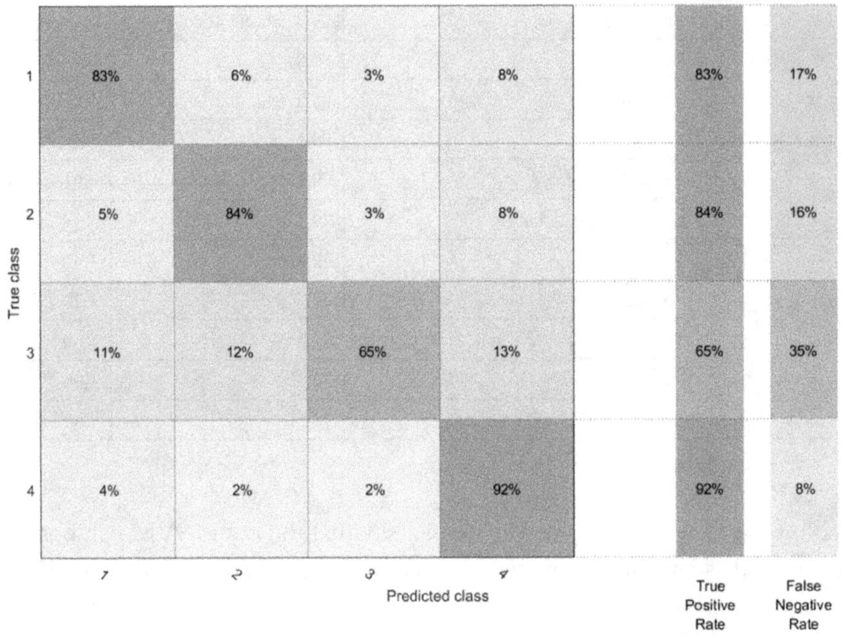

Figure 13. Roc curve for GLCM-based Cubic SVM

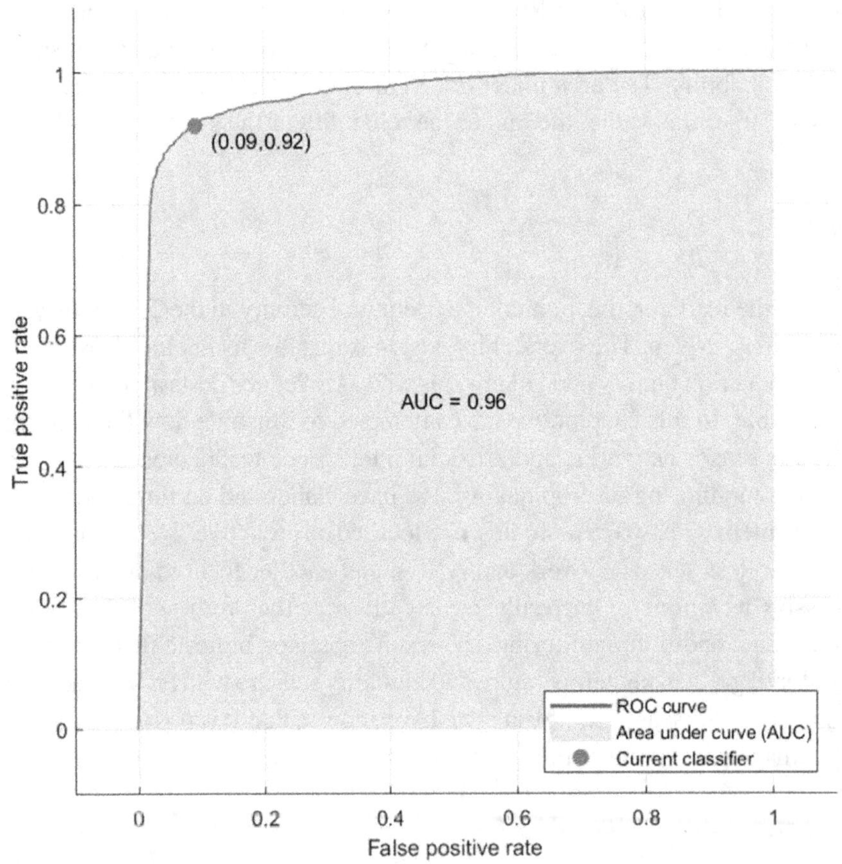

Future research trends with respect to the AI applications in agriculture discussed in this chapter will most likely involve the integration of these systems into more complex infrastructures that will not only monitor agricultural products for disease or other quality deterioriation, but will also proactively take some kind of physical action. This will pave the way for agricultural cyberphysical systems that will automate the farming process for enhanced quality and efficiency. As a potential concrete example of this concept, the silobag could conceivably be redesigned to automatically dry the grain in case the relative humidity gets too high, without explicit human intervention.

Of course, introduction of AI technology in agriculture will not come without challenges. While these technological advances increase agricultural productivity, they have some drawbacks such that their initial establishment costs can be relatively high. Most applications will likely depend on cheap and widespread availability of sensor infrastructure. The relatively simple structure of the sensor modules makes them susceptible to hacking and other kinds of cyber-attacks. Security and privacy in data transfer are also very important since the data will be mostly stored on cloud storage infrastructure. While a decision support system guides the agriculturalists about climate model, plant modeling, using such systems has a steep learning curve. Moreover, since most of the technological applications in the precise agriculture are hardware dependent, some hardware failures can cause delays in the field and lead to inaccurate decisions. Furthermore, farmers' attitudes and lifestyle can affect the success of such systems. All these issues will need to be addressed by the research community before any widespread deployment of AI technology in agriculture takes place. It is without a doubt that we are on the cusp of a technology revolution in agriculture, and hope that this chapter stimulates other researchers to devote their energies to this fast-growing field.

ACKNOWLEDGMENT

This research was supported by Scientific Research Projects Support Office of Erciyes University [grant number FYL-2017-7141].

REFERENCES

Akyildiz, I. F., Su, W., Sankarasubramaniam, Y., & Cayirci, E. (2002). Wireless sensor networks: A survey. *Computer Networks*, *38*(4), 393–422. doi:10.1016/S1389-1286(01)00302-4

Altman, N. S. (1992). An introduction to kernel and nearest-neighbor nonparametric regression. *The American Statistician*, *46*(3), 175–185.

Atmojo, U. D., Salcic, Z., Kevin, I., Wang, K., & Park, H. (2015). System-level approach to the design of ambient intelligence systems based on wireless sensor and actuator networks. *Journal of Ambient Intelligence and Humanized Computing*, *6*(2), 153–169. doi:10.100712652-014-0221-3

Baykara, C. K. (2018). Wireless Network Based Grain Control System Design For Grain Bagging System (Silobag) (Master of Science Master of Science). Erciyes University, Kayseri, Turkey.

Cortes, C., & Vapnik, V. (1995). Support-vector networks. *Machine Learning*, *20*(3), 273–297. doi:10.1007/BF00994018

Dilay, Y., Soy, H., & Bayrak, M. (2012). Employment of the wireless sensor networks in precision agriculture and investigation of the application areas. *Igdir University Journal of the Institute of Science and Technology, 2*(2).

Eisenhauer, M., Rosengren, P., & Antolin, P. (2010). *Hydra: A development platform for integrating wireless devices and sensors into ambient intelligence systems. In The Internet of Things*. Springer.

FAO. (2018). The State of Agricultural Commodity Markets 2018. Agricultural trade, climate change and food security. Rome: FAO.

Food, S. (2016). Global initiative on food loss and waste reduction. In *Key facts on food loss and waste you should know*. Academic Press.

Haralick, R. M., Shanmugam, K., & Dinstein, I. H. (1973). Textural features for image classification. *IEEE Transactions on Systems, Man, and Cybernetics, SMC-3*(6), 610–621. doi:10.1109/TSMC.1973.4309314

Haykin, S. (1994). *Neural networks: a comprehensive foundation*. Prentice Hall PTR.

Hughes, D.P., & Salathé, M. (2015). *An open access repository of images on plant health to enable the development of mobile disease diagnostics*. CoRR abs/1511.08060

Kumar, D., & Kalita, P. (2017). Reducing postharvest losses during storage of grain crops to strengthen food security in developing countries. *Foods, 6*(1), 8. doi:10.3390/foods6010008 PMID:28231087

Li, X., & Zhu, W. (2011). Apple grading method based on features fusion of size, shape and color. *Procedia Engineering, 15*, 2885–2891. doi:10.1016/j.proeng.2011.08.543

Majumdar, S., & Jayas, D. S. (2000). Classification of cereal grains using machine vision: IV. Combined morphology, color, and texture models. *Transactions of the ASAE. American Society of Agricultural Engineers, 43*(6), 1689. doi:10.13031/2013.3069

Ojha, T., Misra, S., & Raghuwanshi, N. S. (2015). Wireless sensor networks for agriculture: The state-of-the-art in practice and future challenges. *Computers and Electronics in Agriculture, 118*, 66–84. doi:10.1016/j.compag.2015.08.011

Paliwal, J., Visen, N., Jayas, D., & White, N. (2003). Comparison of a neural network and a non-parametric classifier for grain kernel identification. *Biosystems Engineering, 85*(4), 405–413. doi:10.1016/S1537-5110(03)00083-7

Panda, S. S., Ames, D. P., & Panigrahi, S. (2010). Application of vegetation indices for agricultural crop yield prediction using neural network techniques. *Remote Sensing, 2*(3), 673–696. doi:10.3390/rs2030673

Park, S., Im, J., Jang, E., & Rhee, J. (2016). Drought assessment and monitoring through blending of multi-sensor indices using machine learning approaches for different climate regions. *Agricultural and Forest Meteorology, 216*, 157–169. doi:10.1016/j.agrformet.2015.10.011

Savary, S., Ficke, A., Aubertot, J.-N., & Hollier, C. (2012). *Crop losses due to diseases and their implications for global food production losses and food security*. Springer. doi:10.100712571-012-0200-5

Singh, K. R., & Chaudhury, S. (2016). Efficient technique for rice grain classification using back-propagation neural network and wavelet decomposition. *IET Computer Vision*, *10*(8), 780–787. doi:10.1049/iet-cvi.2015.0486

Taner, A., Tekgüler, A., & Sauk, H. (2015). Yapay sinir ağları ile makarnalık buğday çeşitlerinin sınıflandırılması. *Anadolu Tarım Bilimleri Dergisi*, *30*(1), 51–59. doi:10.7161/anajas.2015.30.1.51-59

Togola, A., Seck, P., Glitho, I., Diagne, A., Adda, C., Toure, A., & Nwilene, F. (2013). Economic losses from insect pest infestation on rice stored on-farm in Benin. *Journal of Applied Sciences (Faisalabad)*, *13*(2), 278–285. doi:10.3923/jas.2013.278.285

Visen, N., Paliwal, J., Jayas, D., & White, N. (2002). Ae—automation and emerging technologies: Specialist neural networks for cereal grain classification. *Biosystems Engineering*, *82*(2), 151–159. doi:10.1006/bioe.2002.0064

West, J. S., Canning, G. G., Perryman, S. A., & King, K. (2017). Novel Technologies for the detection of Fusarium head blight disease and airborne inoculum. *Tropical Plant Pathology*, *42*(3), 203–209. doi:10.100740858-017-0138-4 PMID:32714500

Wilson, C. L., & Wisniewski, M. E. (1989). Biological control of postharvest diseases of fruits and vegetables: An emerging technology. *Annual Review of Phytopathology*, *27*(1), 425–441. doi:10.1146/annurev.py.27.090189.002233

Yu, X., Wu, P., Han, W., & Zhang, Z. (2013). A survey on wireless sensor network infrastructure for agriculture. *Computer Standards & Interfaces*, *35*(1), 59–64. doi:10.1016/j.csi.2012.05.001

Zhang, M., Qin, Z., Liu, X., & Ustin, S. L. (2003). Detection of stress in tomatoes induced by late blight disease in California, USA, using hyperspectral remote sensing. *International Journal of Applied Earth Observation and Geoinformation*, *4*(4), 295–310. doi:10.1016/S0303-2434(03)00008-4

Zhang, Y. (2011). *Design of the node system of wireless sensor network and its application in digital agriculture.* Paper presented at the 2011 International Conference on Computer Distributed Control and Intelligent Environmental Monitoring. 10.1109/CDCIEM.2011.371

ADDITIONAL READING

Callaway, E. H. Jr. (2003). *Wireless sensor networks: architectures and protocols*. CRC press. doi:10.1201/9780203500705

Jain, A. K., Mao, J., & Mohiuddin, K. M. (1996). Artificial neural networks: A tutorial. *Computer*, *29*(3), 31–44. doi:10.1109/2.485891

Kussul, N., Lavreniuk, M., Skakun, S., & Shelestov, A. (2017). Deep learning classification of land cover and crop types using remote sensing data. *IEEE Geoscience and Remote Sensing Letters*, *14*(5), 778–782. doi:10.1109/LGRS.2017.2681128

Liakos, K. G., Busato, P., Moshou, D., Pearson, S., & Bochtis, D. (2018). Machine learning in agriculture: A review. *Sensors (Basel)*, *18*(8), 2674. doi:10.339018082674 PMID:30110960

Moulin, S., Bondeau, A., & Delecolle, R. (1998). Combining agricultural crop models and satellite observations: From field to regional scales. *International Journal of Remote Sensing*, *19*(6), 1021–1036. doi:10.1080/014311698215586

Reed, T. R., & Dubuf, J. H. (1993). A review of recent texture segmentation and feature extraction techniques. *CVGIP. Image Understanding*, *57*(3), 359–372. doi:10.1006/ciun.1993.1024

Yick, J., Mukherjee, B., & Ghosal, D. (2008). Wireless sensor network survey. *Computer Networks*, *52*(12), 2292–2330. doi:10.1016/j.comnet.2008.04.002

Yu, H., & Kim, S. (2012). SVM Tutorial-Classification, Regression and Ranking. Handbook of Natural computing, 1, 479-506.

KEY TERMS AND DEFINITIONS

Artificial Intelligence: A paradigm of modern-day computing that tries to mimic the workings of the human brain to come up with highly adaptive, self-configuring computing infrastructures that can constantly learn.

Artificial Neural Network: A highly nonlinear computing structure, modeled after the neuronal organization of the human brain, used for solving some of the most complex problems in science and engineering, including regression, classification, and pattern recognition. A typical artificial neural network consists of an input layer, an output layer, as well as one or more hidden layers with connections between the neurons at each layer.

Feature Extraction: A common step in most image processing applications, where the image is pre-processed in order to come up with those aspects of the image that uniquely characterize a given image (for an image classification application, for instance). In most image processing applications, not all parts of the image are equally important. Feature extraction is therefore employed in order to take advantage of this fact and reduce the computational burden.

Grain Condition Monitoring System: A system utilizing sensors, wireless sensor networks and machine learning techniques to monitor the condition of harvested grain that is stored somewhere (e.g. in a silo, or silobag etc.).

Image Segmentation: A step in most image processing applications, where the image is processed to extract a given object.

Supervised Learning: A learning model employed by a neural network, whereby the network is presented with training data that is appropriately labeled. In supervised learning, the network knows what output it is supposed to give in response to a given input, and thus, the network tries to adjust its parameters to approach the desired output as part of the learning process.

Wireless Sensor Network: A network composed of physically small, low-power network nodes that also incorporate sensing functionality.

Chapter 8
Precision Management Practices for Legal Cultivation of Cannabis (*Cannabis sativa L.*):
Precision Management Practices for Cannabis Cultivation

Aitazaz Ahsan Farooque
University of Prince Edward Island, Canada

Farhat Abbas
https://orcid.org/0000-0002-2032-8527
University of Prince Edward Island, Canada

ABSTRACT

Cannabis (Cannabis sativa L.) growers worldwide lack reliable and research-based information about precision management practices (PMP) of cannabis. The history, legal framework, and PMP for cultivation of cannabis have been reviewed with special emphasis on water management, nutrient management, and disease control for optimum cannabis production. The aim is to provide guidelines for precision farming of cannabis to meet fibrous and medicinal needs of the humankind. Therefore, the scope of this chapter is for the potential of hemp cultivation to meet industry needs of fiber and medicine. Methods of irrigation scheduling, nutrient applications, and keeping greenhouse hygienically clean for disease-free (i.e., powdery mildew) hemp production are discussed. Reviewed and recommended application rates of irrigation and nutrients, and environment controls have been tabulated. Chemical, biological, and physical controls of PM control and crop input requirements for disease-free cultivation of hemp are presented.

DOI: 10.4018/978-1-7998-5000-7.ch008

INTRODUCTION

Cannabis can grow in all climate extremes with its life cycle starting from germination time of 7 days and total growing season of approximately 100-120 days. Its cultivation in North America prevails a long history recorded from the Government of the United States of America encouraging its farmers to grow the fibrous form of cannabis for meeting the country's fiber needs for clothing, rope making, and fiber sails products during 1800s (Millery, 2011). However, the immigrants from South of USA introduced amusing use of cannabis during 2000s; this forced USA to enact the Marijuana Tax Act during 1937, to effectively criminalize marijuana consumption in USA; nonetheless, USDA (the United States Department of Agriculture) incentivized farmers to grow fibrous form of cannabis to meet wartime fiber needs of the country and collating partners during World War II (Millery, 2011).

The cannabis international legal framework comprises the 1961 Single Convention on Narcotic Drugs as amended by the 1972 Protocol (UNGA, 1975), the 1971 Convention on Psychotropic Substances (UNGA, 1975), and the 1988 Convention Against Illicit Traffic in Narcotic Drugs and Psychotropic Substances (UNESOC, 1998). The use of cannabis for scientific and thus medical purposes has been permitted under the legal framework of the European Union (EMCDDA, 2018). There has been a recent trend in some countries of decriminalizing the use of cannabis for personal use, but the current international laws still require them to regulate and control the use of cannabis. For an increased public acceptance of recreational marijuana use and challenges in controlling illegal use of cannabis in Canada, the issue of legalizing the illicit use of this drug has been raised publicly (Hajizadeh, 2016). Department of Justice of the Government of Canada on its official website https://www.justice.gc.ca/eng/cj-jp/cannabis/ details legalization and regulation of cannabis in Canada (Government of Canada, 2019).

In this book chapter, history and legal framework of cannabis cultivation, and PMP for legal cultivation of cannabis are presented by putting a special emphasis on water management, nutrient management, and disease control particularly under greenhouse situations. The aim is to promote and provide guidelines to the farmers/growers for precision farming of cannabis to meet fibrous and medicinal needs of the humankind. This chapter is concerned only with medical purpose production of cannabis.

Cannabis (*Cannabis sativa* L.) has a long history of cultivation (Schulteset al., 1974) originating from central Asia (McPartland, 2000) and has been on industrial demand for centuries to produce fibers, oils, textile products, papers, construction materials, energy, cosmetics, medicines and chemicals (Grabowskaet al., 2009; Bakroet al., 2018). Cannabis has also been used medicinally dating back approximately 5000 years (Lemberger, 1980; Pertwee, 2014). The potential use of cannabis and cannabinoids has been termed to be a source for relief medicines as painkillers, anti-emetics, anticholinergic, appetizer, and in treatment of glaucoma and epilepsy by the US and the British bodies specialized in medical services (Morris, 1997).

Cannabis: Hemp vs. Marijuana

Numerous synonyms for the word 'cannabis' have been used in literature including hemp, marijuana, ganja, hash, or hashish in North America, Europe, Caribbean, and Asia, respectively. However, English literature commonly refers to the word "cannabis". Forsyth (2000) enlisted a detailed list of synonyms comprising historic and contemporary, scientific and slang used for cannabis.

Cannabinoids and Strains of Cannabis

Cannabis is classified sativa and *indica*. Hemp is a member of the cannabis *sativa* family. Marijuana is a member of either the *sativa* or *indica* families. Since, the two originate from *sativa*, they have some similar chemical properties. Physically and biologically, the two are different from each other. For example, hemp has thin leaves more intensified in upper parts of the plant and less dense on the lower part of its stem. Its plants grow taller and skinnier than those of Marijuana which has wider leaves and thick buds possessing a short bushy look. A typical plant of hemp is 1.2 to 5 m tall with a stem diameter of 4 to 20 mm. Typically, the male plants are longer with less branches compared to female plants (Clarke, 1999). Kompolti is a Hungarian industrial strain of cannabis with potential of high CBD; over 14% of Cannabidiol is possible when you grow these industrial cannabis seeds. Kompolti has a peculiar growth style. Lissonet al., (2000) modeled development of a Kompolti to simulate the effect of plant density on leaf appearance, expansion and senescence and found that leaf area per node before reaching to a plateau increased linearly with node number followed by a linear declined. They further reported slowed down node production and reduction in the maximum leaf area per node with increases in plant density. Cannabis yield is estimated from floral bud tissues (Hawleet al., 2018); the bud has a relatively high density of oil-rich cells targeted for high and quality yield of cannabis (Happyanaet al., 2013).

MAJOR CROP INPUTS, ENVIRONMENTAL CONTROL AND CHALLENGES

Growth of hemp is affected by nutrient supply, intensity and duration of light hours and water use efficiency. Nitrogen (N) supply can rapidly increase the growth of hemp; however, oversupply of N results in irregular growth of hemp resulting to leafy and juicy plants and stems diameter thickness more than the optimum range (Ranalli, 1999). Long days with more light hours (12-14) support vegetative growth (Sankari&Mela, 1998); whereas, short days with less light hours are supportive environment for flowering and maturation stages. With proper irrigation water supply hemp cultivars reach to the height of 5 m in 4 to 6 monthperiod (Clarke &Merlin, 2016).

One of the major challenges in growing hemp is a disease-free cultivation. Like many other crops, cannabis plants experience threats from a variety of diseases including powdery mildew (PM), which (*Podosphaerapannosa*) is one of the major challenges in greenhouse cultivation. Spores of PM travel through wind and enter into the greenhouse their ventilation systems. They are also carried by pets and greenhouse workers to the greenhouse. Favorable conditions for PM spores to spread and become problematic for disease free cultivation of cannabis become available in warmer conditions like in Spring and Fall months famous for high relative humidity (RH) and changing day-night temperatures.

Open and Controlled Environment Cultivation

Open fields provide ideal conditions for cannabis wild growth. Since these ecosystems provide the required sunlight, light well-drained composted/fertile soil, and enough irrigation, it grows well wildly when exposed to open water bodies including canals and rivers, pastures, and agricultural fields (Millery, 2011). However, open field cultivation of hemp faces soil borne diseases from pathogens including *pythium spp.* and *sclerotiumrolfsii* especially during the warmer spring periods and in the well-drained lighter sandy soils (Hall et al., 2017). Field cultivation of hemp has ever been debated for plant density for the financial importance of single biomass content of the plant. Van der Werfet al., (1995) studied quality and yield of a fiber hemp variety under the influence of density of plants and their thinning and reported that the number of nodes at the onset of flowering decreased with increasing plant density; however, the rate of decrease with age of plant was not sensitive to density of plants.

Greenhouses are used for controlled environment cultivation of hemp. A typical greenhouse has various section for management of irrigation water, nutrients, mother plants, nurseries, growing plants to maturity, harvesting flowers, and processing the final product for packing. The workers follow protocols for dressing, handling items, and performing various duties. The protocols may include wearing prescribed clothing and protective masks/gadgets specified for working in various sections of the greenhouse to ensure safety, security, disinfection, and prevention of plants from physical, chemical and biological damages.

Most of the arrangements in greenhouse are made artificially. The sections where mother or regular plants are kept are lit with artificial lights to mimic sunshine for extended number of hours. Usually 15 to 18 hours of artificial lights are provided to the plants. Most of the greenhouses use dataloggers to record inside temperature, relative humidity, and light intensity.

Precision Management Practices

Precision management practices cover a range of material, input and practices needed to grow cannabis; for example, a medium to grow plants such as soil and inputs such as water and nutrients. The practices include measures to control disease and weeds.

Soilless media

Since the natural soil comes with threats of disease carrying bacteria and pathogens, it is a common practice to use soilless media in greenhouse to grow cannabis. Soilless media, typically called as growing media is commercially available. Different inert materials including, coconut crush, rock wool, peat moss, pine bark, coconut coir, are sometimes synthetic materials are used to form loose media, growing cubes or slabs for plants. The media come with certain porosity to accommodate growth of roots, movement of irrigation water, supply of oxygen to roots, and transport of nutrients. Soilless media are most commonly peat-based, with a wide array of amendments used to create the best media based on crop requirements and production techniques. Adjusting water-holding capacity, pore space within the media, bulk density and even characteristics like pH are possible with the careful selection of media components and amendments.

Irrigation water management

Hawley et al. (2018) reported that bud quality and yield of cannabis was affected by irrigation scheduling in addition to other plant carrying aspects. Pejiceet al., (2018) used Hargreaves equation to calculate evapotranspiration requirements of hemp (320 mm for the whole growing season) at its growth stages comprising i) Stage-I: the stage between sowing of seeds and the point of developing of 3-4 pair leaf pairs, ii) Stage-II: from 3-4 pair leaf pairs to coming out of male flowers, and iii) Stage-III: a period beyond appearing of male flowers. They reported that irrigation water application with drip irrigation system to meet the calculated evapotranspiration resulted in significant improvement in developing of plant stem, emerging of new leaves, flowers and increase in plant height, but not the thickness of stem indicated from its diameter and thus the fiber content. Kleinwächter&Selmar (2015) reported that controlled or moderate drought stress influence the production of certain secondary metabolites of cannabis. Irrigation water requirements collected from literature for various environmental conditions are tabulated as Table 1.

Table 1. Irrigation requirements for cultivation of hemp

Sr. No.	Climatic conditions	Irrigation requirements (mm)	Source
1	Not mentioned	250-350 per growing season	Kišgeci, 1994
2	Ukraine	250-280 per growing season	Pejiceet al., 2018
3	The Netherlands	650 mm per year	Van Dam, 1995
4	Northwest Tasmania in Australia	535 mm during growing season	Lisson& Mendham, 1998
5	Not mentioned	700 mm	Bócsa&Karus 1998
6	Italy	410-460	Di Bari et al. (2004)

Irrigation scheduling

Advanced cultivation of cannabis has used soil moisture content sensors the readings of influenced by location of the sensor within the pot/media and growth media composition (Stemeroff, 2017). This influence of external factors necessitates sensor calibration for specific growth media/systems (Abbas et al., 2011; Fares et al., 2011). Although as a common practice, a network of sensors is used in a field or in a greenhouse setup, but an average value of soil moisture content is used for the entire crop is used to trigger the valve of irrigation system on or off the automated irrigation system. However, numerous factors are considered to develop proper irrigation scheduling (Ehlers &Goss, 2003). Since, it is impractical to expect growers to deploy these research tools on a large scale to apply irrigation management strategies (Stemeroff, 2017), the use of simple methods to schedule irrigation for cannabis can be beneficial. Irrigation scheduling for cannabis can be conducted using Pan Evaporation method (FAO, 1998) and/or Hargreaves equation (Hargreaves & Allen 2003).

The pan evaporation method involves the use of i) a standard "USWB class-A pan" and ii) a standard rain-gauge. The USWB class-A pan is used to determine values for daily pan evaporation (E_p) by monitoring water level drop from its free water surface. The USWB class-A pan can be made with a 20-gauge galvanized iron sheet panted white. The diameter of this top-open and closed bottom pan is to be fixed at 121.5 cm and pan depth at 25.4 cm. The pan is equipped with a vertical pointer in the stilling well accompanied with a hook gauge to see the level of the maintained water. In order to let air circulated beneath the pan, the pan is placed on a hollow frame preferably made of wood or a material that does not warm out with heat or freeze with low temperature ensuring insulating effect to the pan from its bottom. Depending upon irrigation intervals (hourly versus daily), rate of evaporation can be measured, with the help of a hook gauge, from the fall of level of water in the stilling well at specified irrigation intervals. In case of field use and rainfall, necessary evaporation values need to be adjusted. The water in the pan is brought back to its original level by adding the equal amount of water that evaporated to the pan bring back pan's pointer tip to its original position. That is, the drop of water level is maintained by adding water in the pan to the original position of pointer tip level. The amount of water added is divided by the surface areas of pan and stilling well together to find out the depth of water added which is taken as the daily evaporation rate (E_P).

Marketable rainfall gauge with a graduated cylinder wall or conventionally made simple gauge consisting of a deep pan equipped with a ruler can be used to measure rainfall on daily basis. Amount of rainfall water in the gauge can be converted to depth units by dividing the total volume received with surface area of the gauge for effective rainfall values (RE). The following equations can be used to calculate supplemental irrigation water requirements (SIWR).

$$SIWR = \left(E_P \times k_P \times k_C\right) - RE \text{ (for field cultivation)}$$

$$SIWR = \left(E_P \times k_P \times k_C\right) \text{ (for greenhouse cultivation)}$$

Where, k_P is pan coefficient (0.7 for USWB pan as the rate of evaporation from USWB pan is higher than that over a large free water surface; therefore, the E_P value is multiplied by 0.7 to obtain the evaporation rate over the large free water surface), k_C is crop factor, and RE is effective rainfall. The values of k_C for the three stages of crop growth; i.e., Stage-I, Stage-II, and Stage-III can be taken as 0.5, 0.9, and 1.1, respectively (Cosentinoet al., 2012; García-Tejeroet al., 2014).

Hargreaves equation: Daily evapotranspiration (ET_D) from cannabis plants can be calculated using Hargreaves equation as given below (Hargreaves & Allen, 2003):

$$ET_O = 0.0023\left(T_m + 17.8\right)\left(\sqrt{T_{\max} - T_{\min}}\right)R_a$$

where ET_O is reference evapotranspiration (mm day^{-1}), T_m is the daily average temperature of air (°C), T_{max} is the daily maximum temperature of air (°C), T_{min} is the daily minimum temperature of the air (°C), and R_a stands for the extraterrestrial radiation (MJ m^{-2} day^{-1}). Values of R_a vary with coordinates of the location, day of the year and can be found in Chapter 4 of the famous paper 56 of Food and Agricultural Organization of the United Nations (FAO, 1998). The Daily water used on plants evapotranspiration can be calculated as:

$$ET_D = (ET_O)\, k_C$$

where k_C crop coefficient, and (ET_D) is the daily ET.

Application of Supplemental Irrigation

It is a common practice to apply supplemental irrigation to the plants by keeping a margin of 10-15% over-irrigation than the calculated depth/volume - a factor recommended for covering up errors in irrigation application efficiency of the irrigation application systems be that drip or sprinkler irrigation systems. The reason behind this exercise is to avoid plants going under wilting point and to ensure that the media of growth (artificial media used to grow cannabis in greenhouses) remains within the field capacity range. However, excessive application of supplemental irrigation results into salt accumulation in the soil-less growth media

Nutrient Management

Application of N to the seedbeds or during the early stage (vegetative stage) of crop growth benefit growth of industrial hemp since N is important for protein content in the grain and the dry matter yields of fiber crops (Hall et al., 2017). There are numerous plant nutrition sources available in the market including silica that helps strengthening the plants overall by thickening the leaves' cell wall in order to make it resistant for spores of PM to hold on to it and enter into the surface of leaves in search of food. The leaves as well as stem of cannabis can be thickened and hardened with the supply of N that increases the plant biomass for fibrous yield (Jordan et al., 1946; Ivonyiet al., 1997; Bócsa&Karus 1998).

Van der Werfet al. (1995) compared growth of cannabis in treatments applied with 80 kg N/ha, and 200 kg N/ha and reported increase in biomass due to increased stem as radial growth of plants supplied with N at the rate of 200 kg/ha. On the other hand, Van der Werf and Van den Berg (1995) reported that the application of 80 kg N/ha resulted in increased bast content, plant height, proportions of male and female plants and stem dry matter. Struiket al. (2000) tested the effects of rates N application including 100, 160, and 220 kg/ha on stem health and reported that it increased with increases in rates of application of N; however, they concluded that treatments with relatively lower rate of N application produced enough fiber for the tested species. This is complemented by the suggestion of Scheifele (1999) that the requirements of N requirements for fiber sativa species are lower than expected and may be decreased from 140 to 100 kg/ha, the phosphorus (P) and potassium (K) levels should be increased. Ivanyi (2005) related the concentration of leaf nutrient of fiber hemp with its yield and reported concluded that suggested to redesign nutrient management practices of hemp for its advanced production. Forrest et al. (2008) investigated anatomy and morphology of an industrial fiber hemp species sativa "Fédrina" for their cultivation in field and in greenhouse settings in western province of Canada – British Columbia and reported that the application of N at the rate of 150 and/or 300 kg/ha was beneficial for field cultivation and application of 90 kg P_2O_5/ha resulted in maximum fiber yield under greenhouse settings. Nutrient requirements for hemp have been reported in literature as summarized in Table 2.

Salt Accumulation

Fertilization is one of the most important factors for indoor cannabis production. Over-fertilization can lead to salt accumulation in the root zone; whereas, under-fertilization can cause nutrient deficiency and lower yields (Bar-Yosef, 1999). Guttormsen (1969) studied distribution of salts in plastic pots having lettuce plants irrigated under sub-irrigation methods and reported accumulation of salts particularly in the upper soil layers of the growing pots even with normal amount of application of N and K through water application. His detailed results showed a proportion between the amounts of the applied N, K and the amounts of Ca and Mg in water extracts showing a tendency of exchangeable Ca and Mg.

The decision about base fertilizer rates for cannabis is difficult to make due to the differences in growing conditions (Wright &Niemiera, 1987). Nutrient requirements for cannabis vary with growth stage in flowering plants; the flowering stage requires higher concentration of nutrients (Raviv&Lieth, 2007). No reports have been published on nutrient requirements of cannabis at its different growth stages. Cannabis growers can benefit from the knowledge based on nutrient requirements calculated and experimented for different growth stages of other crops. For example, for sweet peppers (Capsicum annum L.), the supply of 30% N from $NH4^+$-N in the vegetative stage and using only NO_3–N in the flowering stage can result in the highest total plant and fruit yield (Xuet al., 2001).

Table 2. Summarized literature on nutrient requirements for production of cannabis

Nature of study	Nutrient requirements	Source
Field study	200 kg/ha N	Van der Werfet al., 1995; Ehrensing, 1998; Vera et al., 2004; Aubinet al., 2015
Field study	80 kg/ha N (for bast content, stem height, and proportion of stem dry matter)	Van der Werf& Van den Berg, 1995
Field study	100 kg/ha N	Scheifele, 1999
Field study	100 kg/ha N	Struiket al., 2000
Laboratory experiments	150 mg/kg AL-P_2O_5; 200-350 mg/kg AL-K_2O 100 kg/ha N	Kadar et al., 2003
Laboratory experiments	110-130 mg/kg AL-P_2O_5; 310-330 mg/kg AL-K_2O; 80 kg/ha N	Ivanyi, 2005
Field study	NPK (at a rate consistent with wheat production)	Reisingeret al., 2005
Field and greenhouse trials	150-300 kg/ha N of any N fertilizer type 90 kg/ha P_2O_5	Forrest et al., 2008
Walk-in growth chamber	283 mg N/L of a liquid organic fertilizer (2.0N–0.87P–3.32K)	Caplanet al., 2017a
Greenhouse	389 mg N/L using a liquid organic fertilizer (4.0N–1.3P–1.7K)	Caplanet al., 2017b

Salt accumulation because of the application of synthetic fertilizers can be avoided by using organic manures that release nutrients at a slower rate providing less chance of salts soil build-up in soil and resultant uptake by plants. Voglet al. (2004) compared hemp seed yields (variety Fedora-19) under an open experimental field comprising small plots with those grown in eleven open hemp fields under practical organic growing conditions in Austria.

Diseases of Cannabis and Their Control

In the face of the quote "hemp has no enemies" (Dewey, 1914), there are plenty of pathogens and pests that act as enemies of hemp causing a huge damage to its growth, production, yield, and quality (Bakroet al., 2018). Bakroet al. (2018) has summarized common fungal disease of cannabis typically caused by more than fungi taxa (McPartland& Hughes, 1994).

The outbreak of PM in the greenhouse can be detected in-situ by developing an image processing-based machine vision system as the machine learning techniques have been utilized in a variety of data-driven applications including image processing in agriculture and food industry (Chen et al., 2010; Savakar&Anami 2015; Rojas-Moraledaet al., 2017). Cameras installed at various angles on a moving buggy or on a robot in a greenhouse can also detect early stage leaf infection but with the added task of processing a large number of images for disease feature detection. Technology advancement in near or far future may solve this technical issue.

Powdery Mildew

Powdery mildew (PM) is a common fungal disease infecting cannabis. A casual organism for PM in hemp was first reported for eastern Europe by Jaczewski (1927). This fungal infection is common in outdoor hemp particularly in the regions of temperate and subtropical environment (Transhelet al., 1933). The top leaf surfaces are easily infected showing dusty look of powdery mycelium that turns leaves chlorotic and dead. Symptoms of PM can be identified by the growth of a white powdery fungus on diseased portions of the affected plants.

Symptoms of Powdery Mildew

Symptoms of PM, in addition to appearing on the upper surface of leaves, get developed on the lower surface of leaves. This sequence is not necessary as the disease symptoms vary randomly in their appearance as irregular purplish spots, followed by the appearance of white powder on the leaf surfaces. The symptoms may include unusual scab-like cuts, witches'-brooms, bendy and distorted shoots, early leaf colouring followed by fallen, reduced or undersized growth and curling of leaves.

Effect of Powdery Mildew on Plant Growth

The powdery appearance results from the superficial growth of the fungus as threadlike strands over the plant surface and the production of chains of spores. In the environment of combination of moderate to high temperatures and high humidity, more or less 300,000 spores are produced each day on the infection site equivalent to the size of 1 cm2. An infected plant becomes producer of spores in a period of 5-10 days with a continued cycle under greenhouse conditions (2019 Kentucky Pest News). Younger and new shoots, stems, other vegetative parts of cannabis easily come under the attack of fungi of PM than the mature parts of plants. Plants of cannabis must be continuously monitored to detect earlier appearance of PM which initially appears on the lower and/or middle leaves before spreading to the top leaves.

Sources of Powdery Mildew

This disease is caused by a range of fungi genera including *Erysiphales*, and *Podosphaeraxanthii*, which are commonly reported ones. For greenhouse cultivated cannabis, the important genera are *Oidium*, *Erysiphe*, *Leveillula*, *Sphaerotheca*, and *Microsphaera*. *Botrytis cinerea* remains and dominant parasite fungi for cannabis. Its spores are commonly present in outdoor air as well as in indoor environments. In the North America its average concentration ranges between 45 and 49 CFU/m^3 (Shelton, 2002). Botrytis *cinerea* is a *necrotrophic* fungus (a type of fungi that attack damaged and injured wine vines producing toxins that then kill the vine; the nutrients of the dead vine thrive the fungi).

Powdery Mildew in Greenhouses

Plant cuttings and clones contaminated with PM are the main source of spread of this disease in other plants in the greenhouse; however, the PM is often easily overlooked in the early stages, especially when one is handling a large number of microscopic spores during activities in greenhouses (Chris, 2017). Most of the times, it is no possible to judge if the spores of PM exist till the moment, they start taking life and spread throughout the greenhouse.

Environmental controls including low temperature, high RH, low light shade, and poor air circulation are the factors responsible for development of PM in a greenhouse. Until these factors are not optimal, the microscopic spores of PM stay dormant meaning that they exist but do not show no visible signs of growth, which does not necessarily mean that there is no potential of this disease to spread; it does as soon as the favourable conditions become available (Chris, 2017).

In greenhouses, PM usually survives between crops as hyphae (a long, branching filamentous structure of a fungus) because of ideal conditions comprising temperatures ranging from 68 to 86 °F and relative humidity beyond 95% followed by low light intensities. With an inverse relationship between RH and temperatures RH increases with lowering of temperatures influencing production as well as spread of PM conidia. High RH stimulates conidia to germinate; a conidium (plural conidia) is an asexual, non-motile spore of a fungus. Non-motile spore cannot move independently, using metabolic energy. Literature has reported the following tabulated optimal conditions suitable for PM to thrive on various crops in greenhouses (Table 3).

Table 3. Optimum conditions for powdery mildew to thrive in a greenhouse

Crop	Temperature (°F)	Relative humidity (%)	Other conditions	Source
Marijuana	68-86	> 95	Low light intensities or shade Light air circulation	Brais, (n.d.)
Rose	68	75 ± 2	Normal day light	Suthaparanet al., 2012
Cannabis	Cool to moderate	high	-	Barloy&Pelhate, 1962

Plant Disease Detection Through Machine Vision

Machine vision is a science that duplicates the effect of human vision by electronically perceiving and understanding an image, and preparing a suitably rapid, consistent, cost effective/economic and objective assessment of objects (Sun, 2000). The approach involves development of studies and algorithms to analyze and automatically extract useful details about an object or a set of objects perceived (Gunasekaran, 1996; Sun, 2016). Machine vision also involves processing images of physical objects for feature detection from them. With the help of good algorithms, machine vision systems progressively improve their performance on a specific task. Machine vision systems are being applied to identify, detect and predict crop diseases and plant stress phenotyping in agricultural research fields.

Several successful machine vision applications have been reviewed in the literature to detect plant diseases in different cropping systems (Story et al., 2010; Pineda et al., 2018). Wspanialy and Moussa (2016) developed and tested a machine vision system for PM detection in greenhouse tomato plants to a success rate of 85%. The PM detection system can reduce expenses associated with manual scouting for monitoring diseased plants in a greenhouse. This can be done through adopting the use of a robotics system to watch out for early signs of diseases in the greenhouse on daily basis and report the observation to the operators to ground truth the observation for further steps (Wspanialy&Moussa, 2016).

Real-Time Detection of Powdery Mildew With Deep Learning and Machine Vision

The outbreak of PM in the greenhouse can be detected in-situ by developing an image processing-based machine vision system using deep learning techniques. The machine vision system can consist of a graphical user interface based PM detection program, two µeye cameras, a real-time kinematic global positioning system, and a laptop computer. A colour co-occurrence matrices-based image texture analysis program can be generated by using C# programming language. The installed cameras inside the greenhouse can be programmed to take images of the plants on intermittent basis. Factor affecting machine vision parameters can be selected by evaluating performance upon collected images from several healthy and PM affected leaves. Mahmud (2019) tested the real-time performance of developed vision system in comparison to manual detection in three strawberry fields targeting a total of 36 randomly selected rows. He reported that the machine vision system was moderately capable of detecting PM disease with acceptable mean absolute errors and root mean square errors.

Challenges in Automated Detection of Disease

In-situ detection of infected leaves in a crop has real challenges and opportunities. Technology has always been on the go for innovation and advancement. The machine vision system introduced by Wspanialy and Moussa (2016) has various innovative features, as claimed by the developers, including i) it does not require specific environment but operates under normal lighting conditions, ii) it learns PM image features using the Hough forest machine learning technique without demanding for clean and ideal background conditions, and iii) its image acquiring and processing speed is very high because of the use of parallel nature of the Hough Forest algorithm. Systems using machine vision and visible light image processing to detect disease and pests attacked leaves under field conditions have attained high detection rates (Velázquez-Lópezet al., 2011). Limitations with such systems include i) input images making them unsuitable for independent detection and ii) uniform background against images taken from the top of a plant, and iii) limiting an input image to only a single centered leaf (Zhenget al., 2009; Zhang &Meng, 2011). Cameras installed at various angles on a moving buggy or on a robot in a greenhouse may help solve this issue but with the added task of processing a large number of images for disease feature detection. Technology advancement in near or far future may solve this technical issue.

Optimal Greenhouse Environments

Greenhouse cannabis cultivation and management practices consume high energy use, operating costs, and uncontrolled greenhouse gas emission; nonetheless, for cannabis producers, energy costs are usually acceptable given the high product value (Mills et al., 2012). Chris (2017) enlisted three precautionary measure to preventing cannabis cultivation from PM in greenhouses; the measures include i) genetics, ii) plant nutrition, and iii) controlled environments.

Seed Selection

Depending upon the source of seed, some of the strains of the same plant get infected and some do not even under identical conditions of growth. Therefore, it is worth surveying prior to deciding for the cannabis strain and picking varieties known best for pest and disease resistance. Only a registered seed distributer must be contacted for quality seeds as well as for detailed information about strain's resistance from pests and diseases.

Control After the Outbreak of Disease

Botrytiscinerea is the most dangerous and tough parasite fungi that dominate cannabis. Its spores are found in inside as well as outside air. Cannabis fungi can be controlled by chemical, biological, physical, or by the combined use of these methods.

Chemical Control

Spores of *Botrytis cinerea* migrate to the cannabis greenhouses through air and through ground carried by workers' clothing and shoes. Therefore, the most of greenhouses have protocols of disinfecting their employees prior to their entry to the facility. The use of traditional fungicides for controlling Botrytis *cinerea*, may include the precautionary measure of using concentration not harmful for seriously ill patients (Cervantes 2006). Usually, the fungicide sprays are recommended in emergency cases since the fungicide affects quality of buds resulting into poor quality taste, taste and value of the products. Small scale spraying of K soap or Neem oil on infected parts are considered the best chemical treatments of plants grown for pharmaceutical purposes (Barco et al., 2018). The attack of PM in greenhouse can be prevented through proper supply of nutrients to the plants. A healthier plant can defend itself better against disease attacks.

Biological Control

In its early stages PM appears as white colored cushions on plant leaves; therefore, it can be detected at earlier stages. Kendrick (1985) has recommended *hyperparasitic* fungi including *Ampelomycesquisqualis* as a biological control of early stage PM. Eladet al. (1998) tested two biocontrol preparations for their ability to control *Sphaerothecafusca*and*Botrytis cinerea*on greenhouse cucumber and reported that *Trichodermaharzianum*T39 (TRICHODEX) spray reduced PM severity by up to 97% but its efficacy declined to 18–55% control as the epidemic progressed. They further reported that unlike on young leaves, on older leaves the control of PM by *T. harzianum*T39 was poor; whereas, *Ampelomycesquisqualis*(AQ10) was very effective against PM, achieving up to 98% of control. Some microorganisms that compete PM have also been tested to serve as an organic control of PM (Schumacher &Tudzynski, 2011; Costa et al., 2013).

Physical Control

Rise of temperature due to sun or due to artificial sources of heat causes lowers RH. The chain of conidia dries under such conditions. Conidia are source of spread of dry PM and can be functional only through air circulation in the greenhouse. Physical entry of conidia is thus through open portions of greenhouse where from the air or the workers enter or exit the facility. The airborne conidia can be removed from the greenhouse through proper air filtration and through the use of UV germicidal disinfection. Air filtration in combination with UV-C wavelength light serves efficiently in removing airborne conidia and other fungal spores from inside of the greenhouse. Because of the reasonable size of *Botrytis cinerea* spores; i.e., diameter = 11-12 microns, these spores can completely be captured with filters rated MERV 12 and higher (Brais, n.d.).

In cases where chemical and biological controls do not effectively control PM in the greenhouse, physical methods remain the only ways to limit PM invasions. The use of germicidal UV as a proven disinfection technology has been recommended in the literature which shows that a well-engineered mix of air coil treatment with germicidal UV and air filtration can be an efficient way of reducing PM outbreaks in greenhouses. Such a system follows continuous UV surface decontamination of all cooling and heating coils of the facility and frequent air changes with filtration complemented by germicidal UV.

Sterilization of Botrytis Cinereasporesl

Literature reports the use of germicidal sterilization of *Botrytis cinerea* spores with varying wavelength UVs including UV-A (315-400 nm), UV-B (28-315 nm), and UV-C (100-280 nm). Mercier et al. (2001) achieved 95% spore disinfection rates using 40-450 mJ/cm^2 of UV-C irradiation dosage for slowing down the decay on Bell peppers caused by *Botrytis cinerea*. Among UV-A, UV-B, and UV-C, Latorreet al. (2012) found UV-C as the most effective wavelength to for irradiation on *Botrytis cinerea*. Their findings showed no survival of PM spores at a UV dose of 110 mJ/cm^2. Using UV susceptibility rate constant ($k = 0.0052\ m^2/J$) a UV lighting system can be designed for *Botrytis cinerea* to apply the sterilization dosage in less than a minute exposure.

Proper filtration of air supported with germicidal disinfection of air and surfaces using ultraviolet (UV) light is an efficient way of eliminating PM outbreaks in the greenhouse. However, UV-C radiation induces metabolic modifications in plants as reported by Marti et al. (2014) for secondary metabolite biosynthesis. Marti et al. (2014) tested UV-C radiation for its effects on leaves of three different selected species of plants including cannabis and reported that the metabolomic modifications were species-specific; for example, no notable changes in the content of cannabinoid were observed in cannabis plants.

Environmentally, following some basic routine can prevent PM to attack greenhouse plants. The routine includes selection of seed genetics (registered safe strain of cannabis), ensuring plant health through proper plant nutrition, and maintaining controlled environments inside the greenhouse. Optimal conditions for growing healthy and PM free cannabis in a greenhouse include i) 70-80 °F inside temperature, ii) 40-50% relative humidity, iii) good fresh blowing air circulation including an exhaust system, iv) a combination of nutrients (150 mg/kg AL-P_2O_5, 310-330 mg/kg AL-K_2O, 150-300 kg/ha N from any fertilizer type, 90 kg/ha P_2O_5), v) UV-C germicidal disinfection, and vi) hygienic indoor conditions.

Fujikawaet al. (2013) tested the use of supplemental UV radiations to controls a rose PM disease under the greenhouse conditions and suggested that UV radiation can suppress PM development through inhibition of fungal growth and activation of host defenses. A system designed for controlling *Botrytis* can also effectively control other fungal contaminants. The UV floor/surfaces disinfection systems and the UV air disinfection systems are the efficient ways to use germicidal UV light for controlling mold in greenhouses. Furthermore, the greenhouse air quality can be upgraded with the use of air filters and positive pressurization.

An air-tight growing facility with positive air pressure minimizes outbreak of PM. In China, the major greenhouses are designed to have a pressurized isolated entry point to control the entrance of spores into the facility. Based on literature review, the optimal conditions for PM free optimum production require control on greenhouse temperature and humidity. Germination and reproduction of PM spores is favored by a combination of temperatures and humidity of certain values. Controlling the greenhouse environments is vital for disease free cultivation of cannabis. This can be achieved through avoiding the favourable environmental conditions for PM to live as well as correct setting of controlled environment for cannabis plants to flourish. The following controlled environmental factors can take away the environment for PM and allow the cannabis plants to thrive.

***Indoor temperature*:** Greenhouse inside temperature in combination with a proper airflow reduces spreading of PM infections to the plant leaves. For optimal production of cannabis, literature suggests 70-80 °F (Mills et al., 2012) and 68 °F (Marti et al., 2014) indoor temperature.

***Humidity control*:** A proper exhaust or ventilation system replaces the stale air with fresh air resultant into reduced humidity. Literature reports different values for RH to maintain during greenhouse cultivation of cannabis; for example, 40-50% for cannabis (Mills et al., 2012), 70% for cannabis (Marti et al., 2014), and 60-70% for seedlings and 40% for flowering stages of marijuana (Online Blog).

***Airflow*:** A constant blow of air prevents tiny spores of PM to land and latch on plant leaves. A proper airflow keeps the greenhouse breezy and fresh. For a small greenhouse (grow room) of 22 m^2 size, Mills et al. (2012) modeled a ventilation system that computes using 7 oscillating fans to keep the place dry and simultaneously equipped with 30 supply and exhaust fans for ventilation. Table 1A annexed with Mills et al. (2012) details several other components features modeled for a highly equipped and controlled greenhouse.

***Lighting*:** Spectral quality of greenhouse lights affect cannabis plant growth. Therefore, making the quality of lights to address the needs of plants can result in increases quality of buds, consistency in plant growth, and the cannabis yield. Spectral quality of lights varies with species, cultivars, and production strategies (Critten, 1991; Nemali& van Iersel, 2004; Fu et al., 2012; Ilicet al., 2012; Blomet al., 2016; Li et al., 2017). Studies have been reported on growth of various plants affected by spectral quality of lights (Loughrin&Kasperbauer 2001; Yorioet al., 2001; Beamanet al., 2009; Chang et al., 2009; Lefsrudet al, 2008). The role of sub-canopy lights on production of cannabis was investigated by Hawley et al. (2018) by growing plants without additionally provided lighting to the below canopy areas, with red/blue, and with red-green-blue supplemental sub-canopy lights and reported that both combinations of lights result in significant increases in cannabis yield as well as THC concentration in tissues of bud present at lower portions of canopy of the plants that were grown in control treatments.

***Consistency*:** Ideal greenhouse climate helps as sudden changes in temperature and/or humidity can create favorable conditions for PM to spread. This can be done through environmental controllers available in the market. Literature report values on the optimal environmental conditions for a greenhouse to grow powdery mildew free cannabis (Table 4).

Table 4. Summarized literature on temperature, relative humidity, day light hours, and air flow for greenhouse production of cannabis

Crop	Temperature (°F)	Relative humidity (%)	Daily photo period (hr)	Air flow	Source
Cannabis	70-80	40-50		7 oscillating fans to keep a 22 m^2 room dry 30 supply and exhaust fans for ventilation	Mills et al., 2012 (See annex Table 1A for further detail)
Cannabis	68	70	16	-	Marti et al., 2014
Marijuana	70-80	60-70 (seedling) 40 (flowering)	Normal day	Fresh air and sun light	Blog

Hygienically maintained greenhouse: A regular protocol of ensuring thorough cleanliness inside the greenhouse can help avoid favorable environment of PM to grow and spread. Cleanliness of a greenhouse starts from the basic steps of cleaning all the tools, picking up and discarding all plant debris immediately, and wiping down all surfaces regularly to greatly reduce chances of a PM outbreak.

The work areas are typically recommended to be cleaned with solution of peroxide and water. If diluted to very low concentration, this solution can be used to spray on plant leaves, branches, and stem. The common equipment under use in greenhouse for plant trimming including scissors, cutters, trimmers, and blades can be cleaned using rubbing alcohol. Care must be taken for not spraying alcohol on plants. Glick (2012) enlisted precautionary measures to take care of hygienic conditions of the greenhouse, which include i) consistently observing the plants during all especially development stages, ii) looking for symptoms of congregation aphids and mites on above and especially under the leaves, iii) removing of the debris of the diseased/infected plants, and iv) looking for any cutworms and soil predators gathering around the base of the plant stem. Table 5 summarizes the optimal greenhouse environmental and crop input requirements for disease free cultivation of cannabis.

Table 5. The optimal greenhouse environmental and crop input requirements for disease free cultivation of cannabis

Temperature (°F)	Relative humidity (%)	Daily light (hr)	Air circulation and sterilization	Nutrients
70-80	40-50	16	To keep a 22 m^2 room dry and ventilated - 7 oscillating fans - 30 supply and exhaust fans UVC (100-280 nm)	150 mg/kg AL-P_2O_5 310-330 mg/kg AL-K_2O 150-300 kg/ha N from any fertilizer type 90 kg/ha P_2O_5

FUTURE SCOPE

Future scope of cannabis research and production activities depend on its legalization in various countries. Although its production for medical or recreational use has been ongoing since 1999 in selective states of North America but full-scale activities still wait for a country and/or worldwide adoption of this practice. Despite the use of cannabis and its effects on population health, many countries have yet to legalize its commercial production. This has resulted limitations on the use of cannabis and cannabinoids and its health effects. Development and testing of best management practices for cannabis production this remain halted. Literature suggest that the research on cannabis and cannabinoids cannot be promoted without removing barriers, which are yet to be identified and addressed.

CONCLUSION

Literature reveals that despite high energy related costs, cannabis cultivation in greenhouse seems profitable because of its high product value. However, regardless of the production costs related to energy, the producers need to pay attention to important measures for disease control in greenhouse that remain i) genetics, ii) plant nutrition, and iii) controlled environments. Genetics deals with seed selection, plant nutrition is all about maintaining optimal concentration of NPK especially N, and controlled environment including controlled supply of irrigation water, light, aeration, and indoor temperature can help efficiently controlling disease outbreak through chemical, biological, physical, or by the combined use of these methods.

REFERENCES

Abbas, F., Fares, A., & Samira, F. (2011). Field Calibrations of Soil Moisture Sensors in a Forested Watershed. *Sensors (Basel)*, *11*(6), 6354–6369. doi:10.3390110606354 PMID:22163959

Ahmed, S. A., Ross, S. A., Slade, D., Radwan, M. M., Zulfiqar, F., & ElSohly, M. A. (2008). Cannabinoid ester constituents from high potency Cannabis sativa L. *Planta Medica*, *74*(03), 536–542. doi:10.1055-2008-1075212

Aubin, M.-P., Seguin, P., Vanasse, A., Tremblay, G. F., Mustafa, A. F., & Charron, J.-B. (2015). Industrial hemp response to nitrogen, phosphorus, and potassium fertilization. *Crop. Forage TurfgrassMgt*, *1*(1), 1–10. doi:10.2134/cftm2015.0159

Bakro, F., Wielgusz, K., Bunalski, M., & Jedryczka, M. (2018). An overview of pathogen and insect threats to fibre and oilseed hemp (Cannabis sativa L.) and methods for their biocontrol. *Integrated Control in Oilseed Crops IOBC-WPRS Bulletin*, *136*, 9–20.

Bar-Yosef, B. (1999). Advances in fertigation. *Advances in Agronomy*, *65*, 1–77. doi:10.1016/S0065-2113(08)60910-4

Beaman, A. R., Gladon, R. J., & Schrader, J. A. (2009). Sweet basil requires an irradiance of 500 m mol·m^{-2}·s^{-1} for greatest edible biomass production. *HortScience*, *44*(1), 64–67. doi:10.21273/HORTSCI.44.1.64

Blom, T. J., Straver, W. A., Ingratta, F. J., Khosla, S., & Brown, W. (2016). *Carbon dioxide in greenhouses*. Ontario Ministry of Agriculture, Food, and Rural Affairs. http://www. omafra.gov.on.ca/english/crops/facts/00-077.htm#suppl

Bócsa, I., & Karus, M. (1998). The cultivation of hemp. Botany, varieties, cultivation and harvesting. Hemptech.

Caplan, D., Dixon, M., & Zheng, Y. (2017a). Optimal Rate of Organic Fertilizer during the Flowering Stage for Cannabis Grown in Two Coir-based Substrates. *HortScience*, *52*(12), 1796–1803. doi:10.21273/HORTSCI12401-17

Caplan, D., Dixon, M., & Zheng, Y. (2017b). Optimal rate of organic fertilizer during the vegetative-stage for cannabis grown in two coir-based substrates. *HortScience*, *52*(9), 1307–1312. doi:10.21273/HORTSCI11903-17

Cervantes, G. (2006). *Marijuana Horticulture: The Indoor/Outdoor Medical Grower's Guide*. Van Patten Publishing.

Chang, X., Alderson, P. G., & Wright, C. J. (2009). Enhanced UV-B radiation alters basil (Ocimum-basilicum L.) growth and stimulates the synthesis of volatile oils. *Journal of Horticulture and Forestry, 12*, 27–31.

Chen, K., Sun, X., Qin, C., & Tang, X. (2010). Color grading of beef fat by using computer vision and support vector machine. *Computers and Electronics in Agriculture, 70*(1), 27–32. doi:10.1016/j.compag.2009.08.006

Chris, C. (2017). *Powdery Mildew: Identify, Prevent and Treat White Powdery Mildew on Cannabis: Part 2 – Prevention of PM*. Cannabis Chris Articles. http://cannabischris.com/2017/08/powdery-mildew-prevention-of-pm/

Clarke, R. C. (1999). Botany of the genus Cannabis. In P. Ranalli (Ed.), *Advances in Hemp Research* (pp. 1–19). Food Products Press, The Haworth Press Inc.

Clarke, R. C., & Merlin, M. D. (2016). Cannabis Domestication, Breeding History, Present-day Genetic Diversity, and Future Prospects. *Critical Reviews in Plant Sciences, 35*(5-6), 293–327. doi:10.1080/07352689.2016.1267498

Cosentino, S. L., Riggi, E., Testa, G., Scordia, D., & Copani, V. (2013). Evaluation of European developed fibre hemp genotypes (Cannabis sativa L.) in semi-arid Mediterranean environment. *Industrial Crops and Products, 50*, 312–324. doi:10.1016/j.indcrop.2013.07.059

Costa, L., Rangel, D. E. N., Morandi, M. A. B., & Bettiol, W. (2013). Effects of UV-B radiation on the antagonistic ability of Clonostachysrosea to Botrytis cinerea on strawberry leaves. *Biological Control, 65*(1), 95–100. doi:10.1016/j.biocontrol.2012.12.007

Critten, D. L. (1991). Optimization of CO_2 concentration in greenhouses: A modelling analysis for the lettuce crop. *Journal of Agricultural Engineering Research, 48*, 261–271. doi:10.1016/0021-8634(91)80020-F

Dewey, L. H. (1914). Hemp in U.S.D.A. Yearbook 1913 United States Department of Agriculture.

Di Bari, V., Campi, P., Colucci, R., & Mastrorilli, M. (2004). Potential productivity of fibre hemp in southern Europe. *Euphytica, 140*(1-2), 25–32. doi:10.100710681-004-4751-1

Ehlers, W., & Goss, M. (2003). *Water Dynamics in Plant Production*. CABI Publishing. doi:10.1079/9780851996943.0000

Ehrensing, D. T. (1998). Feasibility of industrial hemp production in the United States Pacific Northwest. *Oregan State Univ. Agr. Expt. Sta. Bul.*, 681.

Elad, Y., Kirshner, B., Yehuda, N., & Sztejnberg, A. (1998). Management of powdery mildew and gray mold of cucumber by Trichodermaharzianum T39 and Ampelomycesquisqualis AQ10. *BioControl, 43*(2), 241–251. doi:10.1023/A:1009919417481

ElSohly, M. A., & Slade, D. (2005). Chemical constituents of marijuana: The complex mixture of natural cannabinoids. *Life Sciences, 78*(5), 539–548. doi:10.1016/j.lfs.2005.09.011 PMID:16199061

Elzinga, S., Fischedick, J., Podkolinski, R., & Raber, J. C. (2015). Cannabinoids and terpenes as chemotaxonomic markers in cannabis. *Natural Products Chemistry & Research, 3*, 1–9.

European Monitoring Centre for Drugs and Drug Addiction. (2018). Cannabis legislation in Europe: an overview. Publications Office of the European Union. doi:10.2810/566650

FAO. (1998). Crop evapotranspiration - Guidelines for computing crop water requirements. *FAO Irrigation and Drainage Paper, 56*. Retrieved June 01, 2019, from http://www.fao.org/3/X0490E/x0490e00.htm

Farag, S., & Kayser, O. (2015). Cultivation and breeding of Cannabis sativa L. for preparation of standardized extracts for medicinal purposes. In Medicinal and aromatic plants of the world. Springer.

Fares, A., Abbas, F., Domingos, M., & Mair, A. (2011). Improved Calibration Functions of Three Capacitance Probes for the Measurement of Soil Moisture in Tropical Soils. *Sensors (Basel), 11*(5), 4858–4874. doi:10.3390110504858 PMID:22163880

Forsyth, A. (2000). Psychoactive drugs: the pharmacopoeia of substance use. London: Stationery Office.

Fu, W., Li, P., & Wu, Y. (2012). Effects of different light intensities on chlorophyll fluorescence characteristics and yield in lettuce. *Scientia Horticulturae, 135*, 45–51. doi:10.1016/j.scienta.2011.12.004

García-Tejero, I. F., Durán-Zuazo, V. H., Pérez-Álvarez, R., Hernández, A., Casano, S., Morón, M., & Muriel-Fernández, J. L. (2014). Impact of Plant Density and Irrigation on Yield of Hemp (Cannabis sativa L.) in a Mediterranean Semi-arid Environment. *Journal of Agricultural Science and Technology, 16*, 887–895.

Government of Canada. (2019). *Cannabis Legalization and Regulation*. Available at: https://www.justice.gc.ca/eng/cj-jp/cannabis/

Grabowska, L., Rebarz, M., & Chudy, M. (2009). Breeding and cultivation of industrial hemp in Poland. Institute of Natural Fibres and Medicinal Plants.

Graham, D. S. (2002). Medical marijuana: Canada's regulations, pharmacology, and social policy. *Canadian Pharmacists Journal, 137*(902), 23–27.

Guttormsen, G. (1969). Accumulation of salts in the sub-irrigation of pot plants. *Plant and Soil, 31*(3), 425–438. doi:10.1007/BF01373814

Hajizadeh, M. (2016). Legalizing and Regulating Marijuana in Canada: Review of Potential Economic, Social, and Health Impacts. *International Journal of Health Policy and Management, 5*(8), 453–456. doi:10.15171/ijhpm.2016.63 PMID:27694657

Hall, J. (2017). *Industrial hemp and sugarcane rotation – not just a pipe dream*. A report of the Managing Director – Industrial Hemp Australia. Accessed on May 24, 2019 at www.australiansugarcane.com.au/Back issues/133jjsug09/21_Industrial.pdf

Happyana, N., Agnolet, S., Muntendam, R., Van Dam, A., Schneider, B., & Kayser, O. (2013). Analysis of cannabinoids in laser-microdissectedtrichomes of medicinal Cannabis sativa using LCMS and cryogenic NMR. *Phytochemistry, 87*, 51–59. doi:10.1016/j.phytochem.2012.11.001 PMID:23280038

Hargreaves, G. H., & Allen, R. G. (2003). History and Evaluation of Hargreaves Evapotranspiration Equation. *Journal of Irrigation and Drainage Engineering, 129*(1), 53–63. doi:10.1061/(ASCE)0733-9437(2003)129:1(53)

Hawley, D., Graham, T., Stasiak, M., & Dixon, M. (2018). Improving Cannabis Bud Quality and Yield with Subcanopy Lighting. *HortScience*, *53*(11), 1593–1599. doi:10.21273/HORTSCI13173-18

Hough, M., Warburton, H., Few, B., May, T., Man, L.-H., Witton, I., & Turnbull, P. J. (2003). *A growing market: The domestic cultivation of cannabis York*. Joseph Rowntree Foundation.

Ilic, Z. S., Milenkovic, L., Stanojevic, L., Cvetkovic, D., & Fallik, E. (2012). Effects of the modification of light intensity by color shade nets on yield and quality of tomato fruits. *Scientia Horticulturae*, *139*, 90–95. doi:10.1016/j.scienta.2012.03.009

Ivanyi, I. (2005). *Relationship between leaf nutrient concentrations and yield of fibre hemp (Cannabis sativa L.). Alps-Adria Scientific Workshop*, Portoroz, Slovenia. 10.1556/CRC.33.2005.1.23

Ivonyi, I., Izsoki, Z., & Van der Werf, H. M. G. (1997). Influence of nitrogen supply and P K levels of the soil on dry matter and nutrient accumulation of fibre hemp (Cannabis sativa L.). *Journal of the International Hemp Association*, *4*(1), 82–87.

Jaczewski, A. (1927). *Mucnisto-rosjanyegriby* [Powdery mildew fungi] KarmannyiOpreditel'. *Gribov*, *5*(2), 412.

Jordan, H., Lang, A. L., & Enfield, G. H. (1946). Effects of fertilizers on yields and breaking strengths of American hemp, Cannabis sativa. *Journal of the American Society of Agronomists*, *38*(6), 551–563. doi:10.2134/agronj1946.00021962003800060010x

Kadar, J., & Tarkany-Szucs, S. (2003). Effect of mineral fertilization on the yield of fibre hemp (Cannabis sativa L.). *Novenytermeles*, *52*(2), 217–228.

Kendrick, B. (1985). *The fifth kingdom*. Mycologue Publication.

2019 Kentucky Pest News. (n.d.). Available at http://www.kyhempdisease.com/powdery-mildew-of-hemp.html

Kišgeci, J. (1994). *Praise to Hemp*. Nolit. (in Serbian)

Knight, G., Hansen, S., Connor, M., Poulsen, H., McGovern, C., & Stacey, J. (2010). The results of an experimental indoor hydroponic cannabis growing study, using the "Screen of Green" (ScrOG) method-yield, tetrahydrocannabinol (THC) and DNA analysis. *Forensic Science International*, *202*(1-3), 36–44. doi:10.1016/j.forsciint.2010.04.022 PMID:20462712

Kowalski, W. J. (2009). *Ultraviolet Germicidal Irradiation Handbook: UVGI for Air and Surface Disinfection*. Springer. doi:10.1007/978-3-642-01999-9

Latorre, B. A., Rojas, S., Diaz, G. A., & Chuaqui, H. (2012). Germicidal effect of UV light on epiphytic fungi isolated from bleberry. *CienInvAgr*, *39*(3), 473–480.

Lefsrud, M. G., Kopsell, D. A., & Sams, C. E. (2008). Irradiance from distinct wavelength light emitting diodes affect secondary metabolites in kale. *HortScience*, *43*(7), 2243–2244. doi:10.21273/HORTSCI.43.7.2243

Leggett, T. (2006). Review of the world cannabis situation. *Bulletin on Narcotics*, *58*, 1–155. PMID:19066071

Lemberger, L. (1980). Potential therapeutic usefulness of marihuana. *Annual Review of Pharmacology and Toxicology*, *20*(1), 151–172. doi:10.1146/annurev.pa.20.040180.001055 PMID:6104468

Li, T., Ji, Y. H., Zhang, M., & Li, M. Z. (2017). Determining optimal CO_2 concentration of greenhouse tomato based on PSO-SVM. *Applied Engineering in Agriculture*, *33*(2), 157–166. doi:10.13031/aea.11578

Lisson, S., & Mendham, N. (1998). Response of fibre hemp (Cannabis sativa L.) to varying irrigation regimes. *Journal of the International Hemp Association*, *5*(1), 9–15.

Lisson, S. N., Mendham, N. J., & Carberry, P. S. (2000). Development of a hemp (Cannabis sativa L.) simulation model 3. The effect of plant density on leaf appearance, expansion and senescence. *Australian Journal of Experimental Agriculture*, *40*(3), 419–423. doi:10.1071/EA99060

Loughrin, J. H., & Kasperbauer, M. J. (2001). Light reflected from colored mulches affects aroma and phenol content of sweet basil (Ocimumbasilicum L.) leaves. *Journal of Agricultural and Food Chemistry*, *49*(3), 1331–1335. doi:10.1021/jf0012648 PMID:11312860

Mahmud, M. S. (2019). *Development of a machine vision system for strawberry powdery mildew disease detection* (MS thesis). Dalhousie University, Halifax, Nova Scotia, Canada.

Mallery, M. (2011). Marijuana National Forest: Encroachment on California Public Lands for Cannabis Cultivation. *Berkeley Undergraduate Journal*, *23*(2), 51.

Marti, G., Schnee, S., Andrey, Y., Simões-Pires, C., Carrupt, P., & Gindro, J. K. (2014). Study of Leaf Metabolome Modifications Induced by UV-C Radiations in Representative Vitis, Cissus and Cannabis Species by LC-MS Based Metabolomics and Antioxidant Assays. *Molecules (Basel, Switzerland)*, *19*(9), 14004–14021. doi:10.3390/molecules190914004 PMID:25197936

McPartland, J. M. (2000). *Cannabis pathogens XI: Septoria spp. on Cannabis sativa, sensustricto*. Academic Press.

McPartland, J. M., & Hughes, S. (1994). Cannabis pathogens VII: A new species, Schiffnerula cannabis. *Mycologia*, *86*(6), 867–869. doi:10.1080/00275514.1994.12026491

Mercier, J., Baka, M., Reddy, B., Corcuff, R., & Arul, J. (2001). Shortwave Ultraviolet Irradiation for Control of Decay Caused by Botrytis cinerea in Bell Pepper: Induced Resistance and Germicidal Effects. *Journal of the American Society for Horticultural Science*, *126*(1), 128–133. doi:10.21273/JASHS.126.1.128

Milanes-Murcia, M. E. (2017). Cannabis and Water Management: International Regulation and the Legal Framework of the European Union. *University of Pennsylvania Law Review*, *49*(1), 117–153.

Morris, K. (1997). The cannabis remedy-wonder worker or evil weed? *Lancet*, *350*(9094), 1828. doi:10.1016/S0140-6736(05)63651-3 PMID:9454053

Pejic, B., Sikora, V., Milic, S., Mackic, K., Koren, A., & Bajic, I. (2018). *Effect of drip irrigation on yield and evapotranspiration of fibre hemp (Cannabis sativa L.)*. Academic Press.

Pertwee, R. G. (2014). *Handbook of cannabis 781*. doi:10.1093/acprof:oso/9780199662685.001.0001

Pineda, M., Pérez-Bueno, M. L., & Barón, M. (2018). Detection of bacterial infection in melon plants by classification methods based on imaging data. *Frontiers in Plant Science*, *9*, 164. doi:10.3389/fpls.2018.00164 PMID:29491881

Potter, D. J. (2014). A review of the cultivation and processing of cannabis (Cannabis sativa L.) for production of prescription medicines in the UK. *Drug Testing and Analysis*, *6*(1-2), 31–38. doi:10.1002/dta.1531 PMID:24115748

Potter, D. J., & Duncombe, P. (2012). The effect of electrical lighting power and irradiance on indoor-grown cannabis potency and yield. *Journal of Forensic Sciences*, *57*(3), 618–622. doi:10.1111/j.1556-4029.2011.02024.x PMID:22211717

Potter, G. R. (2006). *Weed, Need and Greed: Domestic Marijuana Production and the UK Cannabis Market* (PhD thesis). University of Sheffield.

Radwan, M. M., ElSohly, M. A., El-Alfy, A. T., Ahmed, S. A., Slade, D., Husni, A. S., Manly, S. P., Wilson, L., Seale, S., Cutler, S. J., & Ross, S. A. (2015). Isolation and pharmacological evaluation of minor cannabinoids from highpotency Cannabis sativa. *Journal of Natural Products*, *78*(6), 1271–1276. doi:10.1021/acs.jnatprod.5b00065 PMID:26000707

Ranalli, P. (1999). Agronomical and physiological advances in hemp crops. In P. Ranalli (Ed.), *Advances in Hemp Research* (pp. 61–84). Food Products Press, The Haworth Press Inc. doi:10.1201/9781498705820-11

Raviv, M., & Lieth, J. H. (2007). *Soilless culture: Theory and practice* (1st ed.). Elsevier B.V.

Reisinger, P., Lehoczky, E., & Komives, T. (2005). Competitiveness and precision management of the noxious weed Cannabis sativa L. in winter wheat. *Communications in Soil Science and Plant Analysis*, *36*(4-6), 629–634. doi:10.1081/CSS-200043303

Rojas-Moraleda, R., Valous, N. A., Gowen, A., Esquerre, C., Härtel, S., Salinas, L., & O'Donnell, C. (2017). A frame-based ANN for classification of hyperspectral images: Assessment of mechanical damage in mushrooms. *Neural Computing & Applications*, *28*(1), 969–981. doi:10.100700521-016-2376-7

Sankari, H., & Mela, T. J. N. (1998). Plant development and stem yield of non-domestic fibre hemp (Cannabis sativa L.) cultivars in long-day growth conditions in Finland. *Journal Agronomy & Crop Science*, *181*(3), 153–159. doi:10.1111/j.1439-037X.1998.tb00411.x

Savakar, D. G., & Anami, B. S. (2015). Grading of bulk food grains and fruits using computer vision. *Journal of Agricultural Engineering and Biotechnology*, *3*(1), 1–10. doi:10.18005/JAEB0301001

Scheifele, G. (1999). *Determining the feasibility and potential of field production of low THC industrial hemp (Cannabis sativa) for fibre and seed grain in Northern Ontario*. Ontario Ministry of Agriculture and Food.

Schultes, R. E., Klein, W. M., Plowman, T., & Lockwood, T. E. (1974). Cannabis: An example of taxonomic neglect. *Botanical Museum Leaflets, Harvard University*, *23*, 337–367.

Schumacher, J., & Tudzynski, P. (2011). Morphogenesis and Pathogenicity in Fungi. *Current Genetics*, *22*, 225–241.

Shelton, B. G., Kirkland, K. H., Flanders, W. D., & Morris, G. K. (2002). Profiles of Airborne Fungi in Buildings and Outdoor Environments in the United States. *Applied and Environmental Microbiology*, *68*(4), 1743–1753. doi:10.1128/AEM.68.4.1743-1753.2002 PMID:11916692

Stemeroff, J. (1917). *Irrigation Management Strategies for Medical Cannabis in Controlled Environments* (PhD thesis). University of Guelph.

Story, D., Kacira, M., Kubota, C., Akoglu, A., & An, L. (2010). Lettuce calcium deficiency detection with machine vision computed plant features in controlled environments. *Computers and Electronics in Agriculture*, *74*(2), 238–243. doi:10.1016/j.compag.2010.08.010

Struik, P. C., Amaducci, S., Bullard, M. J., Stutterheim, N. C., Venturi, G., & Cromack, H. T. H. (2000). Agronomy of fibre hemp (Cannabis sativa L.) in Europe. *Industrial Crops and Products*, *11*(2-3), 107–118. doi:10.1016/S0926-6690(99)00048-5

Sun, D. W. (2000). Inspecting pizza topping percentage and distribution by a computer vision method. *Journal of Food Engineering*, *44*(4), 245–249. doi:10.1016/S0260-8774(00)00024-8

Sun, D. W. (Ed.). (2016). *Computer vision technology for food quality evaluation*. Academic Press.

Suthaparan, A., Stensvand, A., Solhaug, K. A., Torre, S., Mortensen, L. M., Gadoury, D. M., Seem, R. C., & Gislerød, H. R. (2012). Suppression of powdery mildew (Podosphaerapannosa) in greenhouse roses by brief exposure to supplemental UV-B radiation. *Plant Disease*, *96*(11), 1653–1660. doi:10.1094/PDIS-01-12-0094-RE PMID:30727454

Transhel, V., Gutner, L., & Khokhryakov, M. (1933). A list of fungi found on new cultivated textile plants. Moscow Inst. Nov. Lubian. Syria. *Trudy*, *4*, 127–140.

UN Economic and Social Council (ECOSOC). (1988). *United Nations Convention Against Illicit Traffic in Narcotic Drugs and Psychotropic Substances*. Available at: https://www.refworld.org/docid/49997af90.html

UN General Assembly. (1971). *Convention on Psychotropic Substances*. Available at: https://www.refworld.org/docid/3b00f1ad4b.html

UN General Assembly. (1972). *Protocol Amending the Single Convention on Narcotic Drugs*. Available at: https://www.refworld.org/docid/3b00f1c118.html

Van Dam, J. E. G. (1995). Potentials of hemp as an industrial fibre crop. *Proceedings of First Bioresource Hemp Symposium, nova-Institute*, 405-422.

Van der Werf, H. M. G., & Van den Berg, W. (1995). Nitrogen fertilization and sex expression affect size variability of fibre hemp (Cannabis sativa L.). *Oecologia*, *103*(4), 462–470. doi:10.1007/BF00328684 PMID:28306994

Van der Werf, H. M. G., Van Geel, W. C. A., Van Gils, L. J. C., & Haverkort, A. J. (1995). Nitrogen fertilization and row width affect self-thinning and productivity of fibre hemp (Cannabis sativa L.). *Field Crops Research*, *42*(1), 27–37. doi:10.1016/0378-4290(95)00017-K

van der Werf, H. M. G., Wijlhuizen, M., & De Schutter, J. (1995). Plant density and self-thinning affect yield and quality of fibre hemp (Cannabis sativa L.). *Field Crops Research, 40*(3), 153–164. doi:10.1016/0378-4290(94)00103-J

Vanhove, W., Van Damme, P., & Meert, N. (2011). Factors determining yield and quality of illicit indoor cannabis (Cannabis spp.) production. *Forensic Science International, 212*(1-3), 158–163. doi:10.1016/j.forsciint.2011.06.006 PMID:21737218

Velázquez-López, N., Sasaki, Y., Nakano, K., Mejía-Muñoz, J. M., & Romanchik, K. E. (2011). Detection of powdery mildew disease on rose using image processing with Open CV. *Revista Chapingo Serie Horticultura, 17*(2), 151–160. doi:10.5154/r.rchsh.2011.17.017

Vemuri, V. K., & Makriyannis, A. (2015). Medicinal chemistry of cannabinoids. *Clinical Pharmacology and Therapeutics, 97*(6), 553–558. doi:10.1002/cpt.115 PMID:25801236

Vera, C. L., Malhi, S. S., Raney, J. P., & Wang, Z. H. (2004). The effect of N and P fertilization on growth, seed yield and quality of industrial hemp in the Parkland region of Saskatchewan. *Canadian Journal of Plant Science, 84*(4), 939–947. doi:10.4141/P04-022

Wright, R. D., & Niemiera, A. X. (1987). Nutrition of container-grown woody nursery crops. *Horticultural Reviews, 9*, 75–101.

Wspanialy, P., & Moussa, M. (2016). Early powdery mildew detection system for application in greenhouse Automation. *Computers and Electronics in Agriculture, 127*, 487–494. doi:10.1016/j.compag.2016.06.027

Xu, G., Wolf, S., & Kafkafi, U. (2001). Effect of varying nitrogen form and concentration during growing season on sweet pepper flowering and fruit yield. *Journal of Plant Nutrition, 24*(7), 1099–1116. doi:10.1081/PLN-100103806

Yorio, N. C., Goins, G. D., Kagie, H. R., Wheeler, R. M., & Sager, J. C. (2001). Improving spinach, radish, and lettuce growth under red light-emitting diodes (LEDs) with blue light supplementation. *HortScience, 36*(2), 380–383. doi:10.21273/HORTSCI.36.2.380 PMID:12542027

Zhang, M., & Meng, Q. (2011). Automatic citrus canker detection from leaf images captured in field. *Pattern Recognition Letters, 32*(15), 2036–2046. doi:10.1016/j.patrec.2011.08.003

Zheng, L., Zhang, J., & Wang, Q. (2009). Mean-shift-based color segmentation of images containing green vegetation. *Computers and Electronics in Agriculture, 65*(1), 93–98. doi:10.1016/j.compag.2008.08.002

Chapter 9
Predictable Scenarios of Fuzzy Logic Analysis for Sprinkler Irrigation Control

Mohamed Elsayed Elhagarey
Desert Research Center, Egypt

Mohamed M. Hushki
Faculty of Mechanical Engineering, SzentIstvan University, Hungary

Szabo E. Istvan
Faculty of Mechanical Engineering, SzentIstvan University, Hungary

ABSTRACT

MATLAB will be utilized to validate the various irrigation systems and report it; the air temperature, wind, and humidity will be member functions to improve the efficiency of irrigation performance before the irrigation process, and the fuzzy information system consists of fuzzy rules, which are derived from information of experts or from input-output learning of the system. Rules mimic human reasoning. Mamdani method is mostly applied in the fuzzy inference technique, and the generalized bell function is used for both of temperature and wind where the triangular function used for humidity. The principles were based on the last experiments and personal experiences, and the output will change into a crisp value from this procedure of defuzzification. There are many different methods to do defuzzification, but the most common technique is centroid method. The resultant surface graphic is enough to monitor, validate, and report the irrigation system efficiency to execute and schedule the irrigation practice management.

INTRODUCTION

Soft computing technology is an interdisciplinary research field in computational sciences. At present, various techniques of soft computing, such as statistics, machine learning, neural network, and fuzzy data analysis, are being used for exploratory data analysis.

DOI: 10.4018/978-1-7998-5000-7.ch009

Proposed a fuzzy set theory in which the set boundaries were not precisely defined; but, boundaries were in fact gradational. Such a set was characterized by the continuum of grades of membership (characteristic) function which assigned a grade of membership ranging between zero and one to each object. The central concept of fuzzy set theory is a membership function, which numerically represents the degree to which an element belongs to a set. In fuzzy set theory, an element can be a member of a particular set and, at the same time, a member of a different set to a certain degree. In fuzzy rule-based systems, knowledge are represented by an if–then rules (Zadeh, 1965).

Fuzzy rules consist of two parts: an antecedent part, which states conditions on the input variable (s), and a consequent part, which describes the corresponding values of the output variable (Allahverdi, 2002; Carman & Seflek, 2004). Fuzzy application areas include estimation, prediction, control, approximate reasoning, intelligent system design, machine learning, image processing, machine vision, pattern recognition, in medical computing, robotics, optimization, civil, chemical and industrial engineering, but fuzzy applications in hydrology and meteorology, are relatively less common (Kindler, 1992).

An automated irrigation system not only allows a better water use efficiency, but it also provides all the necessary information to generate detailed water usage reports which are critical to assess and improve irrigation performance. An automated irrigation system resolves one of the most difficult irrigation problems: when and how much water to apply to ensure thorough wetting of the root zone without loss of water past the roots. The flow front of the irrigation water can easily be detected with soil, water sensors buried in the ground at the required depth (Castilla, 1997).

Agriculture plays a vital role in economy of countries throughout the globe providing raw material to industries and fulfilling the increasing needs of immensely growing population pressure. However, in spite of great agricultural importance, productivity is not up to the mark and farmer's gains are substantial. Several issues are anticipated responsible like high cost of production, inflation, poverty, agricultural risks, inadequate access to finance, inadequate availability of inputs and the most noteworthy climate change; putting huge threat to the water availability, which is prime source of irrigation in agriculture sector (Shabbir & Tahir, 2014).

Once the soil has reached desired moisture content, the sensors send a signal to a controller to turns off the power to a solenoid valve or a pump which controls irrigation. As a result, the automated irrigation system prevents water escaping past the root zone and therefore, improves the efficiency of water use (Sigrimis et al., 2001). The main goal of this investigation is produce a intelligence model using MATALB and Fuzzy logic system to monitoring, validate and reported the irrigation systems efficiency to correct the irrigation practices managements.

Table 1. List of acronyms and nomenclature

TEMP	= Temperature membership function,
WIND	= Wind membership function,
HUM	= Humidity membership function,
RAIN	= Rain fall membership function,
IE	= efficiency of irrigation performance membership outputs,
MF	= Membership function, and
FIS	= Fuzzy information system.
IF	= Irrigation efficiency

MATERIALS AND METHODS

Software

For sprinkler irrigation system, fuzzy logic is used based on the most factors effect of the irrigation performance, such as (air temperature, wind and air humidity) to predict the perfect time to start and shutdown the sprinkler irrigation using the automated control valve.

- **Control surfaces identification**

The variable factors and function membership will estimate for every calculated variable. Figures: 1 and 2.

Figure 1. The generalized bell function

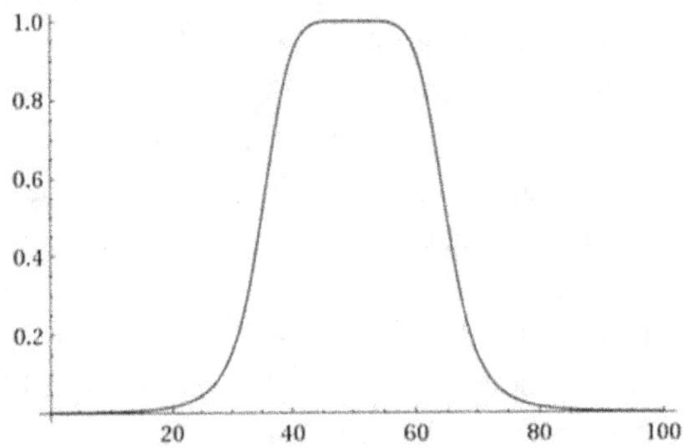

Figure 2. The triangular function

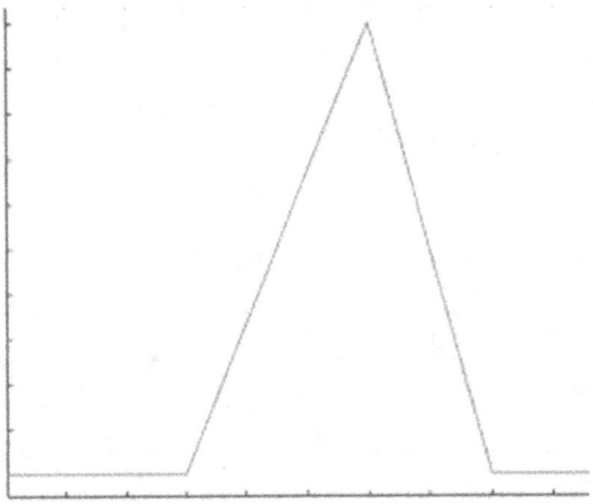

- **Control surfaces attitude**

Fuzzy logic rules will design for input data to achieve a variance procedure, in addition to merge all of the input data and output data through fuzzy logic control. Figure: 3, 4, 5, 6 and 7.

Figure 3. The membership functions for temperature input

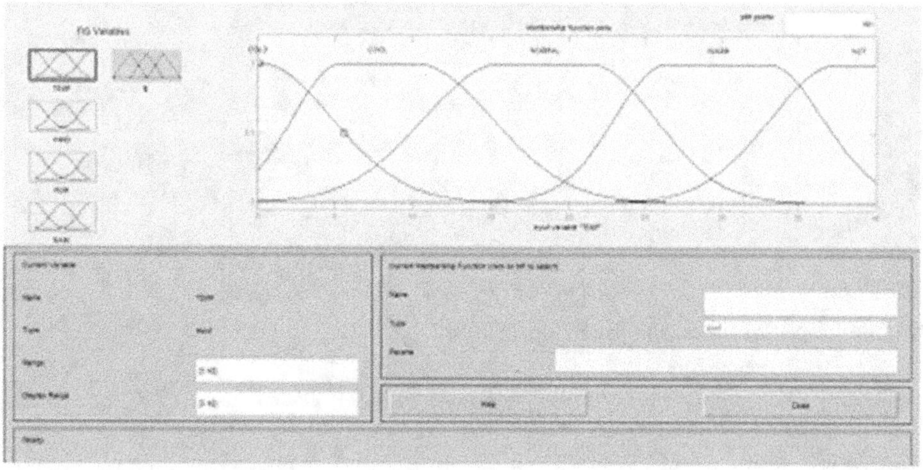

Figure 4. The membership functions for wind input

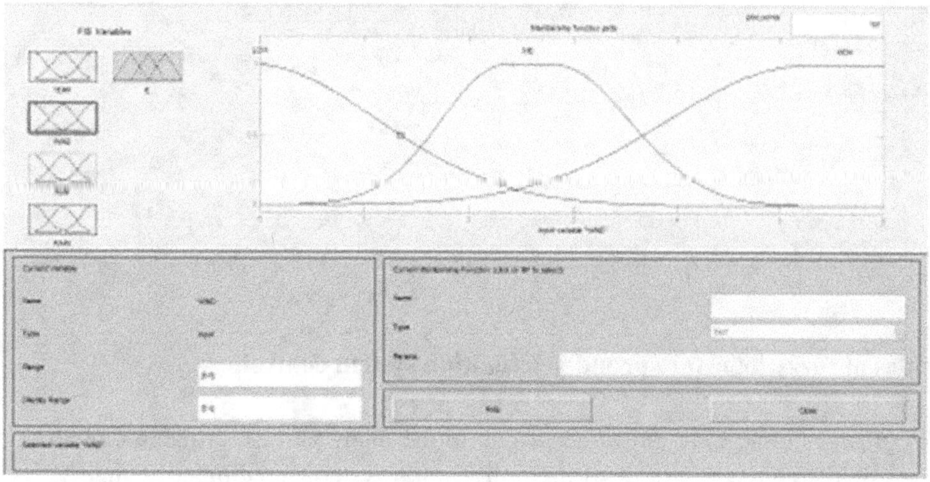

- **Fuzzy logic system and decision action**

Fuzzy rules are used derived from information of experts or from data. The rules mimic human reasoning. Mamdani method is generally used in a fuzzy inference technique. Fuzzy inference system used rules to generate fuzzy outputs, in this scheme, there are 3 inputs against each input there is fuzzy linguistic variables as indicated in Figure 8 and 9.

Figure 5. The membership functions for humidity input

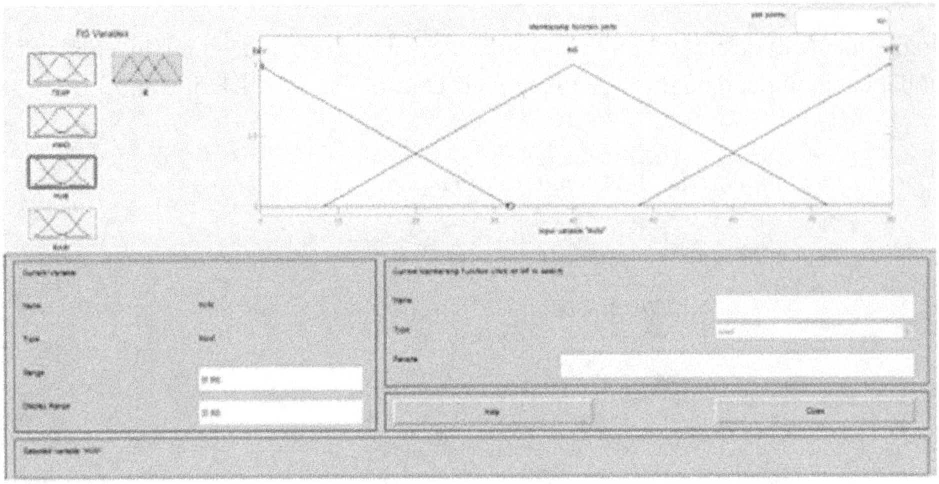

Figure 6. The membership functions for rainfall input

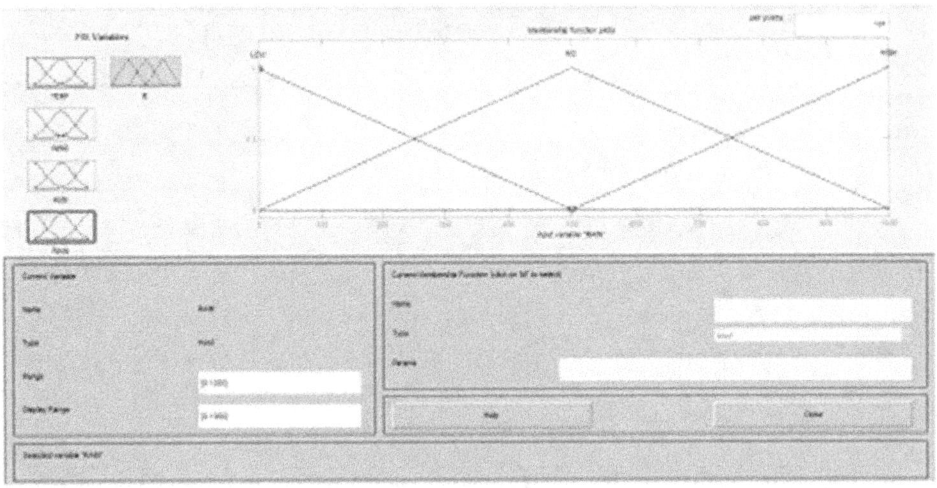

- **The rules of fuzzy logic for sprinkler irrigation system controller**

The rules of fuzzy logic were putted according to the many past experiments in addition to the irrigation expert predictable behavior of sprinkler irrigation system and climate change.

1- If (TEMP is COLD) then (IF is High) (1)
2- If (TEMP is NORMAL) then IF is Med) (1)
3- If (TEMP is HOT) then ((IF is LOW) (1)
4- If (TEMP is HOT) then (IF is LOW) (1)
5- If (WIND is HIGH) then ((IF is LOW) (1)
6- If (WIND is MID) then (IF is Med) (1)

7- If (WIND is LOW) then (IF is High) (1)
8- If (HUM is WET) then (IF is High) (1)
9- If (HUM is MID) then (IF is Med) (1)
10- If (HUM is DRY) then ((IF is LOW) (1)
11- If (RAIN is LOW) then (IF is High) (1)
12- If (RAIN is MED) then (IF is Med) (1)
13- If (RAIN is HIGH) then (IF is LOW) (1)

Figure 7. The membership functions for efficiency of irrigation performance output

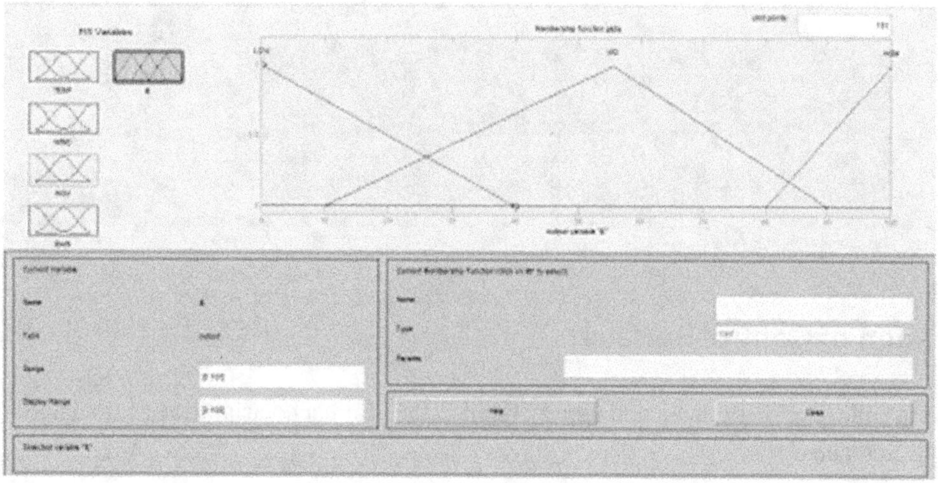

Figure 8. The Fuzzy Logic Tool Box uses simple Mamdani system

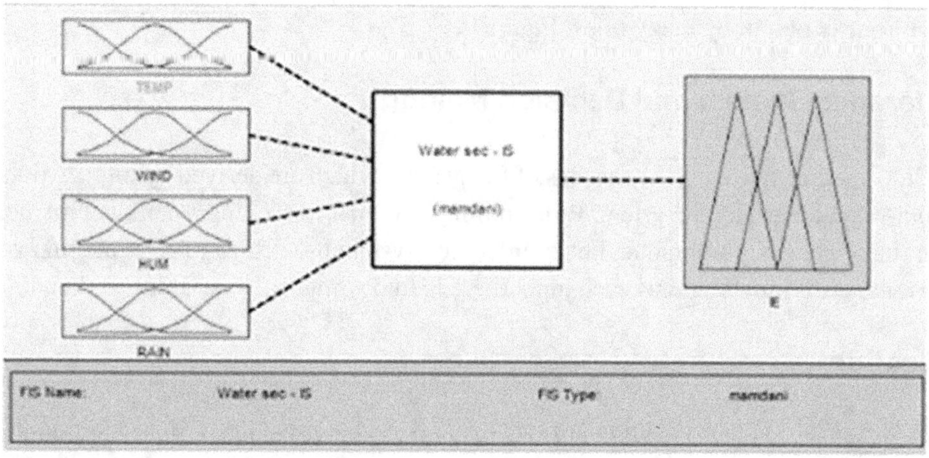

Figure 9. The fuzzy defined rules for irrigation performance

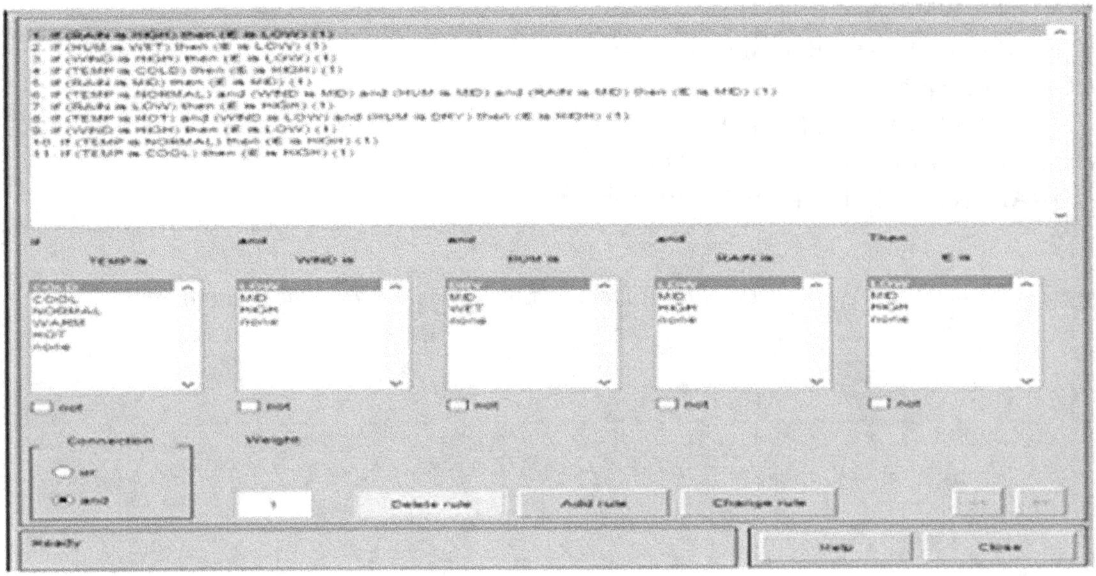

Identification of Control Surfaces

Linguistic variables are recognized and membership values for each variable are calculated in this step. Figures 3, 4, 5, 6 and 7.

Behavior of Control Surfaces

In this step fuzzy rules are constructed for different inputs to perform different actions. Fuzzy inputs associate with fuzzy output by fuzzy rules. Figure 9.

Fuzzy Inference System and Decision Making

The Fuzzy information system (FIS) consists of fuzzy rules which are derived by information of experts or from input-output learning of system. Rules mimics " human reasoning. Mamdani method is generally used in fuzzy inference technique. Fuzzy inference system used rules to generate fuzzy outputs, in this system there are 4 inputs against each input there is fuzzy linguistic variables as shown in Figure 9.

Defuzzification

Defuzzification is a process of conversion from a fuzzy set to a crisp number. For crisp input value, there are fuzzy membership for input variables, and each variable cause different fuzzy outputs cells that will used to activate or to be fired. Output will change into crisp value from this procedure of defuzzification. Defuzzification can be done by different methods but most common technique is centroid method (Patil & Desai, 2013; Figure 10).

Figure 10. Steps involved in Fuzzy Logic

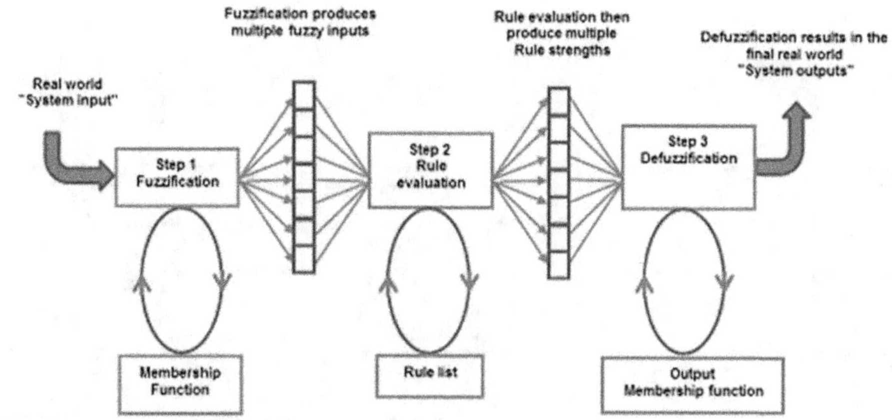

The first membership function input is temperature and the second is wind, they are under the *generalized* bell membership function. The *generalized* bell -membership function is specified by three parameters and has the function name G*bell*-MF. G*bell* MF formula is next:

$$F(x; a, b, c) = 1/(1 + |(x-c)/b|^{20}).$$

The third membership function is humidity and the fourth is rainfall, beside, they are under triangular function. The simplest membership functions are formed using straight lines. Of these, the simplest is the *triangular* membership function, and it has the function name triMF. It's nothing more than a collection of three points forming a triangle. Figures 2.

The data values of temperatures are ranged from zero to 40 Celsius, where the data values of wind MF are ranged from zero to 6 Km/h, and then the data values of humidity are ranged from zero to 80%, finally the data values of rainfall are ranged from zero to 1000 mm/day.

The triangular function formula is next:

$$(x; a, b, c) = \max(\min((x-a)/(b-a).(c-x)/(c-b)),0)$$

A membership function associated with a given fuzzy set maps an input value to its appropriate membership value.

Fuzzy Rules

The basic fuzzy rules were needed to describe a useful relationship between the four inputs and the final output (Efficiency of irrigation performance). The fuzzy rules are given based on published experiments and personal experiences. Figure 9, 10 and 11.

Figure 11. The rules interaction of variables under various membership functions

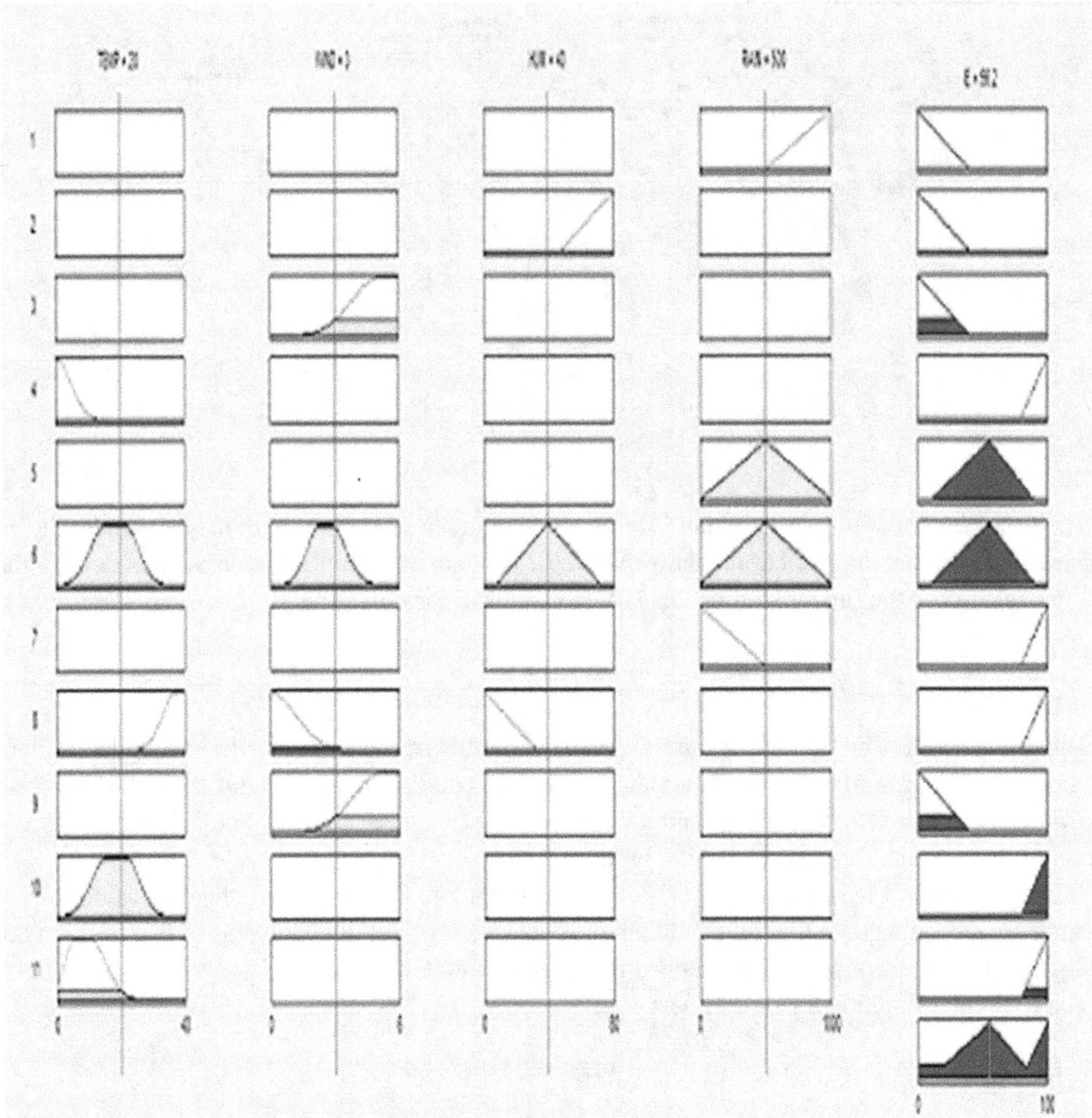

THE FIELD EVALUATION OF FUZZY LOGIC DESIGN FOR SPRINKLER IRRIGATION SYSTEMS

The filed evaluation trials were done on research farm, tests were conducted in various hours during the day and night to study the climate condition on the sprinkler irrigation performance and water losses, the nozzle diameter of sprinkler is 7.32 mm and the riser length is 2 meters above the soil surface, the operating pressure is 4.5 bar, there are 100 containers (water receptacles) as a net design (10 x 10 m) the diameter of the container is 15 cm. The temperature, wind and humidity were measured during the test process, the water volume in the container was measured after test process done, in addition to the water salinity (EC) in water before pumping and water after pumping in containers, by the same way the water amount measured before and after pumping, the time of one test is one hour. By using the next equation, it could be calculated the evaporation of water losses, according to the increasing of saline water, which the salt concentration in the pumped saline solution is exposed to the EC increasing as a direct result of the decreasing of water amount by water evaporation, and using the percentage of both of the pumped water EC and the received water in container to determine the evaporation losses by wind drift, temperature and defect pressure of humidity. According to the next relationship:

$$E = 100\,(EC_A - Ee_B)\,/Ee_B$$

Where

E : The evaporation losses (%),

ECA : The electrical conductivities of the samples of the water in the catch-containers, after pumping, and

EeB : The electrical conductivities of the samples of the supply water, before pumping.

RESULTS

Air Temperature

Fuzzy surface graph of temperature shows the defuzzification relationship of temperature and other membership function such as humidity, rainfall and wind, beside the irrigation performance, the increasing of temperature is followed by evaporation and transpiration of both of soil and plant, the soil evaporation causes the water losses specially in the highest temperatures degrees and reduces the efficiency of irrigation performance and efficiency, As a consequence, there will be a salt stress on the plant according to the soil salt concentration increasing as a result of water losses by evaporation.Figures 14 (A, B, C, D, G and K).

The irrigating efficiency, when the air temperature increased the evapotranspiration increased also, the evaporation is the water which transfer from the liquid water of plant surface and soil surface, transpiration is the consumed water during biological process from plant, and both of water evaporation caused the water losses and reduction of water efficiency, for that when the environment condition is support to evaporate water, in the same token, loss water, it's advised to shut off the valve of irrigation system.

The relationship between increasing of evaporation and temperature, it's crystal clear the positive relationship, according to Lorenzini (2002). See Figure 12.

Air Humidity

The humidity degree ranged from 0 to 80% according the climatology of Hungary beside the green wild area of forest and fields.

Figure 12. Evaporation vs temperature (Dimensionless), cited from G. Lorenzini. 2002

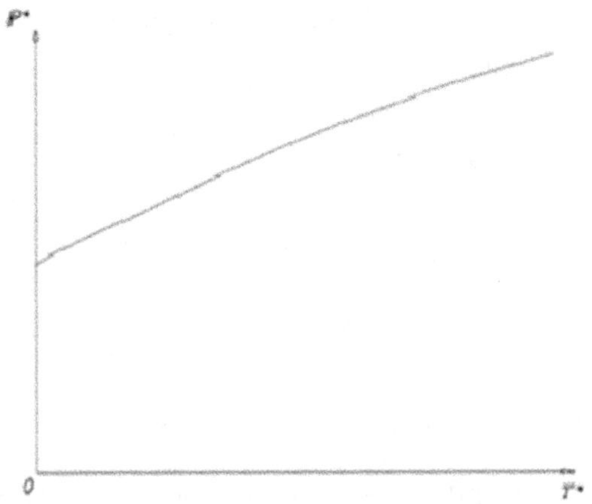

The air humidity is important factor of irrigation performances and influences indirectly on the efficiency of irrigation performance and by the same token on the plant productivity and quality. When the humidity degree increased may be caused a lot of fungus disease so the efficiency of irrigation performance will reduce specially, the irrigation systems which apply water on the plant surface like sprinkler system, beside the high humidity degree influence negatively on the transpiration process. Figures 14 (B, E, G, H, J and M)

When temperatures are high and the air humidity is low, the vapor pressure deficiency (and evaporation rate) will be higher. Note that when temperatures are low, the vapor pressure deficiency will be depressed, despite the humidity level. Likewise, when the air humidity is high, the vapor pressure deficiency will be decreased, despite the temperature. Figure, 13.

In Hungary, the humidity degree ranged from 0 to 80%, because of the climatology beside the green wide area.

The air humidity is an essential membership condition of irrigation operations and influences indirectly on the irrigation operation by the same token on the plant productivity and quality. When the air humidity increased, it causes fungus disease, the sprinkler irrigation systems applied water along the plant surface, beside the high air humidity influence negatively on the transpiration process.

Wind Velocity

This wind variable shows the effect of air flow on irrigation performance, where there's correlation between both of wind and irrigation performance, when the wind velocity increasing this cause the air moisture movement, by the same token the air need to have a lot of evaporation to replace the Humidity reduction and makes evaporation from both of plant and soil making water and salt stress. According to the relationship of both of wind and efficiency of irrigation performance the suitable function is *The generalized bell function.* Figures 14(A, D, E, F, H and L).

Figure 13. Sprinkler evaporation nomograph (cited from Forst and Schwalen, 1955)

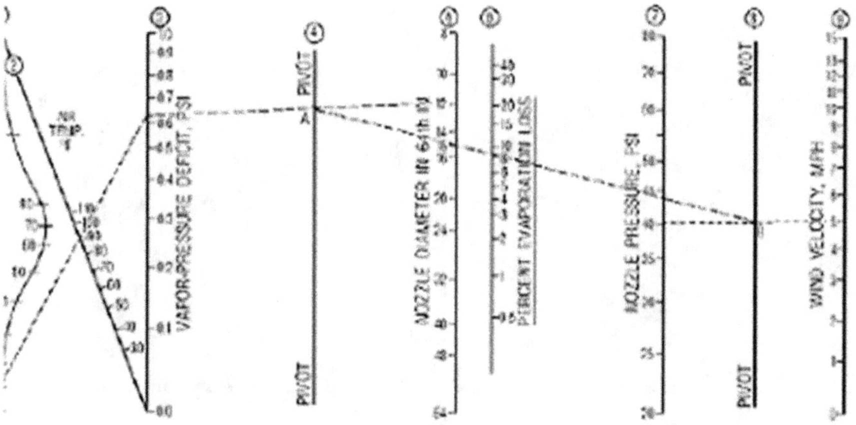

The Most Perfect Time for Evaporation

The most perfect time for evaporation is a short time, when the water droplet leaves the sprinkler hole and end by fall down on the surface soil or plant. When the water is pumped through sprinkler which has more riser and more jet length, this time will be increased and give more chance for water to evaporate, the water evaporation increased for the more riser height according to the wind speed is increased whenever the height of sprinkler riser increased according there is no limitations of wind.

Rain Fall

The average annual precipitation across the country is 600 millimeters. The maximum of rain, nearly 1,000 millimeters, so the rainfall is an important factor in irrigation process, beside control of supplemental irrigation on and off, there is a strong correlation relationship between of rainfall and efficiency of efficiency of irrigation performance as shown in the fuzzy defuzzification of rainfall and other factors versus irrigation efficiency. Figures 14(C, f, J, K, L and M).

THE FIELD EVALUATION OF FUZZY LOGIC OF SPRINKLER IRRIGATION SYSTEM CONTROLLER

Regards to table (There are many tests were done to evaluate fuzzy logic of sprinkler irrigation system, will compare the results to the predictable rules and surface fuzzy surface graphs according to the location weather conditions as follows: Table 2.

If HUM is in the overlap zone of both of MID to WET (51%), WIND is MID (14 mil/h) and TEMP is WARM (29Co) the evaporation is MID (7.1%) so the valve opining hole must be MID according to irrigation efficiency (IF) is MID and also the irrigation time will be recommended to be not long not short according the next day's weather conditions. It's important to mention that the mean of MID for valve is referring to the time of irrigation will be under control and not to be fully recommended or fully

Figure 14. The Fuzzy surface graph of membership functions versus efficiency of irrigation performance

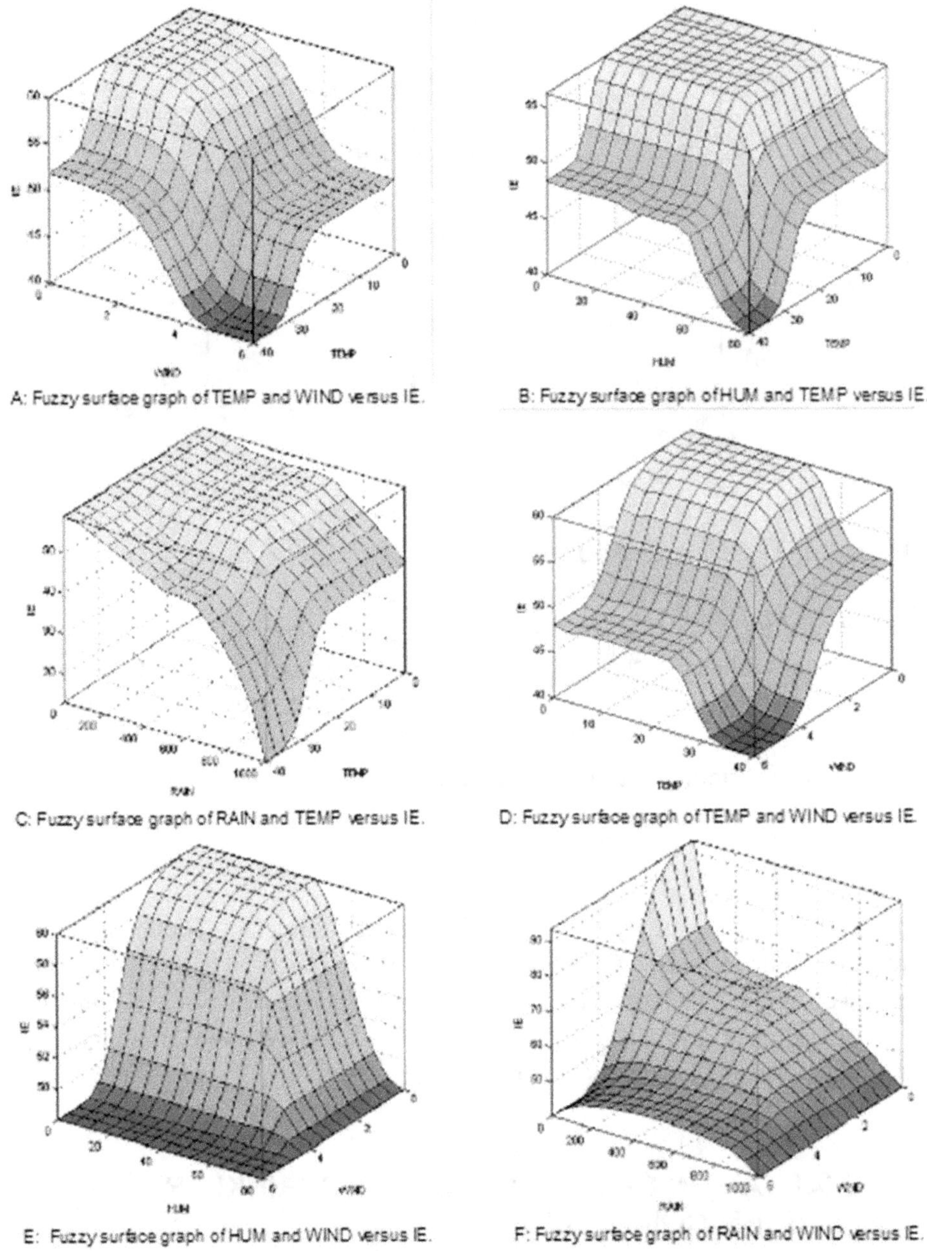

A: Fuzzy surface graph of TEMP and WIND versus IE.

B: Fuzzy surface graph of HUM and TEMP versus IE.

C: Fuzzy surface graph of RAIN and TEMP versus IE.

D: Fuzzy surface graph of TEMP and WIND versus IE.

E: Fuzzy surface graph of HUM and WIND versus IE.

F: Fuzzy surface graph of RAIN and WIND versus IE.

prevented, according to also the next day the weather conditions, in other test If HUM is MID (31%), WIND is MID (23 mil/h) and TEMP is NORMAL (22 Co) the evaporation is MID (8.9%) so the valve opining hole must be MID. Finally, test If HUM is HIGH (78%), WIND is LOW (7 mil/h) and TEMP is HOT (39 Co) the evaporation is LOW (4.4%) so the valve will be open according to the irrigation efficiency (IF) is high and means the time irrigation will be recommended to be lengthier. That clear the fuzzy logic system is able to be the nuclear design of the control system of sprinkler irrigation system to increase the irrigation performance efficiency.

Table 2. The evaporation water losses versus HUM, WIND and TEMP

TEMP (C°)	WIND (mil/h)	HUM (%)	Evaporation losses (%)
29	14	51	7.1
32	11	57	5.8
22	23	31	8.9
22	13	39	7.3
21	13	36	7.4
34	12	64	6.6
39	7	78	4.4
37	9	65	5.4
34	16	51	7.5
39	9	64	5.6
31	13	28	7.4

Sprinkler Irrigation Controller

Flowchart clears the irrigation controller in sprinkler systems, besides; it can monitor and operate the monitor in three crossed phases. The desired climate parameters, this item shows the parameters of both of wind, temperature and humidity (climatic conditions) which it needed to start irrigation process, the input variables in this phase, there is a physical sign impact to start the decision of sprinkler irrigation process under typical climate conditions (wind, temperature and humidity) to save water, the three actual input values compared with the database values, where the dynamic decision is token related to the start time and the time of irrigation process. Figure 15.

Closed Loop Control System

This is based on collected of back controlling concept (feed-forward) and reactions of watching factors, in this kind of control, there is a reaction of necessary data to determine the start time and the irrigation time. There are many of climate factors (wind, temperature and humidity), which impact of irrigation process scheduling (start time and irrigation time). Some of them are variables and must measure before irrigation process, these factors have physical nature, such as (wind temperature and humidity). So that, when the values of these factors change the irrigation start decision must change also, according the climate factor changing, the monitor receives the reaction from one or more remote sensor in a field where continually present a modern data to the control unit of factors of climate, according to the presented sensors measurement and the past historical parameters for programing. The monitor controls of valve open and close process. It's important to note on this system, the system status compared with the desired status and based on this comparison the decision will take to on or off in the irrigation process using the solenoid valve. The control tools of closed loop for irrigation system lead to deiced of irrigation process time and start time. It chooses the most perfect and ideal time and climate condition, which do not cause of water losses by evaporation from sprinkler irrigation system, and increase the irrigation performance efficiency. Figure: (6).

Figure 15. The flow chart of sprinkler system controller design

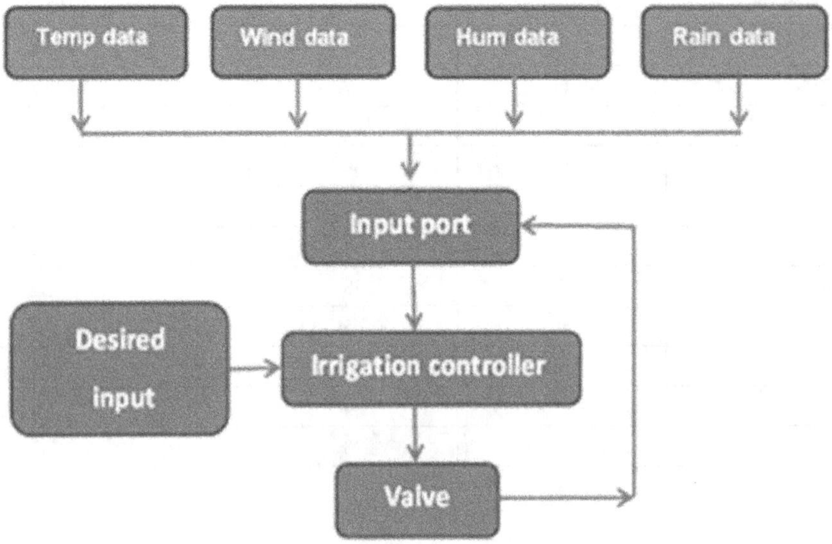

DISCUSSION

Air temperature, wind, humidity and rainfall are basic influence of irrigation efficiency directly, beside the interaction between every factor and the others, and need to monitor and watch irrigation process to report and validate it, the resulted surface graphic are enough to monitor, validate and report the irrigation systems efficiency to exact and schedule the irrigation practices managements

CONCLUSION

The water lost by evaporation in the air during sprinkler irrigation system process, the evaporated water losses depends on (climate, evaporation time, the surface of evaporation area). The basic environmental factors which effect on the sprinkler irrigation systems are (air temperature, wind velocity, air humidity and the interaction of them), the obtained surface graphic shown the last interaction during fuzzy logic rules and introduce the predictable results for every case, and it' base of establishment of a control system of sprinkler irrigation systems using fuzzy logic control, depends on the monitoring and predictable of weather data and does a defuzzification to them to desist the most right decision to on or off the of irrigation systems. It's important to mention that the fuzzy logic system has availability to change the boundary of data of parameter according to the sit weather conditions.

ACKNOWLEDGMENT

The authors are grateful for the financial support of Tempus public foundation, Stipendium Hangaricum programme and the Desert Research Center, DRC, Cairo, Egypt.

REFERENCES

Allahverdi, N. (2002). *Expert Systems: An Artificial Intelligence Application.* Atlas Press.

Carman, K., & Seflek, A. Y. (2004). Fuzzy expert system design for tyre traction performance. *J. Agric. Mach. Sci., 1*(1), 63–68.

Castilla, N. (1997). Greenhouse drip irrigation management and water saving. *Cahiers Options Méditerranéennes, 31*, 189–202.

Dainello, J. F., & Fipps, G. (2001). *Irrigation. Department of horticulture Science. Texas A&M University.*

ElHagarey, M. E., & Hushki, M. M. (2017). Fuzzy Logic Model Approaches for Water Saving in Irrigation Systems. *European Journal of Academic Essays, 4*(4), 157–165.

Evans, R. G., Buchleiter, G. W., Sadler, E. J., King, B. A., & Harting, G. B. (2000). Controls for precision irrigation with self-propelled systems. In *Proceedings of the 4th decennial national irrigation symposium*. ASAE.

FAO. (1960). *Irrigation by sprinkler. FAO Agriculture Development Paper No. 65*. FAO.

Forst, K. R., & Schwalen, H. C. (1955). Sprinkler evaporation losses. *Agricultural Engineering, 36*(8), 526–528. http://www.wac-infonet.org/id/16377

James, G. L. (1988). *Principles of Farm Irrigation System Design*. John Wiley and Sons, Inc.

Kindler, J. (1992). Rationalizing water requirements with aid of fuzzy allocation model. *Journal of Water Resources Planning and Management, 118*(3), 308–323. doi:10.1061/(ASCE)0733-9496(1992)118:3(308)

Lorenzini, G. (2002). Air temperature effect on spray evaporation in sprinkler irrigation. *Irrigation and Drainage, 51*(4), 301–309. doi:10.1002/ird.68

Pair, C.H. (1960). *Sprinkler Irrigation*. Soil and Water Conservation Research Division, Agricultural Research Service, Leaflet-No. 476

Patil, P., & Desai, B. L. (2013). Intelligent Irrigation Control System by Employing Wireless Sensor Network. *International Journal of Computers and Applications, 79*(11), 2–5.

Phocaides, A. (2007). *Handbook om pressurized irrigation technique. Food and Agriculture Organization of the United Nations.*

Scherer, T. F., Kranz, W., Pfost, P., Wermer, H., Wright, J. A., & Yonts, C. D. (1999). Sprinkler irrigation systems. WCAinfoNET.

Shabbir, F., & Tahir, Z. (2014). Impact of Electric power on Agriculture industrialization and value chain. *Proceeding of international Conference entitled "emerging horizons of Agricultural extension for sustainable rural development.*

Sigrimis, N., Arvanitis, K. G., Pasgianos, G. D., & Ferentinos, K. (2001). Hydroponics water management using adaptive scheduling with an on-line optimiser. *Computers and Electronics in Agriculture, 31*(1), 31–46. doi:10.1016/S0168-1699(00)00172-1

Swarner, L. R., Langley, M. N., Maletic, J. T., Phelan, J. T., Lawhon, L. F., Shockley, D. G., & Brownscombe, R. H. (1959). *Irrigation on Western Farms*. Bulletin No. 199. United States Department of the Interior, Bureau of Reclamation. United States Department of Agriculture, Soil Conservation Service Agriculture Information.

APPENDIX: SOME TYPES OF SPRINKLER IRRIGATION SYSTEMS

Sprinkler irrigation system is defined as a pressurized system where water is distributed through a network of pipelines to the field and applied through selected sprinkler heads or water applicators (Dainello and Fipps, 2001). The adoption of sprinkler irrigation technology accelerated after World War II, due to development of light-weight aluminum pipes with the impact-type sprinkler (Scherer *et al.*, 1999). Sprinkling is mans imitation of natural rainfall. Such application is adapted to most soils that can be irrigated and is particularly adapted to sandy soils that take water rapidly. Soils that are too shallow, too steep, or too rolling to be irrigated by surface methods can often be effectively irrigated with sprinklers. Small streams of water can be used efficiently with this type of irrigation and it is adapted to all the major farm crops except rice. Sprinkler systems have certain limitations, whatever, in that water distribution is affected by wind, and power requirements are usually higher than for other methods of irrigation. The water used must be clean and free from debris, and a constant water supply is needed for most economical use of the equipment. Sprinkler systems will give many years of satisfactory service if they are designed specifically to fit the conditions on the farm and are properly operated and maintained **(Swarner *et al.*, 1959).**

a) Advantages of sprinkler irrigation system:
    1. Land with irregular topography can be irrigated with minimum leveling and disturbance of the topsoil. The same applies to shallow soils.
    2. On sloping lands runoff and soil erosion that usually accompany it can be eliminated.
    3. Sandy or other highly permeable soils can be irrigated with out excessive losses by deep percolation, thus reducing the danger of creating drainage problems.
    4. Field ditches are not necessary. Thus, increasing the area available for crop production and minimizing the problem of ditch maintenance.
    5. Effective use of small continuous streams of water,
    6. Nutrients can be applied through the irrigation system (fertigation).
    7. Excessive losses by deep percolation can be eliminated, thus reducing the danger of creating drainage problems.
    8. With proper drainage, sprinklers can be used efficiently to flush accumulated salts own the profile.
    9. Automation is readily available for many sprinkler systems.
    10. Can be effective for weather modification (micro-climate).

b) Disadvantages:
    1. Initial cost can be high (compared to surface irrigation systems).

Also, operating cost (energy) can be high compared to nonpressurized systems, unless sufficient head is available from a gravity-fed supply.

2. Low irrigation efficiencies under high winds and hot conditions.
3. Distribution uniformity can be deteriorating over time.
4. Fluctuating flow rates at the water source can be very problematic.

5. Water quality can be a problem with overhead sprinklers if water is saline, and in terms of clogging and nozzle wear. Also some types of water are corrosive to sprinkler pipes and other hardware.
6. Some fruit crops cannot tolerate wet conditions during maturation (unless fungicides, etc., are used).
7. Low intake rate soils (< 3 mm/h) cannot be irrigated by sprinklerirrigation system due to the runoff loss.
8. Damage of weak branches and flowers near sprinkler head.

Types of Sprinkler Irrigation Systems

Sprinkler systems have been generally classified as pivot, liner mover, permanent,semi- permanent (semi-portable), and portable. The classificationdepends on whether the lateral pipe line (including sprinklers), mainpipeline and pumping unit are movable or fixed. Sprinkler systems maybe more specifically classified according to special mechanical featuresthat are used to move the lateral pipelines (Pair, 1960).

Center Pivot Irrigation Systems:

The first "Self-Propelled Sprinkling Irrigating Apparatus" was invented in 1948 and patented in 1952 by Frank Zybach in eastern Colorado. The early systems were the foundation of the development of modern self-propelled center pivot and linear move irrigation systems. These very adaptable water application methods have experienced tremendous growth around the world in recent years due to: 1) their potential for highly efficient and uniform water applications; 2) their high degree of automation requiring less labor than most other irrigation methods; 3) large areal coverage; and 4) their ability to economically apply water and water soluble nutrients over a wide range of soil, crop and topographic conditions. Most of the following discussion is directed towards center pivot machines although much will also apply to lateral move machines.

Water losses from spray heads near the top of the canopy typically range from 0-2% due to droplet evaporation, wind drift is usually less than 5%, evaporation from crop canopy ranges from 4 to 8%, and soil evaporation less than 2% whereas runoff may range from 0 to 15% or more depending on slope and soil conditions. Spray heads and impacts mounted on top of the pipe lateral may have droplet evaporation and wind drift losses as high as 15%. Evaporation may slightly offset crop water use, but this amount is difficult to measure or calculate and is usually less than 15% of total ET. For spray irrigation on drops over a crop with a full canopy, application efficiencies of about 90 to 92% are attainable with no surface runoff whereas sprinklers on the top of the pipe may attain efficiencies from 80-85% (Evans *et al.*, 2000). See Figures 16 to 19.

a) **Permanent systems:** A fully permanent system has buried main line, sub-main and laterals with stationary pumping plant. Sprinklers (or nozzles) are permanently located on each riser (James, 1988). Equipment and installation cost is higher than that of any other system, but labor requirements are low. Figure 20.
b) **Semi-permanent systems:** Have portable lateral line, permanent main line and stationary water source and pumping plant. The main line is usually buried. The risers are located at suitable intervals on the laterals (James, 1988). Cost of equipment and installation is moderate, also labor requirements are moderate. Figure 21 and 22.

c) **Portable systems:** The fully portable system has portable lateral pipelines with sprinkler and portable pumping plant (FAO, 1960). This type has the lowest cost of any sprinkler system. Labor requirements for operating are higher than those of any other system. Portable systems can be used anywhere sprinklers are used. Sprinkler systems have been generally classified as permanent, semi- permanent (semi-portable), and portable. The classification depends on whether the lateral pipe line (including sprinklers), main pipeline and pumping unit are movable or fixed. Sprinkler systems may be more specifically classified according to special mechanical features that are used to move the lateral pipelines (Pair, 1960). Figure 23.

Figure 16. Primary center pivot

Figure 17. Lateral move

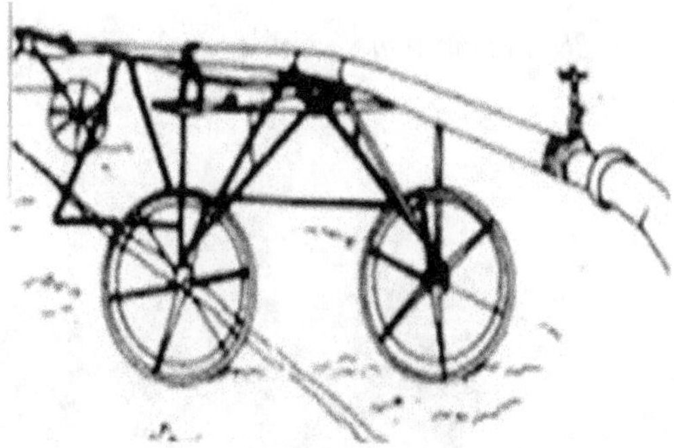

Figure 18. Pivot irrigation systems

Side-Roll Sprinkler System

In this type the lateral is axed carry out on a big diameter wheel and moved along filed for distance ranged from 18-24m according to the desired design and the lateral spaces, a diesel engine is installed in the center of line to move system to irrigated the next site in the field, the lateral length is ranged from 120-400 m and this system is good for feed and pasture, the rise of axe pipe is rise above ground by 0.6-1.0 m, in case of higher crop such as maize, the rise of axe pipe is 1.3 m. Figure 24 to 26.

Figure 19. The water draft by wind during irrigation process using pivot irrigation system

Figure 20. Permanent sprinkler irrigation systems

Figure 21. Semi- Permanent sprinkler irrigation system

Figure 22. Semi-portable sprinkler irrigation systems

Figure 23. Types of portable sprinkler irrigation systems

Figure 24. Side wheel roll sprinkler irrigation systems

Figure 25. Side-roll wheel sprinkler during irrigation process and water losses

Figure 26. Side roll wheel sprinkler system during operating on slope

The Gun Sprinkler

The gun sprinkler system is suitable for wide irrigated area and grass cultivated, where the space between sprinklers ranged from 35 to 50 m, according to the sprinkler characters, and the sprinkler flow is about 30-35 m^3/h. and worked under operating head reached to 5-7 Bar, the water droplet size of gun-sprinkler is big which caused the soil impacting and losses water by run-off according to the poor of infiltration rate happiness. Figure 27 and 28.

Sprinkler Nozzles

Generally, gun sprinkler nozzles are fitted with either tapered or orifice nozzles. Jets produced by tapered nozzles have large drops and are less susceptible to wind than those produced by orifice nozzles. Since orifice nozzles produce smaller drops, they are used on more delicate crops, which are susceptible to large drop impact. However, the jets from orifice nozzles are also more susceptible to wind.

Figure 27. Gun sprinkler system (mounted on tractor)

Figure 28. Gun sprinkler system during irrigation process

Sprinklers Types

Most of the agricultural sprinklers are the hammer-drive, slow rotating impact type, single or twin nozzle. The sprinklers shoot jets of water into the air and spread it to the field in the form of raindrops in a circular pattern. They are available in various nozzle sizes, flow discharges, operating pressures and wetted diameters or diameter coverage, full circle or part circle. They are classified as low, medium and high pressure/capacity, as shown in Table 3; according to the height of the water jet above the nozzle, they are divided into low angle (4°–11°), or high angle (20°–30°). They are made of brass or high engineering plastics with internal or external threaded connections of –1 inch. They are installed vertically on small diameter riser pipes, 60 cm above ground, fitted on the laterals. The sprinkler spacing in the field is rectangular or triangular at distances not exceeding 60 percent of their diameter coverage. Filtration requirements, where necessary, are about 20 mesh. Figure 29, 30 and 31.

Table 3. Sprinkler engineering classifications

Agriculture sprinklers	Nozzle size mm	Head (Bar)	Flow (m^3/h)	Coverage Diameter (m)
Low pressure	3 -4.5 x 2.5 -3.5	1.5- 2.5		12 - 21
Mid pressure	4 - 6 x 2.5 - 4.2	2.5 3.5		24 - 35
High pressure	12 - 25 x 5 - 8	4 – 9		60 - 80

Mini -Sprinklers

These water emitters are small plastic sprinklers of low capacity with flow rates less than 300 litres/h. Their main characteristics are their rapid rotation/whirling, less than a minute per rotation, the very small size of the water drops and the low angle of the water jet above nozzle. They have only one nozzle, of about 2.0 mm. They discharge 150–250 litres/h at 2.0 bars operating pressure. They are full circle and the wetted diameter is only 10–12 m. Mounted at a height of 60 cm onmetallic or plastic rods inserted into the ground, they are connected to PE laterals (25 or 32 mm) through small flexible tubes 7 mm in diameter and 80 cm long. The spacing arrangement in the field is the same as for conventional sprinklers. The spacing does not exceed 6.0 m, i.e. 50 percent of the wetting diameter. The filtration requirements are about 60 mesh (300 microns). Figure 30.

Figure 29. Rotating sprinkler

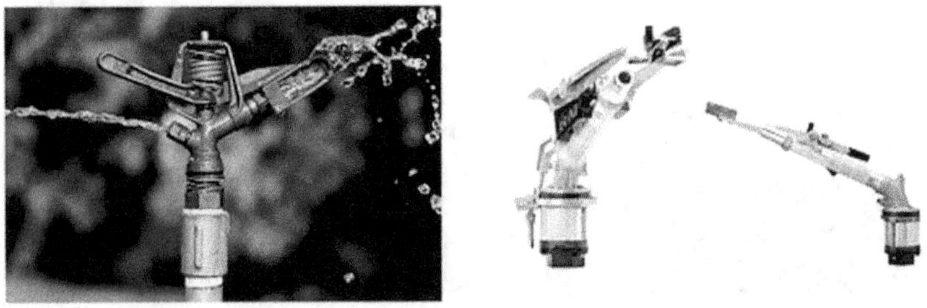

Figure 30. Mini-sprinkler

Spitters, Micro-Jets and Sprayers

These are small plastic emitters with a low water discharge at a low angle in the form of fine drops in a sectorial or full circle pattern. They are mainly used for tree crops. They are of various mechanisms with a wide range of flow rates and water diameters. They have a small passage diameter, thus filtration of the water is essential. Their main performance characteristics are:

- Operating pressure: 1.5–2.0 bars;
- Flow rate: 35–250 litres/h (generally 150 litres/h);
- wetting diameter: 3–6 m;
- Precipitation rate: 2–20 mm/h (generally 4–8 mm/h);
- Filtration requirements: 60–80 mesh (250–200 microns).

Their heads are fixed to small plastic wedges 20–30 cm above ground and they are connected to the PE laterals with 7–9 mm flexible plastic tubes 60–120 cm long and a barbed plunger. They are placed one per tree, 30–50 cm apart. Figure 31.

Figure 31. Micro-jet irrigating

Chapter 10
A Web-Based Platform for Crop-Specific Data Management and Exchange of Farmers' Experiences

Rosa Maria Gonzalez-Amaro
CONACYT, Instituto de Ecología, A. C., Mexico

Miguel Angel Hidalgo-Reyes
https://orcid.org/0000-0001-8303-6030
CONACYT, Comisión Nacional para el Conocimiento y Uso de la Biodiversidad, Mexico

Virginia Lagunes-Barradas
Instituto Tecnológico Superior de Xalapa (ITSX), Mexico & Universidad Veracruzana, Mexico

ABSTRACT

In this chapter, the research theme is focused on the relationship between small farmers and information and communication technologies (ICT). Although there are other previous works that have already analyzed this same relationship, the authors believe that access to information remains a major challenge for farmers. With the application of workshops on agricultural practices of maize, in communities of Oaxaca and Veracruz, they learned about the practices of farmers around the production cycle and applied a survey to find out their opinion regarding the use of ICT. In addition, they used a specialized database to complement the workshops objectives. Next, in collaboration between researchers in the areas of biology and computing, they developed a web platform for access and use of information related to the variables of interest to farmers. Among the main results, they highlight that the community prefers to use cell phones to access such information and that the older generations are looking to transmit experiences and knowledge to the young with the aim of conserving ancestral knowledge.

DOI: 10.4018/978-1-7998-5000-7.ch010

INTRODUCTION

Information and communication technologies (ICT) are having increasing impact on different domains and when it comes to agriculture,this is not an exception. The compiled and managed information could be of interest to a broad range of different actors, such as governmental audience, educational and research institutions, small associations, and certain producers.

In the case of precision agriculture (PA), the agronomic experience is one variable that can greatly influence the agricultural decision-making process, and even more, if technology transfer helps to take the right action in the right place at the right time, the benefits for any producer could be highly positive.

Broadly speaking, technology transfer includes the development of activities aimed at facilitating access to knowledge, products and technological services developed for the agricultural sector(Corpoica, 2015). The concept of technology transfer is evolving towards knowledge transfer, by understanding arguments from personal, social or cultural experience(González Sabater, 2017).

The processes involved in knowledge management are the creation, transmission or socialization, and use of knowledge(EcuRed, 2011), (Canals, 2003). Technology transfer strategies in the agricultural sector are aimed at getting producers to adoptand use the technologies. To the extent that a community makes use or appropriates knowledge or technology, we can speak of innovation; an action that contributes to the social and cultural transformation of a community. Knowledge management is facilitated by ICTs through virtual communities and communities of practice, among others, promoting collective learning through exchange(Badia & García, 2006), (García Yeste, Leena Lastikka, & Petreñas Caballero, 2013).

Information related to these factors is characterized by the analysis and/or control of the behavior (temporal and spatial) of agricultural and social environmental variables associated with the crop. It is therefore of interest to explore studies related to techniques linked to the experimental work of the farmers.

The objectives of this chapter are focused on two main situations that involve collaborative work and complement one another. Firstly, the exchange of information, experiences, culture and traditions of small farmers dedicated to the cultivation of native maize in the different regions ofMexico(Ali *et al.*, 2019); Secondly, this research is based on the interdisciplinary work of experts in the area of biology, mainly in ethnobotany, ecology, sustainable development and agroecology, in combination with researchers from the ICT area, with regard tosoftware development for mobile and web applications.

Although there are some definitions of precision agriculture and several components are continuously added to its conceptual frame, the relevant aspect to consider in the present chapter is the need for information of the small farmers (Ali *et al.*, 2019).

The expected impact of the objectives stated aboveaim to remark, from a research perspective, the conceptual and operational framework used to integrate and distribute the information and knowledge obtained from small maize farmers in specific regions of Mexico, as well as the supported technology. Both aspects can be used as a guide for economic studies within governments and world or local organizations programs.

The chapter begins with an analysis of representatives works from the state of the art and with the specification of the problems related with the understanding of the research context; then it continues with the acquisition of data, first using a maize database from CONABIO and alsofrom the application of a survey among small farmers of specific communities from Oaxaca and Veracruz. Next, there is a specificsection linked to the appropriate use of ICT for the proper visualization and interpretation of the maize crops information by the small farmers. Finally, it concludes and summarizes the main findings as well as a brief description of future work.

BACKGROUND

Extensionism, a term defined by the OECD (2011) as the service provided by staff of education and research institutions, aims to facilitate access to knowledge, information, and technologies for different audiences. Among these audiences are producers, rural economic groups and organizations, and actors in the agricultural, fisheries and aquaculture sector, as it currently occurs in Mexico.

The essential components of extensionism are communication for development and innovation, technology transfer using information and communication technologies, education and training of key actors, and technical support (Cadena Iñiguez *et al.*, 2018).

In 2015, the Inter-institutional Extension Group (IEG) was formed to promote the emergence of different voices and readings to facilitate the exchange of ideas, experiences and visions, with resources such as information and communication technologies (ICTs)(GIE, 2015).

The study developed in this chapter is about getting a better understanding of the factors that affect farmers' decisions, particularly with regard to ICT adoption and, consequently to answer the question, howcouldmaize information systems be improved through the exchange of useful information? The previous research question is based on the three basic elements from Figure 1.

Figure 1. Elements that guide the technological contribution towards small farmers

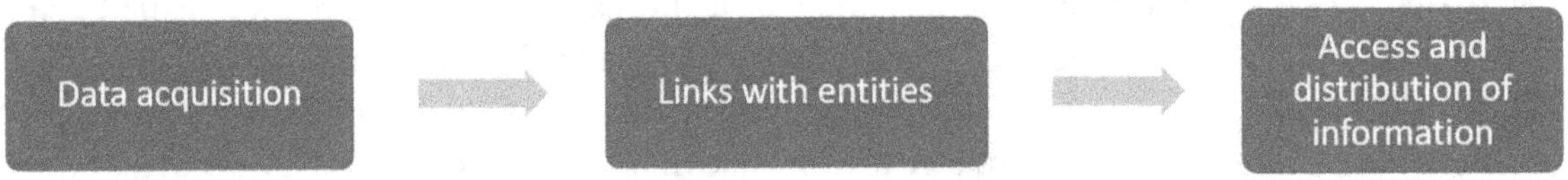

ICT play an important role in precision agriculture (PA), since different parameters that impact the production of crops of given characteristics, can be analyzed and managed using these technologies. The precision agriculture term was first published and defined by the National Research Council as "a management strategy that uses information technology to bring data from multiple sources to bear on decisions associated with crop production".

Also, previous works for PA shows that demographics parameter can affect adoption of ICT(Adrian, Norwood, & Mask, 2005). Some of these parameters are age, farming experience, education level of farm managers/owners, off-farm employment, and farm size, which represent an important subset that could guide a new way of accessing a distributing the information.

With the aim of describing specific topics of Figure 1, the following subsections address each one of the elements (blue boxes).

Data Acquisition

This element is related with the selection of representative works in the maize an ICT domain. The analysis of these studies allows us to know the state of knowledge. In addition, the information collected on maize in Mexico highlights the currently known parameters, as well as those that require further effort to achieve.

Agriculture literature review

The efforts of the government's extension program efforts in Mexico are aimed at three defined sectors of the rural population: self-consumption (smallholder farmer), productive potential, and large producers. For the typology of self-consumption, the perspective is at a territorial scale, with construction of social enterprise, dialogue and exchange of knowledge, capacity building and systemic analysis of life strategies. In this sector, food security and biodiversity conservation are the priority.

The productive potential segment, focuses on production, organization, financing, and integration to the value chain, prioritizing the network of specialists with interdisciplinary vision and biodiversity conservation. For large producers, the focus is on incentives and integration into other export strata, and concentrates on the search for flexible policies and promotion of national seed(GIE, 2015).

According to Raza*et al.* (2020), some demographic attributes of farmers (e.g. age, education, land tenure, tenure status, area cultivated) play an important role in the awareness and adoption of modern production practices. In addition, the authors analyze the most effective information tools among farmers, identifying those that meet their needs, *i.e.*, timely, accurate and quality information to improve their skills in agriculture.

The information needs expressed by farmers depend on the category to which they belong, because some topics are indispensable and others additional. The demanding information is grouped into four areas: technical production, business policy management, trade finance and regulations. Formaize producers in particular, the climate, markets and prices are three areas that demand information in a special way. However, the economic, commercial and climate information areas are the least attended (Nagel Amaro & Martínez Vergara, 2016).

There are other works related to the analysis of production, nutritional quality and other indicators related to agronomic factors of various food products of plant and animal origin. Among these, maize represents one of the most studied because of its importance in the human diet and in the productionof other goods.

Given the importance of this resource in Mexico, there have been numerous studies on this species in various topics such as agronomic uses (Espinosa Tamayo*et al.*, 2019), morphological variability of the native maize(Alvarado-Beltrán *et al.*, 2019),*in situ* and *ex situ* conservation, genetic analysis, pharmaceuticals, among others.

However, as stated, these studies are generally regional and dedicated toresearchers from various disciplinary areas, but rarely focused on providing information for farmers that could support them in making decisions regarding their crops.

ICT literature review

InMéndezLópez (2017) the authors state that the process of ICT adoption by farmers occurs in stages. The starting point is digital literacy, where practice is acquired in basic functionssuch as communication and navigation in simple applications. Then, ICT focus on production processes, including precision agriculture, traceability, and biotechnology. Next comes the use of ICT in administrative management, which involves the use of management software and transaction portals, among others. Also, comprehensive computerized management considers cloud data processing and enterprise resource planning (ERP) systems.

In this process, producer groups are associated, and in the first phases, subsistence farmers are located, followed by family farmers, small and large entrepreneurs. In each phase there is an increase in the quality and use of information for knowledge-based agricultural development(Nagel Amaro & Martínez Vergara, 2016),(Méndez López, 2017).Other studies show that the effective use of ICT among farmers provides easy access to information.

In the experience of ICT use in Latin America, the traditional means used for technology transfer in the agricultural sector, such as radio, television and other face-to-face methodologies, have worked well (Molano Bernal, 2017), (Raza *et al.*, 2020). Text messaging has represented another popular strategy in the agricultural sector, not only in Latin America but also worldwide. For the Food and Agriculture Organization of the United Nations (FAO), the appeal of Short Message Service (SMS) is based on its low cost and ability to work on all types of mobile phones (Palmer, 2012).

The global context requires finding options of different levels of use for each actor or sector regarding ICT. At present, everything leads to a greater availability and access to information, therefore, the type of public to which the information is addressed, the contents and the type of information must be considered. AnICT that has already been used in the agricultural sector isthe *Sowing Platform* in Colombia, which is aimed at providing information, analysis and support for stakeholder coordination, decision-making and knowledge management in science, technology and innovation for the agricultural, livestock, forestry, fisheries and agro-industrial sectors (Corpoica, 2015).

PLATICAR and *Linkata* are two platforms that allow the exchange of knowledge and experiences through blogs, forums, discussion groups and chat. In Mexico, the Ministry of Agriculture, Livestock, Rural Development, Fisheries and Food (SAGARPA) implemented a freemarketing application called "Mercados Mx". This mobile applicationis aimed at producers and buyers in the Mexican Agricultural, Livestock and Fishing sectors. The users are able toregister their products for sale and specify characteristics such as place of origin, pictures, if it has a certification, etc. In addition, it is possible to consult information from the sector like Customs Services, Certifying Agencies, national and international buyers, among others. In Chile, there is the*IDI Plataforma Silvoagropecuaria*, which offers information services for innovation, and promotes planning for small farmers (Nagel Amaro & Martínez Vergara, 2016).

Moreover, there is a special interest in young producers who already are land-workers or community members, belonging to a second or third generation of farmers, so their relationship or roots with the land has another nuance. However, their age is more related to the use and adoption of ICT (Méndez López, 2017). There is also a marked and accelerated process of feminization of the countryside (Concheiro & Grajales, 2005), which representsan opportunity to introduce them to the use of ICT.

The incorporation of farmers in the access and use of ICT, provides them with useful information that supports their decision-making process in the following aspects:

- **Planning:** from choosing production, knowledge about production and determining the productive input (seeds, seedlings, etc.).
- **Technical feasibility**: in terms of soil, water and climate requirements and economic, financial, commercial, and human resources.
- **Production**: data on irrigation practices, pruning, harvesting, storage, nutrition control, weeds, pets, and diseases.
- **Marketing**: national and international market strategies, financing, procedures, and institutional support.

- **Export**: the process of exporting, negotiations, shipment of products, franchises, and special funds.

Shankar*et al.* (2020) point out that, given the primacy of the agricultural sector in India, the analysis of massive data (Big Data), is key to decision making, in favor of understanding the possible ways in which a farmer's income can increase. The authors propose a simulation model based on Bayesian classifiers, which allow the farmer to make a cost-benefit balance between low and high commercial value crops, according to the characteristics of the producer and the conditions of the land. It also provides useful information for policy makers to study and analyze crop growth trends in different states and regions. Finally, this work provides a discussion forum or social network where farmers from different regions can exchange information.

Aditya*et al.* (2020) in their study on the use of Big Data in agriculture, highlight general information that is used in different areas of agriculture: climatology and climate change, land, research on plants and animals, crops, soil, weeds, food availability and security, decision-making by farmers, insurance and financing for farmers and remote sensing. They recognize the importance of data correlation and classification methods for pattern detection in agriculture. In addition, they discuss how critical it is for farmers to understand the importance of technology collecting large amounts of crop data and transforming it into meaningful knowledge.

Other works recognized the different technologies that could be applied in agriculture and its components such as management, nutrients or plant physiology (Ali *et al.*, 2019), and which couldbe enhanced with the use of techniques like geospatial analysis or remote sensing. In fact, authors identified a set of topics that might benefit from ICT adoption, for example, the development of expert systems, the design of government policies as well as detecting information needs among local producers.

In the case of tools for the dissemination of agricultural information, the design and development of webapps represents an increasing popular alternative. The systematic review performed by Tejeda-Castro *et. al.* (2019) found that Spain, Mexico, Colombia and US are the countries that most use web applications. According to their results, most of the webapps are oriented to multiple areas of the crop cycle instead of providing a single or specific function. Another finding showed that web systems have the highest percentage for deployment (71%), but in the case of mobile apps(android and iOS platforms), these represent only the 14%.

For a detailed analysis of the application of ICT in the farming life cycle, Sorensen *et.al.* (2019) established a three-stage process associated to pre-cultivation, cultivation & harvesting and post-harvest.

In the case of Mexico, there are private associations and national organizations that carry out studies related to ICTs. Particularly, the Internet MX Association (AIMX), presents in its web platform different studies about the Internet, for example, electronic banking, online education, and digital economy, to mention a few. However, it is very interesting to know the results of one particular study (AIMX, 2015), which we consider appropriate for the subject of this chapter.

This study is focused on the barriers of access to Internet among people between 13 and 55 years. Although the cities participating in the study are concentrated in 4 geographical areas of Mexico, the labor activity of farmers is not represented.

If the main barriers to Internet access are costs and slow connection, it is to be expected that rural or semi-urban communities will suffer a greater impact. Another important and closely related problem is the little or no supply of service providers in areas where connection problems are constant.

Regarding the search for information, the second barrier linked to greater use corresponds to the existence of other mechanisms to meet the need. This means that among the adult and older population, the presence of traditional media such as radio and television cover their needs. However, it is recognized that specialized information is increasingly being disseminated and given more space, as is the case with the written press.

Collaboration with CONABIO

Due to its great variety of ecosystems, Mexico offers ideal conditions for the cultivation of diverse products. Prominent among these agricultural products is maize, not onlyas an essential part of the Mexican diet, but also because it is the cereal with the greatest production worldwide in order to satisfy the needs of both human and animal consumption, as well as those of industry.Mexico is the center of origin and diversification of 59 breeds of maize. These were created by Mexican farmers, who generate and maintain this diversity, being even greater within each breed (CONABIO, 2012).

However, the most recent and outstanding work in terms of its impact at a national level is the Global Project on Native Corn, coordinated by the National Commission for the Knowledge and Use of Biodiversity (CONABIO).

This study founded the theories of centers of origin and diversification of maize in Mexico and showed the state of the art of diversity and distribution of the species in the country. One of the results that concerns us is the Native Maize Database that contains more than 20 thousand records of native maize populations, with information declared by the farmers.

Native Maize Dataset

The data taken into account for this study are derived from the Global Native Corn Project, coordinated by the CONABIO.The objective of the project, in which a total of 206 researchers from 65 academic institutions participated, is to update the information on maize and its wild relatives in Mexico for the determination of centers of genetic diversity.

A database with 22,931 maize records was built, through the filling out of passport data by farmers and complemented with the collection of germplasm (maize cobs).

The database tables are composed of text, alphanumeric and numeric fields. Of the 103 fields that make up the database, 30 are considered in this platform according to the information that most farmers are interested in consulting.

MAIN OBJECTIVE OF THE CHAPTER

As part of the strategies to combat poverty and generate sustainable regional development, precision agriculture enablesthe efforts to be channeled towards crop management through the use of data analysis and technologies that allow estimation, evaluation and understanding of the variables involved.

This chapter focuses on promoting the use of ICT for the exchange of knowledge and experiences among subsistence farmers. These farmers are characterized by their small agricultural areas, where they carryout low-input farming practices that tend to be sustainable and represent about 70% of the producers in Mexico (FAO & SAGARPA, 2012). These producers are not linked to the market, but according to theFAO (2015) they are the agricultural sector on which the world's food depends and not on intensive farming of large areas.

Public institutions are an influential source of information for small farmers, and those with websites are the most visited. In addition, supplier companies, marketing companies and agrobusinesses assume an informative role towards them. At the same time, private support institutions and technology promotion companies and cooperatives are a strategic factor in transmitting messages across and generating reliable information flows for farmers.

According to Figure 2, the first objective of the technological contribution is data acquisition. In this stage two main activities in green color were achieved: the literature review and the maize dataset analysis; the second stage consisted of the establishment of a partnership with CONABIO, with aninterdisciplinary teamof researchers and with the maize farmers to select the variables from his experience.

Then, the third stage, focuses on access and distribution of information usingthe Native Maize System, (blue color text due to the application of ICT in this research). In addition, the i-Gamma platform appears as a system that benefits from the information shared by the maize system and vice versa. The use and exploitation of ICTs depends on a final training activity. Each of these activities will be described in the following sections.

Figure 2. Research elements differentiated by its analysis & use (green) and construction (blue)

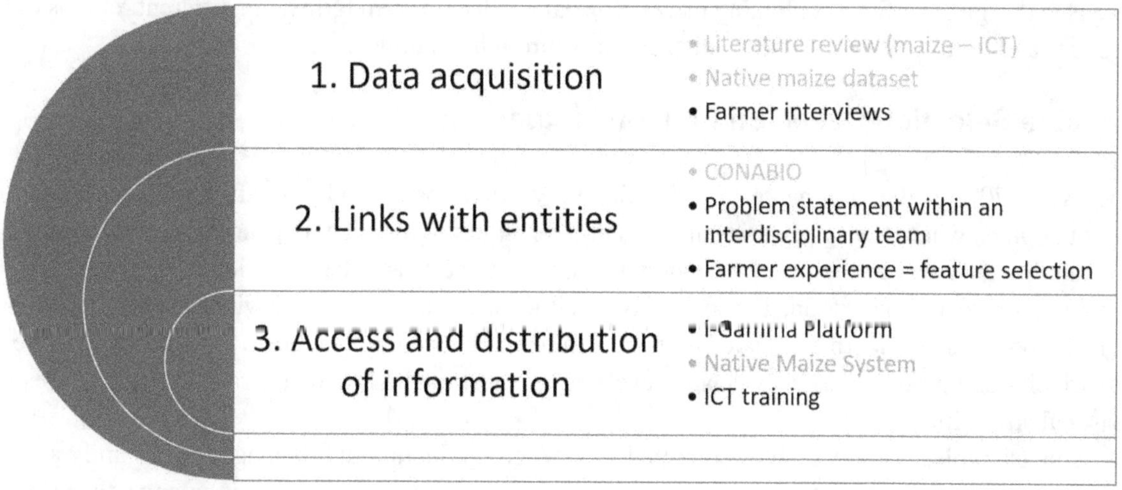

Interviews with Farmers

In order to explore the possibilities of usingICT for consultation and knowledge sharing among maize producers, they were interviewedto determine the ICTs they have access to, the ability to use themand the type of information they are interested in consulting and sharing.

Interviews were conducted with maize producers from the communities of Ocotempa, Oxtotitla, Atlajco and Atempa, in the municipality of Tequila in Veracruz; and San Andrés Huayapam and San Antonino Castillo Velasco in Oaxaca.

Approaching the Problem in an Interdisciplinary Team

This work, inspired by empirical and traditional knowledge, is inevitably collaborative, as there is a need to preserve and share this knowledge through automated mechanisms. Today, the loss of traditional knowledge among younger generations is a reality. The current economic system forces young people to engage in profitable activities outside their communities and therefore prevents their participation in agricultural practices.

Producers, now mostly elderly, have no one to pass their knowledge to and, in the near future, the land will not be cultivated for the same purpose. Having this traditional knowledge at hand could prevent irreparable loss and could represent a source of information available to interested young people as they identify new opportunities in the rural sector.

Deciphering and understanding producers' decision making to grow their crop, in one way or another in specific scenarios, underlines the weight of knowledge that must be disseminated and conserved to contribute to the sustainable development of rural families.

It is important to reflect on the intention of the content to be disseminated, whether for information purposes or knowledge generation purposes, in order to determine the narrative to be used, and the appropriate means for its transfer. The processes involved in knowledge management are the creation, transmission or socialization and use of knowledge (EcuRed, 2011; Canals, 2003).

Therefore, the fundamental proposal is to disseminate knowledge through the use of technology. The idea is to design a platform containing maize crop data collected from farmers in different regions of the country, and to make it accessible to other farmers through a mobile device.

Variable Selection Based on Farmers' Experience

Each year, 70% of the area harvested of maize in Mexico corresponds to native varieties grown on seasonal plots, where the agronomic variables of interest to the producer depend on the environment.

Rainfall determines the time of sowing, while drought and frost affect the yield. The topography of flat areas, or steep slopes define the sowing type, either with land rotation or with conservation tillage. Soil composition and depth are related to the moisture and nutrients available to the crop. This is the set of variables that producers face cycle after cycle as they test methods and generate strategies to resolve seasonal adversities.

Some examples of the variables compiled in a maize production unit managed by a family are: the geographic coordinates, the common name of the maize sown, the race to which it belongs, the origin of the seed (where it originated and how it was obtained), the time during which this seed has been sown, the sowing density, the amount of seed used in sowing and the dates of the main activities during the maize cycle.

The analysis of the above variables is complemented by the exchange of experiences among farmers, where several variables are addressed, such as yield, uses, consumption preferences and the most appreciated characteristics of the maize, as well as its resistance to different weather phenomena such as wind and frost, are addressed.

Similarly, data about the cob, such as size and type of grain, are important variables that allow the identification of the breed. The classification of maize by race allows the grouping of populations that share common characteristics, some of which are morphological, ecological, and genetic. Therefore, some maize crops share cultivation techniques and cares, pest or disease problems.

Relationship Between i-Gamma and the Maize System

The analysis of the data presented in this research, results from the link between two projects developed at the Institute of Ecology, in Xalapa, Veracruz. The first is called "Strategies for Maize conservation and Food Sufficiency" and the second, "Integrity for the Environmental Management of Development Supported by Massive Data and Automated Learning (i-GAMMA)".

Within the i-GAMMA project, ecosystem integrality is explored through the analysis of massive data that provides a measure of ecosystem changes and health status. The agricultural sector depends on the environmental services of ecosystems such as pollination, erosion control, provision of water, soils, nutrients, and pest control, among others. Therefore, it is essential to share data from small farmers, who promote good agricultural practices, with care for the environment, diversity, and food security.

The data set obtained comes from an effort to combine complete and consistent data that will systematize and facilitate the decision-making process of maize farmers in Mexico supported by experts who intend to advise on these crops.

ICT Training

The training of farmers is through community workshops. In this teaching, multimedia resources are used in an important way, thus illustrating each concept in detail.

Because most producers are elderly and most are low-literate, classroom videos and field practices are the most accurate means of learning.

The participation of community technicians is fundamental in the development of workshops. These people know the problems and goodness of the community. They speak the same language as the farmers and are proud to be part ofthe community. They are also the link between the specialists and the producers since their preparation strengthens their performance from the academy in the understanding and transmission of concepts.

Young people in middle and high school represent an important sector for ICT training, since they have skills and relationships with mobile technologies.

FINDINGS AND RECOMMENDATIONS

This section presents the results of the implementation of the framework. After reviewing the literature and the data sets obtained by CONABIO, all the procedures performed by the interdisciplinary team led to the results described below:

Interviews on the use of ICTs by Producers

In order to explore the possibilities of access to the exchange of experiences among maize producers, through the use of ICTs, and also to find out what kind of information they would like to share with other farmers, interviews were conducted with twenty producers from the communities of Ocotempa, Oxtotitla, Atlajco and Atempa in the state of Veracruz and San Andrés Huayapam and San Antonino Castillo Velasco in Oaxaca, to assess the possibilities of access to ICTs, specifically the use of radio, television, telephone, mobile phone, computer, Internet and social networks.

Figure 3. Localization of the communities in Oaxaca and Veracruz

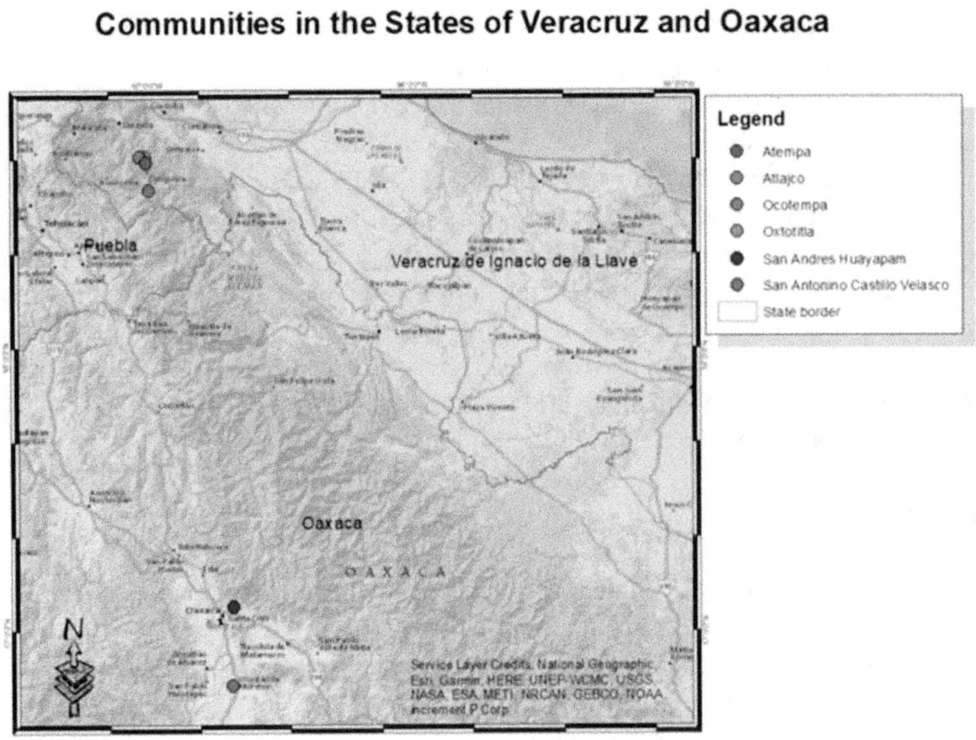

The interview data show that most have radio and television, few have mobile phones and only a few people have access to the internet and social networks. However, they mention the Internet and social networks as means of accurate and timely information. They even consider the cell phone as the main means of exchanging information with other producers.

They were asked about the agricultural information they are interested in consulting and the data they manage and would like to share with other farmers.

The information they are interested in consulting in ICT is mainly, pests and diseases of maize, organic fertilizers, and production methods. They are not interested in knowing means and prices of maize, because the total of their production is for self-consumption. They would also like to share information on yields, pests and diseases, organic fertilizers, types of seeds, maize from other places and best agricultural practices for maize cultivation.

Platform Design with Respect to Workshops

It is believe that, in order to improve the methods of management and operability of national programs of field fomentation directed at the production of basic foods such as maize, it is vital to have a source of information that demonstrates the dynamic of the crop over time, through an informatic tool that provides data to the public, from public policy decision-makers to communication among producers of different regions of Mexico.

Figure 4. Results in the use of ICT resources

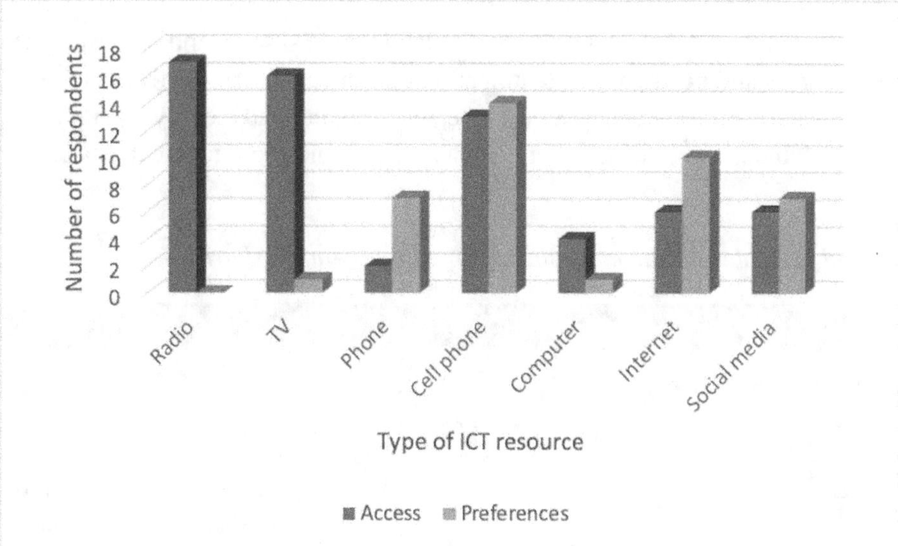

Figure 5. Main topics around maize information

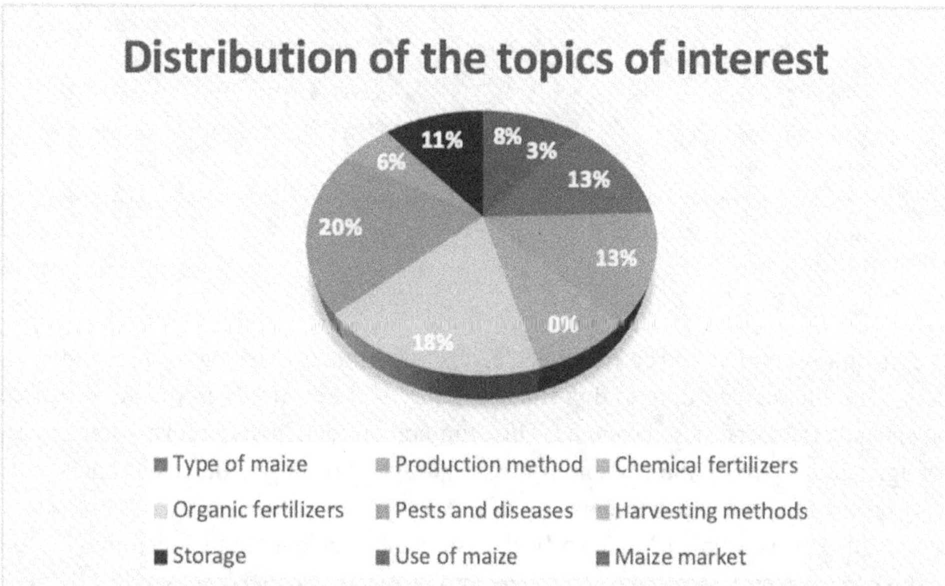

The exchange of experiences regarding the attention given to different phenomena could help resolve problems in similar localities. The experience of working in the field allows us to know the information that is important to share among farmers according to their interests. Here we highlight data such as: soil conservation practices (periods of rest of the soil, types of organic fertilizers), nutritional content of maize (by color, by size) and forms of seed exchange. The exchange occurs with the appropriate technical advice, taking advantage of the experience of Mexican researchers and producers, which requires the use of ICTs to provide a fluid and updated communication of what is happening in the field.

The wireframes shown in Figures 6 to 10, represents graphically the basic operations of the web system. The responsive structure of this page allows to register and consult from different devices, the variables selected from the CONABIO database, which include sowing and harvesting season, control of diseases and plagues and post-harvest treatment that producers carry out from their experience.

Additionally, these records allow queries by state, municipality and locality, identifying the specimen ID, primary race, altitude, uses, yield, variety characteristics and topographical aspects.

Figure 6. Selected variables for the communication with producers, Part A

Sistema de Intercambio de experiencias entre agricultores de maíz

Estado: TODOS   Municipio: TODOS   Localidad: TODOS

ID	Año colecta	Raza primaria	Complejo racial	Influencia de otras razas	Estado	Municipio	Localidad	Altitud	Nom
1	2009	Cónico	Cónico	Serrano	OAXACA	SAN JUAN QUIOTEPEC	San Juan Quiotepec	1967	maí
2	2009	Cónico	Cónico		OAXACA	SAN JUAN QUIOTEPEC	San Juan Quiotepec	1967	maí
3	2009	Cónico	Cónico	Serrano	OAXACA	SAN JUAN QUIOTEPEC	San Juan Quiotepec	1967	maí
4	2009	Nal-tel de Altura	Dentados tropicales		OAXACA	SAN JUAN QUIOTEPEC	San Juan Quiotepec	1967	maí
5	2009	Tuxpeño	Dentados tropicales		OAXACA	SAN JOSÉ CHILTEPEC	San José Chiltepec	35	maí
6	2009	Tuxpeño	Dentados tropicales		OAXACA	SAN JOSÉ CHILTEPEC	San José Chiltepec	35	maí
7	2009	Tuxpeño	Dentados tropicales		OAXACA	SAN JOSÉ CHILTEPEC	San José Chiltepec	35	maí
8	2009	Tuxpeño	Dentados tropicales		OAXACA	SAN JOSÉ CHILTEPEC	San José Chiltepec	36	maí
9	2009	Tuxpeño	Dentados tropicales		OAXACA	SAN JOSÉ CHILTEPEC	San Isidro Naranjal	138	maí
10	2009	Tepecintle	Dentados tropicales		OAXACA	SAN JOSÉ CHILTEPEC	San Isidro Naranjal	138	maí
11	2009	Tepecintle	Dentados tropicales	Tuxpeño	OAXACA	SAN JOSÉ CHILTEPEC	San Isidro Naranjal	138	maí
12	2009	Tuxpeño	Dentados tropicales		OAXACA	SAN JOSÉ CHILTEPEC	San Isidro Naranjal	104	maí
13	2009	Tepecintle	Dentados tropicales		OAXACA	SAN JOSÉ CHILTEPEC	San Isidro Naranjal	104	maí
14	2009	Tepecintle	Dentados tropicales		OAXACA	SAN JOSÉ CHILTEPEC	San Isidro Naranjal	104	maí

In order to acquire complete and consistent data, it is important to drive the systemization of processes in the area of agriculture.According to this study, modern ICTs currently have a higher effective communication capacity than traditional sources. This study recommends that information departments and divisions of agriculture modify the content of modern tools according to the needs and socio-economic conditions of farmers to increase their capacity of use.

This is possible if it is involve Mexican businesses, obtain adequate technical advice and take advantage of the experience of Mexican researchers and producers, which necessitates the use of ICT to provide fluid and up to date communication of that which takes place in the field.

However, such instruments require other elements in order to successfully construct a software-type technological product; i.e., in addition to data collection, the development of an information system must integrate or consult information from other web platforms, such as CONABIO, CONANP or SIAP, only when these are prepared to receive consultations of information from the manual records of focal groups, including the results of systemized interviews applied to farmers.

Figure 7. Selected variables for the communication with producers, Part B

Sistema de intercambio de experiencias entre agricultores de maíz

Estado: TODOS Municipio: TODOS Localidad: TODOS

Nombre común	Usos del grano	Época de siembra 1	Época de siembra 2	Época de cosecha 1	Época de cosecha 2	Densidad de siembra	Indica
maíz amarillo, maíz blanco	Atole, tamal, tortilla	15 de abril		30 de septiembre			Frijol y
maíz amarillo, maíz blanco	Tamal, tortilla	15 de marzo		15 de octubre			Calab
maíz amarillo, maíz blanco	Tamal, tortilla	15 de marzo		15 de octubre			Frijol y
maíz amarillo	Tamal, tortilla	15 de junio	15 de octubre	30 de diciembre	15 de junio		Frijol y
maíz criollo	Tamal, tortilla	15 de mayo	15 de noviembre	15 de noviembre	30 de marzo		Frijol y
maíz criollo	Tamal, tortilla	25 de mayo	25 de noviembre	15 de noviembre	15 de mayo		Calab
maíz criollo	Tamal, tortilla	23 de mayo	15 de noviembre	15 de noviembre	30 de marzo		Frijol y
maíz criollo	Tamal, tortilla	15 de mayo	15 de noviembre	15 de agosto	24 de abril		Frijol,
maíz criollo	Tamal, tortilla	30 de mayo	15 de octubre	15 de septiembre	15 de mayo		Calab
maíz amarillo, maíz morado	Tamal, tortilla	24 de octubre	15 de octubre	15 de septiembre	30 de abril		Calab
maíz amarillo, maíz morado	Tamal, tortilla	15 de julio	1 de noviembre	15 de octubre	15 de mayo		Tomat
maíz criollo	Tamal, tortilla	15 de junio	15 de diciembre	30 de noviembre	15 de mayo		Frijol,
maíz amarillo, maíz morado	Tamal, tortilla	28 de junio	15 de noviembre	30 de octubre	15 de mayo		Calab
maíz amarillo, maíz morado	Tamal, tortilla	28 de junio	15 de noviembre	30 de octubre	15 de mayo		Calab

Figure 8. Selected variables for the communication with producers, Part C

Sistema de intercambio de experiencias entre agricultores de maíz

Estado: TODOS Municipio: TODOS Localidad: TODOS

Indicar cultivos asociados en el policultivo	Rendimiento 1	Rendimiento 2	Enfermedades observadas	Insectos dañinos observados	Malezas observadas	Problemas
Frijol y calabaza					Arvenses	Gorgojo
Calabaza					Arvenses	Gorgojo
Frijol y calabaza					Arvenses	Gorgojo
Frijol y calabaza					Arvenses	Picadura
Frijol y calabaza					Zacates, malva, lengua de vaca	Gorgojo
Calabaza					Zacate, tamalote, zacatón	No
Frijol y calabaza			Espiga, sin control		Zacate y zacatón	No
Frijol, plátano, calabaza				Gusano cogollero	Zacate	Gorgojo y p
Calabaza, chile, frijol			Tallo (Pudrición de tallo)		Zacate y arvenses	No
Calabaza, chile, frijol					Zacate	Picadura
Tomate, calabaza, chile				Gusano cogollero	Zacate	Gorgojo
Frijol, chayote, calabaza			Raíz (Pudrición de raíz)		Zacate	Hongos y g
Calabaza					Zacate y hierba mora	No
Calabaza					Zacate y hierba mora	No

Figure 9. Selected variables for the communication with producers, Part D

Sistema de intercambio de experiencias entre agricultores de maíz

Estado: TODOS Municipio: TODOS Localidad: TODOS

Problemas durante el almacenamiento	Control de plagas, malezas o enfermedades	Control de plagas/Insecticidas	Control de malezas/Herbicidas	Fertilizante usado	Car
Gorgojo			Otro (Manual)		Aca
Gorgojo		Ninguno	Otro (Manual)		Aca
Gorgojo			Otro (Manual)		Poc
Picadura			Otro (Manual)	Químico (Urea)	
Gorgojo			Químico (Fumigación)	Químico (Urea)	Aca
No		Químico (Fumigación)	Químico (Fumigación)	Químico (Urea)	Aca
No			Químico (Fumigación)		Aca
Gorgojo y pudrición		Químico	Químico (Fumigación)	Químico (Urea)	Aca
No	Manual				Aca
Picadura	Manual				Aca
Gorgojo		Ninguno	Químico (hierbamina)		Aca
Hongos y gorgojos			Otro (Manual), Químico (Fumigación)		Aca
No		Químico (Fumigación)	Otro (Manual)		Aca
No		Químico (Fumigación)	Otro (Manual)		Aca

Figure 10. Selected variables for the communication with producers, Part E

Sistema de intercambio de experiencias entre agricultores de maíz

Estado: TODOS Municipio: TODOS Localidad: TODOS

nsecticidas	Control de malezas/Herbicidas	Fertilizante usado	Características que no le gustan de la variedad	La variedad es resistente o tolerante	Aspectos topográficos
	Otro (Manual)		Acame	Otro	Pendiente (%) (25)
	Otro (Manual)		Acame y poca productividad	Otro	Pendiente (%) (25)
	Otro (Manual)		Poca productividad	Otro	Pendiente (%) (25)
	Otro (Manual)	Químico (Urea)		Otro	Pendiente (%) (25)
	Químico (Fumigación)	Químico (Urea)	Acame	Acame, frío, Sequía	Pendiente (%) (0-5)
ón)	Químico (Fumigación)	Químico (Urea)	Acame	Acame, frío, Sequía	Pendiente (%) (0-5)
	Químico (Fumigación)		Acame	Otro	Pendiente (%) (0-5)
	Químico (Fumigación)	Químico (Urea)	Acame	Otro	Pendiente (%) (15-20)
			Acame	Otro	Pendiente (%) (20)
			Acame	Otro	Pendiente (%) (15-20)
	Químico (hierbamina)		Acame	Otro	Pendiente (%) (15-25)
	Otro (Manual), Químico (Fumigación)		Acame y problema de gorgojo	Otro	Pendiente (%) (20-25)
ón)	Otro (Manual)		Acame	Otro	Pendiente (%) (20)
ón)	Otro (Manual)		Acame	Otro	Pendiente (%) (20)

Operational Strengths

The interdisciplinary work of this project has made possible the open exchange of information for analysis and the establishment of a common plan of action. The contributions of each member allowed the development of a whole method from different perspectives and theories.

Different methods of agricultural research were combined, such as study sites, household surveys, preference analysis and estimation of the most likely general classifications of information options, among others. Also, from the area of systems engineering, the technological tools for data management and access were generated, as well as the creation of the necessary interfaces.

The willingness of farmers to learn and participate, as well as the selection of multimedia resources for the system training process, allowed the identification of the resulting data and the most appropriate use of ICT.

Figure 11. Process of information exchange

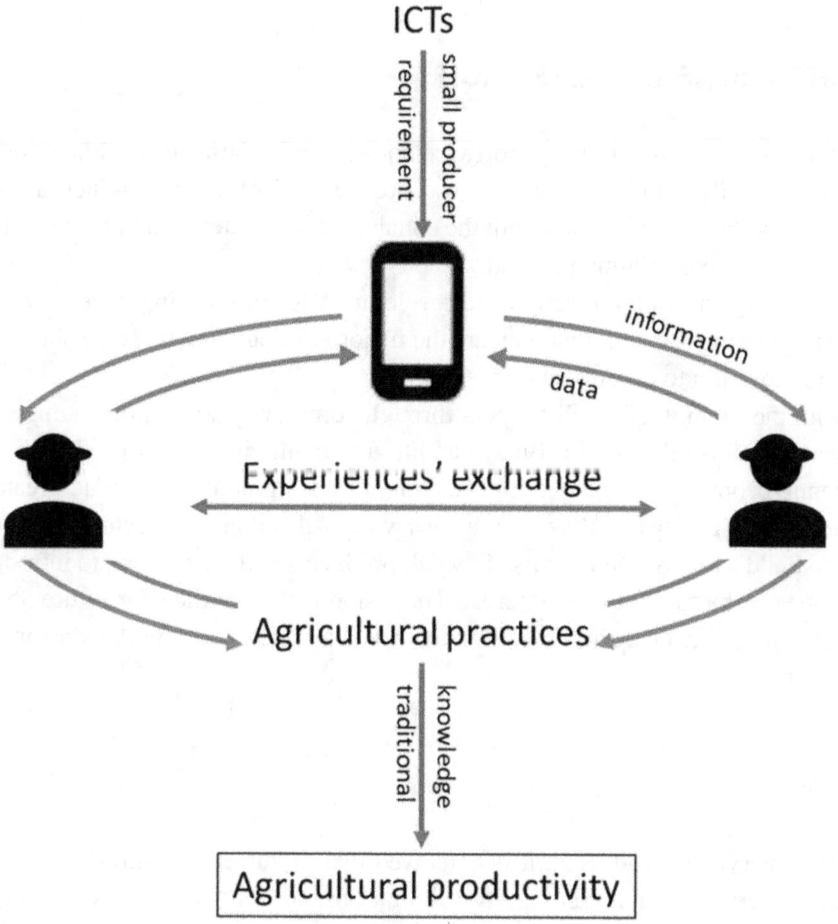

Procedural Shortcomings

A small number of community surveys were achieved; however, these data can be generalized for most producers in Mexico, as they share the same conditions. They live in remote communities, in areas of difficult access, with little infrastructure, where basic water and electricity services are barely covered. The producers are people of low resources, who invest the minimum in agricultural inputs, mainly exercise the family labor and most of them are older adults.

The condition of being older adults and low schooling, can be considered a deficiency in the process, since most do not have someone to inherit the land and share knowledge. However, the few young people who see opportunity in the work of the field, can be considered key elements for the transcendence of this new proposal to share knowledge.

The databases extracted from CONABIO are combined with the data obtained from the interviews, allowing the generation of a semi-automatic system of information exchange between producers.

The correct visualization and interpretation of the data is still in the process of usability and functionality testing, after the training of the farmers.

FUTURE DIRECTIONS OF RESEARCH

Beyond great knowledge in computing or software applications, platforms designed for different types of actors, which are highly intuitive and easy to use, are required. Of course, induction through training, courses and workshops will be necessary with the collaboration of external actors or by taking advantage of the city's infrastructure and human capital.

Advancing the design of communication and development technology in the field, which is accessible and useful for small farmers. Given that they are the majority in our country, they should be included in communication innovation to conserve knowledge.

Continue with the training of small farmers through community workshops, which, for now, is the fast and safe way to inform this sector. But, in addition to being directed at producers, is important to direct to the young people in middle and high school in the community, to create greater vision of the trends in the field, highlighting that they are the ones who easily adopt new techniques and technologies.

This effort to build a platform for the use of small producers is the first step, to introduce them to the use of technologies with topics of their interest. The vision is that, in the near future, they will actively participate in enriching this platform by adding their data and keeping it updated as much as possible, from their location.

CONCLUSION

According to the interviews conducted, it is perceived that producers are interested in having expeditious agricultural information, but there are two limitations: access to current ICTs and the capacity to use them. The adult and elderly producers do not have the ability to use mobile phones and navigate the internet. On the other hand, the limitation of access ICTs is not only an economic issue, but also one of little or no accessibility due to lack of coverage.

Traditional ICT such as radio and television remain the major means of information for producers in rural areas, but they do not prefer them for information exchange. The proposed system should be noted that the assistance of technicians in rural schools is vital for bringing maize producers closer to the use of current ICT for information exchange.

It is considered, that facilitating access to digital ICT among the marginal sectors of the country will not automatically promote its development; however, its intelligent adoption could boost it. The development of technological services designed from the study of the socio-cultural reality of its users is a way to achieve this goal.

The critical limitations of the use of ICT in the agricultural sector are not only the time in which the information arrives due to the quality of internet connection, because it turns out to be inefficient in some areas. Digital literacy in farmers is incipient, especially in the advanced age producers.

The learning obtained from the interdisciplinary work achieved so far, consisted of combining a series of data derived from interviews applied to small maize farmers, among which aspects related to the experience and traditions of these actors stand out. The understanding of these data starts from the identification of their meaning and structure, until the transformation of them into values that can be related to each other. The initial development of the web platform selects and shows the important variables for the small producers. However, the user interface needs a further refinement with the aim of improving its usability. Therefore, a mix of text data and multimedia components such as maps or images will be topics to be analyzed in the short term.

Likewise, extensionism must emphasize the improvement of the quality of life of producers and their families in an integral manner and not only focus their efforts on increasing agricultural production. It should promote that the processes be participatory and long-term, as well as the use of farmer-to-farmer methodologies accompanied by training strategies that use simple and practical language in computer media that highlight the importance and need of youth participation in improve the agricultural impact and guarantee the sustainability of the means of production.

ACKNOWLEDGMENT

We are grateful to the maize producers who participated in this research from the field, by means of answering interviews and attending informative and training workshops. We also thank the Fund for Peace team for facilitating access to the communities and the support of language translation to achieve a better communication with the producers.

We also thank CONABIO for facilitating the Global Corn Project database, which is a raw and crucial piece of data in this review.

This research received no specific grant from any funding agency in the public, commercial, or not-for-profit sectors.

REFERENCES

Aditya Shastry, K., & Sanjay, H. A. (2020). Data Analysis and Prediction Using Big Data Analytics in Agriculture. In *Internet of Things and Analytics for Agriculture* (Vol. 2, pp. 201–224). Springer Nature Singapore., doi:10.1007/978-981-15-0663-5_10

Adrian, A. M., Norwood, S. H., & Mask, P. L. (2005). Producers' perceptions and attitudes toward precision agriculture technologies. *Computers and Electronics in Agriculture*, *48*(3), 256–271. doi:10.1016/j.compag.2005.04.004

AIMX. (2015). *Barreras de acceso a Internet.* Retrieved December 15, 2019, from https://www.asociaciondeinternet.mx/es/estudios

Ali, M., Mubeen, M., Hussain, N., Wajid, A., Farid, H. U., Awais, M., … Nasim, W. (2019). Role of ICT in Crop Management. In M. Hasanuzzaman (Ed.), Agronomic Crops Volume 2: Management Practices (pp. 637–652). Springer Nature Singapore. doi:10.1007/978-981-32-9783-8_28

Alvarado-Beltrán, G., López-Sánchez, H., Santacruz-Varela, A., Muñoz-Orozco, A., Valadez-Moctezuma, E., Gutiérrez-Espinosa, M. A., ... Taboada-Gaytán, O. R. (2019). Morphological variability of native maize (Zea mays L.) of the west highland of puebla and east highland of tlaxcala, Mexico. *Revista de La Facultad de Ciencias Agrarias*, *51*(2), 217–234.

Badia, T., & García, C. (2006). The incorporation of information and communication technologies into teaching and learning based on collaborative project work. *RUSC. Universities and Knowledge Society Journal*, *3*(2). Advance online publication. doi:10.7238/rusc.v3i2.286

Cadena Iñiguez, P., Rendón-Medel, R., Rodríguez-Vázquez, H., Camacho-Villa, C., Santellano-Estrada, E., Guevara-Hernández, F., & Govaerts-Bram, B. (2018). Propuesta metodológica-interinstitucional para un nuevo extensionismo en México. *Revista Mexicana de Ciencias Agrícolas*, *9*(8), 1777–1785. doi:10.29312/remexca.v9i8.826

Canals, A. (2003). Gestión del conocimiento (M. Boisot & A. Cornella, Eds.). Academic Press.

CONABIO. (2012). *Maíces y Razas de Maíces*. Academic Press.

Concheiro, L., & Grajales, S. (2005). Movimientos campesinos e indígenas en México: La lucha por la tierra. *Observatorio Social de América Latina*, *16*(VI), 47–58.

Corpoica. (2015). *Corporación Colombiana de Investigación Agropecuaria*. Retrieved October 13, 2019, from https://web.corpoica.org.co/

EcuRed. (2011). Retrieved from https://www.ecured.cu/Gestión_del_conocimiento

Espinosa Tamayo, L., Rincón Sánchez, F., Ruíz Torres, N., Martínez Reyna, J., & Benavides Mendoza, A. (2019). Respuesta ambiental de poblaciones nativas de maíz del sureste de Coahuila, México. *Nova Scientia*, *2*(23), 108–125. doi:10.21640/ns.v11i23.1931

FAO. (2015). *El estado mundial de la agricultura y la alimentación. La innovación en la agricultura familiar.* Retrieved May 1, 2020, from http://www.fao.org/3/a-i4910s.pdf

FAO & SAGARPA. (2012). *Diagnóstico del Sector Rural y Pesquero en México*. Retrieved May 1, 202AD, from http://smye.info/cuestionario_final/diagnostico/menu_f.php

García Yeste, C., Leena Lastikka, A., & Petreñas Caballero, C. (2013). Comunidades de aprendizaje. *Script Nova. Revista Electrónica de Geografía y Ciencias Sociales, 17*(427).

GIE. (2015). *Grupo Interinstitucional de Extensionismo*. GIE.

González Sabater, J. (2017). *Manual de transferencia de tecnología y conocimiento*. Academic Press.

Grøn Sørensen, C. A., Kateris, D., & Bochtis, D. (2019). ICT Innovations and Smart Farming. In M. Salampasis & T. Bournaris (Eds.), *Information and Communication Technologies in Modern Agricultural Development* (pp. 1–19). Springer Nature Switzerland. doi:10.1007/978-3-030-12998-9_1

Méndez López, J. C. (2017). *Condicionantes y posibilidades en el uso de las TIC. Estudio de caso de dos cooperativas de Chinamperos de Xochimilco*. Beneméritca Universidad Autónoma de Puebla.

Molano Bernal, L. C. (2017). *Estudio del uso de las tecnologías de la información y la comunicación (TIC) como herramientas de transferencia de tecnología en el sector agropecuario colombiano: caso Corpoica*. Universitat Politècnica de València. Retrieved from https://riunet.upv.es/handle/10251/90228

Nagel Amaro, J., & Martínez Vergara, C. (2016). *Acceso y uso de TICs y necesidades de información de los pequeños agricultores de las Regiones de O'Higgins y Maule*. Fundación para la Innovación Agraria.

Palmer, N. (2012). *Las TIC en agricultura. Conectando a pequeños agricultores con el conocimiento, las redes y las instituciones*. Retrieved December 3, 2019, from http://www.fao.org/3/ap561s/ap561s.pdf

Raza, M. H., Khan, G. A., Shahbaz, B., & Saleem, M. F. (2020). Effectiveness of information and communication technologies as information source among farmers in pakistan. *Pakistan Journal of Agricultural Sciences*, *57*(1), 281–288. doi:10.21162/PAKJAS/19.8292

Shankar, S., Rao, M., Shetty, P., Thombre, J., & Manek, H. (2020). A data analytics framework for decision-making in agriculture. In Advances in Data Sciences, Security and Applications (Vol. 612, pp. 85–98). Springer Nature Singapore. doi:10.1007/978-981-15-0372-6_7

Tejada-Castro, M., Delgado-Vera, C., Garzón-Goya, M., Sinche-Guzmam, A., & Cárdenas-Rosales, X. (2019). Trends in the Use of Webapps in Agriculture: A Systematic Review. In R. Valencia García, G. Alcaraz Mármol, J. del Cioppo-Morstadt, N. Vera Lucio, & M. Bucaram Leverone (Eds.), *ICT for Agriculture and Environment* (pp. 130–142). Springer Nature Switzerland. doi:10.1007/978-3-030-10728-4_14

ADDITIONAL READING

CONABIO. (2018). *Origen y diversificación del maíz*. Retrieved from https://www.biodiversidad.gob.mx/genes/origenDiv

Kelleher John. (2018). *Data Science*. The MIT Press.

KEY TERMS AND DEFINITIONS

Data Science: It is a general method that combines theories and techniques from databases, mathematics, statistics, artificial intelligence, and information visualization, with the purpose of analyzing data to obtain patterns that generate insights and novel knowledge.

Interdisciplinary Work: Field of study that crosses the traditional boundaries between various academic disciplines which implies consensus on different priorities, thinking styles, and ethical values.

Native Maize: They are varieties of maize that are given in each locality, also called criollo maize. These maizes have developed appropriate characteristics to the environment where they grow, therefore, they can hardly grow well in places outside the region.

Seasonal Agriculture: It is agriculture in open spaces, which depends on residual humidity and rainfall for the development of the crop. These areas are not susceptible to irrigation technology. In Mexico, 80% of the agricultural production is under these conditions.

Small Farmers: Agricultural producers that are characterized by having a recuded sowing area up to 2Ha. Their parcels are generally distributed in areas of slope or difficult access and they are people with low economic resources and little or no schooling.

Traditional Knowledge: It is the wisdom, attitude, and practices that are developed in a place or region according to the predominant culture. It is important to value and rescue it in order to transmit it to new generations.

Usability: It is a collection of techniques aimed at facilitating the interaction between users and software products.

Chapter 11
Raising the Efficiency of Seohen's Model to Predict Soil Resistance Faced by Chisel Plow

Adil Abd Elsamia Meselhy
Agricultural Mechanization Unit, Department of Soil Conservation, Desert Research Center, Egypt

ABSTRACT

This research was carried out to study the effect of crop factor, which is represented in some plant roots of crops residues (broad bean and wheat) on soil mechanical properties (cohesion strength, C, and internal friction angle, Φ), and thus, the effect of these roots on the power requirements of chisel plow to face the soil resistance was studied. Soil mechanical properties and power requirements of chisel plow (7 blades) were measured in the field directly at various soil depths of (0.05, 0.1, 0.15, and 0.2 m) with and without roots at constant tractor forward speed of about 4 km/h. Moisture content of the soil, broad bean roots, and wheat roots were 21%, 16%, and 14%, respectively, The results showed that the effect of roots of previous crops residues had a significant effect on the soil mechanical properties and power requirements for chisel plow when using the crop factor, which is represented in characteristic of crop residual roots in terms of root mean diameter.

INTRODUCTION

Endo and Tsuruta (1969) reported that soil shear strength increases in proportionto the amount of roots in thesoil. Waldron (1977) reported that when the soil is subject to shear strengths, there is the mobilization of an adding opposition due to the development of tensile strength inside of roots and the whole soil has a greater resistance.

Waldron and Dakessian (1982) mentioned that soil-root shear strength is directly proportional to root cohesion. This means a soil with high root cohesion will increase the soil-shear strength, adding to slope stability. Roots of plants increase soil cohesion by binding to soil particles. Wu (1984) reported that the number, depth, size, and growth patterns of roots affect the soil cohesion. Greenway (1987) reported that in generally accepted that plant roots provide reinforcement to a soil matrix due to the different material

DOI: 10.4018/978-1-7998-5000-7.ch011

properties of soil and roots. He also added that several factors could affect the root-reinforcement of a soil, including root density, root branching and tensile strength of the roots. Fitter (1993) appeared that root systems mechanically reinforce soil by transferring shear stress in the soil to tensile resistance in the roots. Ekwue and Stone (1995) proved that soil shear strength is very important in studying draft of tillage tools. In other words, the draft of farm implements is mainly affected by shear strength of soil. Lawrence et al. (1996) carried out shear tests in soil without roots and soil contents roots. The results showed that the increase in soil strength parameters (C and Φ) in the soil with roots compared soil without roots was included between 48% and 56%. Ekanayake et al. (1997) concluded that Field and laboratory shear tests have quantified the shear strength of soils both with and without roots. Roots increase the soil shear strength and the normal components of soil resistance. Goldsmith (1998) presented preliminary findings of increased soil shear strength due to roots of plants. Greenway (1987) and Collins (2001) mentioned that several factors can affect the root-reinforcement of a soil, including root density, root branching, tensile strength of the roots and root moisture content. Operstein and Frydman (2000), and Consoli et al. (2002) reported an increase in internal friction angle of soil reinforced by root sand, an increase in the apparent cohesion with increasing cross-sectional area and tensile strength of the roots. Singh and Huat (2004) appeared that the roots interact with the soil to produce a composite material in which the roots are fibers of relatively high tensile strength and adhesion inherent in a matrix of lower shear strength soil. The main contributor to the increase in shear strength appears to come from the cohesion (an increase of 22-64%) and the angle of shearing resistance (increase of 24-54%). Thorne (1990) and De Baets et al. (2008) mentioned that the mechanical characteristic of roots is that they are strong in tension. Soils, on the other hand, are strong in compression and weak in tension. A combined effect of soil and roots, producing a composite material in which the roots are fibers of relatively high tensile strength and adhesion embedded in a matrix of lower tensile strength soil mass, resulting in a reinforced soil. Therefore, it is the tensile of the roots, which contribute to the overall strength of the soil-root composite. Gerardand Mehta (1971) reported that root crops increased bulkdensity, permeability and strength of soil. Willatt and Sulistyaningsih (1990) showed that rice roots increased both the bearing capacity and shearing resistance of soils and a linear relationship between these strength parameters was observed. Huat et al. (2005) said that the root in soil the main elements which related to soil shear strength. Ali and Osman (2008) showed that root significantly contribute to the increase in soil shear strength. Abdullah et al. (2011) found that the presence of vegetation roots would result in an overall increase in the soil strength. Wu et al. (2014) said that roots play an important role in soil reinforcement has been accompanied by a growing interest in studying the mechanical strength of soil-root composites. Murielle et al. (2014) reported that soil mechanical properties were most influenced by (i) density of roots crossing the shear plane, (ii) branching density throughout the soil profile, (iii) total length of coarse roots above the shear plane and (iv) total volume of coarse roots and fine root density below the shear plane. During failure, fine, short and branched roots slipped through soil rather than breaking. Conclusion Root morphological traits such as density, branching, length, volume, inclination and orientation influence significantly soil mechanical properties. Cohen and Schwarz (2017) showed that the main contribution related to the presence of small roots in the most superficial soil layers, particularly in the first 2.0 m that mobilize their tensile strength by soil-root friction increasing the compound matrix strength. Such an effect has been largely known as root reinforcement. Bordoni et al. (2016) found that different vegetation types have the capacity to modify significantly the structure of the soil and soil mechanical properties. Bordoni et al. (2020) mentioned that root density in soil and root mechanical properties vary significantly

Figure 1. Deformed root (Waldron, 1977)

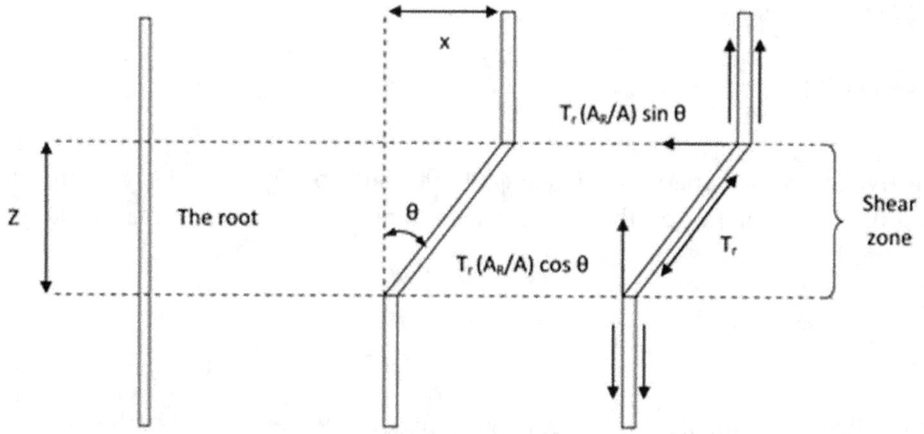

within different land use type, according to the physiological feature of each vegetation species. Root reinforcement is correlated proportionally to root density in soils.

Estimates of root reinforcement of soils have commonly been attained using simple perpendicular root models such as those of Waldron (1977)and Wu et al. (1979) that calculate root reinforcement as an add-on factor to soil strength. The root reinforcement model of Waldron (1977) based on the Coulomb equation in which soil shearing resistance calculated from cohesive and frictional forces:

$$S = C + \sigma_N \tan\varnothing \tag{1}$$

Where S soil-shearing resistance (kPa) σN is the normal stress on the shear plane (kPa), ϕ is soil friction angle (degrees), and C is the cohesion (kPa). Waldron (1977) extended equation (1) for root-permeated soils, by assuming that all roots extended vertically across a horizontal shearing zone, and the roots act like laterally loaded piles, so tension transferred to them as the soil sheared. The modified Coulomb equation becomes:

$$\tau_{sr} = C_s + C_r + \sigma_N \tan\varnothing \tag{2}$$

Where C_r increased shear strength due to roots (kPa). In the **Waldron (1977)** model, the tension developed in the root as the soil is sheared is resolved with a tangential component resisting shear and a normal component increasing the confining pressure on the shear plane. C_r can be represented by:

$$C_r = T_r\left(\sin\theta + \cos\theta\tan\varnothing\right)\left(A_R / A\right) \tag{3}$$

Where T_r average tensile strength of roots per unit area of soil (kPa), (A_R/A) is the root area ratio (no units), A is the area of the soil shear surface, A_R is the total cross-sectional area of all roots crossing the shear surface. Gray (1974) reported the angle of internal friction of the soil appeared to be affected little by the presence of roots. Sensitivity analyses carried out by Wu et al. (1979) showed that the value of the bracketed term in equation (3):

$$\left(\sin\theta + \cos\theta \tan\varnothing\right)$$

fairly insensitive to normal variations in θ and ϕ (40-90° and 25-40° respectively) with values ranging from 1.0 to 1.3. A value of 1.2 was therefore selected to replace the bracketed term and the simplified equation becomes:

$$C_r = 1.2T_r\left(A_R / A\right) \tag{4}$$

Thus, according to the simple perpendicular root model of Wu et al. (1979), the magnitude of reinforcement simply depends on the amount and strength of roots present in the soil. Therefore, the objective of this study is to investigate the effect of crop factor on some soil mechanical properties and power required for operating chisel plow.

MATERIALS AND METHODS

This experiment was conducted in an area of El-QantaraSharq in a sandy loam soil. The experimental area was divided into three experimental pieces. The three pieces were planted in season 2009-2010 as follows: the first with Wheat crop, the second with Broad bean crop and the third left without planting, and taking into account that removal of weeds in all pieces. Wheat and Broad bean were harvested at a height of 5 cm from the soil surface and each piece was divided into four experimental pieces include soil depths (0.05, 0.1, 0.15 and 0.2 m). Soil shear strength was measured in the field immediately at different soil depths for the experimental pieces, which contain the roots of previous crops. Total cohesion force was C_T and total internal friction angle was $\varnothing_T$. Also for the piece without roots soil shear strength were C_S and $\varnothing_S$.Therefore, the root shear strength only was obtained (C_r and $\varnothing_r$) as follows:

$$C_r = C_T - C_S \tag{5}$$

$$\varnothing_r = \varnothing_T - \varnothing_S \tag{6}$$

Root mean diameter and root density in soil were measured at different soil depths and made the equations between root shear strength(C_r and $\varnothing_r$) and root mean diameter, root density andsoil depth as follows:

$$C_{rb} = -119.2d - 1228.571d_{rb} - 1042.857\rho_{rb} + 33.851 \tag{7}$$

$$\varnothing_{rb} = -37d + 157.143d_{rb} - 314.286\rho_{rb} + 11.307 \tag{8}$$

$$C_{rw} = 59.833d + 3750d_{rw} + 291.667\rho_{rw} - 15.171 \quad (9)$$

$$\varnothing_{rw} = 26.467d + 2200d_{rw} + 33.333\rho_{rw} - 6.537 \quad (10)$$

Where: C_{rb}, C_{rw}= cohesion force for Broad bean and Wheat roots respectively (kPa), $\varnothing_{rb}$, $\varnothing_{rw}$= internal friction angle for Broad bean and Wheat roots respectively (degree), d= soil depth (m), d_{rb}, d_{rw}= root mean diameter for Broad bean and Wheat roots respectively (m) and ρ_{rb}, ρ_{rw}= root density in soil for Broad bean and Wheat roots respectively (g/cm^3 of soil). Then replacement the values of (C and Φ) in the Soehne (1956) general equation to predict the power requirements for chisel plow with new values of (C_T and Φ_T) where:

$$C_T = C_S + C_r \quad (11)$$

$$\varnothing_T = \varnothing_S + \varnothing_r \quad (12)$$

It was compensation directly from the values of C_r and Φ_r for roots by terms of root mean diameter, root density and soil depth in Soehne general equation. Power requirements were measured for chisel plow in the field using a hydraulic dynamometer device. The predicted values of power requirements for chisel plow were estimated in the soil with and without roots from Soehne general equation. Prediction of the power requirements in the soil with and without roots was compared with the actual values measured in the field and the efficiency of the Soehne general equation was determined when taking into account the existence roots from non-existence it.

Shear Ring Apparatus

The measurement of soil cohesion and soil internal friction angle was carried out using a shear ring apparatus by (Bekker 1969), Figure 2.

This device consists of a shear ring imbedded in the soil. This enables the operator to apply 105 kPa (15 Ib/in^2) normal pressure comfortably. It also has a calibrated spring, which deflects when the soil is in shear failure, a recording drum on which a recoding pen writes to indicate the values of the shear and normal pressures applied. It also has a handle for the operator to hold and apply the required combination of pressures. The operator forces the circular shear head into the soil so that the shear ring grips the soil and then applies a known amount pressure. Operator twists the handle to apply a shearing force on to the soil. At the point of failure, the shear head starts to turn. The operator takes note this point by making a mark on the recording drum to which is attached a special paper indicating the relationship between normal and shear pressure. Varying amounts of normal pressures are applied and a mark made at each point when the soil starts to fail. A straight line drawn through the points represents the soil failure line or the soil strength. The soil cohesion obtained by interpolating this line so that it cuts the shear stress axis. The intercept is the soil cohesion, C (kPa). The angle between this line and the horizontal gives the value of the soil internal friction angle Φ.

Figure 2. Picture of the shear ring apparatus in the field, (a) and illustration of device parts (b)

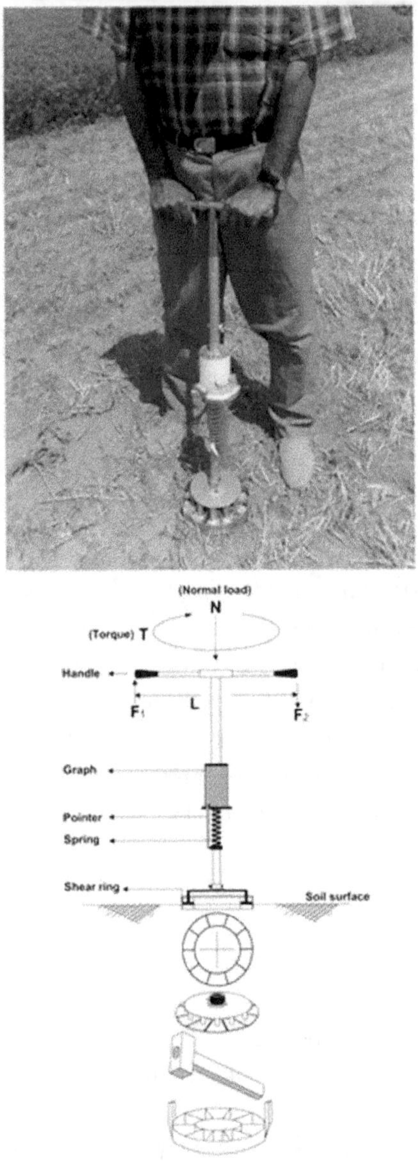

The shear stress τ can be calculate from the torque **T** as:

$$\tau = \frac{3T}{2\pi\left(R_2^3 - R_1^3\right)} \tag{13}$$

Where: R_1 and R_2 are the inner and outer radius of the shear ring. Torque is given by the mean load on the tie rods measured as:

$$T = \frac{(F_1 + F_2)L}{2} \tag{14}$$

Where: F_1 and F_2 are forces separated by the distance L.
The normal stress is given by the vertical load N as:

$$\sigma = \frac{N}{\pi\left(R_2^2 - R_1^2\right)} \tag{15}$$

From measured shear strength for soil without root τ_s obtained:

$$\tau_s = C_s + \sigma_N \tan\varnothing \tag{16}$$

Where: C_s is the effective cohesion of the soil without root.
The soil-root composite shear strength τ_{sr} expressed by:

$$\tau_{sr} = C_s + C_r + \sigma_N \tan\varnothing \tag{17}$$

Where: C_s is the effective cohesion of the soil, C_r is the apparent cohesion provided by roots, σN is the normal stress due to the weight of the soil and $\varnothing$ is the effective internal friction angle of the soil that is unaltered by the presence of roots (O'Loughlin and Ziemer 1982; Wu 1995).

$$C_r = 1.2T_r\left(A_R / A\right) \tag{18}$$

Where: T_r is average tensile strength of roots per unit area of soil (kPa), (A_R/A) is the root area ratio (no units), A is the area of the soil shear surface, A_R is the total cross-sectional area of all roots crossing the shear surface.

Draft Force

Draft force measured by hydraulic dynamometer coupled between the two tractors with the attaching chisel plow to estimate its draught force.

Mathematical Model for Prediction the Power Requirements

The following mathematical model for Soehne (1956) built on Matlab program version-7 to predict the power requirements for chisel plow. The inputs data for the mathematical model were represented with their units in Figure (3: a, b and c) andthe flow chart of the proposed model was as shown in Figure 4.

Figure 3. Inputs data and their measuring units to predict the power requirements for chisel plow in soil, without roots, (a), with Bean roots, (b) and with Wheat roots (c)

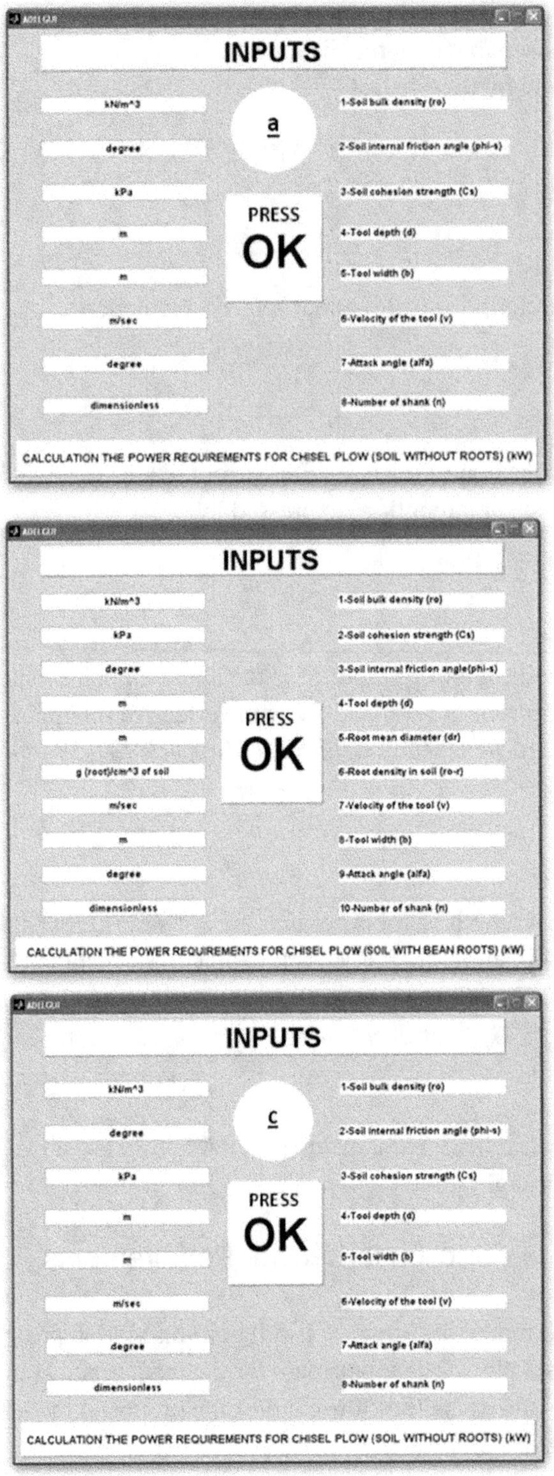

Figure 4. Flow chart of the proposed model to predict the power requirements for chisel plow in soil with and without roots

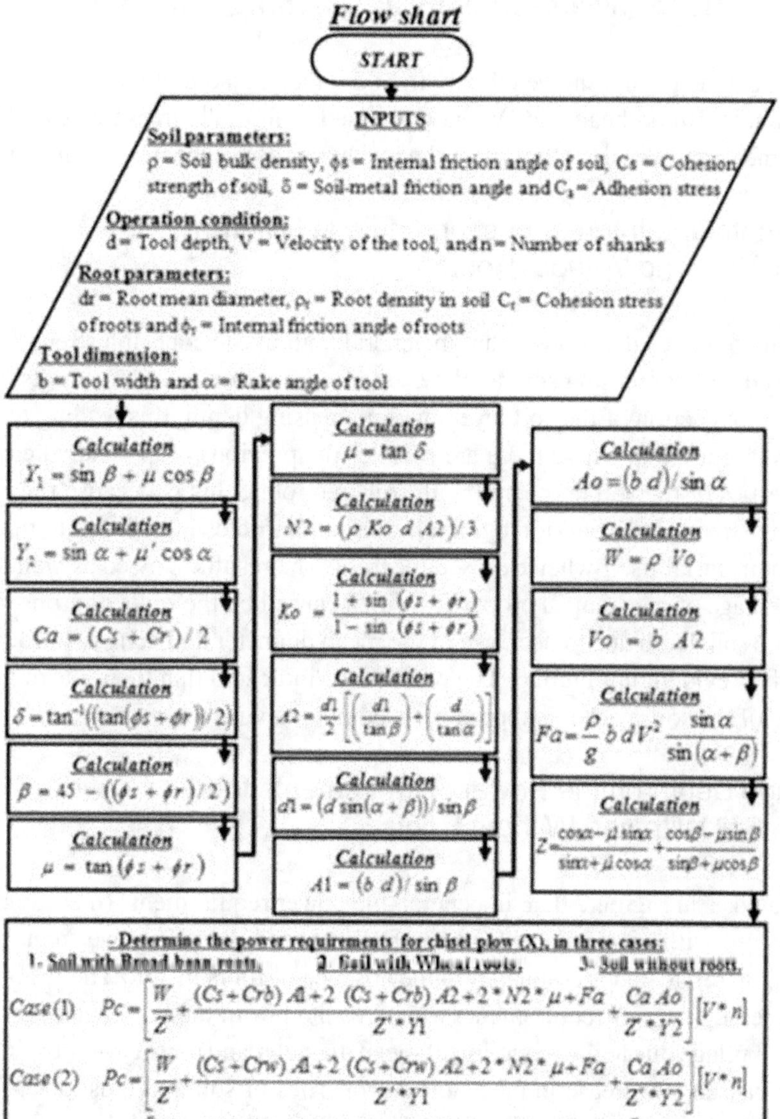

RESULTS AND DISCUSSION

Root Mean Diameter and Root Density at Different Soil Depths

As shown in Figure 5 the results showed that both of the mean diameter and density of the roots in the soil for crops residual, Broad bean and Wheat decreased with the increase of soil depth. Also, found that the mean diameter and density of roots for the residual of Broad beans was higher than Wheat crop.

Cohesion Strength and Internal Friction Angle at Different Depths for Soil With and Without Roots

As shown in Figure 5 the results showed that in general both of the cohesion strength and internal friction angle increased with increasing soil depth whether presence or absence of roots in the soil because of the increase in compaction of the soil layers with increasing depth, this leading to increase the density of the soil. But found that where there are roots of the previous crop residual causes increasing in mechanical properties of the soil compared with soil-free roots, this is because the roots of the previous crop residues increase soil cohesion by binding to soil particles in addition to the resistance which caused by the root material itself when exposed to stress. The results showed as well as in Figure 5 that the increase percentage of mechanical properties which caused by the roots of crops residual decreased with increasing the soil depth due to decrease of the mean diameter and density of roots with increasing soil depth and this is evident in Figure 6. The results also indicated that the roots of broad bean caused a greater increase of the soil mechanical properties compared with wheat roots.

Effects of Tillage Depth on the Power Requirements for Chisel Plow in Soil With and Without Roots

The results indicated as in Figure 7 that, in general, the power requirements (measured or predicted) for chisel plow increased with increasing tillage depth. But found that the power requirements increased in the case of a root crop residues compared in the absence of roots in soil. The results showed that the percentage of increase in power requirements which resulted from the presence of roots decreases with increasing depth of tillage due to the reduction of mean diameter and density of roots in soil with increasing depth, which caused a decrease in mechanical properties of soil and thus lower the soil resistance to shear strength which facing the chisel plow. The results also indicated that the roots of broad bean caused a greater increase of the power requirements compared with wheat roots.

Relationship Between Predicted and Measured Power for Chisel Plow

As in Figure 8 results proved that the introduction of crop factor which represented in this study as the roots in the soil for crops residual (mean diameter and density of roots in soil) as a component of the Soehne mathematical model which used to predict power requirements for chisel plow led to increase the accuracy of the expected values and the model became more a simulation of the practical realities of the measure power requirements for chisel plow in the field. The results showed more accurate for prediction of this model when using crop factor which amounted to about 95% compared with about 91% when non-use of crop factor in the model calculations.

Figure 5. Root mean diameter (m) (a), root density (g/cm3 of soil) (b), average increase in soil cohesion strength with roots (%) (c), average increase in soil internal friction angle with roots (%) (d), soil cohesion strength with roots (kPa) (c) and soil internal friction angle with roots (degree) (d)

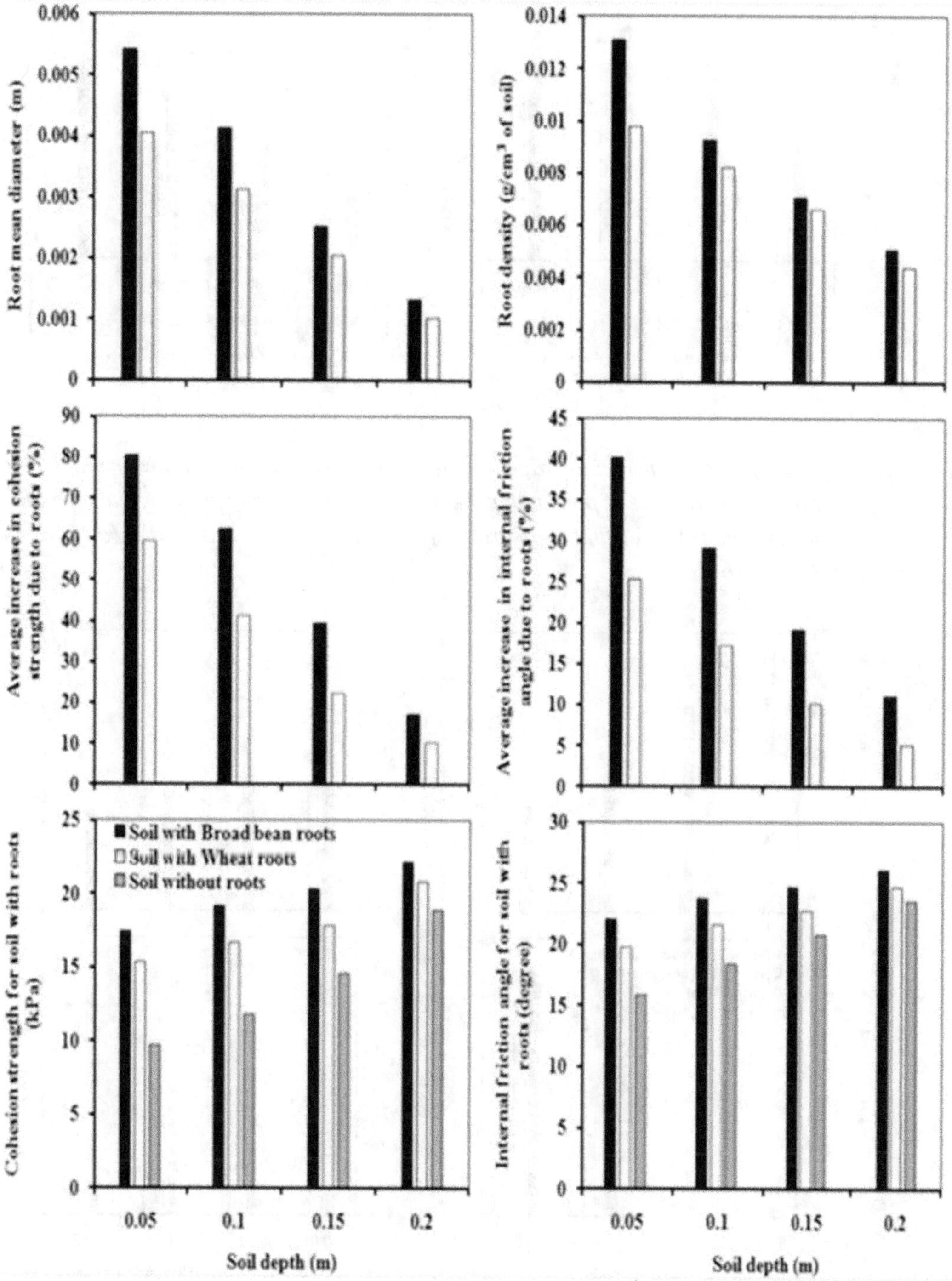

Figure 6. Increasing in soil cohesion strength (kPa) and soil internal friction angle (degree) due to root density (g/cm3 of soil) and root mean diameter (m) for Broad bean (a) and Wheat (b)

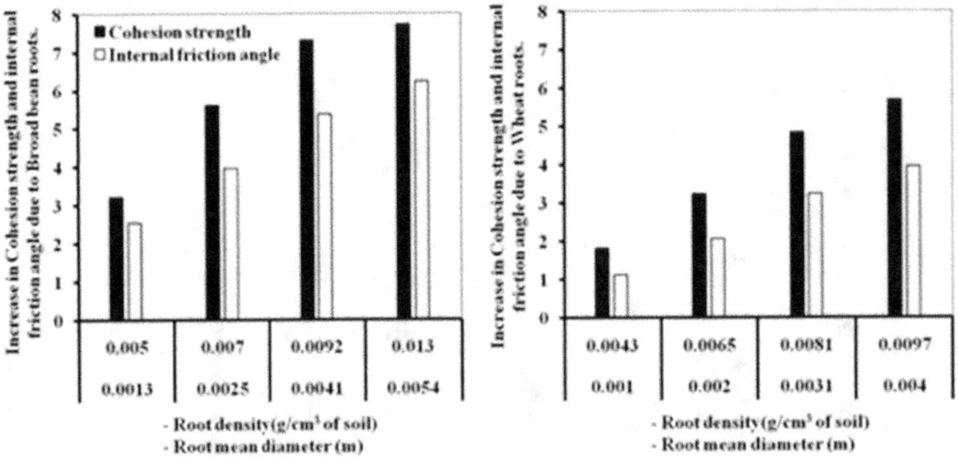

Figure 7. Effect of tillage depth (m) on average increasing in the prediction of power requirements (%) (a), increasing in the measurement of power requirements (%) (b) for chisel plow due to roots, prediction of power requirements (kW) (c) and measurement of power requirements (kW) (d) for chisel plow at soil with and without roots

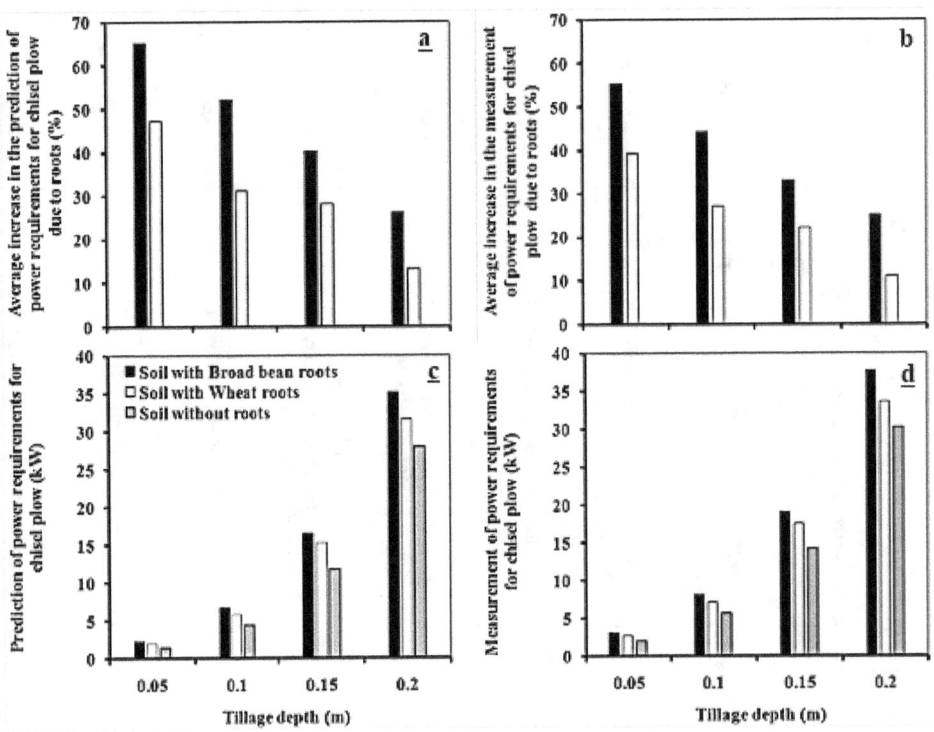

Figure 8. Relationship between predicted and measured power for chisel plow when not taking roots factor (a) and taking roots factor (b) in the mathematical model

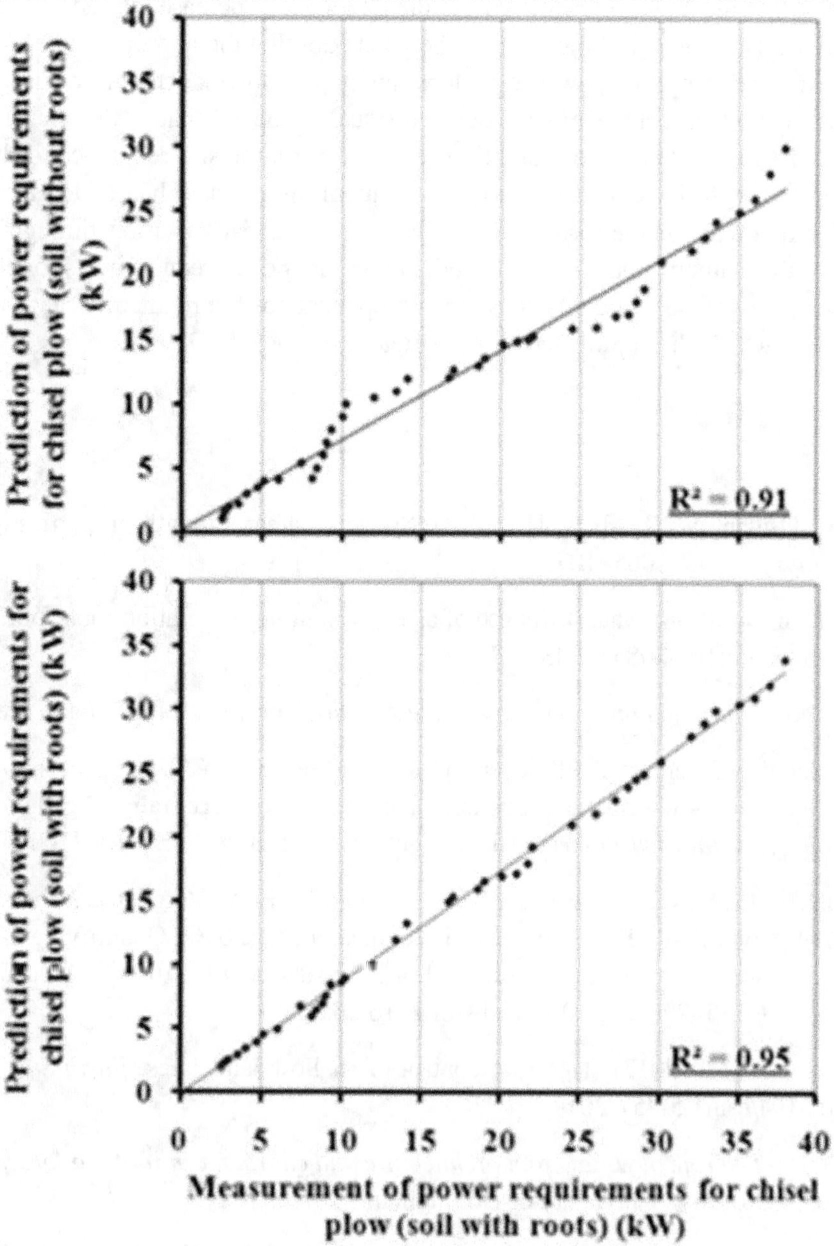

CONCLUSION

This research helps in achieving a precision agriculture system as it increased the accuracy of predicting the power requirements of the chisel plow, which helps in adopting the appropriate decision in providing the tractor with the appropriate power to perform the required work accurately. This study showed that the presence of the roots for the previous crop residual in the soil had a significant and effective impact on the soil mechanical properties and thus roots affect on the soil resistance to shear for chisel plow when compared with the soil, which does not contain the roots. When added the properties of these roots, which represents in the form of mean diameter and density of roots through the soil depth as a component of the mathematical Soehne's model to predict power requirements for chisel plow led to increase the accuracy of this model because the components of the mathematical model closer the simulation of reality which chisel plow faced in the field.

REFERENCES

Abdullah, M. N., Osman, N., & Ali, F. H. (2011). Soil-root shear strength properties of some slope plants. *Sains Malaysiana, 40*, 1065–1073.

Ali, F. H., & Osman, N. (2008). Shear strength of a soil containing vegetation roots. *Soil and Foundation, 48*(4), 587–596. doi:10.3208andf.48.587

Bekker, M. G. (1969). *Introduction to terrain vehicle systems*. The Univ. of Michigan Press.

Bordoni, M., Cislaghi, A., Vercesi, A., Bischetti, G. B., & Meisina, C. (2020). Effects of plant roots on soil shear strength and shallow landslideproneness in an area of northern Italian Apennines. *Bulletin of Engineering Geology and the Environment*. Advance online publication. doi:10.100710064-020-01783-1

Bordoni, M., Meisina, C., Vercesi, A., Bischetti, G. B., Chiaradia, E. A., Vergani, C., Chersich, S., Valentino, R., Bittelli, M., Comolli, R., Persichillo, M. G., & Cislaghi, A. (2016). Quantifying the contribution of grapevine roots to soil mechanical reinforcement in an areasusceptible to shallow landslides. *Soil & Tillage Research, 163*, 195–206. doi:10.1016/j.still.2016.06.004

Cohen, D., & Schwarz, M. (2017). Tree-root control of shallow landslides. *Earth Surf Dynam, 5*(3), 451–477. doi:10.5194/esurf-5-451-2017

Collins, A. J. (2001). *The role of willow root architecture and character in root reinforcement potential* (Unpublished doctoral thesis). University of Nottingham.

Consoli, N. C., Montardo, J. P., Prietto, P. D. M., & Pasa, G. S. (2002). Engineering behavior of a sand reinforced with plastic waste. *Journal of Geotechnical and Geoenvironmental Engineering, 128*(6), 462–472. doi:10.1061/(ASCE)1090-0241(2002)128:6(462)

De Baets, S. D., Poesen, J., Reubens, B., Wemans, K., Baerdemaeker, J. D., & Muys, B. (2008). Root tensile strength and root distribution off typical Mediterranean plant species and their contribution to soil shear strength. *Plant and Soil, 305*(1-2), 207–226. doi:10.100711104-008-9553-0

Ekanayake, J. C., Marden, M., Watson, A. J., & Rowan, D. (1997). Tree roots and slope stability: A comparison between Pinusradiata and Kanuka. *New Zealand Journal of Forestry Science*, *27*, 216–233.

Ekwue, E. I., & Stone, R. J. (1995). Organic matter effects on the strength properties of compacted agricultural soils. *Transactions of the ASAE. American Society of Agricultural Engineers*, *38*(2), 357–365. doi:10.13031/2013.27804

Endo, T., & Tsuruta, T. (1969). A report in regard to the reinforcement action of the vegetational roots upon the tensile strength of the natural soil. 1968 Annu. Rep. Hokkaido Branch, For. Exp. Stn., 183-189.

Fitter, A. H. (1993). Arhitectural analysis of plant root systems. In Methods in comparative plant ecology: A laboratory manual. Chapman and Hall.

Gerard, C.J. & Mehta, H. C. (1971). Influence of a rootcrop on physical properties of a clay soil. *Agronomy*, *63*, 893-895.

Gill, W. R., & Vanden Berg, G. E. (1968). Soil Dynamics in Tillage and Traction. USDA-ARS Agricultural Handbook No. 316. Washington, DC: Government Printing Office.

Goldsmith, W. (1998). Soil reinforcement by river plants: progress results. In *Proceedings of the Conferencon Wetlands Engineering and River Restoration*. American Society of Civil Engineers. 10.1061/40382(1998)30

Gray, D. H. (1974). Reinforcement and Stabilization of Soil by Vegetation. *Journal of Geotechnical Engineering*, (100(GT6)), 695–699.

Greenway, D. T. (1987). Vegetation and slope stability. In Slope Stability. John Wiley.

Huat, B.B.K., Faisal, H. A., & Maail, S. (2005). The effect ofnatural fber on the shear strength of soil. *American Journal of Applied Sciences*, 9-13.

Lawrance, C. J., Rickson, R. J., & Clark, J. E. (1996). The Effect of Grass Roots on the Shear Strength of Colluvial Soils in Nepal. In Advances in Hillslope Processes. John Wiley and Sons.

Murielle, G., Veylon, G., Bernard, A., Vanel, Q., & Stokes, A. (2014). Influence of plant root system morphology and architectural traits on soil shear resistance. *Plant and Soil*, *377*(1-2), 43–61. doi:10.100711104-012-1572-1

O'Loughlin, C., & Ziemer, R. R. (1982). The importance of root strength and deterioration rates upon edaphic stability in steepland forests. In Carbon uptake and allocation in subalpine ecosystems as a key to management. Oregon State University.

Operstein, V., & Frydman, S. (2000). The influence of vegetation on soil strength. *Ground Improvement*, *4*(2), 81–89. doi:10.1680/grim.2000.4.2.81

Singh, H., & Huat, B. B. K. (2004). Origin, formation and occurrence of tropical residual soils. In Tropical Residual Soil Engineering. Taylor and Francis. doi:10.1201/9780203024621.ch1

Soehne, W. (1956). Some principles of soil mechanics as applied to agricultural engineering. *Inst. Agric. Eng. English Translation, 53*, 11-27.

Thorne, C. R. (1990). Effects of vegetation on riverbank erosion and stability. In Vegetation and Erosion. John Wiley and Sons Inc.

Waldron, L. J. (1977). The shear resistance of root-permeated homogeneous and stratified soil. *Soil Science Society of America Journal*, *41*(5), 843–849. doi:10.2136ssaj1977.03615995004100050005x

Waldron, L. J., & Dakessian, S. (1982). Effects of Grass, Legume, and Tree Roots on Soil Shearing Resistance. *Soil Science Society of America Journal*, *46*(5), 894–899. doi:10.2136ssaj1982.03615995004600050002x

Willatt, S. T., & Sulistyaningsih, N. (1990). Effect of plantroots on soil strength. *Soil & Tillage Research*, *16*(4), 329–336. doi:10.1016/0167-1987(90)90068-O

Wu, T. H. (1984). *Effect of Vegetation on Slope Stability. In soil reinforcement and moisture effects on slope stability*. Transportation Research Board.

Wu, T. H. (1995). Slope stabilization. In Slope stabilization and erosion control: a bioengineering approach. Academic Press.

Wu T.H., Kokesh, C.M., Trenner,B. R. & Fox, P. J. (2014). Use of live poles for stabilization of a shallow slope failure. *Journal of Geotechnical and Geoenvironmental Engineering*, *140*(10), 1-13.

Wu, T. H., Mckinnell, W. P., & Swanson, D. N. (1979). Strength of tree roots and landslides on Prince of Wales Island, Alaska. *Canadian Geotechnical Journal*, *16*(1), 19–33. doi:10.1139/t79-003

APPENDIX

List of Symbols for Mathematical Model

Pc	Power requirements for chisel plow (kW)
ρ	Specific weight (kN/m3)
β	Angle of soil shear plane (degree)
α	Rake angle of tool (attack angle) (degree)
A2	Area of the soil slice side (m2)
W	Weight of the soil slice (kN)
d1	Depth of soil disturbed (the effective depth) (m)
b	Chisel blade width (m)
d	Chisel tool depth (the operating depth) (m)
Vo	Volume of the soil slice (m3)
g	Acceleration due to gravity m/sec2
A1	Area of shear failure surface (m2)
Fa	Acceleration force (Inertial force) (kN)
δ	Angle of the soil-metal friction (degree)
μ	Coefficient of internal soil friction
μ'	Coefficient of soil-metal friction
Ao	Chisel tool area (m2)
Ca	Adhesion stress (kPa)
N2	The forces due to earth pressure on the sides of soil slice (kN)
Ko	Coefficient of passive earth pressure
n	Number of shanks
ϕr	Internal friction angle of root (degree)
Cr	Cohesion strength of root (kPa)
ϕrw	Internal friction angle of wheat root (degree)
Crw	Cohesion strength of wheat root (kPa)
ϕrb	Internal friction angle of bean root (degree)
Crb	Cohesion strength of bean root (kPa)
ϕs	Internal friction angle of soil (degree)
Cs	Cohesion strength of soil (kPa)

Section 4

Agricultural Systems Analysis Streamlining, Modeling, and Improvements: Towards Precision Agronomics (Economic and Policy Perspective)

This section addresses the agriculture development and food security from the economic and policy perspective. It presents research on the use of quantitative and qualitative analysis methods to recognize the agricultural export future under the new trends of World Trade Organization taking the Egyptian exports as the case study, and to analyzing the role of institutions, funding/assistance sources, information sources in achieving food security, and extension policy taking the Indonesia food security policy as the case study

Chapter 12
The Role of the Institutional, Assistance, and Source Information Analysis on Food Security and Extension Policy:
Case in Indonesia

Muhamad Rusliyadi
https://orcid.org/0000-0003-4632-5671
Polytechnic of Agricultural Development Yogyakarta Magelang, Indonesia

ABSTRACT

The chapter aims to analyze the role of institutional sources of funding/assistance, source information in term of food security and extension policy. The results of the research show the information on problematic areas in policy implementation with respect to food security and extension policy in term of poverty reduction. The role of the institutional, assistance, and source information is useful in the implementation of other policy sectors in the future, because similar issues or problems may be avoided by taking precautions not to repeat the same mistakes or by applying measures which address the implementation issues. In this way, policy implementation can be carried out effectively, and the outcomes will meet the policy objectives.

INTRODUCTION

The growth in agriculture has been a guiding source of poverty reduction in most developing nations, particularly in the case of rural poverty. The agriculture sector contributes to poverty reduction in terms of employment generation to the poor and supports the development of non-agricultural employment generation in rural fields. Development in agriculture also contributes to a larger supply of food and lower food prices benefiting both rural and urban poor (Grewal et al., 2012).

DOI: 10.4018/978-1-7998-5000-7.ch012

Agricultural development is important in a developing country for achieving industrialization. Farming is the backbone of economic growth in many growing countries, because it provides food security, food self-sufficiency, diversification, and value-added enhancement of competitiveness of exports. Moreover, it increases the wellbeing of farmers and leads directly to poverty reduction.

Indonesia has shifted from an agricultural economy to an industrial one. Nevertheless, the agriculture sector remains a key sector for millions of those who depend directly on agriculture for their livelihood. The agriculture sector has been playing a key role in improving living standards and alleviating poverty. Poverty in rural areas, where most people depend on agriculture for their livelihoods, is one of the immense problems in Indonesia. The poverty level indicates that 16.56% of people in rural areas are poor (CBS, 2010). The increasing numbers of poor people are due to the fact that the bulk of the people cannot satisfy their basic needs and survive at the subsistence level, and therefore ensuring household food security via affordable agricultural workplaces and sustainability scheme policies on the basic demands of the farming sector is vital, especially in terms of food security policy.

According to Darwis and Rusastra (2011), the Food Self-Sufficiency Village Programme (Desa-MandiriPangan) implemented by community groups was supported by agricultural and non-agricultural capital programmes, as well as by rural financial institutions. It was facilitated by agricultural extension workers and expert consultants. This programme is directed at several other objectives: raising the availability of foodstuffs, growing the family unit entitlement, recovering the efficiency of food supply and access programmes, enhancing the local marketplace availability for selling and buying food products, raising capital and industrial options through the introduction of high-yielding varieties of seeds that agriculture community in rural area can develop to increase income, as well as improving the quality of food consumption handling at the household level, and so forth.

The reasons for choosing the Food Self-sufficiency Village Programme (DesaMandiriPangan) as the subject of the study are that: 1) The programme has a comprehensives scheme and assistance; 2) The programme has a direct or an indirect beneficial impact on poverty alleviation and food security efforts; 3) This programme is aimed at resolving the existing problems at the grassroots level; 4) It is expected to contribute importantly to rural growth; and, 5) The programme has been implemented in most of Indonesia's provinces.

Active institutions play a very important role in supporting the DMP Programme. There were institutions in the village sites before the introduction of the DMP Programme, such as the village leader, village councils, farmer groups, rice milling units, a village health station unit, a school, a mosque, a church, private lenders, village shops and traders. New institutions introduced by the Programme are affinity groups, food village teams, rice barns and microfinance institutions. In Kedungdowo, the village leader is serving his second elected term. This shows the trust that villagers have for the village head. The relationship of village leader and village council should be harmonious in order to ensure good consultation with the village council, which typically advises and monitors every village activity.

MATERIALS AND METHODS

This research is focused on the existing food security policy in Indonesia with regard to the DMP Programme that has been implemented since 2006. The methodology used in the study was mainly quantitative, but a qualitative approach was also used to enrich the findings of the quantitative study. Primary data were analysed using Microsoft Excel and Statistical Program for Social Science (SPSS).

Research site in Java island has the largest number of population in Indonesia. It was chosen to represent the research site. Central Java was chosen for this case study because it constitutes 17.0% of the population experiencing poverty (CBS, 2011). Secondly, it has many poor villages when compared to other provinces. Four villages were selected purposively.

Two villages from KebumenRegency, KedungDowo Village at Poncowarno District and SelogiriV-illlage at Karanggayam District, and the others in Magelang Regency at Madukoro Village Kajoran District and Candirejo Village Borobudur District were selected. These villages have been implementing the programme in varying stages with different characteristics and success rate. (Rusliyadi, et al 2019)

Data related to the DMP Programme and extension service were collected using unstructured and semi-structured questionnaire. The survey was conducted in four villages, each of which involved 100 samples. Samples were selected using a purposive and clustered sampling technique based on some particular sample criteria. Criteria of the sample chosen were based on farmer characteristics such as rich farmer, poor farmer, farmer leader, the village leader, man and women. The Scoring o the data presented is by author perception and the investigate scoring exercises using replication, cross-check scoring exercises, and interviews. These activities were undertaken with scoring exercises were developed to assess how local people perceived the importance of various environments and resource (Sheil, D &Liswanti, N 2006).

The purposive sampling technique is a type of non-probability sampling that is most effective when one needs to study a certain cultural domain with knowledgeable experts within. Purposive sampling may also be used with both qualitative and quantitative research techniques. The sample that is selected based on characteristics of a population and the objective of the study. This type of sampling can be very useful in situations when you need to reach a targeted sample quickly, and where sampling for proportionality is not the main concern (Rusliyadi, M and Libin, W, 2018)

RESULT AND DISCUSSION

The Role of Institutions, Organizations, Groups and Individuals (Venn Diagram)

The following Venn diagrams show the institutions, organisations, groups and important individuals in each village site. The institution on the Venn diagram is involved directly with villagers in terms of importance activity in the community. The closeness of the relationship between institutions and communities in the group is shown by placing them in the same area. The Venn diagrams also provide information on external and internal factor such us institutions, organisations, groups or individuals linking in the community. The expectation from this diagram is the provision of information on the direct impact of and influential institutions in the community. This tool helps identify and characterise different institutions in terms of impact, cooperation, and link integration in different circumstancesin a rural village. The output is the information and provision of relationships between the institutions in terms of delivery of programme activities.

Each Venn diagram shows the importance of institutions in a village community. The assessment of the DMP Programme in this analysis is focused on food security and nutrition issues at household level. The institutional collaboration needs adjustments in measurement, especially closeness relative to other institutions. However, each programme has different target groups or institutions in the community, for example, not all member groups receive allocation of aid from the programme. The cooperation is the key to identifying institutional problems in the community and programme implementation.

The Venn diagrams can be used to discover what formal and informal institutions exist in each village and their relationships and benefits for the local community. The distance between the circles in the diagrams indicates the closeness of their relationship, while their sizes indicate their relative sizes or influence in the community.

Kedungdowo Village

As shown in *Figure 1*, institutions within the village all lie within a larger circle, while outside institutions are shown outside the large circle, with their respective estimated distances from Kedungdowo. The distance here is very important, because the process of assisting a programme will be more effective and efficient when the distance between the donor and recipient policies is not too great so that monitoring and implementation can be maximised. The influence and availability of institutions when needed is determined by the symbiotic relationship between villagers and importance institutions in the village.

Each institution shown is an active institution that plays a role in supporting the DMP Programme. There were institutions in the village before the introduction of the Programme. These were village leaders, village councils, farmer groups, rice milling units, the village health station unit, a school, a mosque, a church, private lenders, village shops or traders. New institutions introduced by the Programme are affinity groups, a food village team, rice barn and microfinance institutions. In Kedungdowo, the village leader is serving his second elected term. This shows that the trust that villagers have in the village head is high. The relationship of village leaders and village councils should be harmonious, since activity in the village needs to be consultative with village councils. Normally the village council gives advice and monitors every activity in the village.

The rice milling unit (RMU) in Kedungdowo is relatively small but useful, because it is not far away for villagers to grind their rice. The RMU is an important institution in the village since its function is like a rice barn or financial institution. The RMU assists villagers by providing farming inputs and store for rice stock. The mechanism involves the villagers being given cash or input in the pre-planting stage and they repay this after harvest in the form of rice or money that is equivalent to the value of inputs provided at the pre-planting stage. It is like a soft loan with low interest, but in some cases the RMU charges double the price of the inputs given to farmers, with the result that the villagers have low earnings. Since the implementation of the DMP Programme in Kedungdowo the introduced rice barns are also providing farmers with micro financing (LKD), which reduces the function of theRMU. The role of the RMU has been taken over by the farmer groups who run the operation of the ricebarn and the LKD. Low interest is an important factor for poor farmers in making a decision to seek a loan, which is handled at the regular meeting of farmer groups. The poor members of society find the rice barn beneficial, especially during periods of food shortages.

The nearest institutions outside the village, such as the district office, the agriculture and forestry office, the military/police outpost and the market are about 7km distant. This is relatively far from the village, which makes coordination less optimal. According to villagers, "a far distance from the district office means lack of information for their development". The centres of district municipalities in general are the centres of economic activity for some villages. The point is that economic development in villages depends on the distance from the village to the district centre. The nearest market is about 7km from the village. It is important to understand that the closer the village is to the market, the more it is

Figure 1. Venn Diagram for Kedungdowo Village
Sources: Field work primary data (2013)
NB: RMU = Rice Milling Unit, FVT = Food Village Team, FG = Farmer Gorups, AG = Affinity Groups.

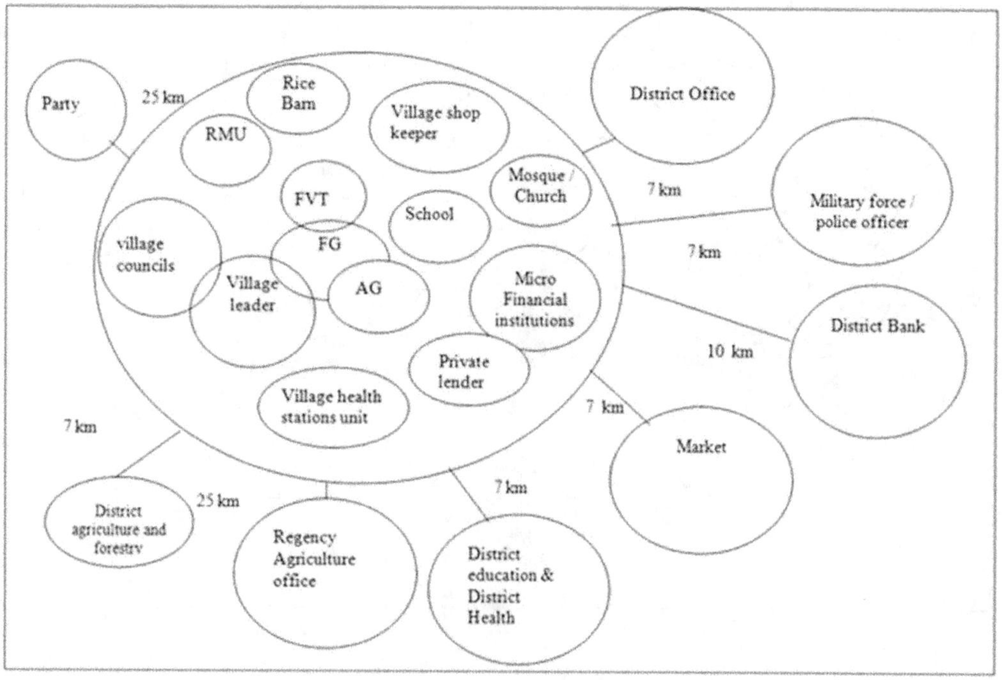

developed. The market is, therefore, central to village activity and employment opportunity. The local daily market that provides the basic needs of the rural community is an economic buffer (food security) for surrounding villages to do business. It is related to the previous observation that the availability of a market makes it easier to start or do business.

Selogiri Village

Institution/group activity in Selogiri is generally not very dynamic. Village institutions would be involved only if there was a government programme for the village. The village leader has uneasy relations with some community leaders, according to the interviews. This has reduced the effectiveness of existing programmes in the village. Affinity groups as drivers in the DMP Programme are quite active, but the effectiveness level is still below the level of effectiveness in other villages.

Figure 2 describes the general institutional Venn diagram for Selogiri. The distance of the village from the district office is less compared to the situation in Kedungdowo, but the distance to the regency office is considerable – about 35 kilometres. The Affinity Group (AG) formed as a result of the DMP activities consists of four groups of farmers with different business areas of agriculture, livestock, fisheries and cassava cracker manufacturers. Fifty percent of funds are channelled to agricultural groups so that many more farmers are involved in the programme. The village food team (FVT) which consists of representatives of each affinity group holds regular meetings every month attended by the head of the village and the village council. The constraints faced in the implementation of activities and future implementation plans are being worked out together so as to encourage the participation of all levels of society.

Figure 2. Venn Diagram at Selogiri Village
Sources: Field work primary data (2013)
NB: RMU = Rice Milling Unit, FVT = Food Village Team, FG = Farmer Gorups, AG = Affinity Groups.

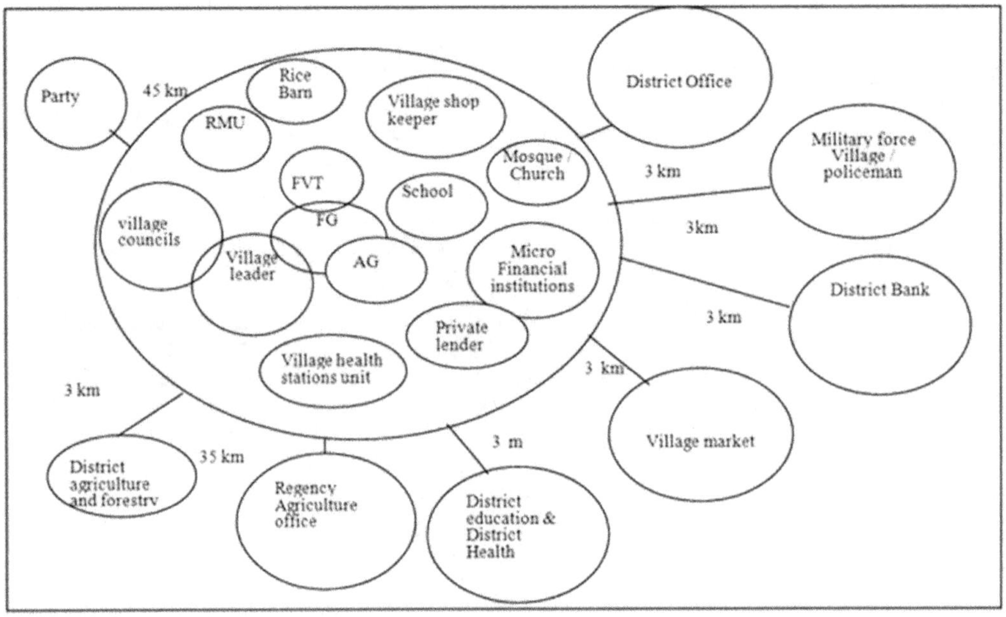

Candirejo Village

The farmer groups in Candirejo are relatively more dynamic than in the other villages. The fundamental difference in the diagram below compared to other villages is, firstly, an absence of rice milling units (RMU) in the village, as farmers grind their rice in a neighbouring village. Secondly, grants are not allocated to the building of rice barns, but are used to purchase livestock in the form of local and *etawa* goats. The allocation of funding is for poor families, which is involved as a farmer group (affinity group). Thirdly, the distance from the village to the district office is 4km. Finally, tourism resources are contributing to the revenue of the village, such as a handicraft store and homestay facilities. Some villagers earn money by selling when local and foreign tourists visit Candirejo.

Madukoro Village

Madukoro Village is located on the border between Purworejo Regency and Magelang Regency, relatively far from the centre of the district (about 7km). DMP activities run well as the farming group in this village is much better in terms of their participation. In Madukoro an individual or group has self-motivation initiated by the village leader. High response of the public to a programme makes it easy to introduce and implement it. The important success factor is the efforts of the village/hamlet leader in driving the activity initiated by the DMP Programme. The influence of the hamlet leader in Madukoro is greater than village's head because the village head is relatively young and pays less attention to villagers. The important point of this case is the role of the village head or chief of the hamlet, which is a critical factor in the village's development. Village leaders should have the ability to mobilise the

Figure 3. Venn Diagram for Candirejo Village
Sources: Field work primary data, (2013).
NB: RMU = Rice Milling Unit, FVT = Food Village Team, FG = Farmer Groups, AG = Affinity Groups.

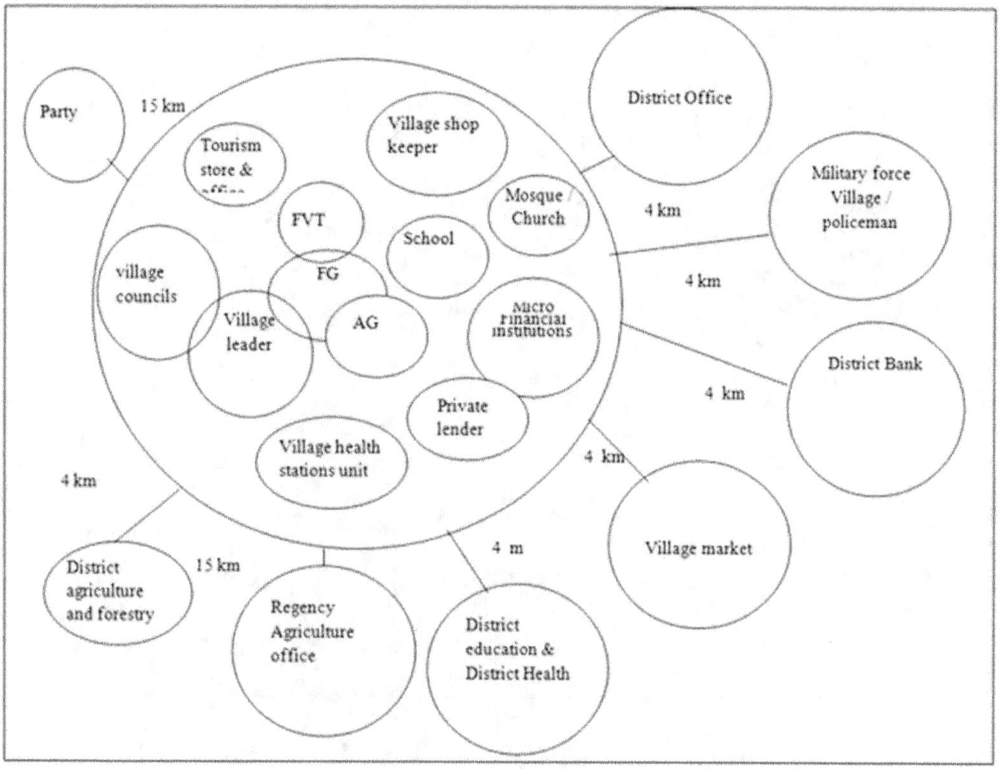

community to be active in most village programmes. Key persons who have great influence in the village have a significant impact on the success of a programme, measured by the highest participation of the community. The policies at village level require effort and involvement from several key persons to achieve implementation impact.

Analysis of the Sources of Information and Assistance

This analysis is the result of focus group discussions (FGD) and randomly surveyed poor households as respondents. Information is a source of knowledge that can be used by farmers to contribute to their income by practical business. To villagers, knowledge is difficult to obtain due to village remoteness. Education alone is not enough for development to take place at village level. Villagers need experience and accurate information. Starting businesses (the opportunity to open/start a business) in the village is difficult because of lack of information and capital. Therefore, information is very helpful for villagers to start a business and information on the latest technologies should also be useful for them to promote business development.

The sources of funding received by the villages are identified and analysed based on perception of the randomly selected villagers. The funding resources are categorised on the basis of where the funding comes from: the State budget (National budget), local budget (Province or Regency budget), International

Figure 4. Venn Diagram at Madukoro Village
Sources: Field work primary data (2013)
NB: RMU = Rice Milling Unit, FVT = Food Village Team, FG = Farmer Gorups, AG = Affinity Groups.

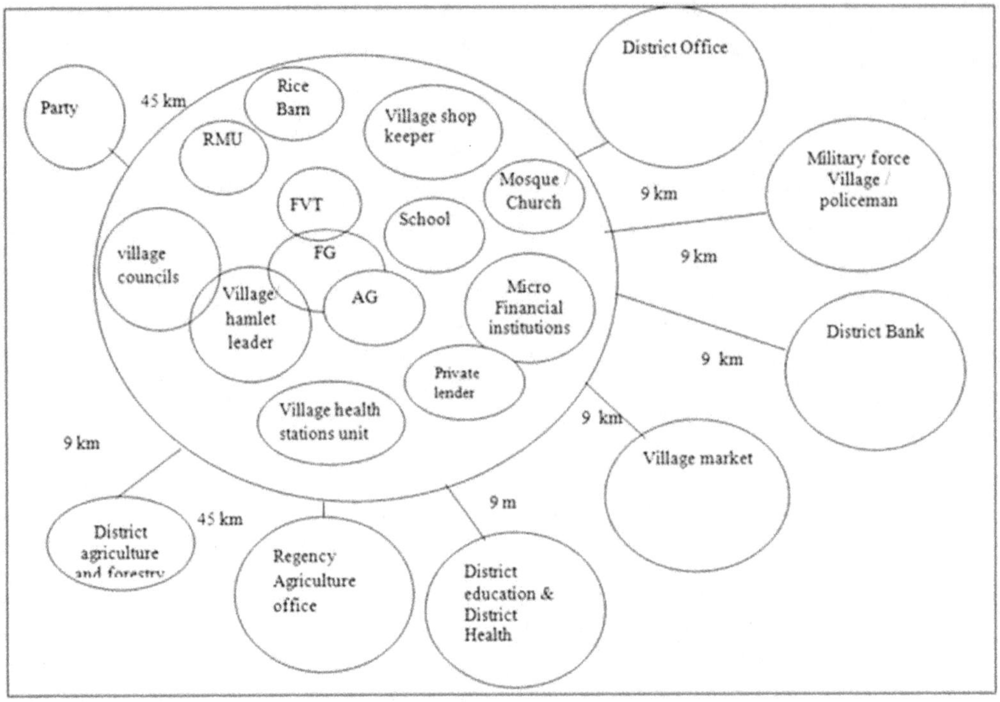

funds, and NGOs. There are four types of assistance made available to the research sites for rural village development: cash funding, physical assistance, aid-type of goods, and types of integrated programmes such as the DMP Programme.

Kedungdowo Village

Based on discussions and analysis of several groups in Kedungdowo (Table 1), villagers receive information from friends, relatives and parents. This shows that verbal information obtained through discussion and conversations is influential. Information obtained through formal schooling is less influential, because it tends to contain theoretical knowledge with little information on practices. Villagers expressed their desire to be continuously trained and exposed to the latest technologies for agriculture, fisheries and husbandry. Extension officers need to be more active at village level in providing information and knowledge-based resources in village workshops.

Funds in Kedungdowo mostly come from the National budget. The majority of the financial resources coming from the central government are distributed through local government. This means that, in the distribution of financial resources, the central government should still consider the local situation by administering a proper allocation of funds. The local government is currently striving to boost revenue by looking to local and foreign investors and persuading them to invest in their areas so that more funding is available for rural development. The local government should be proactive in generating sources

Table 1. Information Source Scoring of Kedungdowo Village

No.	Information type required	Information Source				
		Media (TV, Newspapers, magazines)	Schools	Extension Officers	Friends, relative, parents	Score
1	Agriculture	3	4	8	9	24
2	Livestock	4	3	7	8	22
3	Fisheries	5	4	7	7	23
4	Forestry	7	3	9	8	27
5	Business opportunity	7	6	6	9	28
6	Markets	8	3	6	8	25
7	Employment	8	4	5	8	25
	Score	42	27	48	57	

NB: Scoring is out of 10; the higher the score, the greater is the information obtained
Sources: Field work primary data (2013)

of funds through local revenue – *PendapatanAsli Daerah* (PAD). This point is very important for rural development and the gap between the national and regional budgets can be minimised.

Selogiri Village

In Selogiri, information obtained from friends, relatives and parents also received high score from respondents. Why are these information sources significant for farmers' farming knowledge? This is because other resources, such as the mass media, are rarely accessed by villagers or they must actively obtain this information, while information from friends, relatives and parents is readily available through daily interactions in the village. Also, this information is free of charge. Villagers might have no time to seek information through the media because of working in the fields.

The sources of financial assistance in Selogiri (Table 4) are almost the same as in Kedungdowo, that is, the majority of financial assistance comes from state funds. The biggest aid funds are derived from the National Programme for Community Empowerment (Rural PNPM Mandiri) at 252 million IDR. These funds were directed to the construction of concrete rural roads with partial support employment comes from the community. Villagers in the group discussions said that the national budget has the greatest impact on physical development in PNPM and the highest score (value of 10: see Table 4) in terms of impact on social development through the Health Card/household health scheme. The agricultural extension programme receives assistance from two fund sources with the same pattern and type but different commodities (wetland rice and upland rice). This programme was initiated by School Field of Integrated Rice Crop Management (*SekolahLapangPengelolaanTanamanTerpadu* - SLPTT) funded by APBN. The upland rice SLPTT is funded by the Local Budget APBD. The collaboration programme implementation by national and local budgets in the same time but different commodities is possible, but it needs different targets (villagers) and areas (location).

Table 2. Sources of Assistance Scoring of Kedungdowo Village

No.	Type of poverty Assistance	Sources of Assistance				Score
		APBN National Budget	APBD Provinces or Regency Budget	International Funds	NGO	
1	*Bantuan IDT* - National Aid for underdeveloped villages	6	-	-	-	6
2	*KartuSehat/Gakin*Health Card	9	-	-	-	9
3	*Raskin* (rice for the poor)	9	-	-	-	9
4	*BantuanlangsungTunai BLT* - Direct cash assistance	8	-	-	-	8
5	*Desa mandiri pangan (DMP) Programme*	9	-	-	-	9
6	*PAMSIMAS* - Community-Based Water Supply and Sanitation	0	-	8	-	8
7	*BedahRumah* - House Renovation Programme	5	7			12
8	Agricultural Extension assistance	8	-	-	-	8
9	*Programme NasionalPemberdayaanMasyarakat - PNPM Mandiriperdesaan* -National Programme for Community Empowerment - Rural PNPM Mandiri	10	-	-	-	10
	Score	64	7	8		

NB: Scoring is out of 10; the higher the score, the higher its impact
Sources: Field work primary data, (2013).

Table 3. Information Source Scoring of Selogiri Village

No.	Information type required	Information Source				
		Media (TV, Newspapers, magazines)	Schools	Extension Officers	Friends, relative, parents	Score
1	Agriculture	2	4	7	8	21
2	Livestock	3	4	8	10	25
3	Fisheries	5	5	6	6	22
4	Forestry	6	4	8	9	27
5	Business opportunity	5	7	5	8	25
6	Markets	7	3	5	7	22
7	Employment	7	3	4	9	23
	Score	35	30	43	57	

NB: Scoring is out of 10; the higher the score, the greater the amount of information obtained
Source: Field work primary data (2013)

Table 4. Sources of Assistance Scoring of Selogiri Village

No.	Type of poverty Assistance	Sources of Assistance				Score
		APBN National Budget	APBD Provinces or Regency Budget	International Funds	NGO	
1	*Bantuan IDT* - National Aid for underdeveloped villages	5	-	-	-	5
2	*KartuSehat/Gakin* Health Card	10	-	-	-	10
3	*Raskin* (rice for the poor)	8	-	-	-	8
4	*BantuanlangsungTunai BLT* - Direct cash assistance	7	-	-	-	7
5	*Desamandiripangan (DMP)* - Food Self-Sufficient village Programme	8	-	-	-	8
6	*PAMSIMAS* - Community-Based Water Supply and Sanitation	0	-	9	-	9
7	*BedahRumah* - House Renovation Programme	0	6	-	-	6
8	Agricultural Extension assistance	8	7	-	-	15
9	*Programme NasionalPemberdayaanMasyarakat - PNPM Mandiriperdesaan*National Programme for Community Empowerment	9	-	-	-	9
	Score	55				

NB: Scoring is out of 10; the higher the score, the higher its impact
Source: Field work primary data (2013).

Table 5. Information Source Scoring of Candirejo Village

No.	Information type required	Information Source				
		Media (TV, Newspapers, magazines)	Schools	Extension Officers	Friends, relative, parents	Score
1	Agriculture	4	3	9	10	26
2	Livestock	6	4	8	9	27
3	Fisheries	6	7	9	9	31
4	Forestry	7	3	9	8	27
5	Business opportunity	7	6	6	9	28
6	Markets	7	4	5	9	25
7	Employment	7	6	4	8	25
	Score	40	30	41	52	

NB: Scoring is out of 10; the higher score, the higher the information
Sources: Field work primary data (2013)

Candirejo Village

Candirejo is a village which enjoys good access to information. It is a tourist village that often interacts socially and culturally with people from outside the village. The information obtained by villagers is varied and spreads quickly in the village. One example is about business opportunities in other places and success stories become special information for villagers. However, not all villagers have access to the information. It depends on the distance between the village centre where they live and the other villages where their relatives live. The neighbourhood activity in the village is very important, since information resources are mostly transmitted and received from the interactions of people and relatives. Thus the information obtained by villagers is very limited. The important implication from this case is that rural development policy should be focused on providing information through actual practical training in agriculture and business development opportunities, rather than through information provided through various media sources. An additional solution is training on how to get information from reliable resources such as the Internet and news media.

Table 6. Sources of Assistance Scoring of Candirejo Village

No.	Type of poverty Assistance	Sources of Assistance				Score
		APBN National Budget	**APBD Provinces or Regency Budget**	**International Funds**	**NGO**	
1	*Bantuan IDT* - National Aid for underdeveloped villages	7	-	-	-	7
2	*KartuSehat/Gakin* Health Card/Gakin scheme	8	-	-	-	8
3	Raskin - Rice for the poor	9	-	-	-	9
4	*BantuanlangsungTunai BLT* - Direct cash assistance	6	-	-	-	6
5	*Desa mandiri pangan (DMP) Programme* – Food Self-Sufficient village	8	-	-	-	8
6	*PAMSIMAS* - Community-Based Water Supply and Sanitation	0	-	8	-	0
7	*BedahRumah* - House Renovation Programme	7	6	-	-	13
8	Agricultural Extension assistance	6	7	-	-	13
9	*PNPM Mandiriperdesaan* - National Programme for Community Empowerment	9	-	-	-	9
10	*DesaWisata* Tourism Village	9	-	-	8	14
	Score	69	13	8	8	

NB: Scoring is out of 10; the higher score, the higher its impact
Source: Field work primary data (2013)

The sources of funding assistance in Candirejo can be seen in Table 6. Unlike the previous villages, funds from NGOs are provided. This aid comes in the form of mentoring activities provided near the famous Borobodur Temple in the field of hospitality and tourism. From 2009 until 2011 Candirejo was provided with full assistance by NGOs from Yogyakarta. In 2012 the local government assisted Candirejo through the *Program NasionalPemberdayaanMasyarakat*- PNPM programme to build infrastructure and support facilities to become a tourist village. The community believes that the tourism programme can contribute significantly to their economy if some villagers take the opportunity to make their homes into guesthouses. In terms of the DMP Programme, its activities are more focused on agriculture development, including introduction of livestock farming packaged with management training, and skills training based on local resource availability. The goal of the Programme is to increase the income of poor families living on the village's border where the impact of tourism is less obvious.

Madukoro Village

Based on the results of group discussions (Table 7), extension officers have an important role in every activity in the village. The expected function of the extension officers is to successfully disseminate knowledge and information on programme implementation activities. However, the scoring results show that the sources of information are mostly friends, relatives and parents, with total of 63 points, whereas information disseminated by extension officers only scored 57. This means that extension officers need to improve their performance to the expected level, for example, by frequently visit the villages and visiting their friends or relatives so that they know the problems of villagers. In other words, the extension officers need more creativity and innovation in disseminating information. The villagers at Madukoro suggested that the extension activities should not only come from the district level but also from the regency level and even from the provincial level. The extension officers in this case are from various sectors such as agriculture, husbandry, fisheries and even health. In addition, the variety of information from experienced extension officers is very important. The important issue to be borne in mind is the need for information about the villagers to be known deeply by extension officers through observations. Information to be disseminated should include updated technology and knowledge associated with programme implementation.

Madukoro enjoys considerable aid from the government. Many programmes implemented successfully in a village benefit from cooperation by many agencies in implementation and this gives rise to high impact contribution in terms of village development. In the case of Madukoro, the DMP Programme was not implemented in conjunction with other programmes; they were implemented at different times. The point to be learned is that many programmes should be integrated and implemented at the same time so that the impact is more significant.

CONCLUSION AND RECOMMENDATION

The DMP Programme has resulted in the establishment of a microfinancial institution (*LembagaKeuanganDesa* - LKD) in the village, participated in by farmer groups. However, the funds allocated for the LKD are limited and they cannot cover the needs of all villagers. There remains a lack of capital necessary to start businesses for some farmers. The villagers proposed establishing a farmer cooperative in the village to support 'off-farm' and 'on-farm' activities and to be managed by farmer groups. The aim

Table 7. Information Source Scoring of Madukoro Village

No.	Information type required	Information Source				
		Media (TV, Newspapers, magazines)	Schools	Extension Officers	Friends, relative, parents	Score
1	Agriculture	4	4	7	10	25
2	Livestock	5	4	9	9	27
3	Fisheries	6	5	10	9	30
4	Forestry	6	4	8	9	27
5	Business opportunity	8	6	8	8	30
6	Markets	7	4	7	9	27
7	Employment	7	6	8	9	30
	Score	43	33	57	63	

NB: Scoring is out of 10; the higher score, the higher the information
Source: Field work primary data (2013)

Table 8. Sources of Assistance Scoring of Madukoro Village

No.	Type of poverty Assistance	Sources of Assistance				Score
		APBN National Budget	APBD Provinces or Regency Budget	International Funds	NGO	
1	*Bantuan IDT* - National Aid for underdeveloped villages	6	-	-	-	6
2	*KartuSehat/Gakin* Health Card/Gakin scheme	10	-	-	-	10
3	Raskin/ rice for the poor	9	-	-	-	9
4	*BantuanlangsungTunai BLT* - Direct cash assistance	8	-	-	-	8
5	*Desa mandiri pangan (DMP) Programme* – Food Self-Sufficient village	9	10	-	-	19
6	*PAMSIMAS* - Community-Based Water Supply and Sanitation	0	-	7	-	7
7	*BedahRumah/*House Renovation Programme	5	6	-	-	13
8	Agricultural Extension assistance	7	7	-	-	14
9	*Programme NasionalPemberdayaanMasyarakat - PNPM Mandiriperdesaan* National Programme for Community Empowerment	8	-	-	-	8
10	Fishery fishpond programme	8	-	-	-	8
11	*KawasanRumahPangan Lestari (KRPL)* Sustainable Food House zones	8	-	-	-	8
	Score	62	23	7	0	

NB: Scoring is out of 10; the higher score, the higher its impact
Source: Field work primary data (2013).

of the cooperative is to provide the place for savings and loans with low interest rates and to gain profits for the villagers.

On the issue of information sources, analysis shows that the villagers receive information related to the Programme mostly from friends, relatives and parents. Verbal information obtained through discussion and conversations is influential. Information obtained through formal schooling is less influential because it tends to contain theoretical knowledge with little information on practices. Villagers want to be trained and exposed to the latest technologies for agriculture, fisheries and husbandry. Extension officers need to be more active at the village level in provided information and knowledge-based resources.

Analysis on sources of funding shows that villages receive funding mostly from the national budget, implemented through local government. To ensure sustainability in efforts to alleviate poverty central government should continue to consideration the local situation in allocation of funds. The local government is currently striving to boost revenues by looking to local and foreign investors to invest in their villages so that the budget can be increased. The local government should be proactive in generating financial sources through local revenue. In this way the gap between the national and regional budgets can be reduced.

Extension officers play an important role in every activity in the village. Their function is to disseminate knowledge and information on programme implementation. However, that information is mostly obtained from friends, relatives and parents. If extension officers are to play their role they need to frequently visit villages and come to the problems faced. Extension activities should not be only from the district level but also from regency or province level. Information dissemination is more effective if officers are experienced and come from various sectors, such as agriculture, husbandry, fisheries and health. Extension officers need to know villagers' problems through close observation. Information should include up-to-date technology and information in terms of programme implementation.

REFERENCES

Central Bureau of Statistics (CBS). (2010). Indonesia in Number Book. Statistical Yearbook of Indonesia 2010.BadanPusatStatistik, BPS.

Central Bureau of Statistics (CBS). (2011). Indonesia in Number Book. Statistical Yearbook of Indonesia 2011.BadanPusatStatistik, BPS.

Darwis, V., & Rusastra, I. W. (2011, June). Optimizing Rural Community Empowerment through Integrating PUAP and DesaMandiriPangan Programs. *Agricultural Policy Analysis Journal.*, *9*(2), 125–142.

Gandhi, V. P., & Marsh, R. (2003). *Development and Poverty Reduction: Do Institutions Matter? A Study on the Impact of Local Institutions in Rural India.* The 25th International Conference of Agricultural Economists, Durban, South Africa.

Grewal, B., Grunfeld, H., & Sheehan, P. (2012). *The contribution of agricultural growth to poverty reduction.* ACIAR Impact Assessment Series Report No. 76. Australian Centre for International Agricultural Research, Canberra.

Quisumbing, A. R. (2003). Household Decisions, Gender, and Development A Synthesis of Recent Research. An earlier, expanded version of Chapter 8, "Gender differentials in farm productivity: Implications for household efficiency and agricultural policy." Food Policy, 20, 407–423.

Rusliyadi, M., Bin Hj. Mohd. Jamil, A., Maseleno, A., Tri Kumalasari, R., & Othman, M. (2019). Agricultural extension policy, agricultural growth and poverty reduction in Indonesia. *IACSIT International Journal of Engineering and Technology*, *7*(4), 5539–5550. doi:10.14419/ijet.v7i4.13337

Rusliyadi, M., Jamil, A., Mohd, B., & Kumalasar, R. T. (2019). Analysis of household food security policy: Case of food security village programme, Indonesia. *Asian Journal of Agriculture and Rural Development*, *9*(1), 19–32.

Rusliyadi, M., & Jamil, A. B. (2020). Participatory Poverty Assessment Effort in Food Security and Extension Policy: Evidence From Indonesia. In V. Erokhin & T. Gao (Eds.), *Handbook of Research on Globalized Agricultural Trade and New Challenges for Food Security* (pp. 481–499). IGI Global., doi:10.4018/978-1-7998-1042-1.ch024

Rusliyadi, M., & Jamil, A. B. (2021). Food Security Policy Analysis of Household Level: Case of Food Security Village Programme in Indonesia. In Ordóñez de Pablos, P., & Lytras, M. D. (Ed.), Global Challenges and Strategic Disruptors in Asian Businesses and Economies (pp. 60-75). IGI Global. http://doi:10.4018/978-1-7998-4787-8.ch004

Rusliyadi, M., Jamil, A. B., Kumalasari, R. T., & Rouf, A. A. (2021). The Institutional Impact and Community Perception Implementation of Food Security Policy on Poverty Reduction: Case of Food Self-Sufficiency Village Programme, Indonesia. In Ordóñez de Pablos, P., & Lytras, M. D. (Ed.), Global Challenges and Strategic Disruptors in Asian Businesses and Economies (pp. 29-42). IGI Global. http://doi:10.4018/978-1-7998-4787-8.ch002

Rusliyadi, M., Jamil, A. B., Kumalasari, R. T., & Rouf, A. A. (2021). Institutional Role and Asisstance Service Analysis on Food Security Policy Case in Indonesia. In Ordóñez de Pablos, P., & Lytras, M. D. (Ed.), Global Challenges and Strategic Disruptors in Asian Businesses and Economies (pp. 43-59). IGI Global. http://doi:10.4018/978-1-7998-4787-8.ch003

Rusliyadi, M., & Libin, W. (2018). Agriculture development programs for poverty reduction evidences from Indonesia and china - comparative study case. *Asian Journal of Agriculture and Rural Development*, *8*(2), 104–118.

Rusliyadi, M., Mohd, H., Jamil, A., & Anshari, M. (2020). Analysis of Food Security Policy by Participatory Poverty Assessment (PPA) Effort: Case in Indonesia. In P. Ordoñez de Pablos, M. Almunawar, & M. Abduh (Eds.), Economics, Business, and Islamic Finance in ASEAN Economics Community (pp. 105-127). Hershey, PA: IGI Global. doi:10.4018/978-1-7998-2257-8.ch006

Sheil, D., & Liswanti, N. (2006). Scoring the importance of tropical forest landscapes with local people: Patterns and insights. *Environmental Management*, *38*(1), 126–136. doi:10.100700267-005-0092-7 PMID:16508796

ENDNOTE

1 Purposive sampling is also known as judgmental, selective, or subjective sampling. Based on the recommendation by the Agricultural and Food Security Agency in Regency Level.

Section 5

Agricultural Systems Analysis Streamlining, Modeling, and Improvements: Towards Precision Agronomics (Technological Perspective)

This section takes the technological perspective especially with respect to plant biotechnology subject and its role to secure food and to alleviate the risk of loss of the genetic variability of cultivated plants as a result of environmental changes and human practices. Also, it touches the issue of increasing the efficiency of producing animal-derived food products.

Chapter 13
Importance of Artificial Environment Conditions on Plant Biotechnology, Plant Growth, and Secondary Metabolites

Ahmed M.M. Gabr
National Research Centre, Egypt

Amal A El-Ashry
National Research Centre, Egypt

Hanan S. Ebrahim
National Research Centre, Egypt

Mohamed K. El-Bahr
National Research Centre, Egypt

ABSTRACT

Plant biotechnology (micropropagation, secondary metabolites, etc.) is the science of growing different plant cells under controlled artificial environment conditions. Thence, successful plant biotechnology depends on the techniques of plant tissue culture. Plant tissue culture laboratories have style needs that distinguish them from alternative forms of laboratories and a few wants are distinctive for successfully plant growth as well as production of plant secondary metabolites. The most often manipulated physical parameters are light, temperature, and relative humidity within the culture vessels. The light is most important physical factor which might have an effect on metabolite production. Light not only affects the in vitro growth and developments but also very important factor affecting the production of plant secondary metabolites. The action of light on plants happens in the most part in two ways. In the first way, the light source gives the energy required by the plant for photosynthesis. In the second way, the signal which gotten by photoreceptors it regulates plant metabolism, differentiation, and growth. Temperature is different with the different in vitro life cycle, as well as the aim of production such as plant growth or secondary metabolites production.

DOI: 10.4018/978-1-7998-5000-7.ch013

INTRODUCTION

Plant biotechnology is presently utilized in several fields to attain completely different objectives like: micropropagation, *in vitro* preservation and cryopreservation, secondary metabolites' production etc. (Kozai and Smith, 1995).

The major factors of plant tissue culture, which are interconnected or linked to each other (Figure 1), are a) establishment and sterilized plant material, b) media composition for culture, and c) in vitro physical conditions for the growth room and the culture vessel (DeFossard, 1986).

Figure 1. The main barriers of plant tissue culture

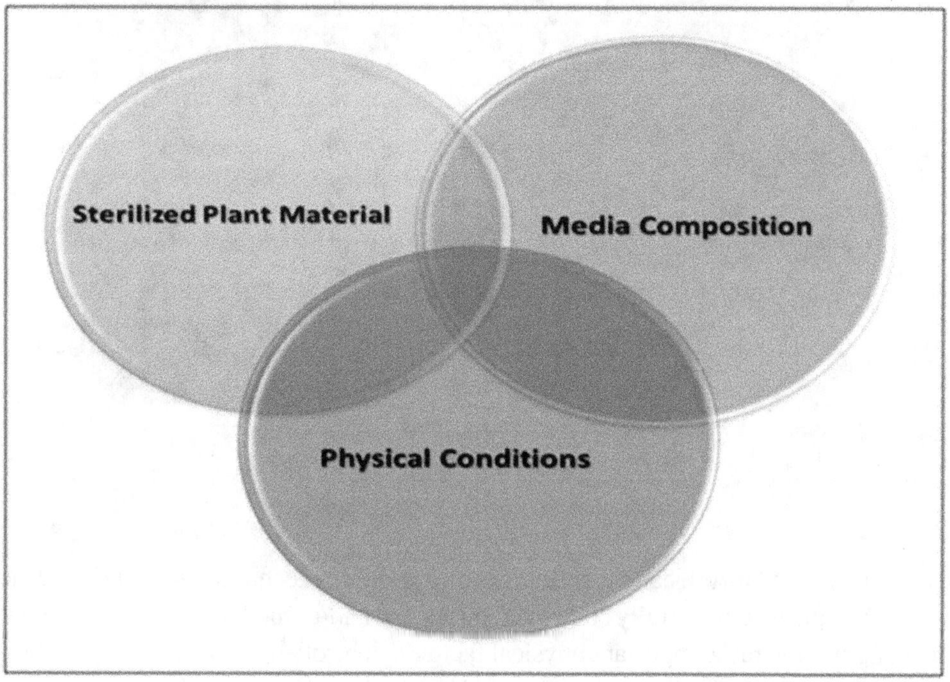

The importance of *in vitro* culture futures, as contrasted to any other plant production system, is the exacting controlled, artificial, of the environment condition surrounded the *in vitro* cultures. All These environmental factors had a significantly affected on the growth characteristics (Lumsden *et al.* 1994; Chen, 2015). Unlike plants in the field, which is growing at the unpredictable environment changes, each factor of the tissue culture condition, surrounding the plant cell cultures within a vessel, can be managed (Kozai and Smith, 1995; Isah, 2015) (Figure 2).

Physical condition, factors including temperature, light (quality, quantity and duration), air movement, and the culture vessels are predetermined and can be maintained as a constant or changing during the growth cycle of plant cell cultures. These physical factors can be modified, controlled, to achieve a desired effect by controlling growth room temperature, changing vessel bottom to influence internal humidity, modified the light quality and quantity to reach the optimum growth of plant cell cultures (Kozai and Smith, 1995; Chen, 2004). This high opportunity for physical conditions control is unique to *in vitro* plant cell growth and gives plant biotechnology distinct advantages for manipulating plant cells under his control. Hence, control the physical conditions, factors, is an awesome responsibility and an advantageous challenge for *in vitro* plant growth and as well as plant secondary metabolites' production. Finally, *in vitro* physical factors are correlated to each other and it must be control all of them in the same growth room (Figure 3).

In this state of the art, physical conditions such as; temperature, culture vessels, and light will discuss.

Figure 2. An overview of plant tissue culture growth room A) empty growth room, and B) culture incubated in growth room

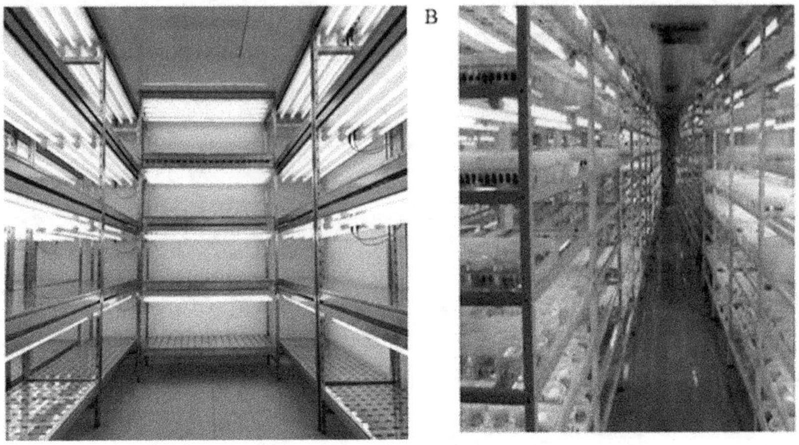

LIGHT

Plant biotechnology laboratories, can be defined as an agriculture technology wherein the surrounding microclimate of any plant cells are fully controlled by normal and/or modern facilities during the cell life cycles. Hence, light is the most important physical factor in controlled environmental conditions, which is directly affect not only plant cell growth and development but also the production of plant secondary metabolites (Kendrick and Kronenberg 2012; Hart 2012; Kozai 2016).

Light Sources and History

In order to decide on the most effective source of illumination for plant growth, it's necessary to know the history, characteristics, light mechanism and applicability of light sources. Light source have three major technologies, which are 1) incandescence lamp, 2) electrical discharge (fluorescent lamb), and 3) solid state light sources (LED; light-emitting diode, organic OLED; light-emitting diode) (Figures 4). Shortly, first two are called "legacy" technologies, and also the third is taken to account modern technologies (Batista *et al.* 2018).

Figure 3. Physical conditions of in vitro plant cell cultures

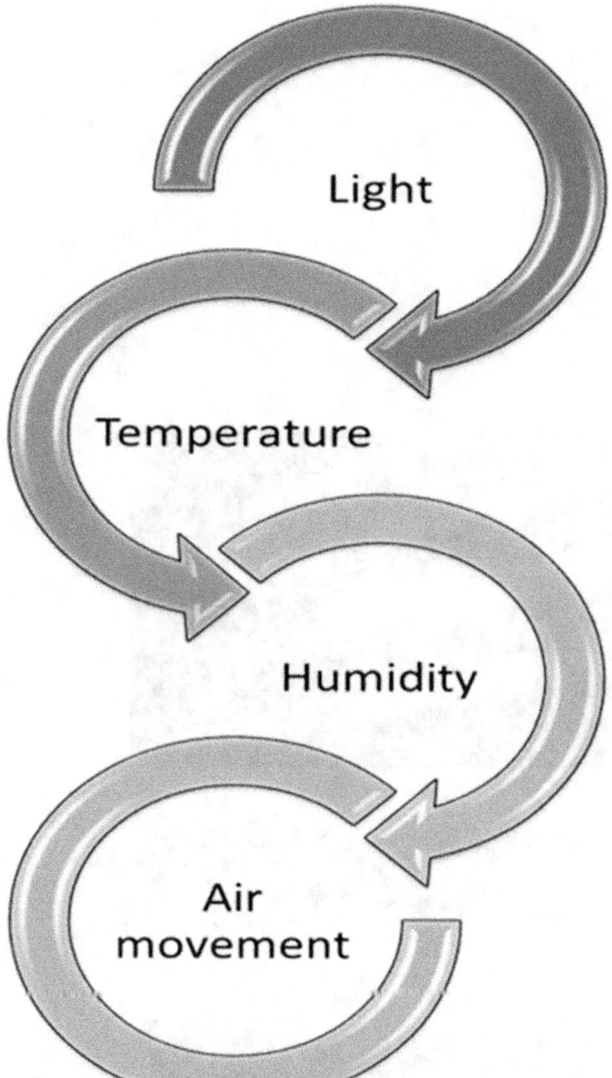

Legacy Light Sources

- **Incandescence lamp**
  Incandescent lamp is the first generation of the artificial light, and was the most common lamb for more than 130 years. While, there has been prohibition to use by several countries in step with the rule: phase out of incandescent lamps in 2008 (Paul, 2010).
- **Fluorescent lamb**
  Fluorescent lamb is the second generation of the artificial light, and it is considered the most common lamb in all plant biotechnology (Schuerger *et al.* 1997; Bantis *et al.* 2016).

Figure 4. Light source: A) incandescence lamp, B) fluorescent lamb, C) light-emitting diode (LED), and D) organic light-emitting diode (OLED)

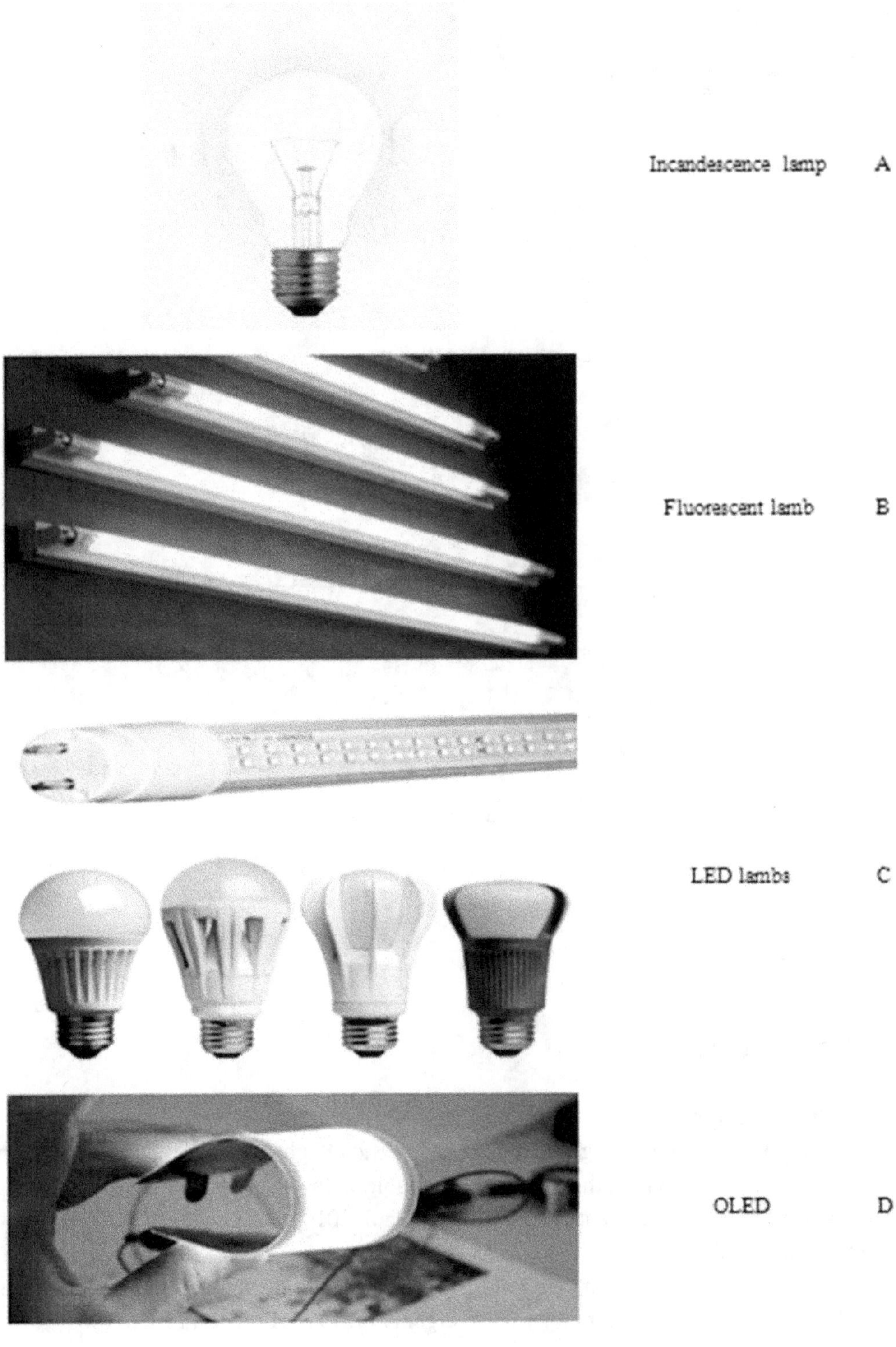

Figure 5. Visible wavelength spectrums from 400 to 700 nm

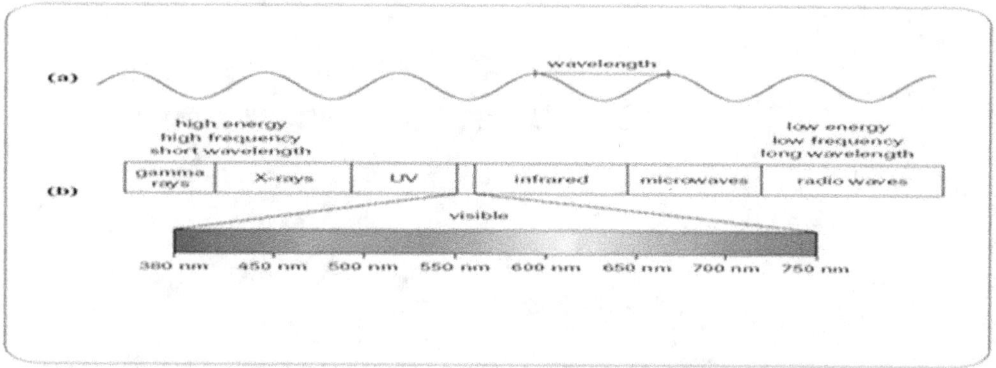

Fluorescent lamps have three main types: cold cathode, hot cathode and electroluminescent lamps (Bagher, 2016).

Solid State Light Sources

- **Light - emitting diode**

  Solid state light sources, including both Light Emitting Diode (LED) and Organic Light Emitting Diode (OLED), are the new generation of artificial lighting sources, which had more than 50 years of history from the first infrared LED patented in 1961 (Bourget, 2008). Reference to Samsung study, the price of a white LED system will decrease by 50% in order to make it more competitive with fluorescent lamps over the next 4–5 years (Chang, 2008; Chang *et al.* 2012). Nowadays, LEDs are used in anywhere, such as TV and computer displays, home lighting, traffic lights, greenhouse lighting, plant biotechnology labs etc. LEDs are available in a multi broadband wavelength across the ultraviolet (250-380 nm), visible (380-760 nm) and infrared (760-1000 nm), which have been becoming common due to their high efficiency and brightness. Additionally, LED light technology has more advantages, which are small volume, the longest life, low heat, high energy conversion and wavelength controls specifications (Massa *et al.* 2008; Morrow 2008; Schuerger *et al.* 1997).

Light and Plant

Light action on plants happens essentially in 2ways: first, light supply by required energy for photosynthesis, especially red and blue light. Second, the signalwhich received via photoreceptors it regulates plant metabolism differentiation, and growth (Wang *et al.* 2001; Hart, 2012; Kozai, 2016).

Light is that the supply of energy for photosynthesis, and gives signals for photomorphogenesis (Smith, 2013). Photosynthesis is gotten by photosynthetically active radiation (400 to 700 nm) (Figure 5). Chlorophyll a, chlorophyll b and a variety of accessory pigments, including carotenoids (plant pigments transfer absorbed energy to photosynthetic reaction centers (Nishio, 2000). Each pigment has a characteristic action spectrum, which is characterized by physiological activity rate of light wavelength. The red and blue light spectra are strongly absorbed by chlorophyll a and b, but green light is weakly absorb (Nishio, 2000) (Figure 6).

Figure 6. Pigments wavelength absorption

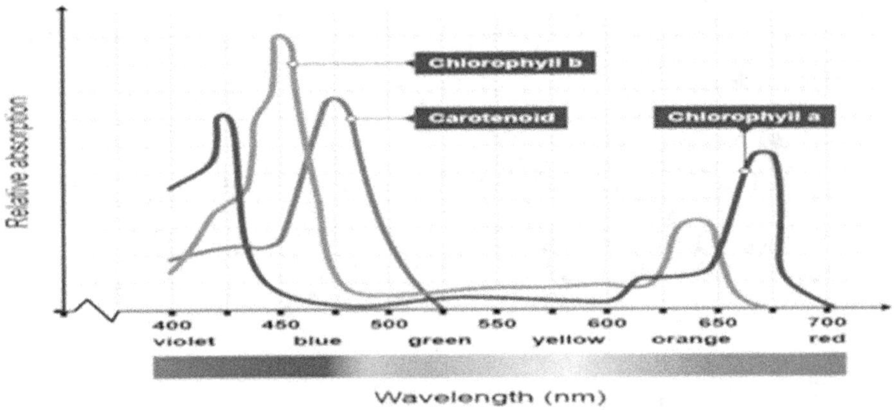

The light reaching plants reflected, transmitted or absorbed. In photosynthesis, part of photons converts into chemical energy, which is in the process of growth and morphogenetic and the other part lost in the form of heat (Kozai, 2013).

Light Applications in *in vitro* Plant Growth

Light, among any other in vitro physical factor, is the most important physical factors in plant biotechnology lab (Gupta and Jatothu, 2013). Light is affecting in vitro plant cells through different ways such as light quantity (intensity), duration (light/dark) and light quality (Gupta and Jatothu, 2013). Moreover, light has different influence on the plant cells during the *in vitro* stage or culture life cycle. For example, potato tissue culture found to produce fewer microtubers/plantlet in the dark than under light conditions for 8 hours (Pruski (2001). Furthermore, microtubers were produced more under shorter photoperiod-with white fluorescent (100 µmol m^{-2} s^{-1}) (8 h light) than under longer days (16 h light) (Wang and Hu, 1982). And, decreasing light photoperiod was increased the size of potato microtubers (Seabrook *et al.* 1993) also resulted in earlier microtuber formation (Garner and Blake, 1989). In order to light intensity, several studies have confirmed that plant cells incubated under low light intensity as well as under dark condition might be favorable for different stages of *in vitro* life cycles (Konstas *et al.* 2003). One of the most important and economical plant propagated by tissue culture is date palm, which is need complete dark conditions in the early stage, callus induction from offshoot (Tisserat, 1979; Huong *et al.* 1999). And for date palm regeneration and growth multiplication, un-differentiated callus cultures were transferred to a new media composition and incubated under normal fluorescent light, white, (100 µmol m^{-2} s^{-1}) (16/8 light/dark) (Aslam and Khan, 2009) (Figure 7). Several studies demonstrated that dark condition has positive impact on plant growth, shoot induction and somatic embryogenesis from some species, like cucumber (Colijn-Hooymans *et al.* 1988), melon (Gray *et al.* 1993; Kintzios *et al.* 1998a, Kintzios and Taravira, 1997) and rose (Kintzios *et al.* 1998b), lavender (*Lavandula angustifolia* Miller subsp. *angustifolia*) callus proliferation was improved with increasing of darkness incubation periodwhereas the somatic embryogenesis was reduced under the identical conditions (Konstas *et al.* 2003). Also, wall germander (*Teucrium chamaedrys*) incubation in darkness didn't have an effect the growth of wall germander cultures, but improved somatic embryogenesis(Kintzios *et al.* 2002). Summart

Figure 7. Date palm growing under dark and light conditions

et al. (2008) reported that the highest growth of rice callus (*Oryza sativa* L.) was obtained under dark condition. Pandino *et al.* (2017) reported that successful established callus cultures from *in vitro* leaves of globe artichoke were inoculated in the dark for the first four weeks, then moved under cool-white fluorescent light (16/8 light/dark). Since, the fluorescent lambs were the most popular in plant biotechnology growth rooms (Economou and Read, 1987). Furthermore, these lambs, light, have a wide range of wide range of wavelengths (350 to 750 nm), such wavelength appears to be unnecessary and gave low quality for promoting plant growth (Gupta and Jatothu, 2003). Other disadvantages of fluorescent lambs are consumption a lot of electrical and generating heat in the growth room, which cause damage or suffer from light stress. Hence, there was a need for improving light source to improve the quality of *in vitro* growing plants as well as reduce the cost of electricity and control the light wavelength.

Recently, LEDs have an applicable alternative source of light for *in vitro* cultures growth and development (as reviewed by Gupta and Jatothu, 2013; Batista *et al.* 2018). LED lighting has high and flexible ability to manage the spectral composition (Hoenecke *et al.* 1992; Brown *et al.* 1995), this ability allows studying the application of LED with specific spectral in the physiological changes of plant growth and development (Yeh and Chung 2009; Nhut and Nam, 2010). Various *in vitro* growth and development are considered as influence under LED lighting as compared to those growings and development under normal fluorescent light. Particularly, red and blue lights have a highly significant impact on *in vitro* plant growth, which used alone or in combination (Nhut *et al.* 2000, 2002; Lian *et al.* 2002; Duong *et al.* 2003; Kim *et al.* 2004a, b, c; Poudel *et al.* 2008; Shin *et al.* 2008; Li *et al.* 2010).Jao and Fang, (2004a,b) studies the effect of LEDs instate of fluorescent light in *in vitro* potato plantlets. They found that LEDs red and blue (ratio 9:4) with photoperiod 16/8 (light/dark) stimulated plant growth as compared with traditional fluorescent light. Similar was with red and blue (1:1) lights were also obtained from *in vitro* cultures of lilium (Lian *et al.* 2002), strawberry (Nhut *et al.* 2003), banana (Duong *et al.* 2003), and chrysanthemum (Kim *et al.* 2004c).

Budiarto, (2010) reported that Anthurium, number of shootes, was promoted by growing under red light that blue light. *In vitro* stem and internods elongation of *Chrysanthemum* (Kim *et al.* 2004c) and grape (Poudel *et al.* 2008) were higher under LED red light. On the other hand stem length of marigold was higher under blue LED light (Heo *etal* . 2002). Also, red with blue led lights enhanced the protocorm-like bodies induction in *Phalaenopsis amabilis* (Nhut *et al.* 2006). Higher growth of *in vitro* plantlets grown under a red/blue (4:1) led ratio, comparing with white light for *Eucalyptus, Musa,* and *Spathiphyllum* (Nhut and Nam 2010). In *Cymbidium*, the highest protocorm-like bodies formation was obtained with red:blue (25%:75%) LEDs (Huan and Tanaka, 2004). Blue diode light was found to be the best condition for shoot promoting of protocorm-like bodies of*Dendrobium officinale in vitro* culture (Lin *et al.* 2010). The results in optimum responses has been suggested that the matching of LEDs wavelengths to plant photoreceptors influence plant morphogenesis (Bourget 2008; Massa *et al.* 2008; Morrow 2008).

Light Applications in Plant Secondary Metabolites

Light influences plant defense stress through regulation of secondary metabolites production (as reviewed Batista *et al.* 2018). Light quality has been involved in the regulation of pigments biosynthetic pathways (carotenoids such as β-carotene and lutein), as well as bioactive compounds such as polyphenols (phenolic acids, anthocyanins, and flavonoids), terpenoids, and glucosinolates (Ballaré 2014; Darko *et al* . 2014; Huché - Thélier *et al.* 2016; Demotes- Mainard *et al.* 2016; Taulavuori *et al* . 2016) (Table 1). The regulation is made through selective activation of different photoreceptors (Ballaré 2014; Huché - Thélier*et al* . 2016; Demotes - Mainard *et al* . 2016). Many studies the influence of light quality on the production of secondary metabolites in *in vitro* plants (Arena *et al* . 2016; Szopa and Ekiert 2016; Gupta and Karmakar 2017).Light qualities gotten by fluorescent lambs have been used to stimulate antioxidant secondary metabolites in shoot and callus cultures of many species (Tariq *et al* . 2014; Fazal *et al* . 2016; Szopa and Ekiert 2016; Gazolla *et al* . 2017; Kawka *et al* . 2017). However, the spectrum of light emitted by fluorescent lamps wasn't specific, can't be controlled and like intensity, moreover, its stability decreased over the time (Darko *et al.* 2014; Taulavuori *et al.* 2016). LEDs light have been used to circumvent problems related to fluorescent lamps and to optimizeplant *in vitro* cultures development (Gupta and Jatothu 2013; Shukla *et al.* 2017). However, the influence on secondary metabolism is underexplored up till now. How and Smith (2003) investigated the effect of four different light/dark cycling treatments on growth and anthocyanin production in callus culture of *Ajuga reptens*. They found that the level of anthocyanin is decreasing with as the length of the duration decrease.Also, the wavelength is significant in case of monochromatic light (Ramawat and Mathur 2007; Mengxi *et al.* 2011).

Favorable impact of light quality on the accumulation of different compounds; flavonoids (Krewzaler and Hahlbrock, 1973), anthocyanin (How and Smith, 2003), betalains (Shin *et al.* 2004), ginsenosides (Yu *et al.* 2005), phenolic acids and linear furanocoumarins (Szopa *et al.* 2012) and lignans (Szopa and Ekiert, 2016) has been shown.Recently, Younas *et al.* (2018) investigate the effect of different LED spectreson silymarin accumulation in callus culture. They found that the level of silychristin, isosilychristin, silydianin, silybin A and silybin B were showed greatest in red light, whereas green light showed the highest accumulation of isosilybin A and isosilybin B. While, the taxifolin amount was the greatest under continuous white light.Khan *et al.* (2018)evaluated the synergistic effects of melatonin and different LEDs in callus cultures of *Fagonia indica*. They investigated that total flavonoids content and Free radical scavenging activity (97%) was found in cultures grown under white LED and melatonin. Inaddition, cultures maintained under white light were also found with highest levels of phenolic and flavonoids production. However, the antioxidant enzymes; Superoxide dismutase and Peroxidase were found optimum in cultures grown under blue LED.

Ultraviolet Radiation-Elicited

Ultraviolet radiation is a small part of the solar radiation, (200 to 400 nm), reaching the Earth's surface but with significant biological impact on all living organisms. These wavelength region can divided into

Table 1. Impact of different LEDs on the accumulation of plant secondary metabolites

Light spectrum	Type of explant	Product	Plants	Ref.
The highest accumulation observed with Red light and the lowest with Blue light	Axillary shoots	phenolic acid and flavonoid contents	*Myrtus communis L.*	**Cioć *et al.* 2018**
LEDs light red, blue, green and yellow	Nodalsegments	Volatile compounds	*Hyptis suaveolens*	**Andrade *et al.* (2017)**
Red led light is higher than fluorescent	Apical segments	Carvacrol γ-Terpinene p-Cymene	*Plectranthus amboinicus*	**Silva *et al.* (2017)**
Blue led	Shoot	Total phenolics and flavonoids	*Swertia chirata*	**Gupta and Karmakar (2017)**
White led	Plantlets (aerial part)	Essential oil profile	*Lippia alba*	**Batista *et al.* (2016)**
White and yellow led	Callus and tissue culture plants	Chlorogenic acid	*Peucedanum japonicum*	**Chen *et al.* (2016)**
Far-red/red/blue/UV-A/white light	Shoot-differentiating callus cultures	Dibenzocyclooctadiene Lignans Phenolic acids	*Schisandra chinensis*	**Szopa and Ekiert (2016)**
Blue, red, green, white LEDs and white fluorescentPhotosynthetic flux density: 13, 27, 35, 47, 69 μmol m^{-2} s^{-1}	Leaves	Monoterpenes Sesquiterpenes	*Achillea millefolium*	**Alvarenga *et al.* (2015)**
Red and far-red LEDs	Shoots	Cucurbitacins	*Aquilaria agallocha*	**Kuo *et al.* (2015)**
red:blue (20:80%)	plantlets	Ginsenoside MR2	*Panax vietnamensis*	**Nhut *et al.* (2015)**
blue led	Shoots	Total phenols	*Rehmannia glutinosa*	**Manivannan *et al.* (2015)**
Red led	shoots	Increase total flavonoids content		
Red and blue led	Adventitious shoot	Harpagoside, (−)-epicatechin caffeic acid, sinapic acid, ferulic acid Trans-cinnamic acid (−)-gallocatechin Ellagic acid Chicoric acid	*Scrophularia takesimensis*	**Jeong and Sivanesan (2015)**
Blue:green:red	Leaves	Amino acids, organic acids, fatty acids, and flavonoid glycosides	*Oryza sativa* cv. Ilmi	**Jung *et al.* (2013)**
Red, blue, and white fluorescent	Adventitious roots	Total phenolic acids, α-tocopherol and β-amyrin	*Panax ginseng*	**Park *et al.* (2013)**

three categories UV-A (315 – 400nm), UV-B (280 – 315nm) and UV-C (100 – 280nm) (Commission Internationale de l'Eclairage, 1987). UV effect is depending upon the plant sensitivity (Lavola *et al.* 2003; Zu *et al.* 2011). UV-B or UV-C with low doses may trigger acclimation responses in plants such as; activation of enzymatic and non-enzymatic defense systems (Loyall *et al.* 2000; Jansen, 2002; Lavola *et al.* 2003; Katerova and Todorova, 2009; Katerova *et al.* 2009; Katerova and Todorova, 2011; Rai *et al.* 2011), while positive response of high UV doses, which could activate repair mechanisms in order to cope with the stress (Frohnmeyer and Staiger, 2003). It is clearly, the defense to UV-B can be related to the stimulation of different secondary metabolite production as well as DNA repair mechanisms (A-H Mackerness, 2000; Brown *et al.* 2005; Ishibashi *et al.* 2006). Low doses UV-C doses (0.5– 9.0kJm^{-2}) has been considered to have commercial prospect by causing beneficial effectsin preventing pathogen diseases and delay senescence during storage of fruits (Shama and Alderson, 2005). In addition, ultraviolet (UV) radiations consider an effective elicitor for boosting the biosynthesis of various plant secondary metabolites (Matsuura *et al.* 2013) (Table 2). UV treatments as an emerging elicitation technology has been applied to different plant and herbs enriched with valuable phytochemicals and causing the increased these health-beneficial substances (Kanazawa *et al*, 2012; Zhang et al., 2013; Nguyen *et al.* 2014; Pandey and Pandey-Rai, 2014 a,b). Hence, UV radiation is considering one of the modern promising elicitation approach the production system and can be easily manipulated (Jiao *et al.* 2015). Therefore, an interesting possibility is to exploit UV radiation for enhancing secondary metabolites production in *in vitro* cultures.

Table 2. Impact of different ultraviolet on the production of secondary metabolites

UV types	Product	plant	Cell Culture	Ref.
UV-A	isoflavonoid	*Astragalus membranaceus*	Hairy root	**Jiao *et al.* 2015**
UV-B	isoflavonoid	*Astragalus membranaceus*	Hairy root	**Jiao *et al.* 2015**
	artemisinin	*Artemisia annua*	hairy roots	**Liu *et al.* 2003**
	vinblastine & vincristine	*Catharanthus roseus*	plants	**Bernard *et al.* 2009**
	Rutin and quercetin	*Fagopyrum tataricum* Gaertn	Hairy root	**Huang *et al.* 2016**
UV-C	isoflavonoid	*Astragalus membranaceus*	Hairy root	**Jiao *et al.* 2015**
	dicaffeoylquinic acids	*Cynara cardunculus* L. var. *scolymus*	Callus	**Menin *et al.* 2013**
	resveratrol	grapevine	Callus	**Ku *et al.* 2005**
	resveratrol	groundnut	Callus	**Liu *et al.* 2010**

TEMPERATURE

Temperature is different for each plant species as well as its cell culture (Smetanska, 2008). Moreover, temperature is different with the different *in vitro* life cycle, as well as the aim of production such as plant growth or secondary metabolites production.

Temperature Application in Plant Growth and Development

Cui *et al.* (2000) examined temperature during light and dark conditions, light period (18°C) and dark period (26°C), and they found increasing the shoot and root weights of *Rehmannia glutinosa* plantlets under light with 26°C. Chen (2015) studied the optimal microclimate conditions for *Phalaenopsis* pot plant *in vitro* growth, results showed that the optimal temperature with light was 29-32°C, while on dark condition the optimal temperature was 22-25°C.

One of the most important applications of temperature in plant biotechnology is *in vitro* preservation (Cruz-Cruz *et al.* 2013; Kaminska *et al.* 2018). Which use for medium-term preservation by slow growth of *in vitro* culture under low temperature or long term-preservation, which call cryopreservation (under -196°C). Slow growth conservation allowssaving the *in vitro* plant cells for several weeks or months depending on the plant species (Cruz-Cruz *et al.* 2013). Son *et al.*(1991) stated that, shoot cultures of *Populus alba* × *P. grandidentata* remained for 5 years at 4°C. Dorion *et al.*(1993) observed that survival rate of *Ulmus sp.* was higher than 85% for 24-30 months at 7°C. Fukui *et al.*(1992) found that *Diospyros kaki* shoot tip cultures of cultivar Fuyu explants survived for 12 weeks at 10°C, while cv. Nishimurawase explants survived for 30 weeks at the same temperature. Hausman *et al.*(1994)reported that *in vitro* shoots cultures of *Populus tremula* × *P. tremuloides*survived by 100% at 4°C for 7, 14 and 28 days periods, Roxas *et al.*(1995) recorded the same results on *Dendranthema grandiflorum* stored at 5°C for 12 and 24 months. Bonga (1997) reported that, explants of *Larix decidua* and *L. xeurolepis* remained viable for at least 9 months at -5°C. Quan Yin *et al.*(1996) observed that, shoot tips and stem segments of citrus survived at a relatively high rate (more than 60%) with storage at 15°C for 12 months. Romano and Martins (1999) on *Quercus suber*, reported that *in vitro* shoots cultures survived by 90% at 5°C for 2 months. Lata *et al.* (2010) found that germplasm of *Podophylum peltatum* preserve at 10°C for 4 and 8 months had higher proliferation than germplasm preserved at 25°C. Similar results showed with Kaminska *et al.* (2016) when successes to prolong the shoot tip preservation of *Taraxacum pieninicum* at 10°C for 9 months. that Slow growth preservation at 4°C showed to be most suitable for several *in vitro* plants such as *Dalbergia sissoo*(Chang and Singh, 2004), *Quercus cerris* and *Quercus robura*(Tsvetkov and Hausman, 2005), *Eclipta alba*(Singh *et al.* 2006), *Deutzia scabra* (Gabr and Sayed, 2010), *Withania somnifera*(Singh *et al.* 2010), *Pistacia vera*(Akdemir *et al.* 2013), *Prunus cerasifera* and *Prunus domestica*(Gianni and Sotile, 2015), Kober 5BB (*Vitis riparia* x *Vitis berlandieri*) (Benelli, 2016),*Taraxcum pieninicum*(Kaminska *et al.* 2018).

Temperature Applications in Plant Secondary Metabolites

The normal temperatures for callus growth and developments as well as cell suspension cultures are in the range of 17- 25°C (Rao and Ravishankar, 2002). And, when the temperature was maintained at 19 °C, biotransformation of digitoxin to digoxin was favored, whereas 32 °C was optimal for the purpureaglycoside-A formation in *Digitalis lanata* cell cultures. A higher yield of ubiquinone in *Nicotiana tabacum* cell cultures has been observed at 32 °C when compared to 24 °C (Smetanska, 2008). Alkaloids in cell cultures of *Catharanthus roseus* at 16 °C had 12 fold higher as compared to the normal 27 °C (Courtois and Guren, 1980). On Additionally, Toivonen *et al.* (1992) reported that 25°C is the optimum temperature for alkaloid production from cell suspension culture of *Catharanthus roseus*. Shohael *et al.* (2006) observed the effect of low and high temperature (12, 16 and 30°C) on biomass, phenolic and flavonoids accumulation in somatic embryo of *Eleutherococcus senticosus*. They found that low temperature increase the accumulation of eleutheroside E and increase the antioxidant capacity, while the biomass was decreased. Yu *et al.* (2005) studied the effect of different temperatures during the *in vitro* daily (light/dark) on growth and ginsenosides accumulation in hairy root of *Panax ginseng*. They reported that the highest hairy root growth was obtained at 20/13° (light/dark), while the accumulation of ginsenosides was obtained in culture incubated at 25/25 (light/dark).

HUMIDITY

Humidity, as *in vitro* physical factor, is considering an important factor for *in vitro* cells grow and development (Chen and Chen, 2002). In vitro plant cells are growing in different types of sterilized vessels (Figure 8). Humidity within the *in vitro* culture vessels is, usually, very high, which cause poor develop of epicuticular wax layer and interrupt stomata function (Chen, 2004). Vessels and their covers are responsible to regulate the air between the internal microenvironment and the external environment of outside air (Huang and Chen, 2005). However, the microenvironment of cultures, in the vessels, are not easy adjusted (Chen and Chen, 2002). *In vitro* plant cells are planted in vessels, which have an important specification, in order to provide uniform and adequate light and to allow the gas exchange, adequate humidity and CO_2 concentrations (Aitken-Chrisie *et al.* 1994; Zimmerman, 1995). In vitro cultures growths are affected by the internal vessels microenvironment such as light, humidity and air temperature (Hung and Chen, 2005). Many studies have been reported that high humidity in the culture vessels cause many physiological and morphological disorder of *in vitro* plantlets (Preece and sutter, 1990; Ghashghaie *et al.* 1992). Under *in vitro* higher relative humidity, the epicuticular and stomata didn't have the normality for evaporation (Ziv, 1990). One of the most common problems in plant tissue culture caused by the high relative humidity is hyperhydricity, (vitrification) (Gribble, 1999; Perez-Tornero *et al.* 2001).

The only or the optimum way to control on the internal microenvironment is to control on the external environmental conditions (Chen, 2003). Chen and Chen, (2002) reported that*in vitro* plantlets grown in small vessel, and the air exchange rate is very low between inside and outside. And medium vessel had sufficient nutrient and water, the internal humidity is very high. These vessels have caps or closures, which have a clear role to isolate the inside environment from the contamination of outside environment and allowance for some gas exchange and provision of sufficient growing area (Smith and Spomer, 1995). Majada *et al.* (2001) were found that *Dianthus caryophyllus* shoots cultured on airtight or ventilated vessel had better performance of stomata function than those grown in sealed vessels. Huang and Chen (2005) studied the effect of seven types of vessels in order to determine their physical properties. These vessels were 1) conical flask glass with black rubber stopper, 2) conical flask plastic with plastic screw-cap closure (flask and closure are made of polycarbonate), 3) Japanese irregular box with polypropylene lids and the wall is made by polycarbonate, 4) GA-7 box with polypropylene lids and the wall is made by polycarbonate, 5) round vessel made by polypropylene (lids and wall), 6) square box made by polypropylene (lids and wall) and 7) rectangular box made by polypropylene (lids and wall). And they measured air exchange rate, transmittance in the culture vessel and the spectral irradiance. References to the gas exchange rate, they had divided the seven vessels to three categories: The first one included conical flasks (glass and plastic) and they had low exchange rate, the second group included the Japanese irregular box, GA-7 box, round vessel and square box had intermediated exchange rates, and the highest exchange rate was shown with the rectangular box. In this point Chen, (2003) reported that the effect of air exchange rate in the internal air temperature of vessels could be neglected. Moreover, air exchange rate is very functional to solve gases accumulation problems.

Humidity Application in Plant Growth and Development

Kozia *et al.* (1995) studied the effect of different relative humidity (80 to 95%) on shoot growth and elongation of potato culture. They reported that potato shooltlet length reduced with decreased the relative humidity. naka*et al.* (1992) reported that the relative humidity in the potato culture vessels had significant

Figure 8. Different types of plant tissue culture vessels

effect on the shootlet evaporation during the first 10 days, while the shootlet evaporation had no effect after 20 days of culture. Sallanon and Maziere, (1992) examined the effect of two levels of humidity on two growth room on *in vitro* rose development. They found that low level of humidity growth room enhance rose growth. Islam *et al.* (2005) studied the effect four different culture vessels (industrial glass jar, magenta vessel, Erlenmeyer flask and culture tube) on the growth of four accessions of mint for 6 weeks at 25°C. They found that the highest losses from the media mint fresh weight were showed with the industrial glass jar while the magenta vessel showed the best *in vitro* performance.

CONCLUSION

Light quantity effect on plant growth, and light quality effect on plant secondary metabolites. Controlled temperature during the in vitro culture life cycle is very urgent to reach the goal of work. Optimize the humidity inside the vessel is very critical to avoid the abnormality growth of in vitro cultures

REFERENCES

Aitken-Chrisie, J., Kozai, T., & Smith, M A L. (1994). *Automation and Environmental Control in Plant Tissue Culture*. Kluwer Academic Publishers.

Akdemir, H., Süzerer, V., Tilkat, E., Yildirim, H., Onay, A., & Çiftçi, Y. O. (2013). In vitro conservation and cryopreservation of mature pistachio (*Pistacia vera* L.) germplasm. *Journal of Plant Biochemistry and Biotechnology*, *22*(1), 43–51. doi:10.100713562-012-0109-2

Alvarenga, I. C. A., Pacheco, F. V., Silva, S. T., Bertolucci, S. K. V., & Pinto, J. E. B. P. (2015). In vitro culture of *Achillea millefolium* L.: Quality and inten- sity of light on growth and production of volatiles. *Plant Cell, Tissue and Organ Culture*, *122*(2), 299–308. doi:10.100711240-015-0766-7

Andrade, H. B., Braga, A. F., Bertolucci, S. K. V., Hsie, B. S., Silva, S. T., & Pinto, J. E. B. P. (2017). Effect of plant growth regulators, light intensity and LED on growth and volatile compound of *Hyptis suaveolens* (L.) Poit in vitro plantlets. *Acta Horticulturae*, (1155), 277–284. doi:10.17660/ActaHortic.2017.1155.40

Arena, C., Tsonev, T., Doneva, D., De Micco, V., Michelozzi, M., Brunetti, C., Centritto, M., Fineschi, S., Velikova, V., & Loreto, F. (2016). The effect of light quality on growth, photosynthesis, leaf anatomy and volatile isoprenoids of a monoterpene-emitting herbaceous species (*Solanum lycopersicum* L.) and an isoprene-emitting tree (*Platanus orientalis* L.). *Environmental and Experimental Botany*, *130*, 122–132. doi:10.1016/j.envexpbot.2016.05.014

Bagher, A.M. (2016). A Detailed Review on Types of lamps and their applications. *J. of Physical and Chemical Science, 4I1*. DOI: doi:10.15297/JPCS.V4I1.02

Ballaré, C.L. (2014). Light regulation of plant defense. *Annu Rev Plant Biol, 65*, 15.1–15.29.

Bantis, F., Ouzounis, T., & Radoglou, K. (2016). Artificial LED lighting enhances growth characteristics and total phenolic content of *Ocimum basilicum*, but variably affects transplant success. *Scientia Horticulturae, 198*, 277–283. doi:10.1016/j.scienta.2015.11.014

Batista, D. S., Castro, K. M., Silva, A. R., Teixeira, M. L., Sales, T. A., Soares, L. I., Cardoso, M. G., Santos, M. O., Viccini, L. F., & Otoni, W. C. (2016). Light quality affects in vitro growth and essential oil profile in *Lippia alba* (Verbenaceae). *In Vitro Cellular & Developmental Biology. Plant, 52*(3), 276–282. doi:10.100711627-016-9761-x

Batista, D. S., Felipe, S. H. S., Silva, T. D., de Castro, K. M., Mamedes-Rodrigues, T. C., Miranda, N. A., Ríos-Ríos, A. M., Faria, D. V., Fortini, E. A., Chagas, K., Torres-Silva, G., Xavier, A., Arencibia, A. D., & Otoni, W. C. (2018). Light quality in plant tissue culture: Does it matter? *In Vitro Cellular & Developmental Biology. Plant, 54*(3), 195–215. doi:10.100711627-018-9902-5

Benelli, C. (2016). Encapsulation of shoot tips and nodal segments for *in vitro* storage of "Kober 5BB" grapevine rootstock. *Horticulturae, 2*(3), 10. doi:10.3390/horticulturae2030010

Bernard, D. G., Cheng, Y., Zhao, Y., & Balk, J. (2009). An allelic mutant series of ATM3 reveals its key role in the biogenesis of cytosolic iron-sulfur proteins in Arabidopsis. *Plant Physiology, 151*(2), 590–602. doi:10.1104/pp.109.143651 PMID:19710232

Bonga, J. M. (1996). Frozen storage stimulates the formation of embryo-like structures and elongating shoots in explants from mature *Larix decidua* and *L.* x *eurolepis* trees. *Plant Cell, Tissue and Organ Culture, 46*(2), 91–101. doi:10.1007/BF00034841

Bonga, J. M. (1997). The effect of collection date and frozen storage on the formation of embryo-like structures and elongating shoots from explants from mature *Larix decidua* and L. x *eurolepis. Plant Cell, Tissue and Organ Culture, 51*(3), 195–200. doi:10.1023/A:1005932628498

Bourget, C. M. (2008). An introduction to light-emitting diodes. *HortScience, 43*(7), 1944–1946. doi:10.21273/HORTSCI.43.7.1944

Brown, C. S., Schuerger, A. C., & Sager, J. C. (1995). Growth and photomorphogenesis of pepper plants under red light-emitting diodes with supplemental blue or far-red lighting. *Journal of the American Society for Horticultural Science, 120*(5), 808–813. doi:10.21273/JASHS.120.5.808 PMID:11540133

Brown, D. M., Zeef, L. A. H., Ellis, J., Goodacre, R., & Turnera, S. R. (2005). Identification of novel genes in Arabidopsis involved in secondary cell wall formation using expression profiling and reverse genetics. *The Plant Cell, 17*(8), 2281–2295. doi:10.1105/tpc.105.031542 PMID:15980264

Budiarto, K. (2010). Spectral quality affects morphogenesis on Anthurium plantlet during in vitro culture. *Agrivita, 32*, 234–240.

Central Bureau of the CIE. (n.d.). *Vienna, Book Review-13*. doi:10.1002/col.5080130115

Chand, S., & Singh, A. K. (2004). Plant regeneration from encapsulated nodal segments of *Dalbergia sissoo* Roxb.—A timber yielding leguminous tree. *Journal of Plant Physiology, 161*(2), 237–243. doi:10.1078/0176-1617-01036 PMID:15022839

Chang, M.-H., Das, D., Varde, P. V., & Pecht, M. (2012). Light emitting diodes reliability review. *Microelectronics and Reliability*, *52*(5), 762–782. doi:10.1016/j.microrel.2011.07.063

Chang, S. (2008). *LED lighting: high efficiency and environmental benefit*. Samsung Economic Research Institute Economic Focus.

Chen, C. (2003). Development of a heat transfer model for plant tissue culture vessels. *Biosystems Engineering*, *85*(1), 67–77. doi:10.1016/S1537-5110(02)00289-1

Chen, C. (2004). Humidity in Plant Tissue Culture Vessels. *Biosystems Engineering*, *88*(2), 231–241. doi:10.1016/j.biosystemseng.2004.02.007

Chen, C. (2004). Humidity in plant tissue culture vessels. *Biosystems Engineering*, *88*(2), 231–241. doi:10.1016/j.biosystemseng.2004.02.007

Chen, C. (2015). Application of growth models to evaluate the microenvironmentalconditions using tissue culture plantlets of *Phalaenopsis sogo* Yukidian'V3'. *Scientia Horticulturae*, *191*, 25–30. doi:10.1016/j.scienta.2015.05.007

Chen, C., & Chen, J. (2002). Measurement of gas exchange rates for plant tissue culture vessels. *Plant Cell, Tissue and Organ Culture*, *71*(2), 103–109. doi:10.1023/A:1019909520132

Chen, X. L., Xue, X. Z., Guo, W. Z., Wang, L. C., & Qiao, X. J. (2016). Growth and nutritional properties of lettuce affected by mixed irradiation of white and supplemental light provided by light–emitting diode. *Scientia Horticulturae*, *200*, 111–118. doi:10.1016/j.scienta.2016.01.007

Cioć, M., Szewczyk, A., Żupnik, M., Kalisz, A., & Pawłowska, B. (2018). LED lighting affects plant growth, morphogenesis and phytochemical contents of *Myrtus communis* L. in vitro. *Plant Cell, Tissue and Organ Culture*, *132*(3), 433–447. doi:10.100711240-017-1340-2

Colijn-Hooymans, C. M., Bouwer, R., & Dons, J. J. M. (1988). Plant Regeneration from Cucumber Protoplasts. *Plant Cell, Tissue and Organ Culture*, *12*(2), 147–150. doi:10.1007/BF00040077

Courtois, D., & Guern, J. (1980). Temperature response of *Catharanthus roseus* cells cultivated in liquid medium. *Plant Science Letters*, *17*(4), 473–482. doi:10.1016/0304-4211(80)90135-2

Cruz-Cruz, C. A., Gonzalez-Arnao, M. T., & Engelmann, F. (2013). Biotechnology and conservation of plant biodiversity. *Resources*, *2*(2), 73–95. doi:10.3390/resources2020073

Cui, Y., Hahn, E., Kozai, T., & Paek, K. (2000). Number of air exchanges, sucrose concentration, photosynthetic photon flux, and differences in photoperiod and dark period temperatures affect growth of *Rehmannia glutinosa* plantlets *in vitro. Plant Cell, Tissue and Organ Culture*, *62*(3), 219–226. doi:10.1023/A:1006412321864

Darko, E., Heydarizadeh, P., Schoefs, B., & Sabzalian, M. R. (2014). Photosynthesis under artificial light: The shift in primary and secondary metabolism. *Philosophical Transactions of the Royal Society of London. Series B, Biological Sciences*, *369*(1640), 20130243. doi:10.1098/rstb.2013.0243 PMID:24591723

De Fossard, R. A. (1986) Principles of plant tissue culture. In Tissue culture as a plant production system for horticultural crops. doi:10.1007/978-94-009-4444-2_1

Dodds, J. H. (1991). Conservation of plant genetic resources.The need for tissue culture. In J. H. Doods (Ed.), *In vitro Methods for conservation of plant genetic resources* (pp. 1–39). Chapman Hall. doi:10.1007/978-94-011-3072-1_1

Dorion, N., Godin, B., & Bigot, C. (1993). Physiological state and clonal variability effects on low temperature storage of in vitro shoot cultures of elms (*Ulmus sp.*). *Scientia Horticulturae*, *56*(1), 51–59. doi:10.1016/0304-4238(93)90101-U

Duong, T. N., Takamura, T., Watanabe, H., Okamoto, K., & Tanaka, M. (2003). Responses of strawberry plantlets cultured in vitro under superbright red and blue lightemitting diodes (LEDs). *Plant Cell, Tissue and Organ Culture*, *73*(1), 43–52. doi:10.1023/A:1022638508007

Fazal, H., Abbasi, B.H., Ahmad, N., Ali, S.S., Akbar, F., & Kanwal, F. (2016) Correlation of different spectral lights with biomass accumulation and production of antioxidant secondary metabolites in callus cultures of medicinally important *Prunella vulgaris L. J Photochem Photobiol B Biol*, *159*, 1–7.

Frohnmeyer, H., & Staiger, D. (2003). Ultraviolet-B radiation-mediated responses in plants, balancing damage and protection. *Plant Physiology*, *133*(4), 1420–1428. doi:10.1104/pp.103.030049 PMID:14681524

Fukui, H., Nishimoto, K., & Nakamura, M. (1992). Varietal differences in rooting ability of *in vitro* subcultured Japanese persimmon shoots. *Journal of the Japanese Society for Horticultural Science*, *60*(4), 821–825. doi:10.2503/jjshs.60.821

Gabr, A. M. M., & Sawsan, S. S. (2010). Slow growth conservation and molecular characterization of *Deutzia scabra* Thunb. *African Journal of Plant Science*, *4*(10), 409–416.

Garner, N., & Blake, J. (1989). The induction and development of potato microtubers *in vitro* on media free of growth regulating substances. *Annali di Botanica*, *63*(6), 663–674. doi:10.1093/oxfordjournals.aob.a087795

Gazolla, A. P., Curado, F. M. L. M. J., Pedroso, R. C. N., Kellner Filho, C., Fátima Sales, J., Rosa, M., Costa, A. C., Pauletti, P. M., Januário, A. H., & Silva, F. G. (2017). The influence of light quality on phenolic acid and biflavonoid production in *Anacardium othonianum* Rizz. seedlings grown in vitro. *Australian Journal of Crop Science*, *11*(05), 528–534. doi:10.21475/ajcs.17.11.05.p314

Ghashghaie, J., Brenckmann, F., & Saugier, B. (1992). Water relations and growth of rose plants cultured in vitro under various relative humidities. *Plant Cell, Tissue and Organ Culture*, *30*(1), 51–57. doi:10.1007/BF00040000

Gianní, S., & Sottile, F. (2015). In vitro storage of plum germplasm by slow growth. *Horticultural Science (Prague)*, *42*(2), 61–69. doi:10.17221/186/2014-HORTSCI

Gray, D.J., McColley, D.W., & Compton, M.E. (1993) High-frequency somatic embryogenesis from quiescent seed cotyledons of *Cucumis melo* cultivars. *J. Am. Soc. Hort. Sci.*, *118*, 425-432.

Gribble, K. (1999). The influence of relative humidity on vitrification, growth and morphology of *Gypsophila paniculata* L. *Plant Growth Regulation*, *27*(3), 179–188. doi:10.1023/A:1006235229848

Gupta, S. D., & Jatothu, B. (2013). Fundamentals and applications of lightemitting diodes (LEDs) in in vitro plant growth and morphogenesis. *Plant Biotechnology Reports*, *7*(3), 211–220. doi:10.100711816-013-0277-0

Gupta, S. D., & Karmakar, A. (2017). Machine vision based evaluation of im- pact of light emitting diodes (LEDs) on shoot regeneration and the effect of spectral quality on phenolic content and antioxidant capacity in *Swertia chirata. Journal of Photochemistry and Photobiology. B, Biology*, *174*, 162–172. doi:10.1016/j.jphotobiol.2017.07.029 PMID:28779689

Hart, J. W. (2012). General introduction. In J. W. Hart (Ed.), *Light and plant growth, Topics in plant physiology: I, v* (Vol. 1, pp. 4–17). Springer Science & Business Media.

Hausman, J., Neys, O., Kevers, C., & Gaspar, T. (1994). Effect of in vitro storage at 4 °C on survival and proliferation of poplar shoots. *Plant Cell, Tissue and Organ Culture*, *38*(1), 65–67. doi:10.1007/BF00034446

Heo, J., Lee, C., Chakrabarty, D., & Paek, K. Y. (2002). Growth responses of marigold and salvia bedding plants as affected by monochromic or mixture radiation provided by a light-emitting diode (LED). *Plant Growth Regulation*, *38*(3), 225–230. doi:10.1023/A:1021523832488

Hoenecke, M. E., Bula, R. J., & Tibbitts, T. W. (1992). Importance of 'Blue' photon levels for lettuce seedlings grown under red-light-emitting diodes. *HortScience*, *27*(5), 427–430. doi:10.21273/HORTSCI.27.5.427 PMID:11537611

How, F.-W., & Smith, M. A. (2003). Effect of Light/Dark cycling on growth and anthocyanin production of *Ajuga reptens* in callus culture. *J. Agric. Res. China*, *52*, 291–296.

Huan, L. V. T., & Tanaka, M. (2004). Effect of red and blue light-emitting diodes (LEDs) on callus induction, callus proliferation, and protocorm-like body formation from callus in *Cymbidium orchid. Environment Control in Biology*, *42*(1), 57–64. doi:10.2525/ecb1963.42.57

Huang, C., & Chen, C. (2005). Physical properties of culture vessels for plant tissue culture. *Biosystems Engineering*, *91*(4), 501–511. doi:10.1016/j.biosystemseng.2005.05.005

Huché-Thélier, L., Crespel, L., Le Gourrierec, J., Morel, P., Sakr, S., & Leduc, N. (2016). Light signaling and plant responses to blue and UV radiations—Perspectives for applications in horticulture. *Environmental and Experimental Botany*, *121*, 22–38. doi:10.1016/j.envexpbot.2015.06.009

Hung, C. D., Hong, C.-H., Kim, S.-K., Lee, K.-H., Park, J.-Y., Nam, M.-W., Choi, D. H., & Lee, H.-I. (2016). LED light for in vitro and ex vitro efficient growth of economically important highbush blueberry (*Vaccinium corymbosum* L.). *Acta Physiologiae Plantarum*, *38*(6), 1–9. doi:10.100711738-016-2164-0

Huong, L.T.L., Baiocco, M., Huy, B.P., Mezzetti, B., Santilocchi, R., & Rosati, P. (1999). Somatic embryogenesis in Canary Island date palm. *Plant Cell Tissue and Organ Culture, 56*(1), 1-7.

Isah, T. (2015). Adjusments to *in vitro* culture conditions and associated anomalies in plants. *Acta Biologica Cracoviensia. Series; Botanica*, *57*(2), 9–28. doi:10.1515/abcsb-2015-0026

Ishibashi, T., Kimura, S., Furukawa, T., & Sakaguchi, K. (2006). DNA repair mechanisms in UV-B tolerant plants. *Japan Agricultural Research Quarterly*, *40*(2), 107–113. doi:10.6090/jarq.40.107

Islam, M. T., Dembele, D. P., & Keller, E. R. J. (2005). Influence of explant, temperature and different culture vessels on in vitro culture for germplasm maintenance of four mint accessions. *Plant Cell, Tissue and Organ Culture, 81*(2), 123–130. doi:10.100711240-004-3307-3

Jansen, M. A. K. (2002). Ultraviolet-B radiation effects on plants: Induction of morphogenic responses. *Physiologia Plantarum, 116*(3), 423–429. doi:10.1034/j.1399-3054.2002.1160319.x

Jao, R. C., & Fang, W. (2004a). Effects of frequency and duty ratio on the growth of potato plantlets in vitro using light emitting diodes. *HortScience, 39*(2), 375–379. doi:10.21273/HORTSCI.39.2.375

Jao, R.C., & Fang, W. (2004b). *Growth of potato plantlets in vitro is different when provided concurrent versus alternating red and blue light photoperiods*. Academic Press.

Jeong, B. R., & Sivanesan, I. (2015). Direct adventitious shoot regeneration, in vitro flowering, fruiting, secondary metabolite content and anti- oxidant activity of *Scrophularia takesimensis* Nakai. *Plant Cell, Tissue and Organ Culture, 123*(3), 607–618. doi:10.100711240-015-0864-6

Jiao, J., Gai, Q. Y., Fu, Y. J., Ma, W., Yao, L. P., Feng, C., & Xia, X. X. (2015). Optimization of *Astragalus membranaceus* hairy roots induction and culture conditions for augmentation production of astragalosides. *Plant Cell, Tissue and Organ Culture, 120*(3), 1117–1130. doi:10.100711240-014-0668-0

Jishi, T., Kimura, K., Matsuda, R., & Fujiwara, K. (2016). Effects of temporally shifted irradiation of blue and red LED light oncos lettuce growth and morphology. *Scientia Horticulturae, 198*, 227–232. doi:10.1016/j.scienta.2015.12.005

Jung, E. S., Lee, S., Lim, S. H., Ha, S. H., Liu, K. H., & Lee, C. H. (2013). Metabolite profiling of the short–term responses of rice leaves (*Oryza sativa* cv. Ilmi) cultivated under different LED lights and its correlations with antioxidant activities. *Plant Science, 210*, 61–69. doi:10.1016/j.plantsci.2013.05.004 PMID:23849114

Kamińska, M., Gołębiewski, M., Tretyn, A., & Trejgell, A. (2018). Efficient long-term conservation of *Taraxacum pieninicum* synthetic seeds in slow growth conditions. *Plant Cell, Tissue and Organ Culture, 132*(3), 469–478. doi:10.100711240-017-1343-z

Kamińska, M., Skrzpek, E., Wilmowicz, E., Tretyn, A., & Trejgell, A. (2016). Effect of light conditions and ABA on cold storage and post-storage propagation of *Taraxacum pieninicum. Plant Cell, Tissue and Organ Culture, 127*(1), 25–34. doi:10.100711240-016-1026-1

Kanazawa, K., Hashimoto, T., Yoshida, S., Sungwon, P., & Fukuda, S. (2012). Short photoirradiation induces flavonoid synthesis and increases its production in postharvest vegetables. *Journal of Agricultural and Food Chemistry, 60*(17), 4359–4368. doi:10.1021/jf300107s PMID:22506664

Katerova & Todorova. (2011). Effect of enhanced UV-C irradiation on growth, some stress markers, polyamines, IAA and IAA-oxidase activity in pea plants (*Pisum sativum* L.). *Compt rend Acad bulg Sci, 64*(11), 1555–1562.

Katerova, Z., Ivanov, S., Mapelli, S., & Alexieva, V. (2009). Phenols, proline and low-molecular thiol levels in pea (*Pisum sativum*) plants respond differently toward prolonged exposure to ultraviolet-B and ultraviolet-C radiations. *Acta Physiologiae Plantarum, 31*(1), 111–117. doi:10.100711738-008-0208-9

Katerova, Z. I., & Todorova, D. (2009). Endogenous polyamines lessen membrane damages in pea plants provoked by enhanced ultraviolet-C radiation. *Plant Growth Regulation*, *57*(2), 145–152. doi:10.100710725-008-9330-3

Kawka, B., Kwiecień, I., & Ekiert, H. (2017). Influence of culture medium composition and light conditions on the accumulation of bioactive compounds in shoot cultures of *Scutellaria lateriflora* L. (*American skullcap*) grown in vitro. *Applied Biochemistry and Biotechnology*, *183*(4), 1414–1425. doi:10.100712010-017-2508-2 PMID:28573603

Kendrick, R. E., & Kronenberg, G. H. (2012). *Photomorphogenesis in plants*. Springer Science & Business Media Kluwer.

Khan, T., Ullah, M. A., Garros, L., Hano, C., & Abbasi, B. H. (2018). Synergistic effects of melatonin and distinct spectral lights for enhanced production of anti-cancerous compounds in callus cultures of *Fagonia indica*. *Journal of Photochemistry and Photobiology. B, Biology*. Advance online publication. doi:10.1016/j.jphotobiol.2018.10.010 PMID:30482427

Kim, H. H., Goins, G. D., Wheeler, R. M., & Sager, J. C. (2004a). Green light supplementation for enhanced lettuce growth under red and blue light emitting diodes. *HortScience*, *39*(7), 1617–1622. doi:10.21273/HORTSCI.39.7.1617 PMID:15770792

Kim, H. H., Goins, G. D., Wheeler, R. M., & Sager, J. C. (2004b). Stomatal conductance of lettuce grown under or exposed to different light qualities. *Annali di Botanica*, *94*(5), 691–697. doi:10.1093/aob/mch192 PMID:15347557

Kim, S. J., Hahn, E. J., Heo, J. W., & Pae, K. K. Y. (2004c). Effects of LEDs on net photosynthetic rate, growth and leaf stomata of *Chrysanthemum* plantlets in vitro. *Scientia Horticulturae*, *101*(1-2), 143–151. doi:10.1016/j.scienta.2003.10.003

Kintzios, S., Katsouri, E., Peppes, D., & Koulocheri, S. (1998a). Somatic embryogenesis and in vitro secondary metabolite production from common mallow (*Malva sylvestris* L.) collected in Greece. *Acta Horticulturae*, (457), 173–178. doi:10.17660/ActaHortic.1998.457.21

Kintzios, S., Manos, C., & Makri, O. (1998b). Somatic embryogenesis from mature leaves of rosa (*Rosa sp.*). *Plant Cell Reports*, *18*(6), 467–472. doi:10.1007002990050605

Kintzios S, Papanastasiou I, Tourgelis P, Papastellatos C, Georgopoulos V, Drossopoulos J (2002). The effects of light on callus growth and somatic embryogenesis from *Lavandula vera* and *Teucrium chamaedrys*:a preliminary study. *Journal of Herbs Spices & Medicinal Plants Spices & Medicinal Plants*, (2), 223-227.

Kintzios, S., & Taravira, N. (1997). Effect of genotype and light intensity on somatic embryogenesis and plant regeneration in melon (*Cucumis melo* L.). *Plant Breed.*, *116*, 359-362.

Konstas, J., Kintzios, S., Drossopoulos, J., & Sarlis, G. (2003). The effect of light intensity and relative exposure under light on the expression of direct or indirect somatic embryogenesis from common mallow (*Malva sylvestris* L.). *Acta Horticulturae*, (597), 315–319. doi:10.17660/ActaHortic.2003.597.45

Kozai, T. (2013). Resource use efficiency of closed plant production system with artificial light: Concept, estimation and application to plant factory. *Proc Japn Acad Series B, 89*(10), 447–461. doi:10.2183/pjab.89.447 PMID:24334509

Kozai, T. (2016). Why LED lighting for urban agriculture? In T. Kozai, K. Fujiwara, & E. Runkle (Eds.), *LED lighting for urban agriculture* (pp. 3–18). Springer. doi:10.1007/978-981-10-1848-0_1

Kozai, T., & Smith, M. A. L. (1995) Environmental control in plant tissue culture-general overview. In Automation and Environmental Control in Plant Tissue Culture. Kluwer Academic Publishers. doi:10.1007/978-94-015-8461-6_14

Krewzaler, F., & Hahlbrock, K. (1973). Flavonoid glycosides from illuminated cell suspension cultures of *Petroselinum hortense. Phytochemistry, 12*(5), 1149–1152. doi:10.1016/0031-9422(73)85031-9

Ku, K.-L., Chang, P.-S., Cheng, Y.-C., & Lien, C.-Y. (2005). Production of stilbenoids from the callus of *Arachis hypogaea*: A novel source of anticancer compound piceatannol. *Journal of Agricultural and Food Chemistry, 53*(10), 3877–3881. doi:10.1021/jf050242o PMID:15884811

Kuo, T. C. Y., Chen, C. H., Chen, S. H., Lu, I. H., Chu, M. J., Huang, L. C., Lin, C. Y., Chen, C. Y., Lo, H. F., Jeng, S. T., & Chen, L. F. O. (2015). The effect of red light and far-red light conditions on secondary metabolism in agarwood. *BMC Plant Biology, 15*(1), 139. doi:10.118612870-015-0537-y PMID:26067652

Lata, H., Moraes, R. M., Bertoni, B., & Pereira, A. M. S. (2010). *In vitro* germplasm conservation of *Podophyllum peltatum* L. under slow growth conditions. *In Vitro Cellular & Developmental Biology. Plant, 46*(1), 22–27. doi:10.100711627-009-9243-5

Lavola, A., Aphalo, P. J., Lahti, M., & Julkunen-Tiitto, R. (2003). Nutrient availability and the effect of increasing UV-B radiation on secondary plant compounds in Scots pine. *Environmental and Experimental Botany, 49*(1), 49–60. doi:10.1016/S0098-8472(02)00057-6

Li, H., Xu, Z., & Tang, C. (2010). Effect of light-emitting diodes on growth and morphogenesis of upland cotton (*Gossypium hirsutum* L.) plantlets in vitro. *Plant Cell, Tissue and Organ Culture, 103*(2), 155–163. doi:10.100711240-010-9763-z

Lian, M. L., Murthy, H. N., & Paek, K. Y. (2002). Effects of light emitting diodes (LED) on the in vitro induction and growth of bulblets of *Lilium* oriental hybrid 'Pesaro'. *Scientia Horticulturae, 94*(3-4), 365–370. doi:10.1016/S0304-4238(01)00385-5

Lin, Y., Li, J., Li, B., He, T., & Chun, Z. (2010). Effects of light quality on growth and development of protocorm-like bodies of *Dendrobium officinale* in vitro. *Plant Cell, Tissue and Organ Culture, 105*(3), 329–335. doi:10.100711240-010-9871-9

Liu, A. R., Chen, S. C., Lin, X. M., Wu, S. Y., Xu, T., Cai, F. M., & Raesh, J. (2010). Endophytic Pestalotiopsis species spp. associated with plants of Palmae, Rhizophoraceae, Planchonellae and Podocarpaceae in Hainan, China. *African Journal of Microbiological Research, 4*, 2661–2669.

Liu, C. Z., Guo, C., Wang, F. O., & Ouyang, F. (2003). Factors influencing artemisinin production from shoot cultures of *Artemisia annua* L. *World Journal of Microbiology & Biotechnology, 19*(5), 535–538. doi:10.1023/A:1025158416832

Loyall, L., Uchida, K., Furuya, M., & Frohnmeyer, H. (2000). Gluthatione and a UV light induced gluthatione S-transferase are involved in signaling to chalcone synthase in cell cultures. *The Plant Cell, 12*, 1939–1950. PMID:11041888

Lumsden, P., Nicholas, J., & Davies, W. (1994). *Physiology, Growth, and Development of Plants in Culture*. Kluwer Academic Publishers. doi:10.1007/978-94-011-0790-7

Majada, J. P., Sierra, M. I., & Sanchez-Tames, R. (2001). Air exchange rate affects the in vitro developed leaf cuticle of carnation. *Scientia Horticulturae, 87*(1), 121–130. doi:10.1016/S0304-4238(00)00162-X

Manivannan, A., Soundararajan, P., Halimah, N., Ko, C. H., & Jeong, B. R. (2015). Blue LED light enhances growth, phytochemical contents, and antioxidant enzyme activities of *Rehmannia glutinosa* cultured in vitro. *Horticulture, Environment and Biotechnology, 56*(1), 105–113. doi:10.100713580-015-0114-1

Massa, G. D., Kim, H. H., Wheeler, R. M., & Mitchell, C. A. (2008). Plant productivity in response to LED lighting. *HortScience, 43*(7), 1951–1956. doi:10.21273/HORTSCI.43.7.1951

Matsuura, H. N., de Costa, F., Yendo, A. C. A., & Fett-Neto, A. G. (2013). Photoelicitation of bioactive secondary metabolites by ultraviolet radiation: mechanisms, strategies, and applications. In S. Chandra, H. Lata, & A. Varma (Eds.), *Biotechnology for Medicinal Plants* (pp. 171–190). Springer-Verlag. doi:10.1007/978-3-642-29974-2_7

Mengxi, L., Zhigang, X., Yang, Y., & Yijie, F. (2011). Effects of different spectral lights on Oncidium PLBs induction, proliferation, and plant regeneration. *Plant Cell, Tissue and Organ Culture, 106*(1), 1–10. doi:10.100711240-010-9887-1

Menin, B., Moglia, A., Comino, C., Hakkert, J. C., Lanteri, S., & Beekwilder, J. (2013). In vitro callus-induction in globe artichoke (*Cynara cardunculus* L. var. scolymus) as a system for the production of caffeoylquinic acids. *The Journal of Horticultural Science & Biotechnology, 88*(5), 537–542. doi:10.1080/14620316.2013.11513003

Morrow, R. C. (2008). LED lighting in horticulture. *HortScience, 43*(7), 1949–1950. doi:10.21273/HORTSCI.43.7.1947

Nguyen, L., Hai, F. I., Yang, S., Kang, J., Leusch, F., Roddick, F., Price, W. E., & Nghiem, L. D. (2014). Removal of pharmaceuticals, steroid hormones, phytoestrogens, UV-filters, industrial chemicals and pesticides by *Trametes versicolor*: Role of biosorption and biodegradation. *International Biodeterioration & Biodegradation, 88*(March), 169–175. doi:10.1016/j.ibiod.2013.12.017

Nhut, D. T., Hong, L. T. A., Watanabe, H., Goi, M., & Tanaka, M. (2000). Growth of banana plantlets cultured in vitro under red and blue lightemitting diode (LED) irradiation source. *Acta Horticulturae*, (575), 7–23.

Nhut, D. T., Huy, N. P., Tai, N. T., Nam, N. B., Luan, V. Q., Hien, V. T., Tung, H. T., Vinh, B. T., & Luan, T. C. (2015). Light-emitting diodes and their potential in callus growth, plantlet development and saponin accumulation during somatic embryogenesis of *Panax vietnamensis* Ha et Grushv. *Biotechnology, Biotechnological Equipment*, *29*(2), 299–308. doi:10.1080/13102818.2014.1000210 PMID:26019644

Nhut, D. T., & Nam, N. B. (2010). Light emitting diodes (LEDs): an artificial lighting source for biological studies. *Proceedings of the 3rd international conference of the development of BME in Vietnam*. 10.1007/978-3-642-12020-6_33

Nhut, D. T., Takamura, T., Watanabe, H., Okamoto, K., & Tanaka, M. (2003). Responses of strawberry plantlets cultured in vitro under superbright red and blue light-emitting diodes (LED). *Plant Cell, Tissue and Organ Culture*, *73*(1), 43–52. doi:10.1023/A:1022638508007

Nhut, T. D., Loan, P. T. L., Thao, P. T. L., Suong, T. T. N., & Lien, H. V. (2006). Effects of some artificial lighting sources on the photomorphogenesis of callus induced embryo and the plantlet regeneration of *Phalaenopsis amabilis* and its applications in micropropagation. *Proceedings of International Workshop on Agricultural Biotechnology*, 20–21.

Nishio, J. N. (2000). Why are higher plants green? Evolution of the higher plant photosynthetic pigment complement. *Plant, Cell & Environment*, *23*(6), 539–548. doi:10.1046/j.1365-3040.2000.00563.x

Pandey, N., & Pandey-Rai, S. (2014a). Short-term UV-B radiation-mediated transcriptional responses and altered secondary metabolism of in vitro propagated plantlets of *Artemisia annua* L. *Plant Cell, Tissue and Organ Culture*, *116*(3), 371–385. doi:10.100711240-013-0413-0

Pandey, N., & Pandey-Rai, S. (2014b). Modulations of physiological responses and possible involvement of defense-related secondary metabolites in acclimation of *Artemisia annua* L. against short-term UV-B radiation. *Planta*, *240*(3), 611–627. doi:10.100700425-014-2114-2 PMID:25023630

Pandino, G., Meneghini, M., Tavazza, R., Lombardo, S., & Mauromicale, G. (2017). Phytochemicals accumulation and antioxidant activity in callus and suspension cultures of *Cynara scolymus* L. *Plant Cell, Tissue and Organ Culture*, *128*(1), 223–230. doi:10.100711240-016-1102-6

Park, S. Y., Lee, J. G., Cho, H. S., Seong, E. S., Kim, H. Y., Yu, C. Y., & Kim, J. K. (2013). Metabolite profiling approach for assessing the effects of colored light–emitting diode lighting on the adventitious roots of ginseng (*Panax ginseng* C. A Mayer). *Plant Omics*, *6*, 224–230.

Pérez-Tornero, O., Egea, J., Olmos, E., & Burgos, L. (2001). Control of hyperhydricity in micropropagated apricot cultivars. *In Vitro Cellular & Developmental Biology – Plant, 37*(2), 250–254.

Perez-Tornero, O., Egea, J., Olmos, E., & Burgos, L. (2001). Control of hyperhydrcity in micropropagated apricot cultivars. *In Vitro Cellular & Developmental Biology. Plant*, *37*(2), 250–254. doi:10.100711627-001-0044-8

Poudel, P. R., Kataoka, I., & Mochioka, R. (2008). Effect of red-and blue-light-emitting diodes on growth and morphogenesis of grapes. *Plant Cell, Tissue and Organ Culture*, *92*(2), 147–153. doi:10.100711240-007-9317-1

Preece, J. E., & Sutter, E. G. (1990). Acclimatization of micropropagated plants to the greenhouse and field. In P. C. Debergh & R. H. Zimmerman (Eds.), *Micropropagation: Technology and Application* (pp. 71–93). Kluwer Academic Publishers.

Pruski, K. (2001). *Micropropagation technology in early phases of commercial seed potato production* (PhD Thesis). Wageningen University, Wageningen, The Netherlands.

Rai, R., Meena, R. P., Smita, S. S., Shukla, A., Rai, S. K., & Pandey-Rai, S. (2011b). UV-b and UV-C pre-treatments induce physilogical changes and artemisinin biosynthesis in *Artemisia annua* L.- An anti-malarial plant. *Journal of Photochemistry and Photobiology. B, Biology, 105*(3), 216–225. doi:10.1016/j.jphotobiol.2011.09.004 PMID:22019553

Rai, R., Pandey, S., & Rai, S. P. (2011a). Arsenic-induced changes in morphological, physiological, and biochemical attributes and artemisinin biosynthesis in *Artemisia annua*, an antimalarial plant. *Ecotoxicology (London, England), 20*(8), 1900–1913. doi:10.100710646-011-0728-8 PMID:21710305

Ramawat, K. G., & Mathur, M. (2007). Factors affecting production of secondary metabolites. In K. G. Ramawat & J. M. Merillon (Eds.), *Biotechnology secondary metabolites* (pp. 59–101). Sci Pub. doi:10.1201/b10756-4

Rao, R. S., & Ravishankar, G. A. (2002). Plant cell cultures: Chemical factories of secondary metabolites. *Biotechnology Advances, 20*(2), 101–153. doi:10.1016/S0734-9750(02)00007-1 PMID:14538059

Romano, A., & Martins-Loução, M. A. (1999). In vitro cold storage of cork oak shoot cultures. *Plant Cell, Tissue and Organ Culture, 591*(2), 155–157. doi:10.1023/A:1006465706905

Roxas, N. J. L., Tashiro, Y., Miyazaki, S., Isshiki, S., & Takeshita, A. (1995). Meiosis and Pollen Fertility in Higo Chrysanthemum *Dendranthema grandiflorum* (Ramat.) Kitam. *Journal of the Japanese Society for Horticultural Science, 64*(1), 161–168. doi:10.2503/jjshs.64.161

Sallanon, H., & Maziere, Y. (1992). Influence of growth room and vessel humidity on the in vitro development of rose plants. *Plant Cell, Tissue and Organ Culture, 30*(2), 121–125. doi:10.1007/BF00034305

Schuerger, A. C., Brown, C. S., & Stryjewski, E. C. (1997). Anatomical features of pepper plants (*Capsium annuum* L.) grown under red light emitting diodes supplemented with blue or far-red light. *Annali di Botanica, 79*(3), 273–282. doi:10.1006/anbo.1996.0341 PMID:11540425

Seabrook, J. E. A., Coleman, S., & Levy, D. (1993). Effect of photoperiod on *in vitro* tuberization of potato (*Solanum tuberosum* L.). *Plant Cell, Tissue and Organ Culture, 34*(1), 43–51. doi:10.1007/BF00048462

Shama, G., & Alderson, P. (2005). UV hormesis in fruits: A concept ripe for commercialization. *Trends in Food Science & TechnologyVolume, 16*(4), 128–136. doi:10.1016/j.tifs.2004.10.001

Shin, S. K., Murthy, H. N., Heo, J. W., & Paek, K. Y. (2004). Induction of betalain pigmentation in hairy roots of red beet under different radiation sources. *Biologia Plantarum, 47*(1), 149–152. doi:10.1023/A:1027313805930

Shin, S. K., Murthy, N. H., Heo, W. J., Hahn, J. E., & Paek, Y. K. (2008). The effect of light quality on the growth and development of in vitro cultured Doritaenopsis plants. *Acta Physiologiae Plantarum, 30*(3), 339–343. doi:10.100711738-007-0128-0

Shohael, A. M., Ali, M. B., Yu, K. W., Hahn, E. J., & Paek, K. Y. (2006). Effect of temperature on secondary metabolites production and antioxidant enzyme activities in Eleutherococcus senticosus somatic embryos. *Plant Cell, Tissue and Organ Culture*, *85*(2), 219–228. doi:10.100711240-005-9075-x

Shukla, M.R., Amritpal, S.S., Kevin, P., Praveen, K.S., & Maxwell, A.P.J. (2017). Application of 3D printing to prototype and develop novel plant tissue culture systems. *Plant Methods.*, 13:6 1:10.

Silva, S. T., Bertolucci, S. K. V., da Cunha, S. H. B., Lazzarini, L. E. S., Tavares, M. C., & Pinto, J. E. B. P. (2017). Effect of light and natural ventilation systems on the growth parameters and carvacrol content in the in vitro cultures of *Plectranthus amboinicus* (Lour.) Spreng. *Plant Cell, Tissue and Organ Culture*, *129*(3), 501–510. doi:10.100711240-017-1195-6

Singh, A. K., Varshney, R., Sharma, M., Agarwal, S. S., & Bansal, K. C. (2006). Regeneration of plants from alginate-encapsulated shoot tips of *Withania somnifera* (L.)Dunal, a medicinally important plant species. *Journal of Plant Physiology*, *163*(2), 220–223. doi:10.1016/j.jplph.2005.06.013 PMID:16399013

Singh, S. K., Rai, M. K., Asthana, P., & Sahoo, L. (2010). Alginate-encapsulation of nodal segments for propagation, short-term conservation and germplasm exchange and distribution of *Eclipta alba* (L.). *Acta Physiologiae Plantarum*, *32*(3), 607–610. doi:10.100711738-009-0444-7

Smetanska, I. (2008). Production of secondary metabolites using plant cell cultures. *Advances in Biochemical Engineering/Biotechnology*, *111*, 187–228. doi:10.1007/10_2008_103 PMID:18594786

Smith, K. (2013). *The science of photobiology*. Springer Science & Business Media.

Smith, M. A. L., & Spomer, L. (1995). Vessels, gels, liquid media, and support systems. In J. Aitken-Chistie (Ed.), *Automation and Environmental Control in Plant Tissue Culture* (pp. 371–404). Kluwer Academic Publishers. doi:10.1007/978-94-015-8461-6_16

Son, S. H., Chun, Y. W., & Hall, R. B. (1991). Cold storage of in vitro cultures of hybrid poplar shoots (*Populus alba* L. x *P. grandidentata* Michx.). *Plant Cell, Tissue and Organ Culture*, *27*(2), 161–168. doi:10.1007/BF00041285

Stearns, E. I. (1987). Commission Internationale de l'Eclairage, Colorimetry (2nd ed.). Publication CIE No. 15.2.

Summart, J., Panichajakul, S., Prathepha, P., & Thanonkeo, P. (2008). Callus Induction and Influence of Culture Condition and Culture Medium on Growth of Thai Aromatic Rice, Khao Dawk Mali 105. *Cell CultureWorld Applied Sciences Journal*, *5*(2), 246–251.

Szopa, A., & Ekiert, H. (2016). The importance of applied light quality on the production of lignans and phenolic acids in *Schisandra chinensis* (Turcz.) Baill.cultures in vitro. *Plant Cell, Tissue and Organ Culture*, *127*(1), 115–121. doi:10.100711240-016-1034-1

Szopa, A., Ekiert, H., Szewczyk, A., & Fugas, E. (2012). The production of therapeutically important phenolic acids and furanocoumarins in in vitro cultures of *Ruta graveolens* and *Ruta graveolens* ssp. divaricata cultured under different light conditions. *Plant Cell, Tissue and Organ Culture*, *110*, 329–336. doi:10.100711240-012-0154-5

Tanaka, K., Fujiwara, K., & Kozai, T. (1992). Effects of relative humidity in the culture vessels on the evaporation and net photosynthetic rates of potato plantlets in vitro. *Acta Horticulturae*, (319), 59–64. doi:10.17660/ActaHortic.1992.319.3

Tariq, U., Ali, M., & Abbasi, B. H. (2014, January 5). Morphogenic and biochemical variations under different spectral lights in callus cultures of *Artemisia absinthium* L. *Journal of Photochemistry and Photobiology. B, Biology*, *130*, 264–271. doi:10.1016/j.jphotobiol.2013.11.026 PMID:24362323

Taulavuoria, K., Hyökya, V., Oksanena, J., Taulavuoria, E., & Tiittob, R. J. (2016). Species-specific differences in synthesis of flavonoids and phenolic acids under increasing periods of enhanced blue light. *Environmental and Experimental Botany*, *121*, 145–150. doi:10.1016/j.envexpbot.2015.04.002

Tisserat, B. (1979). Propagation of Date Palm (*Phoenix dactylifera* L.) *in vitro*. *Journal of Experimental Botany*, *30*(119), 1275–1283. doi:10.1093/jxb/30.6.1275

Toivonen, L., Laakso, S., & Rosenguist, H. (1992). The effect of temperature on growth, indole alkaloid accumulation and lipid composition of *Catharanths roseus* cell suspension culture. *Plant Cell Reports*, *11*, 390–394. PMID:24201539

Trejgell, A., Kamińska, M., & Tretyn, A. (2015). In vitro slow growth storage of Senecio macrophyllus shoots. *Acta Physiologiae Plantarum*, *37*(11), 234. doi:10.100711738-015-1983-8

Waide, P. (2010). *Information paper. Phase out of incandescent lamps. Implications for international supply and demand for regulatory compliant lamps*. https://www.iea.org/publications/freepublications/publication/phase_out.pdf

Wang, H., Ma, L. G., Li, J. M., Zhao, H. Y., & Deng, X. W. (2001). Direct interaction of Arabidopsis cryptochomes with COP1 in light control development. *Science*, *294*(5540), 154–158. doi:10.1126cience.1063630 PMID:11509693

Wang, P., & Hu, C. (1982). In vitro mass suberisation and virus-free seed potato production in Taiwan. *American Potato Journal*, *59*(1), 33–37. doi:10.1007/BF02854881

Yeh, H., & Chung, J. P. (2009). High-brightness LEDs-energy efficient lighting sources and their potential in indoor plant cultivation. *Renewable & Sustainable Energy Reviews*, *13*(8), 2175–2180. doi:10.1016/j.rser.2009.01.027

Younas, M., Drouet, S., Nadeem, M., Giglioli-Guivarc'h, N., Hano, C., & Abbasi, B. H. (2018). Differential accumulation of silymarin induced by exposure of *Silybum marianum* L. callus cultures to several spectres of monochromatic lights. *Journal of Photochemistry and Photobiology. B, Biology*, *184*, 61–70. doi:10.1016/j.jphotobiol.2018.05.018 PMID:29803074

Yu, K. W., Murthy, H. N., Hahn, E. J., & Paek, K. Y. (2005). Ginsenoside production by hairy root cultures of Panax ginseng: Influence of temperature and light quality. *Biochemical Engineering Journal*, *23*(1), 53–56. doi:10.1016/j.bej.2004.07.001

Zhang, Z. Z., Che, X. N., Pan, Q. H., Li, X. X., & Duan, C. Q. (2013). Transcriptional activation of flavan-3-ols biosynthesis in grape berries by UV irradiation depending on developmental stage. *Plant Science*, *208*, 64–74. doi:10.1016/j.plantsci.2013.03.013 PMID:23683931

Zimmerman, R. H. (1995). Environmental effects and their control in plant tissue culture review. *Acta Horticulturae*, (393), 11–14. doi:10.17660/ActaHortic.1995.393.1

Ziv, M. (1990). Vitrification: morphological and physiological disorders of in vitro plants. In P. C. Debergh & R. H. Zimmerman (Eds.), *Micropropagation: Technology and Application* (pp. 45–69). Kluwer Academic Publishers.

Zu, Y. G., Wei, X. X., Yu, J. H., Li, D. W., Pang, H. H., & Tong, L. (2011). Responses in the physiology and biochemistry of Korean pine (*Pinus koraiensis*) under supplementary UV-B radiation. *Photosynthetica*, *49*(3), 448–458. doi:10.100711099-011-0057-4

Chapter 14

Role of Biotechnology in Plant Preservation for Food Security:
In Situ and Ex Situ Preservation

Amal A. El-Ashry
National Research Centre, Egypt

Ahmed M. M. Gabr
National Research Centre, Egypt

Mohamed K. El Bahr
National Research Centre, Egypt

ABSTRACT

Overpopulation and the consequences of urbanization and reduction of agricultural lands represent the most important challenges that face scientists nowadays. In addition, extinction of specific species or reduction in their number occurs continuously in different places of the world at a rapid rate. These challenges urge scientists to use the biotechnological techniques to secure food and to alleviate the risk of loss of the genetic variability of cultivated plants as a result of environmental changes and human practices. These techniques are based on preservation of the genetic materials for long periods. Plants can be stored either in vivo or in vitro. The plant preservation includes in situ and ex situ. One form of the ex situ plant preservation is the in vitro plant preservation. There are different in vitro preservation techniques. However, the two main approaches of in vitro preservation of plants germplasm are slowing the growth and crysoperservation. The former technique could be achieved through either modifying the culture medium or reducing temperature and/or light intensity. The latter is taking place through storing the species between -79 and -196°C, the temperature of liquid nitrogen. Each approach includes several techniques that will be thoroughly discussed with examples in this chapter.

DOI: 10.4018/978-1-7998-5000-7.ch014

INTRODUCTION

Plant biotechnology includes all the promising approaches to develop new plant properties and varieties. These new varieties could be produced in a large scale to be commercialized and to fulfill the human needs (Bhatia, 2015). In additions, there are numerous techniques and routes that are utilized by plant biotechnology for genetic control and screening to improve plant/plant products (Kalia, 2018).

Plant genetic resources describe the plants genetic material of potential value in the present and future. In general speaking, it is specific for crop plants and their wild relatives, yet it can be extended to include all plant species (Maxted & Kell, 2003).

Plant preservation is usually concern with the safety of the plant from any form of injury, destruction or extinction. Preservation of plant genetic resources has become extremely important for improvement of crops so as to face the increasing depletion of natural resources. Preservation of plant genetic resources is necessary for security of food and agro – biodiversity. Genetic diversity provides new and more productive crops which are resistant to different stresses (Rao, 2004).

Preservation of plant genetic resources can be performed *in vivo* or *in vitro,* and either *in- situ* or *ex- situ. In- situ* approach depends on keeping of the plant species in their natural environment, besides maintaining the cultivated species where they have advanced their distinguishing characteristics such as farms (United Nations Conference on Environment and Development, 1992). However, in- situ strategy suffers from some drawbacks as for example, a decline of species and biodiversity loss due to populations and ecosystem composition that result in destroying their habitat. Therefore, the *in- situ* strategies alone are insufficient for saving endangered species.

The *ex- situ* strategies may solve these problems. The *ex -situ* strategies tend to preserve the biological material away from their natural habitats (United Nations Conference on Environment and Development, 1992). The ex-situ strategies include different approaches depends on either they use the whole plant to be conserved like botanical gardens, and field banks or they use parts from the plants to be conserved like seed banks, gene banks or *in- vitro* preservation. There are many botanical gardens all over the word as an example for them; the Royal Botanical Garden in Sidney (Fig. 1). It is considered one of the major botanical gardens and it is located in the middle of Sydney at Australia. It was opened in 1816; the garden is considered the oldest scientific institution in Australia and one of the most important botanical institutions in the world. It covers about 74 acres.

As for the gene banks, the Svalbard Global Seed Vault could be considered as the most famous example for them (Fig. 2). It occupies the deep inside the Norwegian mountain, has the capacity to store about 4.5 million samples. Currently, it holds one million samples. They are collecting samples from all the countries around the world.

Gene banks conserve genetic resources. The most important activity in a gene bank is to treat the samples in a way which will lead for prolonging its viability as long as possible while ensuring its genetic stability. The samples are monitored to be sure that they are not losing viability. Plant samples must periodically regenerated, and new seeds harvested because, even under the optimum conditions of conservation, samples will eventually be lost. However, gene banks are intended to ensure that the genetic resources are adequately collected and used. This means to make sure that the collections are characterized and documented. Gene banks must be able to serve the healthy samples to those who need them (Fig. 3).

Figure 1. The Royal Botanical Garden, Sydney, Australia

Figure 2. The Svalbard Global Seed Vault at the Norwegian Mountain

Figure 3. Schematic diagram of the steps, which are included in gene banks

Ex situ conservation is highly promising route for preserving plants and in some cases it is the only method to keep certain species (Ramsay et al., 2000). We could conclude that in situ and ex situ strategies are complementary approaches, since they offer several alternatives for conservation. While, the choice of the most desirable strategy between them for conservation, usually relies on several factors such as, the biological nature of the species and the feasibility of applying the methods (Altman et al., 2012).

In the past few decades, different *invitro* preservation techniques have substituted for field gene banks based on the required period of storage (Ashmore, 1997; Engelmann, 1991). These techniques usually work well for conserving of species, which have no seeds at all, produce recalcitrant (non- orthodox) seeds, and for plants which are spread vegetatively to preserve specific genotypes (Newburry et al., 1997). *In vitro* preservation process involves storing the explants in a sterile and a pathogen-free environment. Explants miniaturization offers an advantage of reducing the space needed to preserve the plant thus it maintains the germplasm collections with lower labor cost (Newburry et al., 1997). *In vitro* preservation of plants germplasm includes two types. Type I, depends on slowing down the growth by changing the culture medium or decreasing temperature (Withers, 1991). Type II is concerned with freezing the samples, which is called cryopreservation between -79 and -196 °C (Dodds et al., 1991). Cryopreservation allows for preservation of the valued germplasm for long time.

IN VITRO PLANT PRESERVATION USING MINIMAL GROWTH CONDITIONS

In this strategy, preservation is performed via decreasing cell proliferation therefore decreasing the rate of cell growth and consequently extends the lifetime without any genetic alterations. The storage via lowering the frequency of cell growth is performed by decreasing the metabolism of plants; thus increasing the lifetime of the species for months and in some cases for years. In additions, the reduction mechanism of plants metabolism decreasing the storage cost through minimizing the number of subcultures with the benefits of avoiding more contamination that might take place during preparing it. Reduction of the plant metabolism can be achieved *via* several methods among them (a) lowering the temperature of storage medium, (b) decreasing of the light intensity, (c) controlling the water content through addition of osmotic materials i.e. mannitol or sucrose, and (d) adding a growth retardant to the storage environment such as gibberellin biosynthesis inhibitors (Grout & Grout, 1995; Rademacher, 2000; Silva & Scherwinski-Pereira, 2011). These methods are applied individually or collectively. The most conventional way for slowing the growth in order to preserve the plant is by reducing the temperature and light as well. This reduction has a subtle influence on decreasing the respiration, controlling the amount of water loss, the production of ethylene which leads to growth reduction (Jain et al., 2010). Many factors affects the efficiency of the in vitro slow growth storage technique such as the type and physiological state of the explants in the initial stage of storage and culture vessel type and volume, in additions to the type of termination of the culture vessel (8). In the following sections, we will discuss in details the mechanisms of reducing the plant metabolism.

In Vitro Plant Preservation Using Low Temperature

Reduction of the incubation temperature is one of the major tissue culture parameters applied for plant preservation through slowing growth. Under these circumstances, unsaturated lipids are accumulated on the cell membrane resulting in their thickening and impede cell proliferation and their propagation (Newburry et al., 1997). Cold preservation is usually performed in the temperature range of 0 - 5 °c for the cold tolerant species (Ashmore, 1997). While, a higher storage temperature is needed for the tropical and subtropical plants, which have less tolerance towards low temperature (Ford-Lloyd & Jackson, 1986). As an example for low temperature used for minimizing growth, cultured grape cells and organs, which have been stored at temperature (2-10 °C) (Frankel et al., 1975). Also, apple germplasm was successfully stored at 1 - 4 °C (19).Moreover, *in- vitro* raspberry (*Rubus spp.*) and mint (*Mentha spp.*) cultures were kept in twelve hours photoperiod at 4°C(Reed, 1993; Reed, 1999). Furthermore, pear (*Pyrus communis*) was safely maintained at temperature between 4 - 10 °c (Dereuddre et al., 1990). Interestingly, Withers, (Withers, 1982) successfully preserved date palm cultures for 12 - 18 weeks at 5 ° C. Bekheetetal. (Bekheet et al., 2001) stored shoot bud and callus culture of date palm at 5 ºC in dark for 12 months. Later on, Boissonet al., (Boison et al., 2012) had preserved plant cell cultures of both Sycamore (*Acer pseudoplatanus*) and Arabidopsis as cell suspensions cultures on a phosphate free nutrient medium at 5°Cfor six months without subculture. Also, Bekheet et al., (Bekheet et al., 2016) reported that using the low temperature (5°C) is more preferred for the healthy storage of *in vitro* grown jojoba shoot lets till nine months. However, kiwi shoots are conserved at 8 °c (**27**). In the contrary, some plants have no tolerance for cold they needs higher temperature for preservation as for example cassava plantlets are stored at a temperature greater than 20 °c (Roca et al., 1984). In this respect, Banerjee and De Langhe (Banerjee & de Langhe, 1985) Musa *in vitro* plants was saved at 15°C for 15 months. Oil palm plantlets and somatic embryos cannot sustain upon lowering the temperature than 18 °c (Corbineau et al., 1990). Coffee also was preserved at relatively high temperature (20°C) (Bertrand-Desbrunais et al., 1992). Also, Hassan (Hassan, 2002) succeeded in storing somatic embryos of date palm at 18 °C or 27 ° C for 2 -10 months.

In Vitro Plant Preservation at Darkness Condition

This strategy depends on reducing the light intensity and in some cases store in totally darkness. Controlling the light intensity is commonly used in parallel to cold strategy for in vitro preservation of plant germplasm (Engelmann, 1991). Intervals between subcultures can be increased by keeping cultures in a weak light intensity or in the dark and a temperature which is less than the optimal temperature for growth. Whether light or dark storage is preferable and the suitable temperature for storage depends on genotypes (George, 1993). This was obvious in the cases of preservation of banana and plantain, which were preserved for 12 - 15 months at temperature 15 ° C under light intensity of 1000 lux (Banerjee & de Langhe, 1985), while below temperature of 15 ° C, a rapid deterioration occurred causing the death of the culture. It is worth to mention here that variation of the light intensity is important for the plant. Wanas (Wanas, 1999) reported that the leafy or defoliated shoot tips of Citrus spp. cultured on multiplication medium were maintained at 24, 10 and 8 ° C under low light intensity or in complete darkness. The cultures remained viable at 24 ° C for 12 -16 months without any need for subculture. Read and Preece (Read & Preece, 2003) demonstrated the importance of both light and darkness condition for the plant photosynthesis processes to produce the constituents needed for growing. Light is important for the plants to convert its energy into chemical processes, which are converted in to glucose under dark conditions. Glucose is considered as the fuel for the activities needed for the plant growth. Therefore, the performance of the cultures is highly influenced by the light intensity during storage (Romano & Martins, 1999).

In Vitro Preservation by Addition of Osmotica to Culture Media

Osmotic materials are compounds that reduce the water potential of plant cells when added to the culture media causing its slow growth and increase the life time of preservation as seen for a lot of plant species (Shibli, 1991; Wilson et al., 2000).Although, Zimmermann (Zimmermann, 1978)assumed that, the addition of high concentrations of osmotic materials to the culture medium might decrease the turgor pressure that drives cell growth and expansion. According to his hypothesis, this stress would prevent the growth of callus and resist the development of shoots (Brown et al., 1979).

Examples for the osmatic agents that are used for the *in vitro* preservation of grown tissues are the mannitol, sucrose and sorbitol (Shibli et al., 1992).

Sucrose is present in most tissue culture media it acts as source of carbon and energy source. Tarmizi et al., 1993) claimed that using high levels of sucrose could help in reducing growth through decreasing the requirements for subculture of oil plant polyembryonic cultures. This means that the cultures could be maintained for six months duration without the need for subculture. For example, addition of 30 g /l of sucrose in to the tobacco callus culture medium, has declined the cultures ability to produce shoots and increasing the sucrose content to 150 g /l has disabled the shoots production (Brown et al., 1979). In another example, elevating the sucrose level in the media of wild pear and potato has resulted in reduction in the elongation of the microplant (Sarkar & Naik, 1998; Tahtamouni & Shibli, 1999).

Sugar alcohols like (mannitol or sorbitol) are not often metabolized by plant tissues and thus, they cannot be used as carbon sources. For that, they are regularly used as osmotic agents to change the water potential of plant cells. Therefore, sucrose is added together with sugar alcohols to provide the required energy for the tissues. Also, it was noticed that addition of sugar alcohols to the culture medium diminish the boron content (George, 1993). As mentioned above, it was thought that sugar alcohols are not metabolized by cells; however, in recent years it was believed that sugar alcohols exhibit a great antioxidant activity (Williamson et al., 2002).

Mannitol is used as an effective osmotic agent and acts as desiccation. It is produced by a number of plants, which can metabolized it (Moges et al., 2003; Tahtamouni & Shibli, 1999). Mannitol addition to the tobacco cultures prevent the tissue growth (Brown et al., 1979). In additions, it decreases the shoot growth of chrysanthemum (Shibli et al., 1992) and bitter almond (Shibli et al., 1999).

Sorbitol is another type of the sugar alcohol that reduces the shoot growth when added to the culture media of the in *vitro* grown bitter almond (Shibli et al., 1999).Addition of high contents of sucrose, and mannitol or sorbitol at room temperature was found to significantly decrease the growth of bitter almond micro shoots and extended the subculture intervals to four months**(Shibli et al., 1997).**Also, for the *in vitro* preservation of wild pear micro- shoots (*Pyrus syrica)*addition of both sorbitol and mannitol at 0.2 to 0.4 M to the culture media that were kept at 15 ° C, caused a subtle reduction in the growth and lengthen the subcultures intervals to 5 months.

Sorbitol (40 g/l) was used for date palm *in vitro* preservation, resulting in healthy shoot bud cultures after six months that was extended for nine months for callus cultures (Bekheet et al., 2001). Later on, El Bahr et al., (El-Bahr et al., 2016) gave a recommendation by using 40 g/l mannitol or 20 g/l sorbitol for the *in vitro* preservation of Bartamoda date palm while they reported that using 20 g/l of sucrose for is more preferred for the Sakkoty cultivar for 12 months.

Da Silva et al., (Da Silva et al., 2018) had summarized the *In vitro* conservation of ornamental plants using slow growth techniques in the following table.

Cryopreservation

Cryopreservation could be defined as storing the biological material under liquid nitrogen (LN) at -196 °C for a long time. Cryopreservation aims to prevent metabolic processes and biochemical reactions and consequently the cell division, keeping the original properties without change for a long period of time (Niino & Sakai, 1992). This means that cryopreservation technique retain both the phenotype and genotype features of the preserved germ plasm without any genetic alteration. While cryopreservation has various advantages, thawing and freezing injuries of the cell membrane structure and function, which may result in decreasing survival percentages are the main drawbacks of this approach (Ashmore, 1997). Besides, it is too difficult to have the same cryopreservation protocol for all plants as each plant exhibits its own characteristics. The cryopreservation process composed of several steps; (i) addition of osmotic agent to the plant media, (ii) interaction with a cryoprotective agents, (iii) freezing and storing in the temperature of liquid N2 at -196°C, (iv) defrosting, (v) treatments and recovery of growth after melting (Moges et al., 2003). As it is seen in the procedure, the success of the cryopreservation is governed by each step (Razdan et al., 1997). In other words, the pre- and post- freezing manipulations, physiological status and genotype are the most affecting factors in cryopreservation succession (Ashmore, 1997). For example, the explant´s type and its initial physiological status is affecting the storage period (Newburry et al., 1997).Also, cells type govern the ability of cells to tolerate the cold stress (Swan et al., 1999).

Table 1. Slow growth in vitro preservation of ornamental plant

Species	Light/Radiance Used	Temp Used	Storage Period (Months)	Recovery Percentage	Explant Used	References
Acanthostachys strobilacea	12 h (30 µmol m^{-2} s^{-2})	10°C	3	100	Seedling	Carvalho *et al*, 2014(51)
Anthurium andreanum	Darkness	6 -10 °C	8	56	Shoot cultures	Benelli *et al*, 2012 (52)
Camellia japonica (cv Alba Plena)	16h (8 µmol m^{-2} s^{-2})	2- 4 °C	12	100	Shoot cultures	Ballester *et al*, 1997 (53)
Deutzia scabra	Darkness	2- 4 °C	12	96.67	Shoot cultures	Gabr and Sayed,2010 (54)
Dianthus ingoldbyi	Darkness	4 °C	6	58	Shoot cultures	Arda *et al*, 2016(55)
Drimiopsis kirkii	16h (3000 lux)	15 °C	4	64.4	Synthetic seeds	Haque and Ghosh, 2014 (56)
Epidendrum chlorocorymbos	16h (1500 lux)	23 °C	6	100	Seedling	Lopez- Puc 2013 (57)
Heliconia champneianacv	16h LED (100% blue, 100% red, 70% blue + 30% red and control white(25 µmol m^{-2} s^{-2})	25 °C	3	100	Seedling	Rodrigues *et al*, 2018(58)
Metrosideros excelsa Soland exGsertn	Darkness	4 -10 °C	4	58	Synthetic seeds	Benelli *et al*, 2017 (59)
Nandina domestica	Darkness	4 -8 °C	6	100	Shoot cultures	Ozudogra *et al*, 2013 (60)
Polygala myrtifolia	Darkness	4 -10 °C	8	68.8	Synthetic seeds	Benelli *et al*, 2017 (59)
Ranunculus osioticus	Darkness	4 -10 °C	9	100	Shoot cultures	Benelli *et al*, 2012 (52)
Splachnum ampullaceum	Darkness	5°C	30	50	Synthetic seeds	Mallon *et al*, 2007 (61)

The first work on exposing plant tissues to extremely low temperature of liquid nitrogen after dehydration was in 1976 for hardy mulberry twigs (Sakai & Yoshida, 1976). After several years, cryopreservation techniques underwent improvement to be applied to larger range of tissues and organs (Vasil et al., 1994; Wang, Batuman, Bar-Joseph et al, 2002; Altman et al., 1998). There are a lot of ongoing research in order to solve the aforementioned drawbacks of cryopreservation technique. The most critical factor to succeed in cryopreservation is to minimize the intercellular water without affecting the cells survival in order to minimize membrane injuries. Different methods of cryopreservation aim to minimize the intercellular water. Or on another words to cryoprotect the tissue to avoid freezing and thawing injuries. In this respect, Towill, (Dodds & Towill, 1991) suggested that a fast rewarming of the frozen species after cryopreservation is mandatory to prevent recrystallization.

Figure 4. In vitro conservation of N. domestica by slow growth storage. Shoot recovery following 6 months of conservation at 4°C in darkness on storage medium containing 30 (A), 45 (B) or 60 g L-1(C) sucrose, or at 8 °C in darkness on 30 (D), 45 (E) or 60 g L-1 (F) sucrose. A, bar 15 mm; B-F, bar 20 mm (Ozudogruet al., 2013).

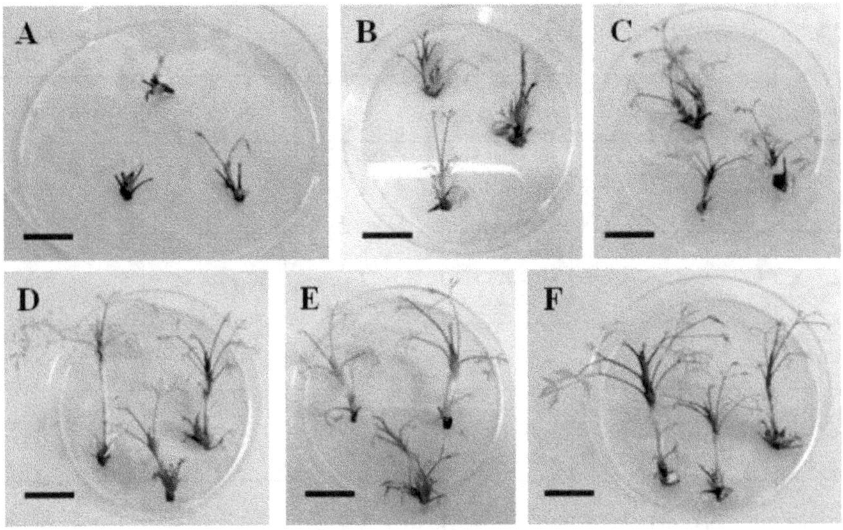

There are two main approaches for cryoprotection; one is termed 'traditional' and is based on the application of cryoprotective agents. More recently, plant cryobiologists have placed greater emphasis on using procedures which avoid the formation of ice during the cryopreservation process and these are based on vitrification techniques. In the following figures some of the equipment's needed for the cryopreservation are represented (Fig 5 and 6).

Pretreatments and Cryoprotection

Cryoprotectant is a treatment that is given before plunging the tissue into the liquid nitrogen to avoid the ice crystal formation in the tissue which leads to avoid occurring of any damage to the tissue which may negatively affect plant regeneration upon thawing of cryo-stored material. A dehydration step, which is necessary to avoid the formation of intracellular ice crystals leads to increase the concentration of solutes in cells and in strong plasmolysis of cells. Cryoprotectants should be non-toxic at proper concentrations, readily miscible with water, with low molecular weight and have the ability to penetrate cells rapidly (Reinhoud et al., 1995). Among those materials the most used cryoprotectants are glycerol, dimethyl sulphoxide (DMSO), ethylene glycol (EG), polyethylene glycol (PEG), amino acids and sugars (Al-Ababneh et al., 2003). In this connection, Tisserat (Tisserat, 1981) succeeded in cryopreservation of date palm embryogenic calli using cryoprotective mixture which consist of polyethylene glycol, glucose and di - methyl sulfoxide (DMSO). Ulrich et al., (Ulrich et al., 1982) reported that the embryogenic calli of date palm var. Medjool cultures were cryopreserved after using different cryoprotective mixtures at -196ºC. They added that the embryos grew and developed normally after cryopreservation when they were pretreated with a cryoprotectant mixture of glycerol and sucrose and then dried to reach a water content of about 0.4 - 0.7 g /g. On one hand, Dereuddre et al. (Dereuddre et al., 1988) declared that the survival rates following cryopreservation in carnation shoot tips increased significantly after the pretreatment with 0.3 to 0.75 molar sucrose for 24 hour. Moreover, Panis et al., (Panis et al., 1996) described a simple cryopreservation method for banana meristem cultures. It involves pre-culturing of the proliferated meristems on MS medium supplemented with 0.3 - 0.5 M sucrose. They found that the sucrose has an important role as a freeze-hardening factor or as a cryoprotictive substance for cryopreservation of banana meristems. On the same hand, Niino et al., (Niino et al., 1997) tested several concentrations of sucrose added to MS medium (0.1, 0.4, 0.7 and 1.0 M) as preculturing treatments for shoot tips of cherry cooled to – 196 °C by vitrification. They reported that pre-culturing for one day on the medium containing 0.7 M sucrose gave the highest level of survival (97%) after cooling to -196 °C. Shibli, (Shibli, 2000) reported that using the sucrose as a pretreatment plays a major role in increasing the plant tissues resistance to both dehydration and freezing in LN. Few years after, cryopreservation methods was performed for nodular tissue of date palm by (Bekheet et al., 2007). Undifferentiated tissue cultures (nodular cultures) were successfully cryopreserved. Among different types of sugars (fructose, glucose, sorbitol and sucrose) used as osmotic agents in pre-culture medium, it was reported that sucrose was the best for date palm tissue cultures. Moreover, cell suspension cryopreservation of date palm cv. Khalas were optimized by (Al-Bahrany & Al-Khayri, 2012). They reported that the highest colony formation, greatest callii growth and highest embryos number was obtained from cell cryoprotected in 10% DMSO supplemented with 0.75 M sucrose.

Figure 5. Samples for the liquid nitrogen containers

Figure 6. The cryotube which contain the explant and plunge it into the liquid nitrogen

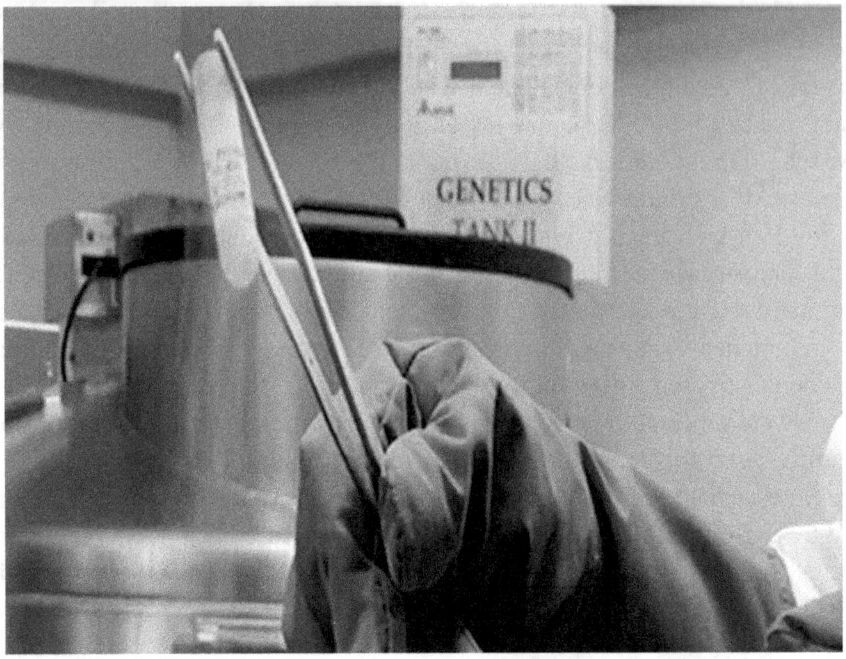

Vitrification

Vitrification is the solidification of liquids without crystallization. It can be defined as «the solidification of a liquid brought about not by crystallization but by an extreme elevation in viscosity during cooling» (Fahy et al., 1984). Vitrification of water in biological systems is dependent on increased cell viscosity, occurring as cell solutes become concentrated. Increased viscosity inhibits the coming together of water molecules to form ice. It can be achieved with different ways but all usually have the end result of increasing solute concentration to a critical viscosity. Vitrification involves treating of samples with cryoprotective substances, dehydration using highly concentrated vitrification solutions, rapid cooling and rewarming are necessary steps then the removal of cryoprotectants and recovery. Plant vitrification solution is an aqueous cryoprotectant solution in which living systems can be cooled slowly to below the glass transition temperature without appreciable ice formation either intra or extra cellularly. DMSO is considered the most common chemical pretreatment used for cryopreservation by vitrification. There are several Vitrification – based cryopreservation techniques like, vitrification, encapsulation- dehydration, encapsulation- vitrification and droplet – freezing. The simple rapid freezing of embryogenic tissues after sucrose cryoprotection and dehydration is considered the simplest vitrification based method which has been successfully used to cryopreserve both meristematic tissues of banana (Panis, 1995) and oil palm embryogenic tissues (Dumet et al., 2000).Matsumoto and Sakai (Matsumoto & Sakai, 2000; Matsumoto & Sakai, 2003) has established a vitrification protocol for grapevine. They used the axillary shoot tips which were excised from 4 to 5 month old *in vitro* stock plantlets then were precultured with 0.3 M sucrose for 3 days and then loaded with a loading solution containing 2 M glycerol and 0.4 M sucrose for 20 min at 25 °C, followed by exposure to half-strength PVS2 at 0 °C for 30 min and then to full-strength PVS2 at 0 °C for 50 min. The PVS2 solution contains (w/v): 30% glycerol, 15% DMSO, and 15% ethylene glycol in 0.4 M sucrose (Sakai et al., 1990), the dehydrated shoot tips were plunged directly into liquid nitrogen for cryostorage. Cryopreserved shoot tips were thawed rapidly in water at 40 °C, and post-thaw cultured for shoot regrowth on a medium composed of half-strength MS supplemented with 1 mg/L BA. This cryoprocedure was applied to ten cultivars of *Vitis*, with an average recovery of about 64% obtained. This technique has been developed for apices, cell suspensions and somatic embryos of different species (Sakai & Engelmann, 2007; Reed et al., 2008). Also, cryopreservation of date palm embryogenic callii via vitrification, encapsulation-dehydration and encapsulation-vitrification was performed by (Subaih et al., 2007). Different components of cryoprotectant like di-methyl sulfoxide (DMSO), sucrose, glycerol and polyethylene glycol (PEG) were examined. In the vitrification-based cryopreservation techniques, precultured explants should be dehydrated by exposure to a vitrification solution like PVS2, before the direct immersion in LN. However, the direct exposing of explants to PVS2 without undergoing any osmoprotection caused harmful effects, because of osmotic stress or chemical toxicity, which is considered as a critical factor in determining the success of cryopreservation by vitrification (Matsumoto et al., 1994); therefore it is essential to induce a high level of osmotoleranceto PVS2. However, preculturing with sugar was insufficient for the shoot tips to establish the required osmotolerance, then a loading treatment was essential (Matsumoto et al., 1994).

Figure 7. Garlic vitrification: 4) explants for vitrification; 5) first regeneration, after two weeks from rewarming; 6) regenerating explants in a petri dish after three weeks from rewarming; 7) Comparative experiment, plantlets taken out of the culture vessels after four months from rewarming (-LN control without liquid nitrogen, +LN after cryopreservation).

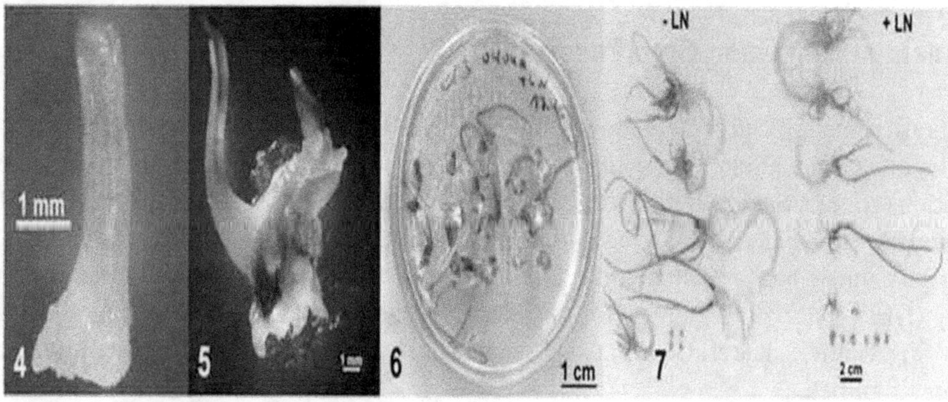

Encapsulation-Dehydration

This technique is based on the artificial seed technology. Encapsulation-dehydration technique includes the encapsulation of plant material in calcium alginate beads, and then the pre- growth treatment using a medium containing high levels of sucrose. The alginate beads must be dehydrated before freezing using either by exposure to silica gel or using air-drying in a laminar flow hood (Ashmore, 1997). This technique has been developed by (Fabre & Dereuddre, 1990), and depends on the inclusion of *solanum spp.* apices in alginate beads and their culturing in a highly concentrated (0.7-1.5 M) sucrose solution followed by physical dehydration then the direct immersion in liquid nitrogen. It was observed that culturing of apices on media supplemented with different sucrose concentrations (0.3-0.7 M), prior to encapsulation, improves the survival percentages after desiccation and freezing. Physical desiccation is carried out either with the exposure to silica gel or in the air flow of the laminar flow cabinet (Paulet et al., 1993). In some temperate species, the incubation of the *in vitro* shoots, nodal segments or apices at low temperature (ranging between 4° to 10ºC), for periods ranging from few days to weeks, elevates the percentage of survival after freezing while using encapsulation-dehydration technique, (Martinez & Revilla, 1999; Niino & Sakai, 1992; Scottez et al., 1992). In this regard, Lynch et al. (Lynch et al., 1996) used the encapsulation-dehydration method for the cryopreservation of *Rosa multiflora* meristems which recorded 25%regenerating explants. While, Pawłowska and Bach (Pawłowska & Bach, 2011) obtained better results with the same method to cryopreserve meristems of *Rosa* 'New Dawn'. In this connection, the encapsulation-dehydration technique was used for the long-term preservation of cocoa (*Theobromacacao* L.) germplasm (Fang et al., 2004). Few years after, N`Nanetal., (N` Nan et al., 2008) used the encapsulation-dehydration successfully with coconut plumules. The apical domes were used as plant materials; pretreatment duration, sugar concentrations and dehydration period were examined as key factors. A moderate survival level around 60% was obtained and 20% of leaf shoots were developed from the cryopreserved explants. In another usage for this technique with roses, Mubbarakh et al., (Mubbarakh et al., 2014) applied it for the *in vitro* fragmented explants of R. hybrid a 'Helmut Schmidt' using TTC (2,3,5-triphenyl-tetrazoliumchloride) to assess the survival. Costeet al., (Coste et al., 2014) used this technique for the cryopreservation of Romanian tomato (*Lycopersicon esculentum* Mill.) cultivars successfully. Shoot tips were excised from *in vitro* grown plants then precultured for24 hours using various sucrose concentrations before dehydrated up to 6 hours in laminar air flow prior to direct immersion in liquid nitrogen(−196 °C) for 24 hours.

Encapsulation-vitrification Cryopreservation

A few years after, another cryopreservation technique based on encapsulation-vitrification was developed, in this technique the precultured shoot tips should be osmoprotected using a loading solution during the encapsulation (Hirai & Sakai, 1999a; Wang, Gafny, Sahae et al, 2002); the encapsulated vitrified shoot tips gave much better and earlier regrowth than encapsulated dehydrated shoot tips, following cooling in LN. This technique has been applied for only a few plants such as wasabi (Matsumoto et al., 1995), strawberry (Hirai et al., 1998), mint (Hirai & Sakai, 1999a), potato (Hirai & Sakai, 1999b), 'Troyer' citrange(Wang, Gafny, Sahae et al, 2002),the procedure of cryopreservation by encapsualtion-vitrification was further simplified.

Droplet Vitrification Cryopreservation

Droplet vitrification cryopreservation technique is one of the techniques that are based on vitrification which based on the usage of cryoprotectants and the high thermal conductivity of aluminum foil. It is done on aluminum foil strips. Initially this technique was developed for banana cryopreservation and raise the regrowth rate by 40–50% compared to vitrification protocol (Panis et al., 2005). In this connection, the shoot tips are dehydrated with a highly concentrated vitrification solution such as PVS2 and then plated on aluminum foil strips for direct immersion into LN. The current technique has been notably used for routine cryopreservation of Musa spp. (Panis et al., 2005).In this regard, Halmagyi and Pinker (Halmagyi & Pinker, 2006a; Halmagyi & Pinker, 2006b) used this technique to develop a procedure for cryostorage the shoot tips of wild roses and 'Rosa × Fairy' hybrid, 'Maidy' and 'Kardinal', however, Pawłowska (Pawłowska, 2011), Pawłowska and Szewczyk-Taranek (Pawłowska & Szewczyk-Taranek, 2015) used this technique for the shoot tips´ cryostorage of Rosa Canina, R. Rubiginosa, R. Agrestis and R. Dumalis. Moreover, Kwasnewska et al., (Kwasnewska et al., 2017) have developed a droplet vitrification technique for Rosa Pomifera. They treated the explants with plant vitrification solution (PVS2) for 10 to 30 minutes. The best results were obtained for shoot tips from dormant buds and a 20-minute in PVS2treatment: the survival reached 84% and regeneration recorded 73%.

Figure 8. Rosa Pomifera 'Karpatia' post-cryopreservation regeneration. Shoot tip survival on 12th day (A) and 4th week (B) after thawing, bar – 5 mm; rose shoots multiplication (12 week after thawing and in vitro rooted plants (D), bar – 10 mm; plant after six weeks of acclimatization (E), potted plant after wintering in the greenhouse and before planting into the soil (F), bar – 100 m

In another study, droplet-vitrification technique has been used with a number of vine and table cultivars, rootstocks and wild grape vine germ plasm some of which, are Chinese wild grape vine germ plasm resistant to grapevine fungal diseases (Bi, 2017).

Figure 9. Plant regeneration from cryopreserved shoot tip of V. vinifera 'Cabernet Sauvignon' by droplet-vitrification. (a) Two weeks- old nodal segments cultured on shoot multiplication medium to promote bud break. (b) Shoot tip excised from (a). (c) PVS2 droplets on an aluminum foil strip. (d) Surviving shoot tip after7 d of post-thaw culture following cryopreservation. (e) Shoot regrowth after 6 weeks of post-thaw culture following cryopreservation. (f) A whole plantlet with well-developed root system after 14 weeks of post-thaw culture following cryopreservation. (Bi 2017).

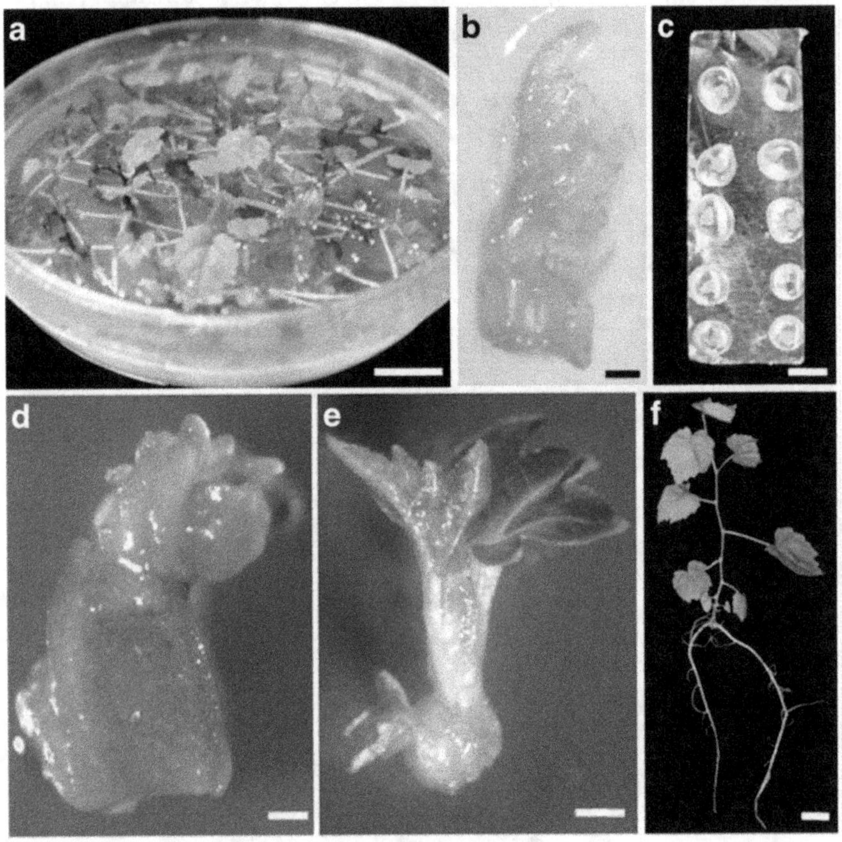

Cryo – Plates Vitrification

Two effective and simple cryopreservation methods using aluminum cryoplates have been developed. The vitrification solution PVS2 for the V cryo-plate process is used for dehydration, whereas the dehydration system D cryo-plate is achieved with the air flow of the laminar fluid cabinet or silica gel. To date, more than 20 articles have been published related to both approaches. The main advantages of the V and D cryo-plate methods are as follows: handling of specimens throughout the procedure is easy and quick because only the cryo-plates are manipulated. Cryo-plates can be filled with a solution (LS) and PVS2/

air flow effectively applied to the sample. Cooling and warming are done easily through the immersion, which results in ultra fast cooling and heating, of the cryo plates respectively in the LN and the 1.0 M sucrose solution. All approaches can achieve high regeneration. The D cryo-plate method can be used for species sensitive to PVS2. The following methods include cryopreservation preparation of material, preconditioning, excision, preculture, explant mounting in cryoplates, osmoprotection, PVS or air flow dehydration, cryo-storage, rewarming and regeneration. Both protocols are promising in order for both herbal and woody plants to be cryopreserved, including tropical plants, following appropriate procedural changes. Optimizing dehydration times, materials pre-conditioning and regeneration conditions after cryopreservation are crucial to a high regrowth levels (Niino et al., 2019). In this regard, cryopreservation using an aluminium cryo-plate (V) was successfully applied to *in vitro*-grown strawberry (*Fragaria* × *ananassa* Duch.) shoot tips (Yamamoto, Fukui, Rafique, Khan et al, 2012). A vitrification procedure using aluminum cryo-plates (V-Cryo-plate procedure) was successfully adopted and adjusted to *in vitro*-grown mulberry shoot tips (*Morus spp.*) (Yamamoto, Fukui, Rafique, Sekizawa et al, 2012).

CONCLUSION

Preservation of plant genetic resources has become extremely important for improvement of crops so as to face the increasing depletion of natural resources. Preservation of plant genetic resources is necessary for food security and agro – biodiversity. Genetic diversity provides new and more productive crops, resistant to biological and environmental stresses.

This chapter has highlighted and demonstrated the importance the two main approaches of the *in vitro* plant preservation as substitutions for the *in vivo* plant preservation. It could be concluded that, although there are some factors that may reduce the success of cryopreservation different techniques application they are still considered the most preferable techniques for the long term storage of plant genetic resources, since the cell division and metabolic and biochemical processes are arrested and thus the cells are allowed to retain their properties unchanged for an indefinite period of time. So, much more efforts must be made to overcome these constraints. Or on another words to cryoprotect the tissue to avoid freezing and thawing injuries.

REFERENCES

Al-Ababneh, S. S., Shibli, R. A., Karam, N. S., & Shatnawi, M. A. (2003). Cryopreservation of bitter almond (Amygdalus communis L.) shoot tips by encapsulation - dehydration and vitrification. *Advanced HortScience*, *17*, 15–20.

Al-Bahrany, A. M., & Al-Khayri, J. M. (2012). Optimizing *in vitro* Cryopreservation of Date Palm (*Phoenix dactylifera* L.). *Biotechnology (Faisalabad)*, *11*(2), 59–66. doi:10.3923/biotech.2012.59.66

Arda, H., Dayan, S., Kartal, C., & Guler, N. (2016). *In vitro* conservation of critically endangered *Dianthus ingoldbyi* Turrill under slow growth conditions. *Trakya UniversityJournal ofNatural Sciences*, *17*, 47–54.

Ashmore, S. E. (1997). *Status report on the development and application of in vitro techniques for the conservation of plant genetic resources*. International Plant Genetic Resources Institute.

Ballester, A., Janeiro, L.V., & Vieitez, A.M. (1997). Cold storage of shoot cultures and alginate encapsulation of shoot tips of *Camellia japonica* L. and *Camellia reticulate* Lindley. *Scientia Horticulturae, 71*, 67-78.

Banerjee, N., & de Langhe, E. (1985). A tissue culture technique for rapid clonal propagation and storage under minimal growth conditions of Musa spp (banana and plantain). *Plant Cell Reports, 4*(6), 341–354. doi:10.1007/BF00269897 PMID:24254081

Bekheet, S. A., Matter, M. A., Taha, H. S., & El-Ashry, A. A. (2016). *In vitro* conservation of Jojoba (*Simmondsia chinensis*) shootlet cultures using osmotic stress and low temperature. *Middle East Journal of Agriculture Research, 5*(4), 396–402.

Bekheet, S. A., Taha, H. S., & Saker, M. M. (2001). *In vitro* long-term storage of date palm. *Biologia Plantarum, 45*(1), 121–124. doi:10.1023/A:1015121111934

Bekheet, S. A., Taha, H. S., Saker, M. M., & Solliman, M. E. (2007). Application of cryopreservation technique for in vitro grown date palm (Phoenix dactylifera L.) cultures. *Journal of Applied Sciences Research, 9*, 859–866.

Benelli, C., Micheli, M., & Carlo, A. (2017). An improved encapsulation protocol for regrowth and conservation of four ornamental species. *Acta Societatis Botanicorum Poloniae, 86*(3), 1–12. doi:10.5586/asbp.3559

Benelli, C., Ozudogru, E. A., Lambardi, M., & Dradi, G. (2012). *In vitro* conservation of ornamental plants by slow growth storage. *Acta Horticulturae*, (961), 89–93. doi:10.17660/ActaHortic.2012.961.8

Bertrand-Desbrunais, ANoirot, MCharrier, A. (1992). Slow growth of *in vitro* conservation of coffee (*Coffea spp.). Plant Cell, Tissue and Organ Culture, 1*, 105–110.

Bhatia, S. (2015). Application of plant biotechnology. In *Modern Applications of Plant Biotechnology in Pharmaceutical Sciences* (pp. 157–207). Elsevier Inc. doi:10.1016/B978-0-12-802221-4.00005-4

Bi, W. L. (2017). *Cryopreservation of Vitis spp. and cryotherapy for eradication of grapevine viruses* (PhD thesis). Northwest A and F University, Yangling, China.

Boison, A. M., Gout, E., Bligny, R., & Rivasseau, C. (2012). A simple and efficient method for the long-term preservation of plant cell suspension cultures. *Plant Methods, 8*(4), 132–144. doi:10.1186/1746-4811-8-4 PMID:22289515

Brown, D. C. W., Leung, D. W. M., & Thrope, T. A. (1979). Osmotic requirements for shoot formation in tobacco callus. *Physiologia Plantarum, 46*(1), 36–41. doi:10.1111/j.1399-3054.1979.tb03182.x

Carvahlo, V., Santos, D. S., & Nievola, C. C. (2014). *In vitro* storage under slow growth and ex vitro acclimatization ofthe ornamental bromeliad *Acanthostachys strobilacea. South African Journal of Botany, 92*, 39–43. doi:10.1016/j.sajb.2014.01.011

Corbineau, F., Engelmann, F., & Côme, D. (1990). Ethylene production as an indicator of chilling injury in oil palm (*Elaeis guineensis* Jacq) somatic embryos. *Plant Science, 71*(1), 29–35. doi:10.1016/0168-9452(90)90065-V

Coste, A., Deliu, C., Valimareanu, S., Mareanu, A., & Halmagyi, A. (2014). Cryopreservation and acclimatization of *Lycopersiconesculentum* (Mill.) genotypes. *Notulae Botanicae Horti Agrobotanici Cluj-Napoca*, *42*(2), 466–471. doi:10.15835/nbha4229579

Da Silva, D. P. C., Ozudogru, E. A., Dos Reis, M. V., & Lambardi, M. (2018). In vitro conservation of ornamental plants. *Ornamental Horticulture (Campinas)*, *24*(1), 28–33. doi:10.14295/oh.v24i1.1163

Dereuddre, J., Fabre, J., & Bassaglia, C. (1988). Resistance to freezing in liquid nitrogen of carnation (Dianthus caryophyllus L. Var eolo) apical and axillary shoot tips excised from different aged in vitro plantlets. *Plant Cell Reports*, *7*(3), 170–176. doi:10.1007/BF00269315 PMID:24241591

Dereuddre, J., Scottez, C., & Duron, Y. A. M. (1990). Effect of cold hardening on cryopreservation of pear (*Pyrus communis* L. cv. Beurre Hardy) shoot tips of *in vitro* plantlets. *C. R. Acad. Sci.*, *30*, 265–272.

Dumet, D., Engelmann, F., Chabrillange, N., Dusseret, S., & Duval, Y. (2000). Cryopreservation of oil palm polyembryogenic cultures. In Current Research, Progress and Applications. Eds. Engelmann, F. and H. Takagi. Cryopreservation of Tropical Plant Germplasm. IPGRI.

El-Bahr, M. K., Abd EL-Hamid, A., Matter, M. A., Shaltout, A., Bekheet, S. A., & El-Ashry, A. A. (2016). In vitro conservation of embryogenic cultures ofdate palm using osmotic mediated growth agents. *Journal of Genetic Engineering and Biotechnology*, *14*(2), 363–370. doi:10.1016/j.jgeb.2016.08.004 PMID:30647634

Engelmann, F. (1991). *In vitro* conservation of horticultural species. *Acta Horticulturae*, (29), 327–334. doi:10.17660/ActaHortic.1991.298.40

Engelmann, F. (1997). *In vitro* conservation methods. In J. H. Newburry & J. A. Callow (Eds.), *Biotechnology and Plant Genetic Resources: Conservation and Use. B. V. Ford – Lloyd* (pp. 119–162). CABI, Wellingford.

Engelmann, F. (2012). Germplasm collection, storage and conservation. In A. Altman & P. M. Hasegawa (Eds.), *Plant Biotechnology and Agriculture* (pp. 255–268). Academic Press. doi:10.1016/B978-0-12-381466-1.00017-1

Fabre, J., & Dereuddre, J. (1990). Encapsulation-dehydration: A new approach to cryopreservation of *Solanum* shoot tips. *Cryo Letters*, *11*, 413–426.

Fahy, G., MacFarlane, D. R., Angell, C. A., & Meryman, H. T. (1984). Vitrification as an approach to cryopreservation. *Cryobiol*, *21*(4), 407–426. doi:10.1016/0011-2240(84)90079-8 PMID:6467964

Fang, J. Y., Wetten, A., & Hadley, P. (2004). Cryopreservation of cocoa (*Theobroma cacao* L.) somatic embryos for long-term germplasm storage. *Plant Science*, *166*(3), 669–675. doi:10.1016/j.plantsci.2003.11.002

Ford-Lloyd, B., & Jackson, M. (1986). Plant Genetic Resources: An Introduction to their Conservation and Use. Edward Arnold Ltd..

Gabr, A. M. M., & Sayed, S. S. (2010). Slow growth conservation and molecular characterization of *Deutzia scabra* Thunb. *African Journal of Plant Science*, *4*, 409–416.

George, E. (1993). *Plant Propagation by Tissue Culture. Part 1: The Technology* (2nd ed.). Exegetics Ltd.

Grout, B. W. W. (1995). Introduction to the *in vitro* preservation of plant cells, tissue and organs. In B. W. W. Grout (Ed.), *Genetic preservation ofplant cellsi n vitro* (pp. 1–20). Springer-Verlag. doi:10.1007/978-3-642-78661-7_1

Halmagyi, A., & Pinker, I. (2006a). Plant regeneration from *Rosa* shoot tips cryopreserved by a combined droplet vitrification method. *Plant Cell, Tissue and Organ Culture*, *84*(2), 145–153. doi:10.100711240-005-9012-z

Halmagyi, A., & Pinker, I. (2006b). Cryopreservation of *Rosa* shoot tips: Importance of preculture conditions. *Acta Horticulturae*, (725), 351–356. doi:10.17660/ActaHortic.2006.725.45

Haque, S. K. M., & Ghosh, B. (2014). Somatic embryogenesis and synthetic seed production - A biotechnological approach for true-to-type propagation and *in vitro* conservation of an ornamental bulbaceous plant *Drimiopsis kirkii* Baker. *Applied Biochemistry and Biotechnology*, *172*(8), 4013–4024. doi:10.100712010-014-0817-2 PMID:24604129

Hassan, M. (2002). *In vitro Studies on Somatic Embryogenesis Conservation of Date Palm* (Ph.D. Thesis). Cairo University.

Hirai, D., & Sakai, A. (1999a). Cryopreservation of *in vitro*-grown axillary shoot-tip meristems of mint (*Mentha spicata* L.) by encapsulation vitrification. *Plant Cell Reports*, *19*(2), 150–155. doi:10.1007002990050725 PMID:30754740

Hirai, D., & Sakai, A. (1999b). Cryopreservation of *in vitro*-grown meristems of potato (*Solanum tuberosum* L.) by encapsulation vitrification. *Potato Research*, *42*(1), 153–160. doi:10.1007/BF02358405

Hirai, D., Shirai, K., Shirai, S., & Sakai, A. (1998). Cryopreservation of *in vitro*-grown meristems of strawberry (*Fragaria ×ananassa* Duch.) by encapsulation-vitrification. *Euphytica*, *101*(1), 109–115. doi:10.1023/A:1018386211577

Kalia, A. (2018). Nanotechnology in Bioengineering: Transmogrifying Plant Biotechnology. Omics Technologies and Bio-Engineering, 211-229.

Kartha, K. K., & Engelmann, F. (1994). Cryopreservation and gemplasm storage. In I. K. Vasil & T. A. Thrope (Eds.), *Plant Cell and Tissue Culture* (pp. 195–230). Kluwer Acadimic Publishers. doi:10.1007/978-94-017-2681-8_9

Kwasnewska, E., Dziedzic, E., & Pawlowska, B. (2017). Integration of cryopreservation and tissue culture for germplasm conservation and ropagation of Rosa pomifera 'Karpatia '. *Notulae Botanicae Horti Agrobotanici Cluj-Napoca*, *45*(1), 208–214. doi:10.15835/nbha45110566

Lopez-Luc, G. (2013). An effective *in vitro* slow growth protocol for conservation of the orchid *Epidendrum chlorocorymbos* Schltr. *Tropical and Subtropical Agroecosystems*, *16*, 61–68.

Lynch, P. T., Harris, W. C., & Chartier-Hollis, J. M. (1996). The cryopreservation of shoot tips of Rosa multiflora. *Plant Growth Regulation*, *20*(1), 43–45. doi:10.1007/BF00024056

Mallon, R., Barros, P., Luzardo, A., & Gonzalez, M.L. (2007). Encapsulation of moss buds: an efficient method for the *in vitro* conservation and regeneration of the endangered moss *Splachnum ampullaceum*. *Plant Cell, Tissue and Organ Culture, 88*, 41–49.

Martinez, D., & Revilla, M. A. (1999). Cold acclimation and thermal transitions in the cryopreservation of hop shoot tips. *Cryo Letters, 19*, 333–342.

Matsumoto, T., & Sakai, A. (2000). Cryopreservation of in vitro-cultured axillary shoot tips of Vitis by vitrification. *Acta Horticulturae*, (538), 177–181. doi:10.17660/ActaHortic.2000.538.28

Matsumoto, T., & Sakai, A. (2003). Cryopreservation of axillary shoot tips of in vitro-grown grape (Vitis) by a two-step vitrification protocol. *Euphytica, 131*(3), 299–304. doi:10.1023/A:1024024909864

Matsumoto, T., Sakai, A., Takahashi, C., & Yamada, K. (1995). Cryopreservation of *in vitro*-grown apical meristems of wasabi (*Wasabijaponica*) by encapsulation-vitrification method. *Cryo Letters, 16*, 189–196.

Matsumoto, T., Sakai, A., & Yamada, K. (1994). Cryopreservation of *in vitro*-grown apical meristems of wasabi (*Wasabia japonica*) by vitrification and subsequent high plant regeneration. *Plant Cell Reports, 13*(8), 442–446. doi:10.1007/BF00231963 PMID:24194022

Maxted, N., & Kell, S. P. (2003). Biodiversity and conservation / Plant diversity, conservation and use. Encyclopedia of Applied Plant Sciences, 25-48.

Moges, A. D., Karam, N. S., & Shibli, R. A. (2003). Slow growth in vitro preservation of African violet (Saintpaulia ionantha Wendl) shoot tips. *Advanced HortScience, 17*, 1–8.

Monette, P. L. (1986). Cold storage of kiwi shoot tips *in vitro*. *HortScience, 21*, 1203–1205.

Morel, G. (1975). Merstim culture techniques for the long-term storage of cultivated plants. In O. H. Frankel & J. G. Hawkes (Eds.), *Crop Genetic Resources for Today and Tomorrow* (pp. 327–332). Cambridge Univ. Press.

Mubbarakh, S. A., Izhar, N. A., Rajasegar, A., & Subramaniam, S. (2014). Establishment of encapsulation-dehydration technique for *in vitro* fragmented explants of *Rosa hybrida* L. cv. Helmut Schmidt. *Emirates Journal of Food and Agriculture, 26*(6), 565–276. doi:10.9755/ejfa.v26i6.18024

N` Nan, O., Hocher, V., & Verdeil, J. L. (2008). Cryopreservation by encapsulation – dehydration of plumules of coconut (*Cocos nucifera* L.). *Cryo Letters, 29*, 339–350. PMID:19137197

Ng, S., & Ng, N. (1991). Reduced – growth storage of germplasm. In J. Dodds (Ed.), *In Vitro Methods for Conservation of Plant Genetic Resources* (pp. 11–39). Chapman and Hall. doi:10.1007/978-94-011-3072-1_2

Niino, T., & Sakai, A. (1992). Cryopreservation of alginate coated *in vitro* – grown shoot tips of apple, Pear and mulberry. *Plant Science, 87*(2), 199–206. doi:10.1016/0168-9452(92)90151-B

Niino, T., Tashino, K., Suzuki, M., Ohuchi, S., Magoshi, J., & Akihama, T. (1997). Cryopreservation of *in vitro* grown shoot tips of cherry and sweet cherry by one – step vetrification. *Scientia Horticulturae, 70*(2-3), 155–163. doi:10.1016/S0304-4238(97)00062-9

Niino, T., Yamamoto, S., Matsumoto, T., Engelmann, F., Valle Arizaga, M., & Tanaka, D. (2019). Development of V and D cryo-plate methods as effective protocols for cryobanking. *Acta Horticulturae*, (1234), 249–262. doi:10.17660/ActaHortic.2019.1234.33

Orlikowska, T. (1992). Effect of *in vitro* storage at 4° C on survival and proliferation of two apple rootstocks. *Plant Cell, Tissue and Organ Culture*, *31*(1), 1–7. doi:10.1007/BF00043468

Ozudogru, E. A., Previati, A., & Lambardi, M. (2010). *In vitro* conservation and cryopreservation of ornamental plants. In S. M. Jain & S. J. Ochatt (Eds.), *Protocols for InVitro Propagation of Ornamental Plants* (pp. 303–324). Humana Press, Springer. doi:10.1007/978-1-60327-114-1_28

Ozudogru, E. A., Silva, D. P. C., Kaya, E., Dradi, G., Paiva, R., & Lambardi, M. (2013). *In vitro* conservation and cryopreservation of *Nandina domestica*, an outdoor ornamental shrub. *Notulae Botanicae Horti Agrobotanici Cluj-Napoca*, *41*(2), 638–645. doi:10.15835/nbha4129335

Panis, B. (1995). *Cryopreservation of Banana (Musa spp) Germplasm* (PhD Thesis). Katholieke University, Leuven, Belgium.

Panis, B., Piette, B., & Swennen, R. (2005). Droplet vitrification of apical meristems: A cryopreservation protocol applicable to all *Musaceae*. *Plant Science*, *168*(1), 45–55. doi:10.1016/j.plantsci.2004.07.022

Panis, B., Totte, N., Nimmen, K., Withers, L., & Swennen, R. (1996). Cryopreservation of banana (*Musa spp*) meristem cultures after preculture on sucrose. *Plant Science*, *121*(1), 95–106. doi:10.1016/S0168-9452(96)04507-4

Paulet, F., Engelmann, F., & Glaszmann, J. C. (1993). Cryopreservation of apices of *in vitro* plantlets of sugarcane (*Saccharum* sp. hybrids) using encapsulation/ dehydration. *Plant Cell Reports*, *12*(9), 525–529. doi:10.1007/BF00236101 PMID:24196115

Pawłowska, B. (2011). The effect of BA and GA3 on shoot multiplication in *in vitro* cultures of Polish wild roses. *Folia Horticulturae*, *23*(2), 145–149. doi:10.2478/v10245-011-0022-5

Pawłowska, B., & Bach, A. (2011). Cryopreservation by encapsulation dehydration of *in vitro* grown shoot buds of *Rosa* 'New Dawn'. *Acta Horticulturae*, (908), 303–307. doi:10.17660/ActaHortic.2011.908.40

Pawłowska, B., & Szewczyk-Taranek, B. (2015). Efficient cryopreservation by droplet vitrification of pentaploid roses and the phenotype of regenerated plants. *Acta Societatis Botanicorum Poloniae*, *84*(4), 439–442. doi:10.5586/asbp.2015.038

Rademacher, W. (2000). Growth retardants: Effects on gibberellin biosynthesis and other metabolic pathways. *Annual Review of Plant Physiology and Plant Molecular Biology*, *51*(1), 501–531. doi:10.1146/annurev.arplant.51.1.501 PMID:15012200

Ramsay, M. M., Jacskon, A. D., & Porley, R. D. (2000). A Pilot study for *ex situ* conservation of UK Bryophytes. *Proceedings of EuroGard 2000—II European Botanic Garden Congress*, 52–57.

Rao, N. K. (2004). Plant genetic resources: Advancing conservation and use through biotechnology. *African Journal of Biotechnology*, *3*(2), 136–145.

Razdan, M. K., & Cockling, E. C. (1997). Biotechnology in conservation of genetic resources. In M. K. Razdan & E. C. Cockling (Eds.), *Conservation of Plant Genetic Resources in vitro, General Aspects* (pp. 1–25). Science Publishers Inc.

Read, P. E., & Preece, J. E. (2003). Environmental management for optimizing micropropagation. *Acta Horticulturae*, (616), 129–133.

Reed, B. (1993). Improved survival of *in vitro* stored rubus germplasm. *Journal of the American Society for Horticultural Science*, *118*(6), 890–895. doi:10.21273/JASHS.118.6.890

Reed, B. (1999). *In vitro* storage conditions for mint germplasm. *HortScience*, *34*(2), 350–352. doi:10.21273/HORTSCI.34.2.350

Reinhoud, P. J., Schrynemakers, E. W. M., Iren, F. V., & Kijne, J. W. (1995). Vitrification and a heat-shock treatment improve cryopreservation of tobacco cell suspensions compared to two-step freezing. *Plant Cell, Tissue and Organ Culture*, *42*(3), 261–267. doi:10.1007/BF00029997

Roca, W. M., Reyes, R., & Beltran, J. (1984). Effect of various factors on minimal growth in tissue culture storage of cassava germplasm. *Proc. 6th Symp. Int. Soc. For Tropical Root Crops*, 441-446.

Rodrigues, P. H. V., Arruda, F., & Forti, V. A. (2018). Slow growth *in vitro* conservation of *Heliconia champneianacv.* Splash under different light spectra. *Scientia Horticulturae*, *75*, 163–166.

Romano, A., & Martins, M. (1999). *In vitro* cold storage of cork oak shoot cultures. *Plant Cell, Tissue and Organ Culture*, *59*(2), 155–157. doi:10.1023/A:1006465706905

Sakai, A., & Engelmann, F. (2007). Vitrification, encapsulation–vitrification and droplet–vitrification: A review. *Cryo Letters*, *28*, 151–172. PMID:17898904

Sakai, A., Hirai, D., & Niino, T. (2008). Development of PVS-based vitrification and encapsulation–vitrification protocols. In B. M. Reed (Ed.), *Plant cryopreservation: a practical guide* (pp. 33–58). Springer. doi:10.1007/978-0-387-72276-4_3

Sakai, A., Kobayashi, S., & Oiyama, I. (1990). Cryopreservation of nucellar cells of navel orange (Citrus sinensis Osb. var. brasiliensis Tanaka) by vitrification. *Plant Cell Reports*, *9*(1), 30–33. doi:10.1007/BF00232130 PMID:24226373

Sakai, A., & Yoshida, S. (1976). Survival of plant tissue at super low temperature VI. Effects of cooling and re- warming rates on survival. *Plant Physiology*, *42*(12), 1695–1701. doi:10.1104/pp.42.12.1695 PMID:16656707

Sarkar, D., & Naik, P. S. (1998). Factors affecting minimal growth conservation of potato microplants *in vitro*. *Euphytic.*, *102*(2), 275–280. doi:10.1023/A:1018309300121

Scottez, C., Chevreau, E., Godard, N., Arnaud, Y., Duron, M., & Dereuddre, J. (1992). Cryopreservation of cold-acclimated shoot tips of pear *in vitro* cultures after encapsulation-dehydration. *Cryobiology*, *29*(6), 691–700. doi:10.1016/0011-2240(92)90073-B

Shibli, R. A. (1991). *In vitro* acclimatization of *(Chrysanthemum morifolium* Ramat.) after *in vitro* water stress. *Plant Cell, Tissue and Organ Culture*, *11*, 15–23.

Shibli, R. A. (2000). Cryopreservation of black iris (*Iris nigricans*) somatic embryos by encapsulation – dehydration. *Cryo Letters, 21*, 39–46. PMID:12148063

Shibli, R. A., Ajlouni, M. M., Jaradat, A., Al Janabi, S., & Shatanawi, M. (1997). Micropropagation of wild pear (*Pyrus syrica*). *Scientia Horticulturae, 69*(1-4), 237–242. doi:10.1016/S0304-4238(96)00972-7

Shibli, R. A., Shatnawi, M. A., Ajlouni, M. M., Jaradat, A., & Adham, Y. (1999). Slow growth *in vitro* conservation of bitter almond (*Amygdalus communis* L.). *Advances in Horticultural Science, 13*, 133–134.

Shibli, R. A., Smith, M. A. L., & Spomer, L. A. (1992). Osmotic adjustment and growth responses of three (*Chrysanthemum morifolium* Ramat.) cultivars to osmotic stress induced *in vitro*. *Journal of Plant Nutrition, 15*(9), 1373–1381. doi:10.1080/01904169209364404

Silva, T. L., & Scherwinski-Pereira, J. E. (2011). *In vitro* Conservation of *Piper aduncum* and *Piper hispidinervum*under slow-growth conditions. *Pesquisa Agropecuária Brasileira, 46*(4), 384–389. doi:10.1590/S0100-204X2011000400007

Subaih, W. S., Shatnawi, M. A., & Shibli, R. A. (2007). Cryopreservation of date palm (*Phoenix dactylifera*) embryogenic callus by encapsulation-dehydration, vitrification and encapsulation-vitrification. *J. J. of Agric. Sci., 3*(2), 156–171.

Swan, T. W., Ohare, D., Gill, R. A., & Lynch, P. T. (1999). Influence of preculture conditions on the post – thaw recovery of suspension cultures of Jerusalem artichoke (*Helianthus tuberosus* L.). *Cryo Letters, 20*, 325–336.

Tahtamouni, R. W., & Shibli, R. A. (1999). Preservation at low temperature and cryopreservation in wild pear (*Pyrus syriaca*). *Advances in Horticultural Science, 13*, 156–160.

Tarmizi, A., Marziah, A., & Halim, A. (1993). Effects of various concentrations of sucrose on growth and proline accumulation in oil palm polyembryogenic cultures. *Biotechnology in Agric.*, 365 – 368.

Tisserat, B. H. (1981). Cryogenic preservation and regeneration of date palm tissue. *HortScience, 16*(1), 47–48.

Towill, L. E. (1991). Cryopreservation. In J. H. Dodds (Ed.), *In vitro methods for conservation of plant genetic resources* (pp. 41–70). Chapman and Hall. doi:10.1007/978-94-011-3072-1_3

Ulrich, J. M., Finkle, B. J., & Tisserat, B. H. (1982). Effects of Cryogenic treatment on plantlet production from frozen and unfrozen date palm callus. *Plant Physiology, 69*(3), 624–627. doi:10.1104/pp.69.3.624 PMID:16662261

United Nations Conference on Environment and Development (UNCED). (1992). *Convention on Biological Diversity*. UNCED.

Wanas, W. (1999). The application of *in vitro* technique for the genetic conservation of some *Citrus spp. Anal. of Agric. Sci., 44*(2), 653–672.

Wang, Q., Gafny, R., Sahae, N., Sela, I., Mawassi, M., Tanne, E., & Perl, A. (2002). Cryopreservation of grapevine (*Vitis vinifera* L.) embryogenic cell suspension by encapsulation – dehydration and subsequent plant regeneration. *Plant Science, 162*(4), 551–558. doi:10.1016/S0168-9452(01)00594-5

Wang, Q. C., Batuman, Ö., Bar-Joseph, P., Li, M., & Gafny, R. (2002). Cryopreservation of *in vitro*-grown shoot tips of 'Troyer' citrange [*Poncirus trifoliata* (L.) Raf. ×*Citrus sinesis* (L.) Osbeck.] by encapsulation-vitrification. *Euphytica*, *128*(1), 135–142. doi:10.1023/A:1020683305690

Williamson, J., Jennings, D., Guo, W., Pharr, M., & Ehrenshaft, M. (2002). Sugar alcohols, salt stress and fungal resistance: Polyols, multifunctional plant protection. *Journal of the American Society for Horticultural Science*, *127*(4), 463–467. doi:10.21273/JASHS.127.4.467

Wilson, S. B., Rajapakse, N. C., & Young, R. E. (2000). Media composition and light affect storability and post storage recovery of micropropagated hosta plantlets. *HortScience*, *34*, 1158–1162.

Withers, L. (1982). Institutes working on tissue culture for genetic conservation. IBGR Secretariat.

Withers, L. A. (1991). Biotechnology and plant genetic resources conservation. In Plant Genetic Resources, Conservation and Management Concepts and Application. International Board for Plant genetic Resources.

Withers, L. A., & Engelmann, F. (1998). *In vitro* conservation of plant genetic resources. In A. Altman (Ed.), *Agricultural Biotechnology* (pp. 57–87). Marcel Dekker.

Yamamoto, S., Fukui, K., Rafique, T., Khan, N., Castillo Martinez, C., Sekizawa, K., & Niino, T. (2012). Cryopreservation of in vitro-grown shoot tips of strawberry by the vitrification method using aluminium cryo-plates. *Plant Genetic Resources*, *10*(1), 14–19. doi:10.1017/S1479262111000906

Yamamoto, S., Fukui, K., Rafique, T., Sekizawa, K., Koyama, A., Ichihashi, I., & Niino, T. (2012). Development of an effective cryopreservation protocol using aluminum cryo-plates for in vitro-grown shoot tips of mulberries (Morus spp.) originated from the tropics and subtropics. *Sanshi-Konchu Biotec*, *81*(1), 57–62.

Zimmermann, U. (1978). Physics of turgor and osmo – regulation. *Annual Review of Plant Physiology*, *29*(1), 121–148. doi:10.1146/annurev.pp.29.060178.001005

Chapter 15
Lipids in Ruminant Nutrition and Its Effect on Human Health

Eman H. Elsabaawy
National Research Centre, Egypt

Sawsan M. Gad
National Research Centre, Egypt

ABSTRACT

Scientific evidence and nutritional guidelines recommend a reduction in total fat intake, particularly of saturated fatty acids, which are associated with an increased risk of obesity, hypercholesterolemia, and cancer. Nutritionists recommend a higher intake of polyunsaturated fatty acids (PUFA), especially n-3 PUFA at the expense of n-6 PUFA. Besides the beneficial effects of n-3 fatty acids on human health, the conjugated linoleic acid (CLA) isomers have attracted increased attention as a result of their health promoting biological properties. As milk and meat are the main sources of CLA for human consumption, increasing such important nutrient in animal products is strongly recommended. Fat supplementation is one of the methods of increasing PUFA content in ruminant products, and it has been shown that PUFA can be increased in milk by supplementation with vegetable oils and oil seeds. Vegetable oils as equivalent to oilseeds show similar effects on CLA content in ruminant products.

INTRODUCTION

An adequate supply of good-quality food is essential for human health and well-being. Ruminant's products represent important sources of nutrients in human diets, providing a nearly ideal pattern of amino acids and energy. In addition, meat and milk contain several compounds of anti-carcinogenic properties. Conjugated linoleic acid (CLA) as one of these components has numerous potential benefits for human health, including potent cancer-fighting properties. This is especially interesting considering that most of natural anti-carcinogens are of plant origin.

DOI: 10.4018/978-1-7998-5000-7.ch015

On the other hand, nutritional quality has an important consideration in food choices because of the growing consumer awareness of the link between diet and health. The goal of increasing the efficiency of animal production has been, and continues to be, an important consideration in producing animal-derived food products. There is also increasing recognition that foods can be contributing factors in the prevention and development of some disease conditions. Many foods contain micro-components that have many effects beyond those associated with their traditional nutrient content, and these are often referred to as ''functional food'' components. "Functional foods" is a generic term used to describe food components that have beneficial effects on human health above those expected on the basis of their nutritive value. In other words, "functional foods" must have a relevant effect on well-being and health or cause a reduction of disease risk. Polyunsaturated fatty acids (PUFA), and especially of n-3 or ώ-3 fatty acids represent one of these microcomponents in animal products.

In times of growing world populations, climate change, and with the damage to critical resources, it is important to take action to develop and sustain the capacity of agricultural and systems of manufacturing to continue to provide food for human. Functional foods" is a generic term used to describe food components that have beneficial effects on human health above those expected on the basis of their nutritive value. It represents an important category for the economic growth of different countries of the world. Concerns about health risks are growing; therefore, increased attention must be given to providing functional food in a sustainable manner to control these concerns. This chapter contends that it is substantial to finds ways to estimate the effects of innovation in this division.

The present chapter focused on:Fat chemistry that related to fatty acid composition of feeds and animal products and its functions. The different polyunsaturated fatty acids and its isomers and the importance of CLA. Factors affecting meat and milk fat content of CLA including the different dietary and animal factors. How to increase CLA content in animal product dietary manipulation and production system. Dietary lipids and human health.

FATTY ACIDS CHEMISTRY

Types of Fatty Acids

Fatty acids are classified as saturated or unsaturated fatty acids. They differ in length and often categorized as short, medium, or long chain fatty acids. Unsaturated fatty acids have one or more double bonds between carbon atoms. The two carbon atoms in the chain that are bound next to either side of the double bond can occur in a *cis* or *trans* configuration.

A *cis* configuration means that adjacent hydrogen atoms are on the same side of the double bond. A *trans* configuration, by contrast, means that the next two hydrogen atoms are bound to opposite sides of the double bond. As a result, they do not cause the chain to bend much, and their shape is similar to straight saturated fatty acids (SFA) **(Fig. 1).** In most naturally occurring unsaturated fatty acids, each double bond has three *n* carbon atoms after it, and all are cis bonds (Bhalla*et al.*, 2009).

Figure 1. Comparison between the trans and the cis-isomers. Adapted from Bhallaet al., (2009).

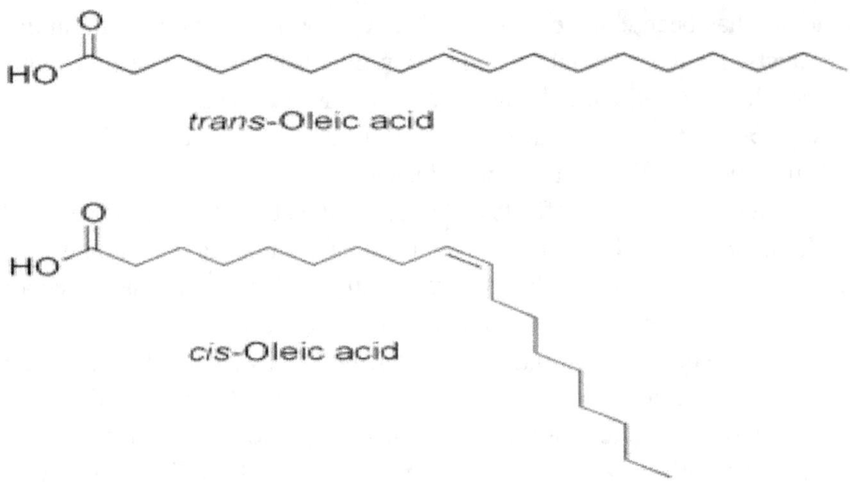

Essential Fatty Acids (Efas)

Essential fatty acids are those fatty acids that humans and other animals must ingest because the body requires them for good health but cannot synthesize them. Also,the term "essential fatty acid" refers to fatty acids required for biological processes, and not those that only act as fuel.

Only two EFAs are known for humans: alpha-linolenic acid (an omega-3 fatty acid) and linoleic acid (an omega-6 fatty acid). Other fatty acids that are only "conditionally essential" include gamma-linolenic acid (an omega-6 fatty acid), lauric acid (a saturated fatty acid), and palmitoleic acid (a monounsaturated fatty acid).

Functions of Essential Fatty Acids

The biological effects of the ω-3 and ω-6 fatty acids are mediated by their mutual interactions. In the body, essential fatty acids serve multiple functions. The balance between dietary ω-3 and ω-6 strongly affects function such as:

- The classic eicosanoids (affecting inflammation and many other cellular functions)
- The endocannabinoids (affecting mood, behavior and inflammation)
- The lipoxins from ω-6 EFAs and resolvins from ω-3 (in the presence of aspirin, down-regulating inflammation).
- The isofurans, neurofurans, isoprostanes, hepoxilins, epoxyeicosatrienoic acids (EETs) and Neuroprotectin D.
    - They form lipid rafts (affecting cellular signaling).
    - They act on DNA (activating or inhibiting transcription factors which is linked to pro-inflammatory cytokine production).

The essential fatty acids start with the short chain polyunsaturated fatty acids:

- ω-3 fatty acids: α-Linolenic acid or ALA (18:3)
- ω-6 fatty acids: Linoleic acid or LA (18:2)

These two fatty acids cannot be synthesized by humans, as humans lack the desaturaseenzymes required for their production. They form the starting point for the creation of longer and more desaturated fatty acids, which are also referred to as long-chain polyunsaturated fatty acids:**ω-3 fatty acid** includeαLinolenic acid or ALA (18:3), Eicosapentaenoic acid or EPA (20:5), and Docosahexaenoic acid or DHA (22:6). **ω-6 fatty acids** include Linoleic acid or LA (18:2), Arachidonic acid or AA (20:4), Gamma-linolenic acid or GLA (18:3), and Dihomo-gamma-linolenic acid or DGLA (20:3). **ω-9fatty acid is** not essential in humans, because humans generally possess all the enzymes required for their synthesis.

Food Sources of Essential Fatty Acids.

Almost all the polyunsaturated fat in the human diet is from EFA. Some of the food sources of ω-3 and ω-6 fatty acids are fish and shellfish, flaxseed (linseed), hemp oil, soya oil, canola (rapeseed) oil, chia seeds, pumpkin seeds, sunflower seeds, leafy vegetables, and walnuts.

Plant sources of ω-3 contain neither eicosapentaenoic acid (EPA) nor docosahexaenoicacid (DHA). The human body can convert α-linolenic acid (ALA) to EPA and subsequently DHA. This however requires more metabolic work, which is thought to be the reason that the absorption of essential fatty acids is much greater from animal rather than plant sources.

CONJUGATED LINOLEIC ACIDS (CLA).

Chemical Composition and Isomers

CLAs are a group of isomers of LA, specially found in meat and milk from ruminants, it is composed of a chain of 18 carbons and ends with a functional group –COOH, with two double bonds in different carbon position (Fig.2). Each double bond can be cis-cis, cis-trans, trans-cis or trans-trans (Eulitz et al., 1999), but those with one trans double bond are bioactive ones (Jensen, 2002). Also they may be in 7-9, 8-10, 9-11, 10-12 carbons and so forth (Kim, 2003). 14 to 19 isomers of CLA where detected in milk fat (Sehat et al., 1998 and Lock and Bauman, 2004).

Benefits of CLA

Most naturally occurring anti-carcinogens are present only at trace amounts and of plant origin. However, CLA is unique among naturally occurring anti-carcinogens, in that it is potent at extremely low levels and present in milk, dairy products and meat from ruminant animals.

As the biomedical studies with CLA expanded, it became apparent that CLA had a range of positive health effects in experimental animal models such as anticarcinogenic, antiatherogenic, altered nutrient partitioning and lipid metabolism, antidiabetic, immunity enhancement, and improved bone mineralization(Bauman et al., 2001). Additional beneficial effects in reducing body fat accretion, delaying

Figure 2. Structures of t-10, c-12-CLA, c-9, t-11-CLA and ordinary linoleic acid, c-9, c-12-octadeca-dienoic acid (bottom)

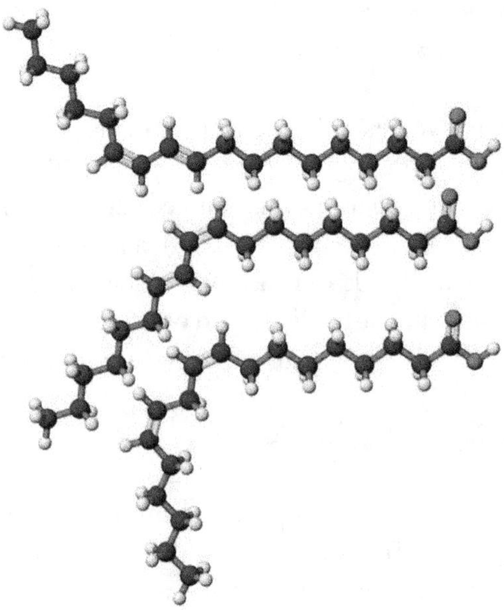

the onset of type II diabetes, retarding the development of atherosclerosis, improving the mineralization of bone and modulating the immune system had also been reported (Belury, 2002).

So, conjugated linoleic acids (CLA) may influence the onset and severity of several chronic diseases, including various cancers, atherosclerosis, obesity, bone density loss, and diabetes. These findings are of special interest to the agriculture community, because dietary sources of CLA are almost exclusively beef and dairy products. Thus, a better understanding of the specific isomers and mechanisms responsible for these positive effects of CLA on human health would be both prudent and economically beneficial.

Data consistently show that relatively low levels of CLA can influence risk of cancer. The predominate CLA isoform (cis-9, trans-11 CLA or rumenic acid) found in beef and milk fat possesses anticarcinogenic effects but does not alter body composition; the trans-10, cis-12 CLA has been shown to inhibit lipogenesis. It may be concluded that foods naturally containing high amounts of CLA (e.g., beef and dairy products) be considered as meeting the definition of functional foods having beneficial effects on human health above those expected on the basis of their nutritive value (Milner, 1999.

Bio Synthesis of CLA

Laura and Hartigh (2019) explained that CLA is synthesized in the rumen as an intermediate in the biohydrogenation of linoleic acid to stearic acid. The supplies of both intermediates and end products of biohydrogenation are affected by the substrate supply and extent of biohydrogenation, thus influencing the CLA content of milk from ruminants. The majority of CLAis present in the rumen in the form of the cis-9,trans-11 isomer. The transfer efficiency of CLA to milk fat is affected by the presence of different isomers of CLA. Ruminant mammary and adipose cells are able to synthesisecis-9,trans-11-CLA from trans-11-18:1 (vaccenic acid) by the action of the _9-desaturase enzyme. Plant oils are high in both linoleic and linolenic acids, which results in increased CLA production in the rumenand in the mammary gland. The CLA content of milk increases when cows are offered grazed grass.

Therefore, it could be emphasized that theCLA found in milk and meat fat of ruminants originates from two sources:

- CLA formed during ruminalbiohydrogenation of linoleic acid.
- CLA synthesized by the animal's tissues from desaturation of trans-fatty acids

Figure 3. Biosynthesis of CLA. (Bauman et al., 1999)

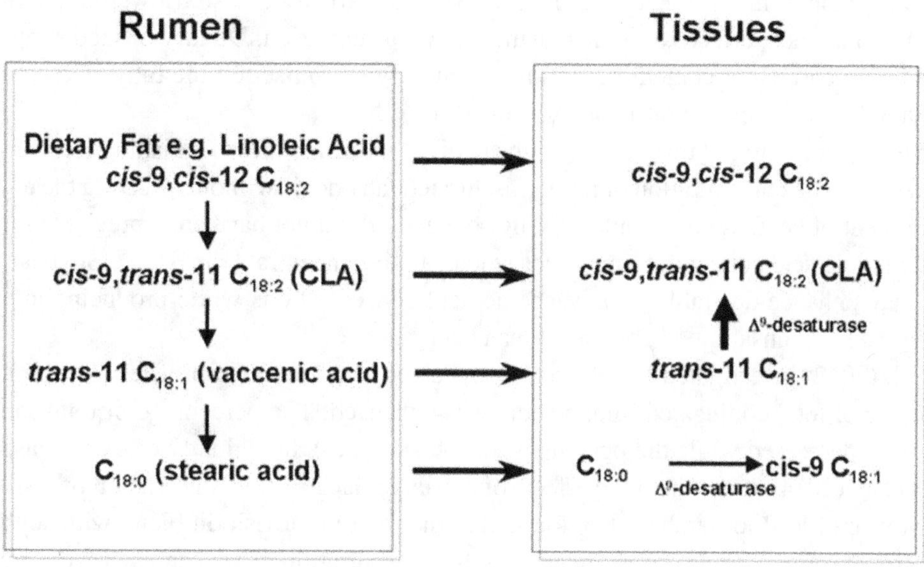

Factors Affecting CLA Content in Animal Products

Factors affecting in CLA content in milk, meat and other food products from various species of animals classified into diet and animal related factors.

Dietary Related Factors

Griinari and Bauman (1999) grouped the dietary effects into three categories relative to the potential mechanism by which they may act in increasing CLA content in ruminant product.

The first category includes dietary factors that provide lipid substrate for rumen biohydrogenation. Plant oils high in linoleic and linolenic acids are particularly effective. These lead to increases in rumen outflow of vaccenic acid and to a lesser extent, CLA isomers.

The second category consists of dietary factors that alter the rumen environment thereby affecting the bacteria involved in rumen biohydrogenation. Forage to concentrate ratio plays an important part and can markedly affect the rumen environment and biohydrogenation.

The third category includes dietary factors that involve a combination of lipid substrate and modification of the rumen population of bacteria. Pasture is one example and typically cows on lush spring pasture will have a ruminant product content of CLA that is 2- to 3-fold greater than corn-based total mixed rations. However, as pasture matures, this difference in CLA diminishes.

Vegetable Oils (Soybean Oil, Rapeseed Oil, Safflower, Sunflower and Linseed).

Scollan et al. (2003) showed that vegetable oils influence CLA content in animal products by supplying PUFA which are substrates for bacterial isomerization or/and biohydrogenation in the rumen. Adding oilseeds to the diet has been proven to be an efficient method to increase the CLA content in the muscle lipids. The inclusion of linoleic acid-rich oilseeds, such as safflower or sunflower, in the diet of ruminants appears to be most effective for increasing CLA concentration. Dietary oilseed supplementation does not only increase CLA content but has also a modifying impact on the fatty acid composition of adipose tissue. However, not all oilseeds exert the same effect.

Ludden et al. (2009) found that adding soybean oil at 5% of the diet of finishing steers for 77 to 189 days may decrease the concentration of monounsaturated fatty acids without affecting the total saturated fatty acid content of beef. While supplementing soybean oil did not have an appreciable impact on the conjugated linoleic acid concentration of beef when a high-concentrate diet is fed. Soybean oil supplementation may enhance desirable trans-vaccenic acid concentrations while producing an undesirable reduction in the monounsaturated content of lean beef.

Mao et al. (2012) investigated the effects of soybean oil on fatty acid profile. The results indicated that percentage of total conjugated linoleic acid was enhanced. The percentage of total saturated fatty acids in muscle decreased while the percentage of total polyunsaturated fatty acids increased.

Ferreira et al. (2014) determined the effects of partial replacement of soybean oil by fish oil on meat fatty acid profile of feed lot lambs. They found that the mixture of fish oil blend with soybean oil improved the lipid profile of the meat by increasing the concentration of vaccenic acid, Eicosapentaenoic acid and Docosahexaenoic acid.

Rapeseed oil did not seem to have positive effects on CLA on meat. Of three studies (two in beef cattle and one with lambs) none showed increased CLA concentrations in the *M. longissimusdorsi* after supplementation with up to 6% of DM rapeseed oil (Stasiniewicz et al., 2000; Strzetelskiet al., 2001andSzumacher-Strabel et al., 2001).

It is well established that the RA and a-linolenic content in milk fat can be increased by feeding oilseeds rich in C18:2 and C18:3 such as sunflower and flax seed, respectively (Dhiman et al., 2005 and Khanal and Dhiman, 2004). Many studies have been performed on the effect that different dietary regimens on the FA profile and CLA of milk fat in cows (Stanton et al., 2003) and goats milk (Chilliard et al., 2006). Despite the fact that the effects of lipid supplements on milk fat synthesis bear some similarities across ruminant species, often goats and ewes respond differently to cows (Chilliard et al., 2003).

Allam et al. (2012) fed goats on rations contained sunflower seed, and linseed, soybean or soy oil. They found that fatty acid of milk was altered by oil supplementation whereas, feeding oils reduced proportion of both short chin (c12:0) and medium chin (c16:0) fatty acid, and feeding oilseed increased the proportion of long chin (> c18:0) fatty acid in milk fat. The increasing of oil seed increased the concentration of CLA in milk fat and the highest value was recorded with SFS ration followed by SBS ration compared to the control of no oil seed inclusion.

Safflower and sunflower are characterized as linoleic acid-rich oilseeds. Studies have shown that safflower seed and sunflower seed supplementation can have positive effects on fatty acid profiles in meat and in improved CLA levels in muscle tissue of lambs and cattle (Boles et al., 2005 and Noci et al., 2007). However, feeding a linoleic acid enriched diet can have a negative effect on the n_6/n_3 ratio (Kott et al., 2003).

Peng et al. (2010) reported that total saturated fatty acid concentrations could be reduced in intramuscular fat when ewes are supplemented with safflower oilseed meal. Fat trimmed from tail fat of ewes supplemented with safflower and sunflower oilseed meal appear to have greater ratios of polyunsaturated fatty acids to saturated fatty acids and greater levels of conjugated linoleic acid in tail fat from ewes supplemented with safflower oilseed meal. The essential fatty acids, linoleic acid and linolenic acid, appear to be generally greater in back fat. Consequently, it appears possible to produce meat products with more beneficial fatty acid composition through differential oilseed supplementation. This may provide more healthful choices of adipose tissue used to extend ground meat products.

Full Fat Soy (Extruded Soybeans)

Aharoni et al. (2005) compared soybean oil with full fat soybeans as supplements over five months in a high forage fattening diet of Friesian bull calves. Extruded full fat soybeans were about 20% more efficient than free oil in increasing the CLA concentration in intramuscular fat. The full fat soybean supplement also resulted in higher PUFA and lower SFA and monounsaturated fatty acids (MUFA) content in the intramuscular fat than supplementation with soybean oil. This may be due to a partial protection of the oils against ruminal biohydrogenation by roughly crushed seeds (Scheeder, 2004). Therefore; using oilseeds instead of free oils may be the preferred option.

Fish Oil

A comparable feeding design used with three lamb breeds documents no effect of fish oil (mean values of 10.0 vs. 11.0 mg/g FAME) but again a significant increase with linseed (Wachira et al., 2002). In both studies combining fish oil with whole linseed let to comparable CLA proportions as with linseed alone but another study (Demirel et al., 2004) with lambs found that the linseed fish oil mixture was more efficient than linseed alone. The comparison of the data is difficult, as the last study reports the CLA content only per 100 g tissue.

The reason for the observed increased CLA levels is not clear yet, as only small amounts of linolenic and linoleic acid are present and the long chain n-3 fatty acids are not isomerized /hydrogenated to CLA or trans-vaccenic acid. Thus long chain n-3 fatty acids present in fish oil may interfere with the biohydrogenation of linolenic or/and linoleic acids or affect D9-desaturase activity (Raes et al., 2004). Chow et al. (2004) postulate that fish oil increases ruminal accumulation of trans-vaccenic acid by inhibiting the final biohydrogenation step to stearic acid. This would supply more substrate for endogenous CLA synthesis. However, further studies are needed to provide an explanation. Feeding fish oil also increases the n-3 long chain PUFA concentration in the intramuscular fat due to the high eicosapentaenoic (EPA) and docosahexaenoic acid (DHA)content in fish oil. Ruminal biohydrogenation of EPA and DHA is limited, and therefore considerable amounts of these fatty acids are available for incorporation into the adipose tissue (Raes et al., 2004 and Ferreira *et al.* 2014).

AbuGhazaleh (2008) studied the effect of adding fish oil (FO) and sunflower oil (SFO) to grazing dairy cows' diets on the temporal changes in milk conjugated linoleic acid (cis-9, trans-11 CLA), Holstein cows were divided into two diet regimen groups. One group (C) was fed a basal diet (7.6 kg DM basis) plus 400 g animal fat. The other group (FOSFO) was fed a basal diet plus 100 g of FO and 300 g of SFO (FOSFO). It is observed that concentrations of cis-9 trans-11 CLA and vaccenic acid (VA) in milk fat were higher for cows fed the FOSFO over 3 week of lipid supplementation.

The concentration of cis-9 trans-11 CLA in milk fat reached maximum on day 3 with both diets and remained relatively constant thereafter. Milk production, milk fat percentages, milk fat yield, milk protein percentages, and milk protein yield were not affected by treatment diets. The concentration of VA in milk fat followed the same pattern of temporal changes as cis-9 trans-11 CLA. The concentration of cis-9, trans-11 CLA in milk fat reached a plateau on day 3 for both the C and FOSFO fed cows at approximately 0.84 and 1.59 g/100 g fatty acids, respectively and remained relatively steady to the last day of feeding.

The effect of fish oil and sunflower oil either alone or in combination in milk cis 9 Trans 11 CLA, milk yield and Milk composition, vary in different articles as shown in Table 1.

Table 1. Effect of fish oil and sunflower oil, either alone or in combination on milk cis-9, trans-11 CLA content, and milk yield and composition

Factors	Fish oil	Sunflower oil	Fish oil+ sunflower oil
cis 9 trans 11 CLA	(+)[1,2,3,4,5,6,7]	(+)[1,2,5]	(++)[1,5]
Milk yield	(+)[1,2] or (0)[5]	(+)[1,2] or (0)[5]	(+)[1] or (0)[5]
Protein yield	(+)[1] or (0)[5]	(+)[1] or (0)[5]	(+)[1] or (0)[5]
lactose yield	(+)[1]	(+)[1]	(+)[1] or (-)
fat yield	(-)[1] or (0)[2,5]	(-)[1] (0)[2,5]	(-)[1] (0)[5]
Milk composition:			
○ protein(concentration)	(-)[1] or (0)[2,5]	(-)[1] or (0)[2]	(+)
○ fat (concentration)	(-)[1] or (0)[5]	(-)[1] or (0)[5]	(0)[5]
○ lactose(concentration)	(+)[1]	(+)[1]	(+)[1]
○ C6:0–C12:0 and C16:0	(-)[1]	(-)[1]	(-)[1]

(+) = increased (++) = increased more (-) = decreased (0) = not effected.

[1]Murphy et al. 2008. [2]AbuGhazaleh et al. 2007. [3] Donovan et al., 2000. [4]Whitlock et al., 2002. [5]AbuGhazaleh.2008. [6]AbuGhazaleh et al. 2003. [7]Kay et al., 2003.

AbuGhazaleh et al. (2002) observed that supplementing lactating cow diets with a blend of FO and extruded soybeans was more efficient in enhancing milk content of *cis*-9, *trans*-11 CLA than when these supplements were fed separately.

Whitlock et al. (2002) reported that, the increase in milk cis-9,trans-11 CLA and TVA concentrations when cows were fed a blend of FO and extruded soybeans was greater than the additive effect of feeding both fat sources separately, Because FO has a very low concentration (<7%) of oleic, linoleic, and linolenic acids, so a component in FO may serve as a ruminal modifier causing greater than expected increases in milk cis-9,trans-11 CLA and TVA concentrations from dietary linoleic and linolenic acids.

The percentage of increasing CLA content in milk as result for fat and oil supplementation varied in the different experiments as shown in Table 2.

Table 2. Effect of different sources of oil and fat supplementations on the percentage of increasing of cis-9, trans-11 CLA

Source	% cis-9,trans-11 CLA increasing	Reference
Fish oil+ sunflower (50 g FO+ 225 g SFO)/d/cow	34%	Murphy et al,2008
Fish oil only (100g /d/cow)	23%	Murphy et al,2008
Fish oil (20g/kg DM)	3.6-fold	Donovan et al. (2000)
full-fat rapeseeds	61%	Lawless et al. (1998)
Ca-salts of fatty acids (FA; 30% linoleic acid)	173%	Schroeder et al. (2003)
soybean oil (500 g/d)	49%	Boken et al., 2005).
Full fat soybeans	24.3%	Lawless et al. (1998)
soybean oil (20 g/Kg. DM)	two-fold	(Whitlock et al., 2002)
Soybean oil (2.0%)	190%	Dhiman et al. (2000)
Extruded soybeans(2% fat)	200	(Whitlock et al., 2002)
fish oil (fed at 2% of diet DM)	360	Donovan et al., 2000
FO+extruded soybeans	321%	AbuGhazaleh et al., 2002).
peanut oil (high oleic acid)	(1.33 g/100 g fat),	Kelly et al. (1998)
sunflower oil (high linoleic acid)	(2.44 g/100 g fat)	
linseed oil (high linolenic acid) at 5.3%	(1.67 g/100 g fat)	

Using Unconventional Feed Sources

Vasta et al. (2008) and Rayudika et al. (2019) reviewed the effects of alternative feed resources, AFR (tannin containing feeds, legume grains, agro-industrial by products) as unconventional feeds on quality of meat and milk of small ruminant. They concluded that feeding sheep and goats alternative feed resources can maintain animal productivity and meat and milk quality. However, to achieve this purpose, diets supplemented with AFR should be carefully formulated, in order to guarantee small ruminants their nutritional requirements and to avoid the presence of anti-nutritional factors on the rations. For instance, tannin-containing feeds cause converse effects on meat and milk fatty acid composition, depending on the amounts of condensed tannins ingested by the animals, while meat colour is evidently paler when condensed tannins are present in the diet. Legume seeds have a high nutritional value and do not cause any detrimental changes in meat quality and milk composition. Meat and milk fatty acid profile can be manipulated by the inclusion of some agro-industrial by-products in the diet depending on the energy value, the fatty acid composition and the fiber content of these alternative feedstuffs.

Dietary Manipulation

As a result of the fatty acid imbalance in human diets, dietary strategies have been used to improve the nutritional and health value of the intramuscular fat of cattle. Thus, manipulation of the fatty acid composition in ruminant meat to reduce SFA content and the n_6/n_3 ratio whilst, simultaneously increasing the PUFA and CLA contents, is of major importance in meat research. It has been shown that in ruminants grazing have potential beneficial effects on PUFA/SFA and n_6/n_3 ratios, increasing the PUFA and CLA contents and decreasing the SFA concentration of beef (Enser et al., 1998 and French et al., 2000).

Although several factors influence the fatty acid composition and the CLA content of beef (e.g. seasonal variation, animal genetics and production practices), diet plays the most important role (Schmid et al., 2006). Dietary factors are often linked with particular production systems (Geay et al., 2001).

Feeding System

Alfaia et al. (2009) studied the effect four feeding system on intramuscular fatty acids and conjugated linoleic acid isomers of beef cattle. The results indicated that feeding systems have a major impact on the fatty acid profile, including CLA isomers, which is independent of the degree of intramuscular fat deposition. a-Linolenic acid, long chain n_3 PUFA, and t11,t13, t11,c13 and t12,t14 CLA isomers in meat were the most sensitive grass intake indicators. In addition, the data reinforced the evidence that beef from pasture-fed animals has a higher nutritional quality (mainly due to the higher levels of n_3 PUFA and CLA) when compared to that from concentrate-fed bulls, as a result of the beneficial effects of grass on meat fatty acid profiles. Meat from bulls fed with pasture only and from animals with 2 months of finishing on concentrate showed PUFA/SFA and n_6/ n_3 ratios of intramuscular fat inside the recommended values for the human diet, in contrast to that from animals exposed to longer finishing periods on concentrate (except for PUFA/SFA ratio in CCC diet). Finally, meat fatty acid profiles seems to be an efficient chemical marker to discriminate between the four feeding systems analyzed, including the different finishing periods of animals on concentrate.

Initial Diet

Bessa et al. (2008) used forty Merino Branco ram lambs to study the effects of initial diet and duration of supplementation with a conjugated linoleic acid (CLA) promoting diet, on carcass composition, meat quality and fatty acid composition of intramuscular fat. The experimental period was 6 weeks. The experimental design involved 2 initial diets (commercial concentrate (C); dehydrated lucerne (L)), and 2 finishing periods(2 and 4 weeks) on dehydrated lucerne plus 10% soybean oil (O). Data were analyzed as a 2 _ 2 factorial arrangement with initial diet and time on finishing (CLA promoting) diet as the main factors. The lambs were randomly assigned to four groups: CCO; COO; LLO; LOO according to the lamb's diet fed in each period.

Lambs initially fed with concentrate showed higher hot carcass weights (11.2 vs 9.6 kg) than lambs fed initially with lucerne. The increase of the duration of finishing period reduced the carcass muscle percentage (57.4% vs 55.5%) and increased the subcutaneous fat percentage (5.67% vs 7.03%). Meat colour was affected by initial diet. Lambs initially fed with concentrate showed a lower proportion of CLA(18:2cis-9, trans-11 isomer) (0.98% vs 1.38% of total fatty acids) and most of n_3 polyunsaturated fatty acids than lambs initially fed with lucerne. Initial diet did not compromise the response to the CLA-promotingdiet and the proportion of 18:2cis-9, trans-11 in intramuscular fat increased with the duration of time on the CLA-promoting diet (1.02% vs 1.34% of total fatty acids).

Pastures, Forages and Grass Feeding

CLA levels in milk of grazing animals are reported to be higher than those of cows, goats or ewes fed dried herbage (Nudda et al., 2005). Several studies in cattle had demonstrated that decreasing the proportion of concentrate in the diet and increasing grass intake, caused a decrease in the concentrations of intramuscular fat and the n_6/n_3 fatty acid ratio (French et al., 2000 and Garcı´a et al., 2005).

French et al. (2000) determined in the intramuscular fat of steers (longissimusdorsi muscle) increasing CLA contents consistent with increasing intakes of grass. Levels of 5.4, 6.6, and 10.8 mg CLA/g FAME were detected in grazing steers with increasing grass intake compared to 3.7 mg/gFAME in animals fed concentrate. Grass silage also positively influenced CLA content (4.7 g/g FAME) but not to the same extent.

Poulson et al. (2004) reported a 6.6 times higher CLA conten tin the longissimus and semitendinosus muscle from steers raised only on forages compared to steers fed a common high grain feedlot diet (13.1 vs. 2.0 mg/g FAME). Steers fed a grain based diet in the growing period and grazed on pasture during the finishing period still had a 4 times higher CLA tissue content compared to those fed only the grain based diet (8.0 vs. 2.0 mg/g FAME). That finishing steers on pasture instead of concentrate feeding leads to higher CLA contents in intramuscular fat (5.3 vs.2.5 mg/g FAME in longissimusdorsi muscle) was confirmed by another study (Realini, et al., 2004).

Pasture feeding does not only cause higher CLA concentrations but also influences fatty acid composition. A decrease in the n-6:n-3 PUFA ratio as well as an increase in the PUFA: saturated fatty acids (SFA) is described in beef adipose and muscle tissue by inclusion of grass in the diet (Nuernberg et al., 2002 and Realini et al., 2004). The decrease in n-6:n-3 PUFA are favorable in regard to current human dietary guidelines (USDA, 2005). Schmid et al. (2006) reported that switching from concentrate-based diet to pasture has been shown to increase CLA content of milk and meet. AbuGhazaleh et al. (2008) studied the response of milk cis 9, trans 11 CLA of cows managed in pasture or in confinement and the results of this study are summarized in Table 3.

Buccioni et al. (2008) reported that little information is available about the effect of the kind of forage on milk fat composition. In fact the modulation of milk fatty acids is generally achieved by lipid supplements and not by the choice of herbage. Different geometrical and positional isomers are synthesized by rumen bacteria during the biohydrogenation of PUFA with respect to those formed in the mammary tissue by action of stearoyl CoA desaturase enzyme.

Table 3. The differences pasture and confinement systems in milk CLA, yield and composition

Parameters	Pasture system[1] Pas	Confinement system[2] lot
$C_{18:2}$ cis 9, trans 11 CLA	1.52	0.84
Vaccenic acid (g/100g total fatty acids.)	4.52	2.15
$C_{18:1}$		
g/100g fatty acids	9.82	10.64
yield (g/d)	64.01	60.65
Milk yield (kg /d)	19.4	23.1
Milk fat		
%	2.95	2.51
Yield (kg/d)	0.51	0.75
Milk protein		
%	3.35	3.34
Yield (kg/d)	0.61	0.75

[1]alfalfa grass pasture, 2 mix of alfalfa hay and corn silage ad libitum.

Feeding Duration

Ludden et al. (2009) examined duration of soybean oil (SBO) supplementation needed to enhance carcass conjugated linoleic acid (CLA) and trans-vaccenic (TVA) content using 96 beef steers fed a 78% corn based diet supplemented with SBO for 0, 77, 137, or 189 days before slaughter. Duration of SBO supplementation had no effect on animal performance or carcass traits, total saturated, or total polyunsaturated fatty acids of Longissimusdorsi (LD). Concentrations of CLA in LD were not affected by SBO supplementation. Concentrations of monounsaturated fatty acids (MUFA) decreased linearly in LD, whereas TVA increased in adipose tissue and tended to increase in LD with increasing duration of SBO supplementation. Supplementing SBO to a concentrate-based diet may enhance TVA without impacting CLA, while reducing the MUFA content of lean beef.

Animal Related Factors

Animal variation is also a major source of differences in milk and meat fat content of CLA.

Animal Species

CLA content is much higher in foods derived from ruminants than those from non-ruminants because of the ability of ruminants to biohydrogenate dietary unsaturated fatty acids (FA) with the help of bacteria present in the rumen. Cows' milk is the richest source of CLA, which has been found to vary from as low as 0.2% to as high as 3.7% of total milk fat. Non-ruminants Only small amounts of CLA have been found in the milk, meat, or egg from non-ruminants (Bee, 2000).

Animal Breed

Ruminant product CLA concentrations between the four breeds ranged from 14.7 to 18.6 mg/g fat, while the variation in milk fat CLA between individual cows within the four breeds ranged from 4.8 to 35.6 mg/g fat. There were significant correlations between milk fat CLA concentrations of individual cows within NM, MB and IH and the correlation was close to being significant for DH cows at two sampling dates. In terms of total fatty acid profiles, DH had higher C16:0 concentrations than the other breeds. The IH had a higher C16:0 than the MB. The C18:0content was higher in milk fat from NM and MB than from DH and total C18:1 concentration was close to being significantly higher in the MB than in the IH.

The data indicated that animal breed has some influence over the CLA content of milk and that cows yielding high levels of milk fat CLA sustain this production over time. While there were some significant differences between the breeds in the concentrations of C16:0, C18:0 and C18:1 it is unlikely that they are large enough to be of practical importance.

Wachira et al. (2002) reported that while there were large differences between sheep breeds in their performance, the effects of breed on fatty acid composition were relatively small in comparison with the effects of diet. The deposition of CLA in muscle and adipose tissue of lambs was affected by both breed and dietary lipid source.

Interaction Between Breed and Diet

Wachira et al. (2002) concluded that the deposition of CLA in muscle and adipose tissue of lambs was affected by both breed and dietary lipid source.

Tsiplakou et al. (2008) conducted an experiment to study the interaction between breed and diet on conjugated linoleic acid (CLA) and fatty acids profile (FA) content of milk fat of four pure of dairy sheep breeds (Awassi, Lacaune, Friesland and Chios). The results of this experiment showed that the sheep breed had no effect on milk FA profile. The only significant effect was on Δ−9 desaturase activity index expressed as C18:1/C18:0. On the contrary, the diet affected significantly the FA profile of all ewes' milk fat, with pasture to cause lower proportions of saturated and higher proportions of unsaturated FA, compared with those of sheep kept and fed indoors. The interaction between breed and diet was significant for the FA C10:0, C14:0, C15:0, C16:0, C17:1, C18:1, C18:2n6c, C18:2n6t, C18:3n6 and CLA and for FA groups SCFA, MCFA, LCFAand MUFA. The S/U FA ratio and the Δ−9 desaturase indexes (C14:1/C14:0 and C18:1/C18:0) were also affected by the breed × diet interaction. In conclusion, pasture feeding when compared with supplementary diet, induce large variation in the milk FA profile.In addition pasture increases FA in milk fat which has potential interest for human health. Breed has small effect on milk FA profile,but it can be ignored as it is shown from the interaction breed × diet.

LIPIDS AND HUMAN HEALTH WITH REFERENCE TO CLA

Almost all the polyunsaturated fats in the human diet are essential fatty acids which play an important role in the life and death of cardiac cells.

Dietary Lipids and Plasma Cholesterol

Fatty acid composition of meat is an important factor in the definition of meat quality due to its relationship to meat odor and flavor and the relationship of fatty acids to nutritional value of fat for human consumption. The ratio of polyunsaturated (PUFA) and saturated (SFA) fatty acids (P/S) and the ratio of n_6 and n_3 PUFA (n_6/n_3) are important indexes for nutritional evaluation of fat with the recommendation that P/S be at least 0.40 and n_6/n_3 less than 4 (Department of Health, 1994).

Nutritionists have focused on the benefits of conjugated linoleic acid (CLA), associated with evidence that CLA is involved in reduction of fat deposition and increased lipolysis in animals (Park et al., 1997) as well as lowering serum cholesterol and reducing the incidence of atherosclerosis. Mensink (2005) confirmed a strong relationship between dietary lipids and plasma cholesterol.

Treatment for Depression

Research suggests that high intakes of fish and omega-3 fatty acids are linked to decreased rates of major depression. Low plasma concentrations of DHA predict low concentrations of cerebrospinal fluid 5-hydroxyindoleacetic acid (5-HIAA). It is found that low concentrations of 5-HIAA in the brain is associated with depression and suicide.

There are high concentrations of DHA in synaptic membranes of the brain. This is critical for synaptic transmission and membrane fluidity. The omega-6 fatty acid to omega-3 fatty acid ratio is important to avoid imbalance of membrane fluidity. Membrane fluidity affects function of enzymes such as adenylate cyclase and ion channels such as calcium, potassium, and sodium, which in turn affects receptor numbers and functioning, as well as serotonin neurotransmitter levels. It is evident that western diets are deficient in omega-3 and excessive in omega-6, and balancing of this ratio would confer numerous health benefits (Givens, 2005).

Cancer

Pariza et al. (2001) reported that milk fat is an important source of potential anticarcinogens as the conjugated linoleic acid (CLA) isomers. A 2006 report in the Journal of the American Medical Association, in their review of literature covering cohorts from many countries with a wide variety of demographics, concluded that there was no link between $n-3$ fatty acids and cancer (MacLean et al., 2006). This is similar to the findings of a review by the British Medical Journal of studies up to February 2002 that failed to find clear effects of long and shorter chain $n-3$ fats on total mortality, combined cardiovascular events and cancer (Lee Hooper, 2006). In those with advanced cancer and cachexia, n-3 fatty acids supplements may be ofbenefit, improving appetite, weight, and quality of life (Colomer et al., 2007).

Patterson et al. (2010) demonstrated that the consumption of high amounts of long chain omega-3 polyunsaturated fats from food produced a 25% reduced risk of additional breast cancer events. These women were also shown to have reduced risk of "all-cause mortality." On the other hand, additional studies performed by different cohorts of women failed to show any correlation between incidence of breast cancer and CLA (Laura and Hartigh, 2019).

Cardiovascular Disease

Omega-3 fatty acids in algal oil, fish oil, fish and seafood have been shown to lower the risk of heart attacks. Omega-6 fatty acids in sunflower oil and safflower oil may also reduce the risk of cardiovascular disease (National Institute of Health, 2006). However, some evidence does not support a beneficial role for omega-3 fatty acid supplementation in preventing cardiovascular disease (including myocardial infarction and cardiac death) or stroke (Evangelos et al., 2012and Kwak et al., 2012).

Inflammation

Some potential benefits for omega-3 fatty acid have been reported in conditions such as rheumatoid arthritis (Fortin et al., 1995).Current research suggests that the anti-inflammatory activity of long-chain *n*−3 fatty acids may translate into clinical effects (Wall et al., 2010). For example, there is evidence that rheumatoid arthritis sufferers taking long-chain n−3 fatty acids from sources such as fish have reduced pain compared to those receiving standard NSAIDs (Ruggiero et al., 2009).

Developmental Disorders

Although not supported by current scientific evidence as a primary treatment for ADHD, autism spectrum disorders, and other developmental differences, omega-3 fatty acids have gained popularity for children with these conditions (Richardson, 2006).

Fish oils appear to reduce ADHD-related symptoms in some children (Richardson, 2006). Double blind studies have shown "medium to strong treatment effects of omega 3 fatty acids on symptoms of ADHD" (Richardson and Montgomery, 2005). There is not enough scientific evidence to support the effectiveness of n-3 fatty acids for autism spectrum disorders (Bent et al., 2009). Fish oil has only a small benefit on the risk of early birth (Secher, 2007 and Jensen, 2006).

Psychiatric Disorders

Though there is some evidence that *n*-3 fatty acids are connected to a variety of mental disorders (Perica and Delas, 2011) there is limited evidence that may be useful as an add-on for the treatment of depression associated with bipolar disorder (Montgomery and Richardson, 2008). There is preliminary evidence that EPA supplementation, either with DHA or medication, is helpful in cases of depression (Naliwaiko et al.,2004).

Cognitive Aging

Epidemiological studies suggest that consumption of omega-3 fatty acids can reduce the risk of dementia, but evidence of a treatment effect in dementia patients is inconclusive (Kawakita et al., 2006 and Cederholm and Palmblad, 2010). However, clinical evidence suggests benefits of treatment specifically in patients who show signs of cognitive decline but who are not sufficiently impaired to meet criteria for dementia (Mazereeuw et al., 2012).

CONCLUSION AND RECOMMENDATION

Omega-3 fatty acids offer a promising complementary approach to standard treatments for several diseases. It has been recommended by the Health organizations that the intake of saturated fatty acids should be reduced, while that of polyunsaturated fatty acids (PUFA) be increased.

REFERENCES

AbuGhazaleh, A. A. (2008). Effect of fish oil and sunflower oil supplementation on milk conjugated linoleic acid content for grazing dairy cows. J. *Animal Feed Science and Technology*, *141*(3-4), 220–232. doi:10.1016/j.anifeedsci.2007.06.027

AbuGhazaleh, A. A., Schingoethe, D. J., Hippen, A. R., & Whitlock, L. A. (2002). Feeding fish meal and extruded soybeans enhances the conjugated linoleic acid (CLA) content of milk. *Journal of Dairy Science*, *85*(3), 624–631. doi:10.3168/jds.S0022-0302(02)74116-7 PMID:11949867

Aharoni, Y., Orlov, A., Brosh, A., Granit, R., & Kanner, J. (2005). Effects of soybean oil supplementation of high forage fattening diet on fatty acid profiles in lipid depots of fattening bull calves, and their levels of blood vitamin E. *Animal Feed Science and Technology*, *119*(3-4), 191–202. doi:10.1016/j.anifeedsci.2005.01.002

Alfaia, P. M. C., Susana, P. A., Susana, I. V. M., Ana, S. H. C., Carlos, M. G. A. F., José, P. C. L., Rui, J. B. B., & José, A. M. P. (2009). Effect of the feeding system on intramuscular fatty acids and conjugated linoleic acid isomers of beef cattle, with emphasis on their nutritional value and discriminatory ability. *Food Chemistry*, *114*(3), 939–946. doi:10.1016/j.foodchem.2008.10.041

Allam, S. M., Ali, M. M., Mohamed, A. H., & Sayed, H. A. (2012). Impact of different forms of oils on productive performance and conjugated linoleic acid content in Goat's milk. *Egyp. J. Nutr. and Food*, *15*(1), 63–64.

Bauman, D.E., Baumgard, L.H., Corl, B.A., & Griinari, J.M. (1999). Biosynthesis of conjugated linoleic acid in ruminant. *Proc. Am. Soc. An. Sci.*, 15–77.

Bauman, D. E., & Griinari, J. M. (2001). Regulation and nutritional manipulation of milk fat: Low-fat milk syndrome. *Livestock Production Science*, *70*(1-2), 15–29. doi:10.1016/S0301-6226(01)00195-6

Bee, G. (2000). Dietary conjugated linoleic acid consumption alter adipose tissue and milk lipids of pregnant and lactating sows. *The Journal of Nutrition*, *130*(9), 2292–2298. doi:10.1093/jn/130.9.2292 PMID:10958826

Belury, M. A. (2002). Dietary conjugated linoleic acid in health: Physiological effects and mechanisms of action1. *Annual Review of Nutrition*, *22*(1), 505–531. doi:10.1146/annurev.nutr.22.021302.121842 PMID:12055356

Bent, S., Bertoglio, K., & Hendren, R. L. (2009). Omega-3 Fatty Acids for Autistic Spectrum Disorder: A Systematic Review. *Journal of Autism and Developmental Disorders*, *39*(8), 1145–1154. doi:10.100710803-009-0724-5 PMID:19333748

Bhalla, R., Singh, A. K., Pradhan, S., & Unnikumar, K. R. (2009). Lipids: structure, function and biotechnology aspects. In K. Chauhan & A. Varma (Eds.), *A Textbook of Molecular Biotechnology* (Vol. A, pp. 173–209). I K International Publishing House.

Boles, J. A., Kott, R. W., Hatfield, P. G., Bergman, J. W., & Flynn, C. R. (2005). Supplemental safflower oil affects the fatty acid profile, including conjugated linoleic acid, of lamb. *Journal of Animal Science, 83*(9), 2175–2181. doi:10.2527/2005.8392175x PMID:16100073

Bolte, M. R., Hess, B. W., Means, W. J., Moss, G. E., & Rule, D. C. (2002). Feeding lambs high-oleate or high-linoleate safflower seeds differentially influences carcass fatty acid composition. *Journal of Animal Science, 80*(3), 609–616. doi:10.2527/2002.803609x PMID:11890397

Buccioni, A., Antongiovanni, M., Petacchi, F., Meleb, M., Serra, A., Secchiari, P., & Minieri, S. (2008). Effect of dried or green herbage on vaccenic acid and conjugated linoleic acid production during in vitro rumen fermentation. *Animal Feed Science and Technology, 140*(1-2), 207–213. doi:10.1016/j.anifeedsci.2007.01.017

Cederholm, T., & Palmblad, J. (2010). Are omega-3 fatty acids options for prevention and treatment of cognitive decline and dementia? *Current Opinion in Clinical Nutrition and Metabolic Care, 13*(2), 150–155. doi:10.1097/MCO.0b013e328335c40b PMID:20019606

Chilliard, Y., Ferlay, A., & Doreau, M. (2001). Effect of different type of forages, animal fat or marine oils in cow's diet on milk fat secretion and composition, expecially conjugated linoleic acid (CLA) and polyunsaturated fatty acids. *Livestock Production Science, 70*(1-2), 31–48. doi:10.1016/S0301-6226(01)00196-8

Chilliard, Y., Ferlay, A., Rouel, J., & Lamberet, G. (2003). A review of nutritional and physiological factors affecting goat milk lipid synthesis and lipolysis. *Journal of Dairy Science, 86*(5), 1751–1770. doi:10.3168/jds.S0022-0302(03)73761-8 PMID:12778586

Chilliard, Y., Rouel, J., Ferlay, A., Bernard, L., Gaborit, P., & Raynal-Ljutovac, K. (2006). Optimising goat's milk and cheese fatty acid composition. In C. Williams & J. Buttriss (Eds.), Improving the fat content of foods (pp. 281–312). Cambridge, UK: Woodhead Publishing Limited.

Chin, S. F., Liu, W., Storkson, J. M., Ha, Y. L., & Pariza, M. W. (1992). Dietary sources of conjugated dienoic isomers of linoleic acid, a newly recognised class of anticarcinogens. *Journal of Food Composition and Analysis, 5*(3), 185–197. doi:10.1016/0889-1575(92)90037-K

Chouinard, P. Y., Corneau, L., Butler, W. R., Chiliard, Y., Drackley, J. K., & Bauman, D. E. (2001). Effect of dietary lipid source on conjugated linoleic acid concentrations in milk fat. *Journal of Dairy Science, 84*(3), 680–690. doi:10.3168/jds.S0022-0302(01)74522-5 PMID:11286421

Colomer, R., Moreno-Nogueira, J. M., García-Luna, P. P., García-Peris, P., García-de-Lorenzo, A., Zarazaga, A., Quecedo, L., del Llano, J., Usán, L., & Casimiro, C. (2007). N-3 fatty acids, cancer and cachexia: A systematic review of the literature. *British Journal of Nutrition, 97*(5), 823–831. doi:10.1017/S000711450765795X PMID:17408522

Demirel, G., Wood, J. D., & Enser, M. (2004). Conjugated linoleic acid content of the lamb muscle and liver fed different supplements. *Small Ruminant Research*, *53*(1-2), 23–28. doi:10.1016/j.smallrumres.2003.07.006

Department of Health. (1994). *Nutritional aspects of cardiovascular disease. Report on Health and Social Subjects No. 46*. HMSO.

Dhiman, T. R., Nam, S. H., & Ure, A. L. (2005). Factors affecting conjugated linoleic acid content in milk and meat. *Critical Reviews in Food Science and Nutrition*, *45*(6), 463–482. doi:10.1080/10408390591034463 PMID:16183568

Dhiman, T. R., Satter, L. D., Pariza, M. W., Galli, M. P., Albright, K., & Tolosa, M. X. (2000). Conjugated linoleic acid (CLA) content of milk from cows offered diets rich in linoleic and linolenic acid. *Journal of Dairy Science*, *83*(5), 1016–1027. doi:10.3168/jds.S0022-0302(00)74966-6 PMID:10821577

Enser, M., Hallet, B., Hewett, G. A., Fursey, G., Wood, J. D., & Harrington, G. (1998). Fatty acid content and composition of UK beef and lamb muscle in relation to production system and implications for human nutrition. *Meat Science*, *49*(3), 329–341. doi:10.1016/S0309-1740(97)00144-7 PMID:22060583

Enser, M., Scollan, N. D., Choi, N. J., Kurt, E., Hallett, K., & Wood, J. D. (1999). Effect of dietary lipid on the content of conjugated linoleic acid (CLA) in beef muscle. *Animal Science (Penicuik, Scotland)*, *69*(1), 143–146. doi:10.1017/S1357729800051171

Evangelos, C. R., Evangelia, E., Eftychia, B., Michael, S. K., & Moses, S. E. (2012). Association Between Omega-3 Fatty Acid Supplementation and Risk of Major Cardiovascular Disease Events A Systematic Review and Meta-analysis. *Journal of the American Medical Association*, *308*(10), 1024–1033. doi:10.1001/2012.jama.11374 PMID:22968891

Ferreira, E. M., Pires, A. V., Susin, I., Gentil, R. S., Parente, M. O. M., Nolli, C. P., Meneghini, R. C. M., Mendes, C. Q., & Ribeiro, C. V. D. M. (2014). Growth, feed intake, carcass characteristics, and meat fatty acid profile of lambs fed soybean oil partially replacedby fish oil blend. *Animal Feed Science and Technology*, *187*, 9–18. doi:10.1016/j.anifeedsci.2013.09.016

Ferreira, E. M., Pires, A. V., Susin, I., Gentil, R. S., Parente, M. O. M., Nolli, C. P., Meneghini, R. C. M., Mendes, C. Q., & Ribeiro, C. V. D. M. (2014). Growth, feed intake, carcass characteristics, and meat fatty acid profile of lambs fed soybean oil partially replacedby fish oil blend. *Animal Feed Science and Technology*, *187*, 9–18. doi:10.1016/j.anifeedsci.2013.09.016

Fortin, P. R., Lew, R. A., Liang, M. H., Wright, E. A., Beckett, L. A., Chalmers, T. C., & Sperling, R. I. (1995). Validation of a meta-analysis: The effects of fish oil in rheumatoid arthritis. *Journal of Clinical Epidemiology*, *48*(11), 1379–1390. doi:10.1016/0895-4356(95)00028-3 PMID:7490601

French, P., Stanton, C., Lawless, F., O'Riordan, E. G., Monahan, F. J., Caffrey, P. J., & Moloney, A. P. (2000). Fatty acid composition, including conjugated linoleic acid, of intramuscular fat from steers offered grazed grass, grass silage, or concentrate-based diets. *Journal of Animal Science*, *78*(11), 2849–2855. doi:10.2527/2000.78112849x PMID:11063308

Garcı'a, P. T., Pensel, N. A., Latimori, N. J., Kloster, A. M., Amigone, M. A., & Casal, J. J. (2005). Intramuscular lipids in steers under different grass and grain regimen. *Fleischwirtschaft International, 73*, 755–758.

Geay, Y., Bauchart, D., Hocquette, J. F., & Culioli, J. (2001). Effect of nutritional factors on biochemical, structural and metabolic characteristics of muscles in ruminants, consequences on dietetic value and sensorial qualities of meat. *Reproduction, Nutrition, Development, 41*(1), 1–26. doi:10.1051/rnd:2001108 PMID:11368241

Givens, D. I. (2005). The role of animal nutrition in improving the nutritive value of animal-derived foods in relation to chronic disease. *The Proceedings of the Nutrition Society, 64*(3), 395–402. doi:10.1079/PNS2005448 PMID:16048674

Griinari, J. M., & Bauman, D. E. (1999). Biosynthesis of conjugated linoleic acid and its incorporation into meat and milk in ruminants. In M. P. Yurawecz, M. M. Mossoba, J. K. G. Kramer, M. W. Pariza, & G. J. Nelson (Eds.), *Advances in Conjugated Linoleic Acid Research* (Vol. 1, pp. 180–200). AOCS press.

Hooper, L., Thompson, R. L., Harrison, R. A., Summerbell, C. D., Ness, A. R., Moore, H. J., Worthington, H. V., Durrington, P. N., Higgins, J. P. T., Capps, N. E., Riemersma, R. A., Ebrahim, S. B. J., & Smith, G. D. (2006). Risks and benefits of omega 3 fats for mortality, cardiovascular disease, and cancer: Systematic review. *BMJ (Clinical Research Ed.), 332*(7544), 752–760. doi:10.1136/bmj.38755.366331.2F PMID:16565093

Husvéth, F., Galamb, E., Gaal, T., Dublecz, K., Wagner, L., & Pál, L. (2010). Milk production, milk composition, liver lipid contents and C18 fatty acid composition of milk and liver lipids in Awassi ewes fed a diet supplemented with protected cis-9, trans-11 and trans-10, cis-12 conjugated linoleic acid (CLA) isomers. *Small Ruminant Research, 94*(1-3), 25–31. doi:10.1016/j.smallrumres.2010.06.003

Ivan, M., Mir, P. S., Koenig, K. M., Rode, L. M., Neill, L., Entz, T., & Mir, Z. (2001). Effects of dietary sunflower seed oil on rumen protozoa population and tissue concentration of conjugated linoleic acid in sheep. *Small Ruminant Research, 41*(3), 215–227. doi:10.1016/S0921-4488(01)00220-6

Jensen, C. L. (2006). Effects of n-3 fatty acids during pregnancy and lactation. *The American Journal of Clinical Nutrition, 83*(6), 1452–1457. doi:10.1093/ajcn/83.6.1452S PMID:16841854

Kawakita, E., Hashimoto, M., & Shido, O. (2006). Docosahexaenoic acid promotes neurogenesis in vitro and in vivo. *Neuroscience, 139*(3), 991–997. doi:10.1016/j.neuroscience.2006.01.021 PMID:16527422

Kay, J. K., Roche, J. R., Thomson, N. A., Griinari, J. M., & Shingfield, K. J. (2003). Increasing milk fat cis-9, trans-11 conjugated linoleic acid content in pasture-fed cows. *Journal of Dairy Science, 86*(Suppl. 1), 146.

Kelly, M. L., Kolver, E. S., Bauman, D. E., VanAmburgh, M. E., & Muller, L. D. (1998). Effect of intake of pasture on concentrations of conjugated linoleic acid in milk of lactating cows. *Journal of Dairy Science, 81*(6), 1630–1636. doi:10.3168/jds.S0022-0302(98)75730-3 PMID:9684170

Khanal, R. C., & Dhiman, T. R. (2004). Biosynthesis of conjugated linoleic acid (CLA): A review. *Pakistan Journal of Nutrition, 3*(2), 72–81. doi:10.3923/pjn.2004.72.81

Kott, R. W., Hatfield, P. G., Bergman, J. W., Flynn, C. R., Van Wagoner, H., & Boles, J. A. (2003). Feedlot performance, carcass composition, and muscle and fat CLA concentrations of lambs fed diets supplemented with safflower seeds. *Small Ruminant Research*, *49*(1), 11–17. doi:10.1016/S0921-4488(03)00052-X

Kwak, S. M.; Myung, S. K.; Lee, Y. J., & Seo, H.G. (2012). Efficacy of Omega-3 Fatty Acid Supplements (Eicosapentaenoic Acid and Docosahexaenoic Acid) in the Secondary Prevention of Cardiovascular Disease: A Meta-analysis of Randomized, Double-blind, Placebo-Controlled Trials. *Archives of Internal Medicine*.

Laura, J., & denHartigh. (2019). Conjugated Linoleic Acid Effects on Cancer, Obesity, and Atherosclerosis: A Review of Pre-Clinical and Human Trials with Current Perspectives. *Nutrients*, *2019*(11), 370.

Lawless, F., Stanton, C., Escop, P. L., Devery, R., Dillon, P., & Murphy, J. J. (1999). Influence of breed on bovine milk cis-9, trans-11-conjugated linoleic acid content. *Livestock Production Science*, *62*(1), 43–49. doi:10.1016/S0301-6226(99)00053-6

Lawson, R. E., Moss, A. R., & Givens, D. I. (2001). The role of dairy products in supplying conjugated linoleic acid to man's diet: A review. *Nutrition Research Reviews*, *14*(1), 153–172. doi:10.1079/095442201108729178 PMID:19087420

Lock, A. L., & Bauman, D. E. (2004). Modifying milk fat composition of dairy cows to enhance fatty acids beneficial to human health. *Lipids*, *39*(12), 1197–1206. doi:10.100711745-004-1348-6 PMID:15736916

Ludden, P. A., Kucukb, O., Rule, D. C., & Hess, B. W. (2009). Growth and carcass fatty acid composition of beef steers fed soybean oil for increasing duration before slaughter. *Meat Science*, *82*(2), 185–192. doi:10.1016/j.meatsci.2009.01.009 PMID:20416764

MacLean, C. H., Newberry, S. J., Mojica, W. A., Khanna, P., Issa, A. M., Suttorp, M. J., Lim, Y. W., Traina, S. B., Hilton, L., Garland, R., & Morton, S. C. (2006). Effects of omega-3 fatty acids on cancer risk: A systematic review. *Journal of the American Medical Association*, *295*(4), 403–415. doi:10.1001/jama.295.4.403 PMID:16434631

Mao, H. L., Wang, J. K., Lin, J., & Liu, J. X. (2012). Fatty Acid Profiles and Stearoyl-CoA Desaturase Gene Expression in Longissimusdorsi Muscle of Growing Lambs Influenced by Addition of Tea Saponins and Soybean Oil. *Journal of Animal Science*, *25*(5), 648–652. PMID:25049609

Mazereeuw, G., Lanctôt, K. L., Chau, S. A., Swardfager, W., & Herrmann, N. (2012). Effects of omega-3 fatty acids on cognitive performance: A meta-analysis. *Neurobiology of Aging*, *33*(7), 17–29. doi:10.1016/j.neurobiolaging.2011.12.014 PMID:22305186

Mensink, R. P. (2005). Effects of stearic acid on plasma lipid and lipoproteins in humans. *Lipids*, *40*(12), 1201–1205. doi:10.100711745-005-1486-x PMID:16477803

Mir, P. S., McAllister, T. A., Zaman, S., Jones, S. D. M., & He, M. L. (2003). Effect of dietary sunflower oil and vitamin E on beef cattle performance, carcass characteristics and meat quality. *Canadian Journal of Animal Science*, *83*(1), 53–66. doi:10.4141/A02-014

Moloney, A. P., Mooney, M. T., Kerry, J. P., & Troy, D. J. (2001). Producing tender and flavoursome beef with enhanced nutritional characteristics. *The Proceedings of the Nutrition Society*, *60*(2), 221–229. doi:10.1079/PNS200077 PMID:11681638

Montgomery, P., & Richardson, A. J. (2008). Omega-3 fatty acids for bipolar disorder. *Cochrane Database of Systematic Reviews*, (2).

Naliwaiko, K., Araújo, R. L., da Fonseca, R. V., Castilho, J. C., Andreatini, R., Bellissimo, M. I., Oliveira, B. H., Martins, E. F., Curi, R., Fernandes, L. C., & Ferraz, A. C. (2004). Effects of fish oil on the central nervous system: A new potential antidepressant? *Nutritional Neuroscience*, *7*(2), 91–99. doi:10.1080/10284150410001704525 PMID:15279495

National Institute of Health. (2005). *Omega-3 fatty acids, fish oil, alpha-linolenic acid*. Author.

Noci, F., Kiely, P. O., Monahan, F. J., Stanton, C., & Moloney, A. P. (2005). Conjugated linoleic acid concentration in M. Longissimusdorsi from heifers offered sunflower oil-based concentrates and conserved forages. *Meat Science*, *69*(3), 509–518. doi:10.1016/j.meatsci.2004.09.006 PMID:22062990

Nudda, A., Mc Guire, M. A., Battacone, G., & Pulina, G. (2005). Seasonal variation in conjugated linoleic acid and vaccenic acid in milk fat of sheep and its transfer to cheese and ricotta. *Journal of Dairy Science*, *88*(4), 1311–1319. doi:10.3168/jds.S0022-0302(05)72797-1 PMID:15778298

Nuernberg, K., Nuernberg, G., Ender, K., Lorenz, S., Winkler, K., & Rickert, R. (2002). n-3 fatty acids, and conjugated linoleic acids of longissimus muscle in beef cattle. *European Journal of Lipid Science and Technology*, *104*, 463–471.

Pariza, M. W., Park, Y., & Cook, M. E. (2001). The biologically active isomers of conjugated linoleic acid. *Progress in Lipid Research*, *40*(4), 283–298. doi:10.1016/S0163-7827(01)00008-X PMID:11412893

Park, Y., Albright, K. J., Liu, W., Storkson, J. M., Cook, M. E., & Pariza, M. W. (1997). Effect of conjugated linoleic acid on body composition in mice. *Lipids*, *32*(8), 853–858. doi:10.100711745-997-0109-x PMID:9270977

Park, Y., McGuire, M. K., Behr, R., McGuire, M. A., Evans, M. A., & Schultz, T. D. (1999). High-fat dairy product consumption increases delta 9c, 11t-18: 2 (rumenic acid) and total lipid concentrations of human milk. *Lipids*, *34*(6), 543–549. doi:10.100711745-999-0396-2 PMID:10405966

Patterson, R. E., Flatt, S. W., Newman, V. A., Natarajan, L., Rock, C. L., Thomson, C. A., Caan, B. J., Parker, B. A., & Pierce, J. P. (2010). Marine Fatty Acid Intake is Associated with Breast Cancer Prognosis. *The Journal of Nutrition*, *141*(2), 201–206. doi:10.3945/jn.110.128777 PMID:21178081

Peng, Y. S., Brown, M. A., Wua, J. P., & Liu, Z. (2010). Different oilseed supplements alter fatty acid composition of different adipose tissues of adult ewes. *Meat Science*, *85*(3), 542–549. doi:10.1016/j.meatsci.2010.03.003 PMID:20416838

Perica, M. M., & Delas, I. (2011). Essential fatty acids and psychiatric disorders". Nutrition in clinical practice. *Official Publication of the American Society for Parenteral and Enteral Nutrition*, *26*(4), 409–425. doi:10.1177/0884533611411306

Poulson, C. S., Dhiman, T. R., Ure, A. L., Cornforth, D., & Olson, K. C. (2004). *Conjugated linoleic acid content of beef from cattle fed diets containing high grain, CLA, or raised on forages*. Livestock Production. doi:10.1016/j.livprodsci.2004.07.012

Purba, Paengkoum, & Paengkoum. (2019). *The link between supplementary tannin level and conjugated linoleic acid (CLA) formation in ruminants: A meta-analysis A meta-analysis of tannin's level prediction on ruminalbiohydrogenation*. bioRxiv preprint first posted online Apr. 17, 2019; doi:10.1101/612523

Raes, K., de Smet, S., & Demeyer, D. (2004). Effect of dietary fatty acids on incorporation of long chain polyunsaturated fatty acids and conjugated linoleic acid in lamb, beef and pork meat: A review. *Animal Feed Science and Technology*, *113*(1-4), 199–221. doi:10.1016/j.anifeedsci.2003.09.001

Realini, C. E., Duckett, S. K., Brito, G. W., Dalla Rizza, M., & De Mattos, D. (2004). Effect of pasture vs. concentrate feeding with or without antioxidants on carcass characteristics, fatty acid composition, and quality of Uruguayan beef. *Meat Science*, *66*(3), 567–577. doi:10.1016/S0309-1740(03)00160-8 PMID:22060866

Richardson, A. J. (2006). Omega-3 fatty acids in ADHD and related neurodevelopmental disorders. *International Review of Psychiatry (Abingdon, England)*, *18*(2), 155–172. doi:10.1080/09540260600583031 PMID:16777670

Richardson, A. J., & Montgomery, P. (2005). The Oxford-Durham study: A randomized, controlled trial of dietary supplementation with fatty acids in children with developmental coordination disorder. *Pediatrics*, *115*(5), 1360–1366. doi:10.1542/peds.2004-2164 PMID:15867048

Ruggiero, C., Lattanzio, F., Lauretani, F., Gasperini, B., Andres-Lacueva, C., & Cherubini, A. (2009). Omega-3 polyunsaturated fatty acids and immune-mediated diseases: Inflammatory bowel disease and rheumatoid arthritis. *Current Pharmaceutical Design*, *15*(36), 4135–4148. doi:10.2174/138161209789909746 PMID:20041815

Scheeder, M. R. (2004). MarkanterZusatznutzenmitfunktionellenFettsauren. In M. Kreuzer, C. Wenk, & T. Lanzini (Eds.), *Lipide in Fleisch, Milch und Ei - Herausforderungfu¨ r die Tiererna¨hrung* (pp. 52–68). ETH Zurich.

Schmid, A. M., Sieber, R., & Bee, G. (2006). Conjugated linoleic acid in meat and meat products: A review. *Meat Science*, *73*(1), 29–41. doi:10.1016/j.meatsci.2005.10.010 PMID:22062051

Scollan, N. D., Enser, M., Gulati, S. K., Richardson, I., & Wood, J. D. (2003). Effects of including a ruminally protected lipid supplement in the diet on the fatty acid composition of beef muscle. *British Journal of Nutrition*, *90*(3), 709–716. doi:10.1079/BJN2003933 PMID:13129478

Secher, N. J. (2007). Does fish oil prevent preterm birth? *Journal of Perinatal Medicine*, *35*(Suppl 1), S25–S27. PMID:17302537

Stanton, C., Murphy, J., McGrath, E., & Devery, R. (2003). *Animal feeding strategies for conjugated linoleic acid enrichment of milk*. Academic Press.

Stasiniewicz, T., Strzetelski, J., Kowalczyk, J., Osieglowski, S., & Pustkowiak, H. (2000). Performance and meat quality of fattening bulls fed complete feed with rapeseed oil cake or linseed. *Journal of Animal and Feed Sciences*, *9*(2), 283–296. doi:10.22358/jafs/68046/2000

Strzetelski, J., Kowalczyk, J., Osiegowski, S., Stasiniewicz, T., Lipiarska, E., & Pustkowiak, H. (2001). Fattening bulls on maize silage and concentrate supplemented with vegetable oils. *Journal of Animal and Feed Sciences*, *10*(2), 259–271. doi:10.22358/jafs/67982/2001

Szumacher-Strabel, M., Potkanski, A., Cieslak, A., Kowalczyk, J., & Czauderna, M. (2001). The effects of different amounts and types of fat on the level of conjugated linoleic acid in the meat and milk of sheep. *Journal of Animal and Feed Sciences*, *10*(Suppl. 2), 103–108. doi:10.22358/jafs/70041/2001

Tsiplakou, E., Kominakis, A., & Zervas, G. (2008). The interaction between breed and diet on CLA and fatty acids content of milk fat of four sheep breeds kept indoors or at grass. *Small Ruminant Research*, *74*(1-3), 179–187. doi:10.1016/j.smallrumres.2007.06.008

United States Department of Agriculture (USDA). (2005). *Dietary guidelines for Americans*. Available from http://www.health.gov/dietaryguidelines/dga2005/document/

Vasta, V., Nuddab, A., Cannas, A., Lanza, M., & Priolo, A. (2008). Alternative feed resources and their effects on the quality of meat and milk from small ruminants. *Animal Feed Science and Technology*, *147*(1-3), 223–246. doi:10.1016/j.anifeedsci.2007.09.020

Wachira, A. M., Sinclair, L. A., Wilkinson, R. G., Enser, M., Wood, J. D., & Fisher, A. V. (2002). Effects of dietary fat source and breed on the carcass composition, n-3 polyunsaturated fatty acid and conjugated linoleic acid content of sheep meat and adipose tissue. *British Journal of Nutrition*, *88*(6), 697–709. doi:10.1079/BJN2002727 PMID:12493092

Wall, R., Ross, R. P., Fitzgerald, G. F., & Stanton, C. (2010). Fatty acids from fish: The anti-inflammatory potential of long-chain omega-3 fatty acids. *Nutrition Reviews*, *68*(5), 280–289. doi:10.1111/j.1753-4887.2010.00287.x PMID:20500789

Whitlock, L., Schingoethe, D.J., Hippen, A.R., Kalscheur, K.F., Baer, R.J., & Ramaswamy, N., & Kasperson, K.M. (2002). Fish oil and extruded soybeans fed in combination increase CLA in milk of dairy cows more than when fed separately. *Journal of Dairy Science*, *85*, 234–243. doi:10.3168/jds.S0022-0302(02)74072-1 PMID:11860116

Compilation of References

Abbas, F., Fares, A., & Samira, F. (2011). Field Calibrations of Soil Moisture Sensors in a Forested Watershed. *Sensors (Basel), 11*(6), 6354–6369. doi:10.3390110606354 PMID:22163959

Abbas, Z., You, K. Y., Shaari, A. H., Khalid, K., Hassan, J., & Saion, E. (2005). Complex permittivity and moisture measurements of oil palm fruits using an open-ended coaxial sensor. *IEEE Sensors Journal, 5*(6), 1281–1287. doi:10.1109/JSEN.2005.859249

Abdullah, M. N., Osman, N., & Ali, F. H. (2011). Soil-root shear strength properties of some slope plants. *Sains Malaysiana, 40*, 1065–1073.

Abdulwaheed, A. (2019). Benefits Of Precision Agriculture in Nigeria. London Journal of Research in Science: Natural and Formal. 19(2). *Compilation, 1*(0), 19–34.

Abegaonkar, M. P., Karekar, R. N., & Aiyer, R. C. (1999). A microwave microstrip ring resonator as a moisture sensor for biomaterials: Application to wheat grains. *Measurement Science & Technology, 10*(3), 195–200. doi:10.1088/0957-0233/10/3/014

Abobatta, W. F. (2020) Precision Agriculture Age. *Open Acc J Agri Res.*

Abobatta, W. F. (2018). Nanotechnology Application in Agriculture. *Acta Scientific Agriculture, 2*(6), 99–102.

AbuGhazaleh, A. A. (2008). Effect of fish oil and sunflower oil supplementation on milk conjugated linoleic acid content for grazing dairy cows. J. *Animal Feed Science and Technology, 141*(3-4), 220–232. doi:10.1016/j.anifeedsci.2007.06.027

AbuGhazaleh, A. A., Schingoethe, D. J., Hippen, A. R., & Whitlock, L. A. (2002). Feeding fish meal and extruded soybeans enhances the conjugated linoleic acid (CLA) content of milk. *Journal of Dairy Science, 85*(3), 624–631. doi:10.3168/jds.S0022-0302(02)74116-7 PMID:11949867

Adamchuk, V. I., Hummel, J. W., Morgan, M. T., & Upadhyaya, S. K. (2004). On-the-go soil sensors for precision agriculture. *Computers and Electronics in Agriculture, 44*(1), 71–91. doi:10.1016/j.compag.2004.03.002

Adamchuk, V. I., Morgan, M. T., & Ess, D. R. (1999). An automated sampling system for measuring soil pH. *Transactions of the ASAE. American Society of Agricultural Engineers, 42*(4), 885–891. doi:10.13031/2013.13268

Aditya Shastry, K., & Sanjay, H. A. (2020). Data Analysis and Prediction Using Big Data Analytics in Agriculture. In *Internet of Things and Analytics for Agriculture* (Vol. 2, pp. 201–224). Springer Nature Singapore., doi:10.1007/978-981-15-0663-5_10

Adrian, A. M., Norwood, S. H., & Mask, P. L. (2005). Producers' perceptions and attitudes toward precision agriculture technologies. *Computers and Electronics in Agriculture, 48*(3), 256–271. doi:10.1016/j.compag.2005.04.004

Aggelopoulou, K. D., Wulfsohn, D., Fountas, S., Gemtos, T. A., Nanos, G. D., & Blackmore, S. (2010). Spatial variation in yield and quality in a small apple orchard. *Precision Agriculture*, *11*(5), 538–556. doi:10.100711119-009-9146-9

Aharoni, Y., Orlov, A., Brosh, A., Granit, R., & Kanner, J. (2005). Effects of soybean oil supplementation of high forage fattening diet on fatty acid profiles in lipid depots of fattening bull calves, and their levels of blood vitamin E. *Animal Feed Science and Technology*, *119*(3-4), 191–202. doi:10.1016/j.anifeedsci.2005.01.002

Ahmad, A. F., Abbas, Z., Abrass, H. A., & You, K. Y. (2019). Optimum design of a microstrip ring resonator sensor to determine the moisture content in oil palm fruits and seeds. *BioResources*, *14*(1), 1819–1837.

Ahmad, L., & Mahdi, S. S. (2018). Recent Advances in Precision Agriculture. In *Satellite Farming* (pp. 129–138). Springer. doi:10.1007/978-3-030-03448-1_10

Ahmed, J., Ramaswamy, H. S., & Raghavan, V. G. S. (2007). Dielectric properties of Indian Basmati rice flour slurry. *Journal of Food Engineering*, *80*(4), 1125–1133. doi:10.1016/j.jfoodeng.2006.09.004

Ahmed, S. A., Ross, S. A., Slade, D., Radwan, M. M., Zulfiqar, F., & ElSohly, M. A. (2008). Cannabinoid ester constituents from high potency Cannabis sativa L. *Planta Medica*, *74*(03), 536–542. doi:10.1055-2008-1075212

AIMX. (2015). *Barreras de acceso a Internet.* Retrieved December 15, 2019, from https://www.asociaciondeinternet.mx/es/estudios

Aitken-Chrisie, J., Kozai, T., & Smith, M A L. (1994). *Automation and Environmental Control in Plant Tissue Culture.* Kluwer Academic Publishers.

Akdemir, H., Süzerer, V., Tilkat, E., Yildirim, H., Onay, A., & Çiftçi, Y. O. (2013). In vitro conservation and cryopreservation of mature pistachio (*Pistacia vera* L.) germplasm. *Journal of Plant Biochemistry and Biotechnology*, *22*(1), 43–51. doi:10.100713562-012-0109-2

Akyildiz, I. F., Su, W., Sankarasubramaniam, Y., & Cayirci, E. (2002). Wireless sensor networks: A survey. *Computer Networks*, *38*(4), 393–422. doi:10.1016/S1389-1286(01)00302-4

Al-Ababneh, S. S., Shibli, R. A., Karam, N. S., & Shatnawi, M. A. (2003). Cryopreservation of bitter almond (Amygdalus communis L.) shoot tips by encapsulation - dehydration and vitrification. *Advanced HortScience*, *17*, 15–20.

Al-Bahrany, A. M., & Al-Khayri, J. M. (2012). Optimizing *in vitro* Cryopreservation of Date Palm (*Phoenix dactylifera* L.). *Biotechnology (Faisalabad)*, *11*(2), 59–66. doi:10.3923/biotech.2012.59.66

Alexandratos, N., & Bruinsma, J. (2012). *World Agriculture Towards 2030/2050.* Academic Press.

Alfaia, P. M. C., Susana, P. A., Susana, I. V. M., Ana, S. H. C., Carlos, M. G. A. F., José, P. C. L., Rui, J. B. B., & José, A. M. P. (2009). Effect of the feeding system on intramuscular fatty acids and conjugated linoleic acid isomers of beef cattle, with emphasis on their nutritional value and discriminatory ability. *Food Chemistry*, *114*(3), 939–946. doi:10.1016/j.foodchem.2008.10.041

Al-Hamdani, A. S., & Al-Shimmary, H. M. (2020). Characterization and Classification of Soil Map Units by Using Remote Sensing and GIS in Bahar Al-Najaf, Iraq. In *Environmental Remote Sensing and GIS in Iraq* (pp. 61–92). Springer. doi:10.1007/978-3-030-21344-2_4

Ali, M., Mubeen, M., Hussain, N., Wajid, A., Farid, H. U., Awais, M., … Nasim, W. (2019). Role of ICT in Crop Management. In M. Hasanuzzaman (Ed.), Agronomic Crops Volume 2: Management Practices (pp. 637–652). Springer Nature Singapore. doi:10.1007/978-981-32-9783-8_28

Ali, F. H., & Osman, N. (2008). Shear strength of a soil containing vegetation roots. *Soil and Foundation, 48*(4), 587–596. doi:10.3208andf.48.587

Allahverdi, N. (2002). *Expert Systems: An Artificial Intelligence Application.* Atlas Press.

Allam, S. M., Ali, M. M., Mohamed, A. H., & Sayed, H. A. (2012). Impact of different forms of oils on productive performance and conjugated linoleic acid content in Goat's milk. *Egyp. J. Nutr. and Food, 15*(1), 63–64.

Allen, E. W. Jr, & Garlan, H. (1962). Evolution of regulatory standards of interference. *Proceedings of the IRE, 50*(5), 1306–1311. doi:10.1109/JRPROC.1962.288087

Altman, N. S. (1992). An introduction to kernel and nearest-neighbor nonparametric regression. *The American Statistician, 46*(3), 175–185.

Alvarado-Beltrán, G., López-Sánchez, H., Santacruz-Varela, A., Muñoz-Orozco, A., Valadez-Moctezuma, E., Gutiérrez-Espinosa, M. A., ... Taboada-Gaytán, O. R. (2019). Morphological variability of native maize (Zea mays L.) of the west highland of puebla and east highland of tlaxcala, Mexico. *Revista de La Facultad de Ciencias Agrarias, 51*(2), 217–234.

Alvarenga, I. C. A., Pacheco, F. V., Silva, S. T., Bertolucci, S. K. V., & Pinto, J. E. B. P. (2015). In vitro culture of *Achillea millefolium* L.: Quality and inten- sity of light on growth and production of volatiles. *Plant Cell, Tissue and Organ Culture, 122*(2), 299–308. doi:10.100711240-015-0766-7

Amacher, M. C., O'Neill, K. P., & Perry, C. H. (2007). Soil vital signs: A new soil quality index (SQI) for assessing forest soil health. *USDA Forest Service - Research Paper RMRS-RP*, (65 RMRS-RP), 1–14. doi:10.2737/RMRS-RP-65

Ameen & Raza. (2017). Green Revolution: A Review. *International Journal of Advances in Scientific Research, 3*(12), 129–137.

Andrade, H. B., Braga, A. F., Bertolucci, S. K. V., Hsie, B. S., Silva, S. T., & Pinto, J. E. B. P. (2017). Effect of plant growth regulators, light intensity and LED on growth and volatile compound of *Hyptis suaveolens* (L.) Poit in vitro plantlets. *Acta Horticulturae*, (1155), 277–284. doi:10.17660/ActaHortic.2017.1155.40

Angelopoulou, T., Tziolas, N., Balafoutis, A., Zalidis, G., & Bochtis, D. (2019). Remote sensing techniques for soil organic carbon estimation: A review. *Remote Sensing, 11*(6), 1–18. doi:10.3390/rs11060676

Anjum, M., & Pradhan, S. N. (2018). Application of nanotechnology in precision farming: A review. *International Journal of Chemical Studies, 6*(5), 755–760.

Ansarudin, F., Abbas, Z., Hassan, J., Yahaya, N. Z., & Ismail, M. A. (2012). A simple insulated monopole sensor technique for determination of moisture content in Hevea rubber latex. *Measurement Science Review, 12*(6), 249–254. doi:10.2478/v10048-012-0034-5

Arda, H., Dayan, S., Kartal, C., & Guler, N. (2016). *In vitro* conservation of critically endangered *Dianthus ingoldbyi*Turrill under slow growth conditions. *Trakya UniversityJournal of Natural Sciences, 17*, 47–54.

Arena, C., Tsonev, T., Doneva, D., De Micco, V., Michelozzi, M., Brunetti, C., Centritto, M., Fineschi, S., Velikova, V., & Loreto, F. (2016). The effect of light quality on growth, photosynthesis, leaf anatomy and volatile isoprenoids of a monoterpene-emitting herbaceous species (*Solanum lycopersicum* L.) and an isoprene-emitting tree (*Platanus orientalis* L.). *Environmental and Experimental Botany, 130*, 122–132. doi:10.1016/j.envexpbot.2016.05.014

Ariffin, A., Mat, S. R., Banjari, M., & Wan, O. W. E. (1990). Morphological changes of the cellular component of the developing palm fruit (Tenera: Elaeisguineensis). *Palm Oil Research Institute of Malaysia Bulletin, 21*, 30–34.

Ashmore, S. E. (1997). *Status report on the development and application of in vitro techniques for the conservation of plant genetic resources*. International Plant Genetic Resources Institute.

Atkinson, J. K., Glanc, M., Prakorbjanya, M., Sophocleous, M., Sion, R., & Garcia Breijo, E. (2013). Thick film screen printed environmental and chemical sensor array reference electrodes suitable for subterranean and subaqueous deployments. *Microelectronics International, 30*(2), 92–98. doi:10.1108/13565361311314485

Atmojo, U. D., Salcic, Z., Kevin, I., Wang, K., & Park, H. (2015). System-level approach to the design of ambient intelligence systems based on wireless sensor and actuator networks. *Journal of Ambient Intelligence and Humanized Computing, 6*(2), 153–169. doi:10.100712652-014-0221-3

Aubert, B. A., Schroeder, A., & Grimaudo, J. (2012). IT as enabler of sustainable farming: An empirical analysis of farmers' adoption decision of precision agriculture technology. *Decision Support Systems, 54*(1), 510–520. doi:10.1016/j.dss.2012.07.002

Aubin, M.-P., Seguin, P., Vanasse, A., Tremblay, G. F., Mustafa, A. F., & Charron, J.-B. (2015). Industrial hemp response to nitrogen, phosphorus, and potassium fertilization. *Crop. Forage TurfgrassMgt, 1*(1), 1–10. doi:10.2134/cftm2015.0159

Auernhammer, H., & Muhr, T. 1991. GPS in a basic rule for environment protection in agriculture. In G. Kranzler (Ed.), Automated Agriculture for the 21st Century. ASAE Publ.

Auernhammer, H., Demmel, M., Muhr, T., Rottmeier, J., & Wild, K. (1994). GPS for yield mapping on combines. *Computers and Electronics in Agriculture, 11*(1), 53–68. doi:10.1016/0168-1699(94)90052-3

Augustin, A., Yi, J., Clausen, T., & Townsley, W. M. (2016). A study of LoRa: Long range & low power networks for the internet of things. *Sensors (Basel), 16*(9), 1466. doi:10.339016091466 PMID:27618064

Badia, T., & García, C. (2006). The incorporation of information and communication technologies into teaching and learning based on collaborative project work. *RUSC. Universities and Knowledge Society Journal, 3*(2). Advance online publication. doi:10.7238/rusc.v3i2.286

Bae, Borgelt, Searcy, Schueller, & Stout. (1987). Determination of spatially variable yield maps. *Am. Soc. Agric. Eng.*

Bagher, A.M. (2016). A Detailed Review on Types of lamps and their applications. *J. of Physical and Chemical Science, 4I1*. Doi:10.15297/JPCS.V4I1.02

Bagheri, N., & Bordbar, M. (2014). Solutions for fast development of precision agriculture in Iran. *Agricultural Engineering International, 16*(3), 119–123.

Bakro, F., Wielgusz, K., Bunalski, M., & Jedryczka, M. (2018). An overview of pathogen and insect threats to fibre and oilseed hemp (Cannabis sativa L.) and methods for their biocontrol. *Integrated Control in Oilseed Crops IOBC-WPRS Bulletin, 136*, 9–20.

Balafoutis, A., Beck, B., Fountas, S., Vangeyte, J., van der Wal, T., Soto, I., ... Eory, V. (2017). Precision Agriculture Technologies Positively Contributing to GHG Emissions Mitigation, Farm Productivity and Economics. *Sustainability, 9*(8), 1339. doi:10.3390u9081339

Ballaré, C.L. (2014). Light regulation of plant defense. *Annu Rev Plant Biol, 65*, 15.1–15.29.

Ballester, A., Janeiro, L.V., & Vieitez, A.M. (1997). Cold storage of shoot cultures and alginate encapsulation of shoot tips of *Camellia japonica* L. and *Camellia reticulate* Lindley. *Scientia Horticulturae, 71*, 67-78.

Banerjee, N., & de Langhe, E. (1985). A tissue culture technique for rapid clonal propagation and storage under minimal growth conditions of Musa spp (banana and plantain). *Plant Cell Reports*, *4*(6), 341–354. doi:10.1007/BF00269897 PMID:24254081

Banhazi, T. M., Babinszky, L., Halas, V., & Tscharke, M. (2012). Precision Livestock Farming: Precision feeding technologies and sustainable animal production. *International Journal of Agricultural and Biological Engineering*, *5*(4), 1–7.

Bantis, F., Ouzounis, T., & Radoglou, K. (2016). Artificial LED lighting enhances growth characteristics and total phenolic content of *Ocimum basilicum*, but variably affects transplant success. *Scientia Horticulturae*, *198*, 277–283. doi:10.1016/j.scienta.2015.11.014

Barnes, E. M., Moran, M. S., Pinter, P. J., & Clarke, T. R. (1996). Multispectral remote sensing and site-specific agriculture: Examples of current technology and future possibilities. *Precision Agriculture*, *3*, 845–854.

Bar-on, L., Jog, A., & Shacham-Diamand, Y. (2019). Four Point Probe Electrical Spectroscopy Based System for Plant Monitoring. In *Proceedings of ISCAS 2019, Japan* (pp. 1–5). 10.1109/ISCAS.2019.8702623

Bar-Yosef, B. (1999). Advances in fertigation. *Advances in Agronomy*, *65*, 1–77. doi:10.1016/S0065-2113(08)60910-4

Batista, D. S., Castro, K. M., Silva, A. R., Teixeira, M. L., Sales, T. A., Soares, L. I., Cardoso, M. G., Santos, M. O., Viccini, L. F., & Otoni, W. C. (2016). Light quality affects in vitro growth and essential oil profile in *Lippia alba* (Verbenaceae). *In Vitro Cellular & Developmental Biology. Plant*, *52*(3), 276–282. doi:10.100711627-016-9761-x

Batista, D. S., Felipe, S. H. S., Silva, T. D., de Castro, K. M., Mamedes-Rodrigues, T. C., Miranda, N. A., Ríos-Ríos, A. M., Faria, D. V., Fortini, E. A., Chagas, K., Torres-Silva, G., Xavier, A., Arencibia, A. D., & Otoni, W. C. (2018). Light quality in plant tissue culture: Does it matter? *In Vitro Cellular & Developmental Biology. Plant*, *54*(3), 195–215. doi:10.100711627-018-9902-5

Bauman, D.E., Baumgard, L.H., Corl, B.A., & Griinari, J.M. (1999). Biosynthesis of conjugated linoleic acid in ruminant. *Proc. Am. Soc. An. Sci.*, 15–77.

Bauman, D. E., & Griinari, J. M. (2001). Regulation and nutritional manipulation of milk fat: Low-fat milk syndrome. *Livestock Production Science*, *70*(1-2), 15–29. doi:10.1016/S0301-6226(01)00195-6

Bausch, W. C., & Duke, H. R. (1996). Remote sensing of plant nitrogen status in corn. *Transactions of the ASAE. American Society of Agricultural Engineers*, *39*(5), 1869–1875. doi:10.13031/2013.27665

Baykara, C. K. (2018). Wireless Network Based Grain Control System Design For Grain Bagging System (Silobag) (Master of Science Master of Science). Erciyes University, Kayseri, Turkey.

Beaman, A. R., Gladon, R. J., & Schrader, J. A. (2009). Sweet basil requires an irradiance of 500 m mol·m^{-2}·s^{-1} for greatest edible biomass production. *HortScience*, *44*(1), 64–67. doi:10.21273/HORTSCI.44.1.64

Beck, J. L., & Kinter, M. L. (1998a). *Photodetector circuit for an electronic sprayer*. U.S. Patent No. 5,763,873, issued June 9, 1998.

Beck, J. L., & Kinter, M. L. (1998b). *Agricultural implement having multiple agents for mapping fields*. U.S. Patent No. 5,809,440, issued September 15, 1998.

Bee, G. (2000). Dietary conjugated linoleic acid consumption alter adipose tissue and milk lipids of pregnant and lactating sows. *The Journal of Nutrition*, *130*(9), 2292–2298. doi:10.1093/jn/130.9.2292 PMID:10958826

Bekheet, S. A., Matter, M. A., Taha, H. S., & El-Ashry, A. A. (2016). *In vitro* conservation of Jojoba (*Simmondsia chinensis*) shootlet cultures using osmotic stress and low temperature. *Middle East Journal of Agriculture Research*, *5*(4), 396–402.

Bekheet, S. A., Taha, H. S., & Saker, M. M. (2001). *In vitro* long-term storage of date palm. *Biologia Plantarum*, *45*(1), 121–124. doi:10.1023/A:1015121111934

Bekheet, S. A., Taha, H. S., Saker, M. M., & Solliman, M. E. (2007). Application of cryopreservation technique for in vitro grown date palm (Phoenix dactylifera L.) cultures. *Journal of Applied Sciences Research*, *9*, 859–866.

Bekker, M. G. (1969). *Introduction to terrain vehicle systems*. The Univ. of Michigan Press.

Belury, M. A. (2002). Dietary conjugated linoleic acid in health: Physiological effects and mechanisms of action1. *Annual Review of Nutrition*, *22*(1), 505–531. doi:10.1146/annurev.nutr.22.021302.121842 PMID:12055356

Benelli, C. (2016). Encapsulation of shoot tips and nodal segments for *in vitro* storage of "Kober 5BB" grapevine rootstock. *Horticulturae*, *2*(3), 10. doi:10.3390/horticulturae2030010

Benelli, C., Micheli, M., & Carlo, A. (2017). An improved encapsulation protocol for regrowth and conservation of four ornamental species. *Acta Societatis Botanicorum Poloniae*, *86*(3), 1–12. doi:10.5586/asbp.3559

Benelli, C., Ozudogru, E. A., Lambardi, M., & Dradi, G. (2012). *In vitro* conservation of ornamental plants by slow growth storage. *Acta Horticulturae*, (961), 89–93. doi:10.17660/ActaHortic.2012.961.8

Benton Jones, J. J. (2001). *Laboratory Guide for Conducting Soil Tests and Plant Analysis*. CRC Press. doi:10.1201/9781420025293

Bent, S., Bertoglio, K., & Hendren, R. L. (2009). Omega-3 Fatty Acids for Autistic Spectrum Disorder: A Systematic Review. *Journal of Autism and Developmental Disorders*, *39*(8), 1145–1154. doi:10.100710803-009-0724-5 PMID:19333748

Bernard, D. G., Cheng, Y., Zhao, Y., & Balk, J. (2009). An allelic mutant series of ATM3 reveals its key role in the biogenesis of cytosolic iron-sulfur proteins in Arabidopsis. *Plant Physiology*, *151*(2), 590–602. doi:10.1104/pp.109.143651 PMID:19710232

Berthold Technologies. (2007). *Contactless Measurement of Water Content and Moisture-Micro-Moist LB 456*. Berthold Technologies.

Bertrand-Desbrunais, ANoirot, MCharrier, A. (1992). Slow growth of *in vitro* conservation of coffee (*Coffea spp.*). *Plant Cell, Tissue and Organ Culture*, *1*, 105–110.

Bhalla, R., Singh, A. K., Pradhan, S., & Unnikumar, K. R. (2009). Lipids: structure, function and biotechnology aspects. In K. Chauhan & A. Varma (Eds.), *A Textbook of Molecular Biotechnology* (Vol. A, pp. 173–209). I K International Publishing House.

Bhandari, S., Raheja, A., Green, R. L., & Do, D. (2017, May). Towards collaboration between unmanned aerial and ground vehicles for precision agriculture. In *Autonomous Air and Ground Sensing Systems for Agricultural Optimization and Phenotyping II*. International Society for OpticsandPhotonics.

Bhatia, S. (2015). Application of plant biotechnology. In *Modern Applications of Plant Biotechnology in Pharmaceutical Sciences* (pp. 157–207). Elsevier Inc. doi:10.1016/B978-0-12-802221-4.00005-4

Bhatti, A. U., Mulla, D. J., & Frazier, B. E. (1991). Estimation of soil properties and wheat yields on complex eroded hills using geostatistics and Thematic Mapper images. *Remote Sensing of Environment*, *37*(3), 181–191. doi:10.1016/0034-4257(91)90080-P

Bhunia, G. S. (2019). Horticulture Development in India: Issues and Scenario of Space Technology. *World J Agri& Soil Sci.*, *3*(1), 1–5. doi:10.33552/WJASS.2019.03.000555

Bi, W. L. (2017). *Cryopreservation of Vitis spp. and cryotherapy for eradication of grapevine viruses* (PhD thesis). Northwest A and F University, Yangling, China.

Bian, Z., Guo, X., Wang, S., Zhuang, Q., Jin, X., Wang, Q., & Jia, S. (2020). Applying statistical methods to map soil organic carbon of agricultural lands in northeastern coastal areas of China. *Archives of Agronomy and Soil Science*, *66*(4), 532–544. doi:10.1080/03650340.2019.1626983

Blackmer, T. M., Schepers, J. S., & Meyer, G. E. (1995). Remote sensing to detect nitrogen deficiency in corn. In P. C. Robert, R. H. Rust, & W. E. Larson (Eds.), *Site-Specific Management for Agricultural Systems* (pp. 505–512). ASA-CSSA-SSSA. doi:10.2134/1995.site-specificmanagement.c35

Blackmer, T. M., Schepers, J. S., Varvel, G. E., & Meyer, G. E. (1996a). Analysis of aerial photography for nitrogen stress within corn fields. *Agronomy Journal*, *88*(5), 729–733. doi:10.2134/agronj1996.00021962008800050008x

Bligaard, J. (2013). *Identified user requirements for precision farming in Germany, Finland and Denmark*. Project Report. Available online at https://www.landbrugsinfo.dk/Planteavl/Praecisionsjordbrug-og-GIS/Filer/pl_13_1370_b1.pdf

Blom, T. J., Straver, W. A., Ingratta, F. J., Khosla, S., & Brown, W. (2016). *Carbon dioxide in greenhouses*. Ontario Ministry of Agriculture, Food, and Rural Affairs. http://www. omafra.gov.on.ca/english/crops/facts/00-077.htm#suppl

Bócsa, I., & Karus, M. (1998). The cultivation of hemp. Botany, varieties, cultivation and harvesting. Hemptech.

Boison, A. M., Gout, E., Bligny, R., & Rivasseau, C. (2012). A simple and efficient method for the long-term preservation of plant cell suspension cultures. *Plant Methods*, *8*(4), 132–144. doi:10.1186/1746-4811-8-4 PMID:22289515

Boles, J. A., Kott, R. W., Hatfield, P. G., Bergman, J. W., & Flynn, C. R. (2005). Supplemental safflower oil affects the fatty acid profile, including conjugated linoleic acid, of lamb. *Journal of Animal Science*, *83*(9), 2175–2181. doi:10.2527/2005.8392175x PMID:16100073

Bolte, M. R., Hess, B. W., Means, W. J., Moss, G. E., & Rule, D. C. (2002). Feeding lambs high-oleate or high-linoleate safflower seeds differentially influences carcass fatty acid composition. *Journal of Animal Science*, *80*(3), 609–616. doi:10.2527/2002.803609x PMID:11890397

Bonfante, A. (2019). LCIS DSS—an irrigation supporting system for water use efficiency improvement in precision agriculture: A maize case study. *Network Protocols and Algorithms.*, *176*, 102646.

Bonga, J. M. (1996). Frozen storage stimulates the formation of embryo-like structures and elongating shoots in explants from mature *Larix decidua* and *L.* x *eurolepis* trees. *Plant Cell, Tissue and Organ Culture*, *46*(2), 91–101. doi:10.1007/BF00034841

Bonga, J. M. (1997). The effect of collection date and frozen storage on the formation of embryo-like structures and elongating shoots from explants from mature *Larix decidua* and L. x *eurolepis*. *Plant Cell, Tissue and Organ Culture*, *51*(3), 195–200. doi:10.1023/A:1005932628498

Bongiovanni, R., & Lowenberg-DeBoer, J. (2004). Precision agriculture and sustainability. *Precision Agriculture*, *5*(4), 359–387. doi:10.1023/B:PRAG.0000040806.39604.aa

Bordoni, M., Cislaghi, A., Vercesi, A., Bischetti, G. B., & Meisina, C. (2020). Effects of plant roots on soil shear strength and shallow landslideproneness in an area of northern Italian Apennines. *Bulletin of Engineering Geology and the Environment*. Advance online publication. doi:10.100710064-020-01783-1

Bordoni, M., Meisina, C., Vercesi, A., Bischetti, G. B., Chiaradia, E. A., Vergani, C., Chersich, S., Valentino, R., Bittelli, M., Comolli, R., Persichillo, M. G., & Cislaghi, A. (2016). Quantifying the contribution of grapevine roots to soil mechanical reinforcement in an areasusceptible to shallow landslides. *Soil & Tillage Research, 163*, 195–206. doi:10.1016/j.still.2016.06.004

Borgelt, S. C., Searcy, S. W., Stout, B. A., & Mulla, D. J. (1989). *A method for determining spatially variable liming rates*. Paper No. 89-1034, International Summer Meeting, American Society of Agricultural Engineers, Quebec City, Quebec, Canada.

Borgelt, S. C., Searcy, S. W., Stout, B. A., & Mulla, D. J. (1994). Spatially variable liming rates: A method for determination. *Transactions of the ASAE. American Society of Agricultural Engineers, 37*(5), 1499–1507. doi:10.13031/2013.28233

Borges, L. F., Bazzi, C. L., Souza, E. G. D., Magalhães, P. S. G., & Michelon, G. K. (2020). Web software to create thematic maps for precision agriculture. *Pesquisa Agropecuária Brasileira, 55*, 55. doi:10.15901678-3921.pab2020.v55.00735

Borghi, E., Avanzi, J. C., Bortolon, L., Junior, A. L., & Bortolon, E. S. O. (2016). Adoption and use of precision agriculture in Brazil: Perception of growers and service dealership. *The Journal of Agricultural Science, 8*(11), 89–104. doi:10.5539/jas.v8n11p89

Bourget, C. M. (2008). An introduction to light-emitting diodes. *HortScience, 43*(7), 1944–1946. doi:10.21273/HORTSCI.43.7.1944

Brader, B., Lee, R. C., Plarre, R., Burkholder, W., Kitto, G. B., Kao, C., Polston, L., Dorneanu, E., Szabo, I., Mead, B., Rouse, B., Sullins, D., & Denning, R. (2002). A comparison of screening methods for insect contamination in wheat. *Journal of Stored Products Research, 38*(1), 75–86. doi:10.1016/S0022-474X(01)00006-6

Bramley, R. G. V., Ouzman, J., & Gobbett, D. L. (2019). Regional scale application of the precision agriculture thought process to promote improved fertilizer management in the Australian sugar industry. *Precision Agriculture, 20*(2), 362–378. doi:10.100711119-018-9571-8

Brevik, E. C., Calzolari, C., Miller, B. A., Pereira, P., Kabala, C., Baumgarten, A., & Jordán, A. (2016). Soil mapping, classification, and pedologic modeling: History and future directions. *Geoderma, 264*, 256–274. doi:10.1016/j.geoderma.2015.05.017

Brown, C. S., Schuerger, A. C., & Sager, J. C. (1995). Growth and photomorphogenesis of pepper plants under red light-emitting diodes with supplemental blue or far-red lighting. *Journal of the American Society for Horticultural Science, 120*(5), 808–813. doi:10.21273/JASHS.120.5.808 PMID:11540133

Brown, D. C. W., Leung, D. W. M., & Thrope, T. A. (1979). Osmotic requirements for shoot formation in tobacco callus. *Physiologia Plantarum, 46*(1), 36–41. doi:10.1111/j.1399-3054.1979.tb03182.x

Brown, D. M., Zeef, L. A. H., Ellis, J., Goodacre, R., & Turnera, S. R. (2005). Identification of novel genes in Arabidopsis involved in secondary cell wall formation using expression profiling and reverse genetics. *The Plant Cell, 17*(8), 2281–2295. doi:10.1105/tpc.105.031542 PMID:15980264

Buccioni, A., Antongiovanni, M., Petacchi, F., Meleb, M., Serra, A., Secchiari, P., & Minieri, S. (2008). Effect of dried or green herbage on vaccenic acid and conjugated linoleic acid production during in vitro rumen fermentation. *Animal Feed Science and Technology, 140*(1-2), 207–213. doi:10.1016/j.anifeedsci.2007.01.017

Budiarto, K. (2010). Spectral quality affects morphogenesis on Anthurium plantlet during in vitro culture. *Agrivita, 32*, 234–240.

Cadena Iñiguez, P., Rendón-Medel, R., Rodríguez-Vázquez, H., Camacho-Villa, C., Santellano-Estrada, E., Guevara-Hernández, F., & Govaerts-Bram, B. (2018). Propuesta metodológica-interinstitucional para un nuevo extensionismo en México. *Revista Mexicana de Ciencias Agrícolas*, *9*(8), 1777–1785. doi:10.29312/remexca.v9i8.826

Canals, A. (2003). Gestión del conocimiento (M. Boisot & A. Cornella, Eds.). Academic Press.

Caplan, D., Dixon, M., & Zheng, Y. (2017a). Optimal Rate of Organic Fertilizer during the Flowering Stage for Cannabis Grown in Two Coir-based Substrates. *HortScience*, *52*(12), 1796–1803. doi:10.21273/HORTSCI12401-17

Caplan, D., Dixon, M., & Zheng, Y. (2017b). Optimal rate of organic fertilizer during the vegetative-stage for cannabis grown in two coir-based substrates. *HortScience*, *52*(9), 1307–1312. doi:10.21273/HORTSCI11903-17

Carman, K., & Seflek, A. Y. (2004). Fuzzy expert system design for tyre traction performance. *J. Agric. Mach. Sci.*, *1*(1), 63–68.

Carr, P. M., Carlson, G. R., Jacobsen, J. S., Nielsen, G. A., & Skogley, E. O. (1991). Farming soils, not fields: A strategy for increasing fertilizer profitability. *Journal of Production Agriculture*, *4*(1), 57–61. doi:10.2134/jpa1991.0057

Carter, Rhoades, & Chesson. (1993). *Mechanization of soil salinity assessment for mapping*. ASAE Paper 931557.

Carvahlo, V., Santos, D. S., & Nievola, C. C. (2014). *In vitro* storage under slow growth and ex vitro acclimatization ofthe ornamental bromeliad *Acanthostachys strobilacea. South African Journal of Botany*, *92*, 39–43. doi:10.1016/j.sajb.2014.01.011

Castilla, N. (1997). Greenhouse drip irrigation management and water saving. *Cahiers Options Méditerranéennes*, *31*, 189–202.

Cederholm, T., & Palmblad, J. (2010). Are omega-3 fatty acids options for prevention and treatment of cognitive decline and dementia? *Current Opinion in Clinical Nutrition and Metabolic Care*, *13*(2), 150–155. doi:10.1097/MCO.0b013e328335c40b PMID:20019606

Ceinsys, T. L. (2018). *Applications of satellite imagery & remote sensing data*. Retrieved from https://www.ceinsys.com/blog/applications-of-satellite-imagery-remote-sensing-data/

Central Bureau of Statistics (CBS). (2010). Indonesia in Number Book. Statistical Yearbook of Indonesia 2010.Badan-PusatStatistik, BPS.

Central Bureau of Statistics (CBS). (2011). Indonesia in Number Book. Statistical Yearbook of Indonesia 2011.Badan-PusatStatistik, BPS.

Central Bureau of the CIE. (n.d.). *Vienna, Book Review-13*. doi:10.1002/col.5080130115

Cervantes, G. (2006). *Marijuana Horticulture: The Indoor/Outdoor Medical Grower's Guide*. Van Patten Publishing.

Chakraborty, M., Khot, L. R., & Peters, R. T. (2020). Assessing suitability of modified center pivot irrigation systems in corn production using low altitude aerial imaging techniques. *Information Processing in Agriculture*, *7*(1), 41–49. doi:10.1016/j.inpa.2019.06.001

Chand, S., & Singh, A. K. (2004). Plant regeneration from encapsulated nodal segments of *Dalbergia sissoo* Roxb.—A timber yielding leguminous tree. *Journal of Plant Physiology*, *161*(2), 237–243. doi:10.1078/0176-1617-01036 PMID:15022839

Chang, M.-H., Das, D., Varde, P. V., & Pecht, M. (2012). Light emitting diodes reliability review. *Microelectronics and Reliability*, *52*(5), 762–782. doi:10.1016/j.microrel.2011.07.063

Chang, S. (2008). *LED lighting: high efficiency and environmental benefit.* Samsung Economic Research Institute Economic Focus.

Chang, X., Alderson, P. G., & Wright, C. J. (2009). Enhanced UV-B radiation alters basil (Ocimumbasilicum L.) growth and stimulates the synthesis of volatile oils. *Journal of Horticulture and Forestry*, *12*, 27–31.

Chappelle, E. W., Kim, M. S., & McMurtrey, F. E. III. (1992). Ratio analysis of reflectance spectra (RARS): An algorithm for the remote estimation of the concentrations of chlorophyll a, chlorophyll b, and carotenoids in soybean leaves. *Remote Sensing of Environment*, *39*(3), 239–247. doi:10.1016/0034-4257(92)90089-3

Charles, W. W. N., Anthony, K. L., & Junjun, N. (2019). *Plant-Soil Slope Interaction.* CRC Press.

Chattopadhyay, A., Singh, A. P., Singh, S. K., Barman, A., Patra, A., Mondal, B. P., & Banerjee, K. (2020). Spatial variability of arsenic in Indo-Gangetic basin of Varanasi and its cancer risk assessment. *Chemosphere*, *238*, 124623. doi:10.1016/j.chemosphere.2019.124623 PMID:31545212

Chen, A., Chen, H. Y., & Chen, C. C. (2014). Use of temperature and humidity sensors to determine moisture content of Oolong tea. *Sensors (Basel)*, *14*(8), 15593–15609. doi:10.3390140815593 PMID:25153142

Chen, C. (2003). Development of a heat transfer model for plant tissue culture vessels. *Biosystems Engineering*, *85*(1), 67–77. doi:10.1016/S1537-5110(02)00289-1

Chen, C. (2004). Humidity in Plant Tissue Culture Vessels. *Biosystems Engineering*, *88*(2), 231–241. doi:10.1016/j.biosystemseng.2004.02.007

Chen, C. (2015). Application of growth models to evaluate the microenvironmentalconditions using tissue culture plantlets of *Phalaenopsis sogo* Yukidian'V3'. *Scientia Horticulturae*, *191*, 25–30. doi:10.1016/j.scienta.2015.05.007

Chen, C., & Chen, J. (2002). Measurement of gas exchange rates for plant tissue culture vessels. *Plant Cell, Tissue and Organ Culture*, *71*(2), 103–109. doi:10.1023/A:1019909520132

Chen, K., Sun, X., Qin, C., & Tang, X. (2010). Color grading of beef fat by using computer vision and support vector machine. *Computers and Electronics in Agriculture*, *70*(1), 27–32. doi:10.1016/j.compag.2009.08.006

Chen, S., Lin, B., Li, Y., & Zhou, S. (2020). Spatial and temporal changes of soil properties and soil fertility evaluation in a large grain-production area of subtropical plain, China. *Geoderma*, *357*, 113937. doi:10.1016/j.geoderma.2019.113937

Chen, X. L., Xue, X. Z., Guo, W. Z., Wang, L. C., & Qiao, X. J. (2016). Growth and nutritional properties of lettuce affected by mixed irradiation of white and supplemental light provided by light–emitting diode. *Scientia Horticulturae*, *200*, 111–118. doi:10.1016/j.scienta.2016.01.007

Chiang, Y. Y., Duan, W., Leyk, S., Uhl, J. H., & Knoblock, C. A. (2020). Creating Structured, Linked Geographic Data from Historical Maps: Challenges and Trends. In *Using Historical Maps in Scientific Studies* (pp. 37–63). Springer. doi:10.1007/978-3-319-66908-3_3

Chilliard, Y., Rouel, J., Ferlay, A., Bernard, L., Gaborit, P., & Raynal-Ljutovac, K. (2006). Optimising goat's milk and cheese fatty acid composition. In C. Williams & J. Buttriss (Eds.), Improving the fat content of foods (pp. 281–312). Cambridge, UK: Woodhead Publishing Limited.

Chilliard, Y., Ferlay, A., & Doreau, M. (2001). Effect of different type of forages, animal fat or marine oils in cow's diet on milk fat secretion and composition, expecially conjugated linoleic acid (CLA) and polyunsaturated fatty acids. *Livestock Production Science*, *70*(1-2), 31–48. doi:10.1016/S0301-6226(01)00196-8

Chilliard, Y., Ferlay, A., Rouel, J., & Lamberet, G. (2003). A review of nutritional and physiological factors affecting goat milk lipid synthesis and lipolysis. *Journal of Dairy Science, 86*(5), 1751–1770. doi:10.3168/jds.S0022-0302(03)73761-8 PMID:12778586

Chin, S. F., Liu, W., Storkson, J. M., Ha, Y. L., & Pariza, M. W. (1992). Dietary sources of conjugated dienoic isomers of linoleic acid, a newly recognised class of anticarcinogens. *Journal of Food Composition and Analysis, 5*(3), 185–197. doi:10.1016/0889-1575(92)90037-K

Chouinard, P. Y., Corneau, L., Butler, W. R., Chiliard, Y., Drackley, J. K., & Bauman, D. E. (2001). Effect of dietary lipid source on conjugated linoleic acid concentrations in milk fat. *Journal of Dairy Science, 84*(3), 680–690. doi:10.3168/jds.S0022-0302(01)74522-5 PMID:11286421

Chow, H. C. (1979). *Methods for measuring the dry rubber content of field latex.RRIM Training Manual on Analytical Chemistry*. Rubber Research Institute of Malaysia.

Chris, C. (2017). *Powdery Mildew: Identify, Prevent and Treat White Powdery Mildew on Cannabis: Part 2 – Prevention of PM*. Cannabis Chris Articles. http://cannabischris.com/2017/08/powdery-mildew-prevention-of-pm/

Christy, C., & Lund, E. (1998). *Device for measuring soil conductivity*. U.S. Patent No. 5,841,282, issued November 24, 1998.

Cioć, M., Szewczyk, A., Żupnik, M., Kalisz, A., & Pawłowska, B. (2018). LED lighting affects plant growth, morphogenesis and phytochemical contents of *Myrtus communis* L. in vitro. *Plant Cell, Tissue and Organ Culture, 132*(3), 433–447. doi:10.100711240-017-1340-2

Clarke, R. C. (1999). Botany of the genus Cannabis. In P. Ranalli (Ed.), *Advances in Hemp Research* (pp. 1–19). Food Products Press, The Haworth Press Inc.

Clarke, R. C., & Merlin, M. D. (2016). Cannabis Domestication, Breeding History, Present-day Genetic Diversity, and Future Prospects. *Critical Reviews in Plant Sciences, 35*(5-6), 293–327. doi:10.1080/07352689.2016.1267498

Cohen, D., & Schwarz, M. (2017). Tree-root control of shallow landslides. *Earth Surf Dynam, 5*(3), 451–477. doi:10.5194/esurf-5-451-2017

Colaco, A. F., Trevisan, R. G., Karp, F. H. S., & Molin, J. P. (2015). Yield mapping methods for manually harvested crops. *Precision Agriculture, 15*, 225–232. doi:10.3920/978-90-8686- 814-8_27

Colburn, J. W., Jr. (1991). *Soil chemical sensor and precision agricultural chemical delivery system and method*. U.S. Patent No. 5,033,397, issued July 23, 1991.

Colijn-Hooymans, C. M., Bouwer, R., & Dons, J. J. M. (1988). Plant Regeneration from Cucumber Protoplasts. *Plant Cell, Tissue and Organ Culture, 12*(2), 147–150. doi:10.1007/BF00040077

Collins, A. J. (2001). *The role of willow root architecture and character in root reinforcement potential* (Unpublished doctoral thesis). University of Nottingham.

Colomer, R., Moreno-Nogueira, J. M., García-Luna, P. P., García-Peris, P., García-de-Lorenzo, A., Zarazaga, A., Quecedo, L., del Llano, J., Usán, L., & Casimiro, C. (2007). N-3 fatty acids, cancer and cachexia: A systematic review of the literature. *British Journal of Nutrition, 97*(5), 823–831. doi:10.1017/S000711450765795X PMID:17408522

CONABIO. (2012). *Maíces y Razas de Maíces*. Academic Press.

Concheiro, L., & Grajales, S. (2005). Movimientos campesinos e indígenas en México: La lucha por la tierra. *Observatorio Social de América Latina, 16*(VI), 47–58.

Consoli, N. C., Montardo, J. P., Prietto, P. D. M., & Pasa, G. S. (2002). Engineering behavior of a sand reinforced with plastic waste. *Journal of Geotechnical and Geoenvironmental Engineering*, *128*(6), 462–472. doi:10.1061/(ASCE)1090-0241(2002)128:6(462)

Corbineau, F., Engelmann, F., & Côme, D. (1990). Ethylene production as an indicator of chilling injury in oil palm *(Elaeis guineensis* Jacq) somatic embryos. *Plant Science*, *71*(1), 29–35. doi:10.1016/0168-9452(90)90065-V

Corpoica. (2015). *Corporación Colombiana de Investigación Agropecuaria*. Retrieved October 13, 2019, from https://web.corpoica.org.co/

Cortes, C., & Vapnik, V. (1995). Support-vector networks. *Machine Learning*, *20*(3), 273–297. doi:10.1007/BF00994018

Cosentino, S. L., Riggi, E., Testa, G., Scordia, D., & Copani, V. (2013). Evaluation of European developed fibre hemp genotypes (Cannabis sativa L.) in semi-arid Mediterranean environment. *Industrial Crops and Products*, *50*, 312–324. doi:10.1016/j.indcrop.2013.07.059

Costa, L., Rangel, D. E. N., Morandi, M. A. B., & Bettiol, W. (2013). Effects of UV-B radiation on the antagonistic ability of Clonostachysrosea to Botrytis cinerea on strawberry leaves. *Biological Control*, *65*(1), 95–100. doi:10.1016/j.biocontrol.2012.12.007

Coste, A., Deliu, C., Valimareanu, S., Mareanu, A., & Halmagyi, A. (2014). Cryopreservation and acclimatization of *Lycopersiconesculentum* (Mill.) genotypes. *Notulae Botanicae Horti Agrobotanici Cluj-Napoca*, *42*(2), 466–471. doi:10.15835/nbha4229579

Courtois, D., & Guern, J. (1980). Temperature response of *Catharanthus roseus* cells cultivated in liquid medium. *Plant Science Letters*, *17*(4), 473–482. doi:10.1016/0304-4211(80)90135-2

Critten, D. L. (1991). Optimization of CO_2 concentration in greenhouses: A modelling analysis for the lettuce crop. *Journal of Agricultural Engineering Research*, *48*, 261–271. doi:10.1016/0021-8634(91)80020-F

Crookston, K. (2006). A top 10 list of developments and issues impacting crop management and ecology during the past 50 years. *Crop Science*, *46*(5), 2253–2262. doi:10.2135/cropsci2005.11.0416gas

Cruz-Cruz, C. A., Gonzalez-Arnao, M. T., & Engelmann, F. (2013). Biotechnology and conservation of plant biodiversity. *Resources*, *2*(2), 73–95. doi:10.3390/resources2020073

Cubeddu, R., D'Andrea, C., Pifferi, A., Taroni, P., Torricelli, A., Valentini, G., Dover, C., Johnson, D., Ruiz-Altisent, M., & Valero, C. (2001). Nondestructive quantification of chemical and physical properties of fruits by time-resolved reflectance spectroscopy in the wavelength range 650–1000 nm. *Applied Optics*, *40*(4), 538–543. doi:10.1364/AO.40.000538 PMID:18357029

Cui, Y., Hahn, E., Kozai, T., & Paek, K. (2000). Number of air exchanges, sucrose concentration, photosynthetic photon flux, and differences in photoperiod and dark period temperatures affect growth of *Rehmannia glutinosa* plantlets *in vitro. Plant Cell, Tissue and Organ Culture*, *62*(3), 219–226. doi:10.1023/A:1006412321864

da Cunha, J. A., Rolim, P. M., & da Silva Chaves, K. S. F. (2020). From seed to flour: Sowing sustainability in the use of cantaloupe melon residue (Cucumis melo L. var. reticulatus). *PLoS One*, *15*(1), e0219229. doi:10.1371/journal.pone.0219229 PMID:31895921

Da Silva, D. P. C., Ozudogru, E. A., Dos Reis, M. V., & Lambardi, M. (2018). In vitro conservation of ornamental plants. *Ornamental Horticulture (Campinas)*, *24*(1), 28–33. doi:10.14295/oh.v24i1.1163

Daberkow, S. G., & McBride, W. D. (2000). Adoption of Precision Agriculture Technologies by U.S. Farmers. *5th International Conference on Precision Agriculture*, Bloomington, IN.

Dahane, A., Berrached, N. E. (2019). *Mobile wireless and sensor, A clustering algorithm for energy efficiency and safety*. Apple Academic Press.

Dahane, A., Berrached, N. E., & Loukil, A. (2015). A Virtual Laboratory to Practice Mobile Wireless Sensor Networks: A Case Study on Energy Efficient and Safe Weighted Clustering Algorithm. *Journal of Information Processing Systems*, *11*(2), 205–228.

Dahane, A., & Kechar, B. (2019). Remote Sensing Data: Automated Irrigation Management Platform Using a WSNs. *International Conference on Embedded Systems in Telecommunications and Instrumentation (ICESTI'19)*, 28-30.

Dahane, A., Kechar, B., Benyamina, A., & Benameur, R. (2020). An IoT Based Smart Farming System Using Machine Learning. *International Symposium on Networks, Computers and Communications (ISNCC'20)*, Montreal, Canada.

Dahane, A., Kechar, B., Meddah, Y., & Benabdellah, O. (2019). Automated Irrigation Management Platform using a Wireless Sensor Network. *2019 Sixth International Conference on Internet of Things: Systems, Management and Security (IOTSMS)*, 610-615. 10.1109/IOTSMS48152.2019.8939170

Dainello, J. F., & Fipps, G. (2001). *Irrigation. Department of horticulture Science. Texas A&M University.*

Daponte, P., De Vito, L., Glielmo, L., Iannelli, L., Liuzza, D., Picariello, F., & Silano, G. (2019). A review on the use of drones for precision agriculture. *IOP Conference Series: Earth and Environmental Science*, *275*(1). 10.1088/1755-1315/275/1/012022

Darko, E., Heydarizadeh, P., Schoefs, B., & Sabzalian, M. R. (2014). Photosynthesis under artificial light: The shift in primary and secondary metabolism. *Philosophical Transactions of the Royal Society of London. Series B, Biological Sciences*, *369*(1640), 20130243. doi:10.1098/rstb.2013.0243 PMID:24591723

Darwis, V., & Rusastra, I. W. (2011, June). Optimizing Rural Community Empowerment through Integrating PUAP and DesaMandiriPangan Programs. *Agricultural Policy Analysis Journal.*, *9*(2), 125–142.

Das, U., Pathak, P., Meena, M. K., & Mallikarjun, N. (2018). Precision Farming a Promising Technology in Horticulture: A Review. *Int. J. Pure App. Biosci.*, *6*(1), 1596–1606. doi:10.18782/2320-7051.3088

Daughtry, C. S. T., Bauer, M. E., Crecelius, D. W., & Hixson, M. M. (1980). Effects of management practices on reflectance of spring wheat canopies. *Agronomy Journal*, *72*(6), 1055–1066. doi:10.2134/agronj1980.00021962007200060045x

Davcev, D., Mitreski, K., Trajkovic, S., Nikolovski, V., & Koteli, N. (2018, June). IoT agriculture system based on LoRaWAN. In *2018 14th IEEE International Workshop on Factory Communication Systems (WFCS)* (pp. 1-4). IEEE. 10.1109/WFCS.2018.8402368

Davis, R. (1974). Pests of stored products. *Bulletin of the Entomological Society of America*, *20*(4), 335–335. doi:10.1093/besa/20.4.335a

De Baets, S. D., Poesen, J., Reubens, B., Wemans, K., Baerdemaeker, J. D., & Muys, B. (2008). Root tensile strength and root distribution off typical Mediterranean plant species and their contribution to soil shear strength. *Plant and Soil*, *305*(1-2), 207–226. doi:10.100711104-008-9553-0

De Fossard, R. A. (1986) Principles of plant tissue culture. In Tissue culture as a plant production system for horticultural crops. doi:10.1007/978-94-009-4444-2_1

De Ollas, C., Morillon, R., Fotopoulos, V., Puertolas, J., Ollitrault, P., Gomez-Cadenas, A., & Arbona, V. (2019). Facing Climate Change: Biotechnology of Iconic Mediterranean Woody Crops. *Front. Plant Sci*, *10*, 427. doi:10.3389/fpls.2019.00427 PMID:31057569

Deepa, G., Singh, V., & Naidu, K. A. (2008). Nutrient compostion and physiochemical properties of Indian medicinal rice –Njavara. *Food Chemistry, 106*(1), 165–171. doi:10.1016/j.foodchem.2007.05.062

Demirel, G., Wood, J. D., & Enser, M. (2004). Conjugated linoleic acid content of the lamb muscle and liver fed different supplements. *Small Ruminant Research, 53*(1-2), 23–28. doi:10.1016/j.smallrumres.2003.07.006

Department of Agriculture. (2009). *Ground Penetrating Radar soil suitability maps*. Retrieved from https://www.nrcs.usda.gov/wps/portal/nrcs/detail/soils/survey/geo/?cid=nrcs142p2_053622

Department of Health. (1994). *Nutritional aspects of cardiovascular disease. Report on Health and Social Subjects No. 46*. HMSO.

Dereuddre, J., Fabre, J., & Bassaglia, C. (1988). Resistance to freezing in liquid nitrogen of carnation (Dianthus caryophyllus L. Var eolo) apical and axillary shoot tips excised from different aged in vitro plantlets. *Plant Cell Reports, 7*(3), 170–176. doi:10.1007/BF00269315 PMID:24241591

Dereuddre, J., Scottez, C., & Duron, Y. A. M. (1990). Effect of cold hardening on cryopreservation of pear (*Pyrus communis* L. cv. Beurre Hardy) shoot tips of *in vitro* plantlets. *C. R. Acad. Sci., 30*, 265–272.

Devices Ltd. D.-T. (2017). *ML3ThetaProbe*. Retrieved from www.delta-t.co.uk

Devitt, K. S. (2018). Cognitive factors that affect the adoption of autonomous agriculture. *Farm Policy Journal, 15*(2), 49–60.

Dewey, L. H. (1914). Hemp in U.S.D.A. Yearbook 1913 United States Department of Agriculture.

Dhiman, T. R., Nam, S. H., & Ure, A. L. (2005). Factors affecting conjugated linoleic acid content in milk and meat. *Critical Reviews in Food Science and Nutrition, 45*(6), 463–482. doi:10.1080/10408390591034463 PMID:16183568

Dhiman, T. R., Satter, L. D., Pariza, M. W., Galli, M. P., Albright, K., & Tolosa, M. X. (2000). Conjugated linoleic acid (CLA) content of milk from cows offered diets rich in linoleic and linolenic acid. *Journal of Dairy Science, 83*(5), 1016–1027. doi:10.3168/jds.S0022-0302(00)74966-6 PMID:10821577

Di Bari, V., Campi, P., Colucci, R., & Mastrorilli, M. (2004). Potential productivity of fibre hemp in southern Europe. *Euphytica, 140*(1-2), 25–32. doi:10.100710681-004-4751-1

Dılay, Y., Soy, H., & Bayrak, M. (2012). Employment of the wireless sensor networks in precision agriculture and investigation of the application areas. *Igdir University Journal of the Institute of Science and Technology, 2*(2).

DJI. (2019). *DJI introduces P4 Multispectral for Precision agriculture and land management*. Retrieved from https://www.suasnews.com/2019/09/dji-introduces-p4-multispectral-for-precision-agriculture-and-land-management/

Dodds, J. H. (1991). Conservation of plant genetic resources.The need for tissue culture. In J. H. Doods (Ed.), *In vitro Methods for conservation of plant genetic resources* (pp. 1–39). Chapman Hall. doi:10.1007/978-94-011-3072-1_1

Doerge, T. (1999). Defining management zones for precision farming. *Crop Insights, 8*, 1–5.

Dokin, B. D., Aletdinova, A. A., & Kravchenko, M. S. (2016). Prospects and features of robotics in russian crop farming. *IOP Conf. Series: Journal of Physics: Conf. Series, 803*(2017). https://iopscience.iop.org/article/10.1088/1742-6596/803/1/012032/

Doolittle, J. A., Sudduth, K. A., Kitchen, N. R., & Indorante, S. J. (1994). Estimating depths to claypans using electromagnetic induction methods. *Journal of Soil and Water Conservation, 49*, 572–575.

Dorion, N., Godin, B., & Bigot, C. (1993). Physiological state and clonal variability effects on low temperature storage of in vitro shoot cultures of elms (*Ulmus sp.*). *Scientia Horticulturae*, *56*(1), 51–59. doi:10.1016/0304-4238(93)90101-U

Dow, A. I., James, D. W., & Russell, T. S. (1973b). *Soil Variability in Central Washington and Sampling for Soil Fertility Tests. Washington Agricultural Experiment Station Bulletin 788*. Washington State University.

Duhan, J. S., Kumar, R., Kumar, N., Kaur, P., Nehra, K., & Duhan, S. (2017). Nanotechnology: The new perspective in precision agriculture. *Biotechnology Reports (Amsterdam, Netherlands)*, *15*, 11–23. doi:10.1016/j.btre.2017.03.002 PMID:28603692

Dumet, D., Engelmann, F., Chabrillange, N., Dusseret, S., & Duval, Y. (2000). Cryopreservation of oil palm polyembryogenic cultures. In Current Research, Progress and Applications. Eds. Engelmann, F. and H. Takagi. Cryopreservation of Tropical Plant Germplasm. IPGRI.

Duong, T. N., Takamura, T., Watanabe, H., Okamoto, K., & Tanaka, M. (2003). Responses of strawberry plantlets cultured in vitro under superbright red and blue lightemitting diodes (LEDs). *Plant Cell, Tissue and Organ Culture*, *73*(1), 43–52. doi:10.1023/A:1022638508007

Dwivedi, A., Naresh, R. K., Kumar, R., & Yadav, R. S. (2017). *Precision Agriculture. Promoting agri-hortucultural, technological innovations*. Parmar Publishers & Distributors.

EcuRed. (2011). Retrieved from https://www.ecured.cu/Gestión_del_conocimiento

Edaphic, S. (2019). *Soil pH Sensor & Meter*. Retrieved from https://www.edaphic.com.au/products/soils/soil-ph-meter/

Eden, T., & Maskell, E. J. (1928). The influence of soil heterogeneity on the growth and yield of successive crops. *The Journal of Agricultural Science*, *18*(2), 163–185. doi:10.1017/S0021859600009023

Ehlers, W., & Goss, M. (2003). *Water Dynamics in Plant Production*. CABI Publishing. doi:10.1079/9780851996943.0000

Ehrensing, D. T. (1998). Feasibility of industrial hemp production in the United States Pacific Northwest. *Oregan State Univ. Agr. Expt. Sta. Bul.*, 681.

Ehsani, R. (2011). Precision Agriculture for Small Growers. *Resource Magazine*, *18*, 11–12.

Eisenhauer, M., Rosengren, P., & Antolin, P. (2010). *Hydra: A development platform for integrating wireless devices and sensors into ambient intelligence systems. In The Internet of Things*. Springer.

Ekanayake, J. C., Marden, M., Watson, A. J., & Rowan, D. (1997). Tree roots and slope stability: A comparison between Pinusradiata and Kanuka. *New Zealand Journal of Forestry Science*, *27*, 216–233.

Ekwue, E. I., & Stone, R. J. (1995). Organic matter effects on the strength properties of compacted agricultural soils. *Transactions of the ASAE. American Society of Agricultural Engineers*, *38*(2), 357–365. doi:10.13031/2013.27804

Elad, Y., Kirshner, B., Yehuda, N., & Sztejnberg, A. (1998). Management of powdery mildew and gray mold of cucumber by Trichodermaharzianum T39 and Ampelomycesquisqualis AQ10. *BioControl*, *43*(2), 241–251. doi:10.1023/A:1009919417481

El-Bahr, M. K., Abd EL-Hamid, A., Matter, M. A., Shaltout, A., Bekheet, S. A., & El-Ashry, A. A. (2016). In vitro conservation of embryogenic cultures ofdate palm using osmotic mediated growth agents. *Journal of Genetic Engineering and Biotechnology*, *14*(2), 363–370. doi:10.1016/j.jgeb.2016.08.004 PMID:30647634

ElHagarey, M. E., & Hushki, M. M. (2017). Fuzzy Logic Model Approaches for Water Saving in Irrigation Systems. *European Journal of Academic Essays*, *4*(4), 157–165.

Elijah, O., Rahman, T. A., Orikumhi, I., Leow, C. Y., & Hindia, M. N. (2018). An overview of Internet of Things (IoT) and data analytics in agriculture: Benefits and challenges. *IEEE Internet of Things Journal*, *5*(5), 3758–3773. doi:10.1109/JIOT.2018.2844296

ElSohly, M. A., & Slade, D. (2005). Chemical constituents of marijuana: The complex mixture of natural cannabinoids. *Life Sciences*, *78*(5), 539–548. doi:10.1016/j.lfs.2005.09.011 PMID:16199061

Elzinga, S., Fischedick, J., Podkolinski, R., & Raber, J. C. (2015). Cannabinoids and terpenes as chemotaxonomic markers in cannabis. *Natural Products Chemistry & Research*, *3*, 1–9.

Endo, T., & Tsuruta, T. (1969). A report in regard to the reinforcement action of the vegetational roots upon the tensile strength of the natural soil. 1968 Annu. Rep. Hokkaido Branch, For. Exp. Stn., 183-189.

Engelmann, F. (1991). *In vitro* conservation of horticultural species. *Acta Horticulturae*, (29), 327–334. doi:10.17660/ActaHortic.1991.298.40

Engelmann, F. (1997). *In vitro* conservation methods. In J. H. Newburry & J. A. Callow (Eds.), *Biotechnology and Plant Genetic Resources: Conservation and Use. B. V. Ford – Lloyd* (pp. 119–162). CABI, Wellingford.

Engelmann, F. (2012). Germplasm collection, storage and conservation. In A. Altman & P. M. Hasegawa (Eds.), *Plant Biotechnology and Agriculture* (pp. 255–268). Academic Press. doi:10.1016/B978-0-12-381466-1.00017-1

Enser, M., Hallet, B., Hewett, G. A., Fursey, G., Wood, J. D., & Harrington, G. (1998). Fatty acid content and composition of UK beef and lamb muscle in relation to production system and implications for human nutrition. *Meat Science*, *49*(3), 329–341. doi:10.1016/S0309-1740(97)00144-7 PMID:22060583

Enser, M., Scollan, N. D., Choi, N. J., Kurt, E., Hallett, K., & Wood, J. D. (1999). Effect of dietary lipid on the content of conjugated linoleic acid (CLA) in beef muscle. *Animal Science (Penicuik, Scotland)*, *69*(1), 143–146. doi:10.1017/S1357729800051171

Espinosa Tamayo, L., Rincón Sánchez, F., Ruíz Torres, N., Martínez Reyna, J., & Benavides Mendoza, A. (2019). Respuesta ambiental de poblaciones nativas de maíz del sureste de Coahuila, México. *Nova Scientia*, *2*(23), 108–125. doi:10.21640/ns.v11i23.1931

Ess, D., & Morgan, M. (2003). *The Precision-Farming Guide for Agriculturists.* Deere & Company.

European Monitoring Centre for Drugs and Drug Addiction. (2018). Cannabis legislation in Europe: an overview. Publications Office of the European Union. doi:10.2810/566650

Evangelos, C. R., Evangelia, E., Eftychia, B., Michael, S. K., & Moses, S. E. (2012). Association Between Omega-3 Fatty Acid Supplementation and Risk of Major Cardiovascular Disease Events A Systematic Review and Meta-analysis. *Journal of the American Medical Association*, *308*(10), 1024–1033. doi:10.1001/2012.jama.11374 PMID:22968891

Evans, R. G., Buchleiter, G. W., Sadler, E. J., King, B. A., & Harting, G. B. (2000). Controls for precision irrigation with self-propelled systems. In *Proceedings of the 4th decennial national irrigation symposium.* ASAE.

Evans, R. G., Han, S., Kroeger, M. W., & Schneider, S. M. 1996. Precision center pivot irrigation for efficient use of water and nitrogen. In P. C. Robert, R. H. Rust, & W. E. Larson (Eds.), *Precision Agriculture. Proc. 3rd Intl. Conf.* ASA-CSSA-SSSA. 10.2134/1996.precisionagproc3.c8

Fabre, J., & Dereuddre, J. (1990). Encapsulation-dehydration: A new approach to cryopreservation of *Solanum* shoot tips. *Cryo Letters*, *11*, 413–426.

Fahy, G., MacFarlane, D. R., Angell, C. A., & Meryman, H. T. (1984). Vitrification as an approach to cryopreservation. *Cryobiol*, *21*(4), 407–426. doi:10.1016/0011-2240(84)90079-8 PMID:6467964

Fang, J. Y., Wetten, A., & Hadley, P. (2004). Cryopreservation of cocoa (*Theobroma cacao* L.) somatic embryos for long-term germplasm storage. *Plant Science*, *166*(3), 669–675. doi:10.1016/j.plantsci.2003.11.002

FAO & SAGARPA. (2012). *Diagnóstico del Sector Rural y Pesquero en México*. Retrieved May 1, 202AD, from http://smye.info/cuestionario_final/diagnostico/menu_f.php

FAO. (1960). *Irrigation by sprinkler. FAO Agriculture Development Paper No. 65*. FAO.

FAO. (1998). Crop evapotranspiration - Guidelines for computing crop water requirements. *FAO Irrigation and Drainage Paper*, *56*. Retrieved June 01, 2019, from http://www.fao.org/3/X0490E/x0490e00.htm

FAO. (2009). How to feed the world 2050 - Global Agriculture towards 2050. Rome, Italy: FAO.

FAO. (2015). *El estado mundial de la agricultura y la alimentación. La innovación en la agricultura familiar.* Retrieved May 1, 2020, from http://www.fao.org/3/a-i4910s.pdf

FAO. (2018). The State of Agricultural Commodity Markets 2018. Agricultural trade, climate change and food security. Rome: FAO.

Farag, S., & Kayser, O. (2015). Cultivation and breeding of Cannabis sativa L. for preparation of standardized extracts for medicinal purposes. In Medicinal and aromatic plants of the world. Springer.

Fares, A., Abbas, F., Domingos, M., & Mair, A. (2011). Improved Calibration Functions of Three Capacitance Probes for the Measurement of Soil Moisture in Tropical Soils. *Sensors (Basel)*, *11*(5), 4858–4874. doi:10.3390110504858 PMID:22163880

Fazal, H., Abbasi, B.H., Ahmad, N., Ali, S.S., Akbar, F., & Kanwal, F. (2016) Correlation of different spectral lights with biomass accumulation and production of antioxidant secondary metabolites in callus cultures of medicinally important *Prunella vulgaris L. J Photochem Photobiol B Biol, 159*, 1–7.

Felton, W., & McCloy, K. (1992). *Controller for agricultural sprays*. U.S. Patent No. 5,144,767, issued September 8, 1992.

Fernandez-Pacheco, D. G., Molina-Martinez, J. M., Jimenez, M., Pagan, F. J., & Ruiz-Canales, A. (2014). SCADA Platform for regulated deficit irrigation management of almond trees. *Journal of Irrigation and Drainage Engineering*, *140*(5), 04014008. doi:10.1061/(ASCE)IR.1943-4774.0000718

Ferreira, E. M., Pires, A. V., Susin, I., Gentil, R. S., Parente, M. O. M., Nolli, C. P., Meneghini, R. C. M., Mendes, C. Q., & Ribeiro, C. V. D. M. (2014). Growth, feed intake, carcass characteristics, and meat fatty acid profile of lambs fed soybean oil partially replacedby fish oil blend. *Animal Feed Science and Technology*, *187*, 9–18. doi:10.1016/j.anifeedsci.2013.09.016

Ferrell, S. L. (2016). Legal Issues on the Farm Data Frontier, Part I: Managing First-Degree Relationships in Farm Data Transfers. *The Drake Journal of Agricultural Law*, *21*, 13.

Finger, R., Swinton, S. M., El Benni, N., & Walter, A. (2019). Precision Farming at the Nexus of Agricultural Production and the Environment. *Annual Review of Resource Economics*, *11*(1), 313–335. doi:10.1146/annurev-resource-100518-093929

Fitter, A. H. (1993). Arhitectural analysis of plant root systems. In Methods in comparative plant ecology: A laboratory manual. Chapman and Hall.

Food and Agriculture Organization. (2017). *The future of food and agriculture – Trends and challenges*. Available online at http://www.fao.org/3/a-i6583e.pdf

Food and Agriculture Organization. (2019). *Digital technologies in agriculture and rural areas – Status report*. Available online at http://www.fao.org/3/ca4985en/ca4985en.pdf

Food, S. (2016). Global initiative on food loss and waste reduction. In *Key facts on food loss and waste you should know*. Academic Press.

Foote, W. (2015). *To Feed the World in 2050, We Need to View Small-Scale Farming as a Business*. Skoll World Forum. Available online at https://www.forbes.com/sites/skollworldforum/2015/04/06/to-feed-the-world-in-2050-we-need-to-view-small-scale-farming-as-a-business/

Ford-Lloyd, B., & Jackson, M. (1986). Plant Genetic Resources: An Introduction to their Conservation and Use. Edward Arnold Ltd..

Forst, K. R., & Schwalen, H. C. (1955). Sprinkler evaporation losses. *Agricultural Engineering*, *36*(8), 526–528. http://www.wac-infonet.org/id/16377

Forsyth, A. (2000). Psychoactive drugs: the pharmacopoeia of substance use. London: Stationery Office.

Fortin, P. R., Lew, R. A., Liang, M. H., Wright, E. A., Beckett, L. A., Chalmers, T. C., & Sperling, R. I. (1995). Validation of a meta-analysis: The effects of fish oil in rheumatoid arthritis. *Journal of Clinical Epidemiology*, *48*(11), 1379–1390. doi:10.1016/0895-4356(95)00028-3 PMID:7490601

Fountas, S., Van der Wal, T., Eory, V., & Gómez-Barbero, M. (2019). *The contribution of precision agriculture technologies to farm productivity and the mitigation of greenhouse gas emissions in the EU*. Available online at: https://ec.europa.eu/jrc/en/publication/contribution-precision-agriculture-technologies-farm-productivity-and-mitigation-greenhouse-gas

Fountas, S., Aggelopoulou, K., Bouloulis, C., Nanos, G. D., Wulfsohn, D., Gemtos, T. A., Paraskevopoulos, A., & Galanis, M. (2011). Site-specific management in an olive tree plantation. *Precision Agriculture*, *12*(2), 179–195. doi:10.100711119-010-9167-4

Fraisse, C. W. (1994). *Variable water application with moving irrigation systems* (PhD dissertation). Colorado State University, Ft. Collins, CO.

Fraisse, C. W., Sudduth, K. A., & Kitchen, N. R. (2001). Delineation of site-specific management zones by unsupervised classification of topographic attributes and soil electrical conductivity. *Transactions of the ASAE. American Society of Agricultural Engineers*, *44*(1), 155–166. doi:10.13031/2013.2296

França, R. P., Iano, Y., Monteiro, A. C. B., & Arthur, R. (2020). A Proposal of Improvement for Transmission Channels in Cloud Environments Using the CBEDE Methodology. In Modern Principles, Practices, and Algorithms for Cloud Security (pp. 184-202). IGI Global. doi:10.4018/978-1-7998-1082-7.ch009

França, R. P., Iano, Y., Monteiro, A. C. B., & Arthur, R. (2020). Improvement for Channels With Multipath Fading (MF) Through the Methodology CBEDE. In Fundamental and Supportive Technologies for 5G Mobile Networks (pp. 25-43). IGI Global.

França, R. P., Iano, Y., Monteiro, A. C. B., & Arthur, R. (2020). Improvement of the Transmission of Information for ICT Techniques Through CBEDE Methodology. In Utilizing Educational Data Mining Techniques for Improved Learning: Emerging Research and Opportunities (pp. 13-34). IGI Global. doi:10.4018/978-1-7998-0010-1.ch002

França, R. P., Iano, Y., Monteiro, A. C. B., & Arthur, R. (2020). Potential proposal to improve data transmission in healthcare systems. In *Deep Learning Techniques for Biomedical and Health Informatics* (pp. 267–283). Academic Press. doi:10.1016/B978-0-12-819061-6.00011-2

Franzen, D. W. (2007). Lessons learned from 40 years of grid-sampling in Illinois. *2007 Indiana CCA Conference Proceedings*, 1–11.

French, P., Stanton, C., Lawless, F., O'Riordan, E. G., Monahan, F. J., Caffrey, P. J., & Moloney, A. P. (2000). Fatty acid composition, including conjugated linoleic acid, of intramuscular fat from steers offered grazed grass, grass silage, or concentrate-based diets. *Journal of Animal Science*, *78*(11), 2849–2855. doi:10.2527/2000.78112849x PMID:11063308

Frohnmeyer, H., & Staiger, D. (2003). Ultraviolet-B radiation-mediated responses in plants, balancing damage and protection. *Plant Physiology*, *133*(4), 1420–1428. doi:10.1104/pp.103.030049 PMID:14681524

Fry, J. E., & Guber, A. K. (2020). Temporal stability of field-scale patterns in soil water content across topographically diverse agricultural landscapes. *Journal of Hydrology (Amsterdam)*, *580*, 124260. doi:10.1016/j.jhydrol.2019.124260

Fuglie, K. (2016). The growing role of the private sector in agricultural research and development world-wide. *Global Food Security*, *10*, 29–38. doi:10.1016/j.gfs.2016.07.005

Fukui, H., Nishimoto, K., & Nakamura, M. (1992). Varietal differences in rooting ability of *in vitro* subcultured Japanese persimmon shoots. *Journal of the Japanese Society for Horticultural Science*, *60*(4), 821–825. doi:10.2503/jjshs.60.821

Fu, W., Li, P., & Wu, Y. (2012). Effects of different light intensities on chlorophyll fluorescence characteristics and yield in lettuce. *Scientia Horticulturae*, *135*, 45–51. doi:10.1016/j.scienta.2011.12.004

Gabr, A. M. M., & Sawsan, S. S. (2010). Slow growth conservation and molecular characterization of *Deutzia scabra* Thunb. *African Journal of Plant Science*, *4*(10), 409–416.

Gandhi, V. P., & Marsh, R. (2003). *Development and Poverty Reduction: Do Institutions Matter? A Study on the Impact of Local Institutions in Rural India.* The 25th International Conference of Agricultural Economists, Durban, South Africa.

Garcı'a, P. T., Pensel, N. A., Latimori, N. J., Kloster, A. M., Amigone, M. A., & Casal, J. J. (2005). Intramuscular lipids in steers under different grass and grain regimen. *Fleischwirtschaft International*, *73*, 755–758.

García Yeste, C., Leena Lastikka, A., & Petreñas Caballero, C. (2013). Comunidades de aprendizaje. *Script Nova. Revista Electrónica de Geografía y Ciencias Sociales, 17*(427).

García, L., Parra, L., Jimenez, J. M., Lloret, J., & Lorenz, P. (2020). IoT-Based Smart Irrigation Systems: An Overview on the Recent Trends on Sensors and IoT Systems for Irrigation in Precision Agriculture. *Sensors (Basel)*, *20*(4), 1042. doi:10.339020041042 PMID:32075172

García-Tejero, I. F., Durán-Zuazo, V. H., Pérez-Álvarez, R., Hernández, A., Casano, S., Morón, M., & Muriel-Fernández, J. L. (2014). Impact of Plant Density and Irrigation on Yield of Hemp (Cannabis sativa L.) in a Mediterranean Semi-arid Environment. *Journal of Agricultural Science and Technology*, *16*, 887–895.

Garner, N., & Blake, J. (1989). The induction and development of potato microtubers *in vitro* on media free of growth regulating substances. *Annali di Botanica*, *63*(6), 663–674. doi:10.1093/oxfordjournals.aob.a087795

Gaultney, L. D., Schulze, D. G., Shonk, J. L., & Van Scoyoc, G. E. (1991). *Real-time soil organic matter sensor.* U.S. Patent No. 5,044,756, issued September 3, 1991.

Gazolla, A. P., Curado, F. M. L. M. J., Pedroso, R. C. N., Kellner Filho, C., Fátima Sales, J., Rosa, M., Costa, A. C., Pauletti, P. M., Januário, A. H., & Silva, F. G. (2017). The influence of light quality on phenolic acid and biflavonoid production in *Anacardium othonianum* Rizz. seedlings grown in vitro. *Australian Journal of Crop Science*, *11*(05), 528–534. doi:10.21475/ajcs.17.11.05.p314

Ge, Y., Thomasson, A., & Sui, R. (2011). Remote sensing of soil properties in precision agriculture: A review. *Frontiers of Earth Science*, *5*(3), 229–238. doi:10.100711707-011-0175-0

Geay, Y., Bauchart, D., Hocquette, J. F., & Culioli, J. (2001). Effect of nutritional factors on biochemical, structural and metabolic characteristics of muscles in ruminants, consequences on dietetic value and sensorial qualities of meat. *Reproduction, Nutrition, Development*, *41*(1), 1–26. doi:10.1051/rnd:2001108 PMID:11368241

Gemtos, T., Fountas, S., Tagarakis, A., & Liakos, V. (2013). Precision Agriculture Application in Fruit Crops: Experience in Handpicked Fruits. *Procedia Technology*, *8*, 324–332. doi:10.1016/j.protcy.2013.11.043

George, E. (1993). *Plant Propagation by Tissue Culture. Part 1: The Technology* (2nd ed.). Exegetics Ltd.

Gerard, C.J. & Mehta, H. C. (1971). Influence of a rootcrop on physical properties of a clay soil. *Agronomy*, *63*, 893-895.

Getinet, A., & Getachew, A. (2019). Precision Agriculture and the Need to Introduce in Ethiopia. *Ethiop. J. Agric. Sci*, *29*(3), 139–158.

Ghashghaie, J., Brenckmann, F., & Saugier, B. (1992). Water relations and growth of rose plants cultured in vitro under various relative humidities. *Plant Cell, Tissue and Organ Culture*, *30*(1), 51–57. doi:10.1007/BF00040000

Gianní, S., & Sottile, F. (2015). In vitro storage of plum germplasm by slow growth. *Horticultural Science (Prague)*, *42*(2), 61–69. doi:10.17221/186/2014-HORTSCI

GIE. (2015). *Grupo Interinstitucional de Extensionismo*. GIE.

Gilchrist, A. (2016). *Industry 4.0: the industrial internet of things*. Apress. doi:10.1007/978-1-4842-2047-4

Gill, W. R., & Vanden Berg, G. E. (1968). Soil Dynamics in Tillage and Traction. USDA-ARS Agricultural Handbook No. 316. Washington, DC: Government Printing Office.

Givens, D. I. (2005). The role of animal nutrition in improving the nutritive value of animal-derived foods in relation to chronic disease. *The Proceedings of the Nutrition Society*, *64*(3), 395–402. doi:10.1079/PNS2005448 PMID:16048674

Goap, A., Sharma, D., Shukla, A. K., & Rama Krishna, C. (2018). An IoT based smart irrigation management system using Machine learning and open source technologies. *Computers and Electronics in Agriculture*, *155*(May), 41–49.

Goldsmith, W. (1998). Soil reinforcement by river plants: progress results. In *Proceedings of the Conferencon Wetlands Engineering and River Restoration*. American Society of Civil Engineers. 10.1061/40382(1998)30

González Sabater, J. (2017). *Manual de transferencia de tecnología y conocimiento*. Academic Press.

González-González, M. G., Gómez-Sanchis, J., Blasco, J., Soria-Olivas, E., & Chueca, P. (2020). CitrusYield: A Dashboard for Mapping Yield and Fruit Quality of Citrus in Precision Agriculture. *Agronomy (Basel)*, *10*(1), 128. doi:10.3390/agronomy10010128

Government of Canada. (2019). *Cannabis Legalization and Regulation*. Available at: https://www.justice.gc.ca/eng/cj-jp/cannabis/

Grabowska, L., Rebarz, M., & Chudy, M. (2009). Breeding and cultivation of industrial hemp in Poland. Institute of Natural Fibres and Medicinal Plants.

Graham, D. S. (2002). Medical marijuana: Canada's regulations, pharmacology, and social policy. *Canadian Pharmacists Journal*, *137*(902), 23–27.

Gray, D.J., McColley, D.W., & Compton, M.E. (1993) High-frequency somatic embryogenesis from quiescent seed cotyledons of *Cucumis melo* cultivars. *J. Am. Soc. Hort. Sci.*, *118*, 425-432.

Gray, D. H. (1974). Reinforcement and Stabilization of Soil by Vegetation. *Journal of Geotechnical Engineering*, (100(GT6)), 695–699.

Greatline, M. W., & Greatline, S. E. (1999). *Method and apparatus for prescription application of products to an agricultural field*. U.S. Patent No. 5,919,242, issued July 6, 1999.

Greenway, D. T. (1987). Vegetation and slope stability. In Slope Stability. John Wiley.

Gregory, P. J., & George, T. S. (2011). Feeding nine billion: The challenge to sustainable crop production. *Journal of Experimental Botany*, *62*(15), 5233–5239. doi:10.1093/jxb/err232 PMID:21841178

Grewal, B., Grunfeld, H., & Sheehan, P. (2012). *The contribution of agricultural growth to poverty reduction*. ACIAR Impact Assessment Series Report No. 76. Australian Centre for International Agricultural Research, Canberra.

Gribble, K. (1999). The influence of relative humidity on vitrification, growth and morphology of *Gypsophila paniculata* L. *Plant Growth Regulation*, *27*(3), 179–188. doi:10.1023/A:1006235229848

Griinari, J. M., & Bauman, D. E. (1999). Biosynthesis of conjugated linoleic acid and its incorporation into meat and milk in ruminants. In M. P. Yurawecz, M. M. Mossoba, J. K. G. Kramer, M. W. Pariza, & G. J. Nelson (Eds.), *Advances in Conjugated Linoleic Acid Research* (Vol. 1, pp. 180–200). AOCS press.

Grisso, R., Alley, M., & Groover, G. (2009). *Precision Farming Tools: GPS Navigation*. Virginia Cooperative Extension. Publication No 442-501.

Grøn Sørensen, C. A., Kateris, D., & Bochtis, D. (2019). ICT Innovations and Smart Farming. In M. Salampasis & T. Bournaris (Eds.), *Information and Communication Technologies in Modern Agricultural Development* (pp. 1–19). Springer Nature Switzerland. doi:10.1007/978-3-030-12998-9_1

Grout, B. W. W. (1995). Introduction to the *in vitro* preservation of plant cells, tissue and organs. In B. W. W. Grout (Ed.), *Genetic preservation of plant cellsi n vitro* (pp. 1–20). Springer-Verlag. doi:10.1007/978-3-642-78661-7_1

Guo, W. C., Nelson, S. O., Trabelsi, S., Kays, S. J. (2007). Dielectric properties of honeydew melons and correlation with quality. *Journal of Microwave & Electromagnetic Energy*, *41*(2), 41-2-44 – 41-2-54.

Guo, W. C., Zhu, X. H., Yue, R., Liu, H., & Liu, Y. (2011). Dielectric properties of Fuji apples from 10 to 4500 MHz during storage. *Journal of Processing and Preservation*, *35*(6), 884–890. doi:10.1111/j.1745-4549.2011.00541.x

Gupta, S. D., & Jatothu, B. (2013). Fundamentals and applications of lightemitting diodes (LEDs) in in vitro plant growth and morphogenesis. *Plant Biotechnology Reports*, *7*(3), 211–220. doi:10.100711816-013-0277-0

Gupta, S. D., & Karmakar, A. (2017). Machine vision based evaluation of im- pact of light emitting diodes (LEDs) on shoot regeneration and the effect of spectral quality on phenolic content and antioxidant capacity in *Swertia chirata*. *Journal of Photochemistry and Photobiology. B, Biology*, *174*, 162–172. doi:10.1016/j.jphotobiol.2017.07.029 PMID:28779689

Gutiérrez, J., Francisco, J., Villa-Medina, A., & Nieto-Garibay, M. Á. P.-G, (2014). *Automated Irrigation System Based on Wireless Sensor Network and GPRS Module*. Academic Press.

Guttormsen, G. (1969). Accumulation of salts in the sub-irrigation of pot plants. *Plant and Soil*, *31*(3), 425–438. doi:10.1007/BF01373814

Guyer, D., Miles, G., Schreiber, M., Mitchell, O., & Vanderbilt, V. (1986). Machine vision and image processing for plant identification. *Transactions of the ASAE. American Society of Agricultural Engineers*, *29*(6), 1500–1506. doi:10.13031/2013.30344

Haboudane, D., Miller, J. R., Tremblay, N., Zarco-Tejada, P. J., & Dextraze, L. (2002). Integrated narrow-band vegetation indices for prediction of crop chlorophyll content for application to precision agriculture. *Remote Sensing of Environment, 81*(2-3), 416–426. doi:10.1016/S0034-4257(02)00018-4

Haggar, R., Stent, C., & Isaac, S. (1983). A prototype hand-held patch sprayer for killing weeds, activated by spectral differences in crop/weed canopies. *Journal of Agricultural Engineering Research, 28*(4), 349–358. doi:10.1016/0021-8634(83)90066-5

Hajizadeh, M. (2016). Legalizing and Regulating Marijuana in Canada: Review of Potential Economic, Social, and Health Impacts. *International Journal of Health Policy and Management, 5*(8), 453–456. doi:10.15171/ijhpm.2016.63 PMID:27694657

Hakkim, V., Joseph, E., Gokul, A., & Mufeedha, K. (2016). Precision farming: The future of Indian agriculture. *Journal of Applied Biology and Biotechnology*, 68–72. doi:10.7324/JABB.2016.40609

Hall, J. (2017). *Industrial hemp and sugarcane rotation – not just a pipe dream.* A report of the Managing Director – Industrial Hemp Australia. Accessed on May 24, 2019 at www.australiansugarcane.com.au/Back issues/133jjsug09/21_Industrial.pdf

Hall, M. N., Robertson, A., & Scotter, C. N. G. (1988). Near-infrared reflectance prediction of quality, theaflavin content and moisture content of black tea. *Food Chemistry, 27*(1), 61–75. doi:10.1016/0308-8146(88)90036-2

Halmagyi, A., & Pinker, I. (2006a). Plant regeneration from *Rosa* shoot tips cryopreserved by a combined droplet vitrification method. *Plant Cell, Tissue and Organ Culture, 84*(2), 145–153. doi:10.100711240-005-9012-z

Halmagyi, A., & Pinker, I. (2006b). Cryopreservation of *Rosa* shoot tips: Importance of preculture conditions. *Acta Horticulturae*, (725), 351–356. doi:10.17660/ActaHortic.2006.725.45

Hamlin, Q. F., Kendall, A. D., Martin, S. L., Whitenack, H. D., Roush, J. A., Hannah, B. A., & Hyndman, D. W. (2020). Quantifying Landscape Nutrient Inputs With Spatially Explicit Nutrient Source Estimate Maps. *Journal of Geophysical Research: Biogeosciences, 125*(2).

Hammond, M. W., & Mulla, D. J. (1989). Field variation in soil fertility: Its assessment and management for potato production. *Proc. 28th Annual Washington State Potato Conference.*

Hammond, M. W., Mulla, D. J., & Fairchild, D. S. (1988). Development of management maps for soil variability. *Proc. 39th Annual Far West Regional Fertilizer Conference.*

Hanan, J. J. (2017). *Greenhouses: Advanced technology for protected horticulture.* CRC press. doi:10.1201/9780203719824

Happyana, N., Agnolet, S., Muntendam, R., Van Dam, A., Schneider, B., & Kayser, O. (2013). Analysis of cannabinoids in laser-microdissectedtrichomes of medicinal Cannabis sativa using LCMS and cryogenic NMR. *Phytochemistry, 87*, 51–59. doi:10.1016/j.phytochem.2012.11.001 PMID:23280038

Haque, S. K. M., & Ghosh, B. (2014). Somatic embryogenesis and synthetic seed production - A biotechnological approach for true-to-type propagation and *in vitro* conservation of an ornamental bulbaceous plant *Drimiopsis kirkii* Baker. *Applied Biochemistry and Biotechnology, 172*(8), 4013–4024. doi:10.100712010-014-0817-2 PMID:24604129

Haralick, R. M., Shanmugam, K., & Dinstein, I. H. (1973). Textural features for image classification. *IEEE Transactions on Systems, Man, and Cybernetics, SMC-3*(6), 610–621. doi:10.1109/TSMC.1973.4309314

Hargreaves, G. H., & Allen, R. G. (2003). History and Evaluation of Hargreaves Evapotranspiration Equation. *Journal of Irrigation and Drainage Engineering, 129*(1), 53–63. doi:10.1061/(ASCE)0733-9437(2003)129:1(53)

Harris, T. (2019). IoT: A Business Perspective. In *Fundamentals of Internet of Things for Non-Engineers* (pp. 279–301). Auerbach Publications. doi:10.1201/9780429274992-15

Hart, J. W. (2012). General introduction. In J. W. Hart (Ed.), *Light and plant growth, Topics in plant physiology: I, v* (Vol. 1, pp. 4–17). Springer Science & Business Media.

Hartley, C. W. S. (1977). *The Oil Palm*. Longman Group Limited.

Hartono, A., Barus, B., Tejo, R. K., Rostiana, O., Manohara, D., & Wahyuno, D. (2020, January). Smart farming on pepper for fertilizers recommendation in Bangka Belitung. In *IOP Conference Series: Earth and Environmental Science* (Vol. 418, No. 1, p. 012049). IOP Publishing. 10.1088/1755-1315/418/1/012049

Hassan, M. (2002). *In vitro Studies on Somatic Embryogenesis Conservation of Date Palm* (Ph.D. Thesis). Cairo University.

Hasted, J. B. (1973). *Aqueous Dielectric*. Chapman and Hall.

Hausman, J., Neys, O., Kevers, C., & Gaspar, T. (1994). Effect of in vitro storage at 4 °C on survival and proliferation of poplar shoots. *Plant Cell, Tissue and Organ Culture*, *38*(1), 65–67. doi:10.1007/BF00034446

Hawley, D., Graham, T., Stasiak, M., & Dixon, M. (2018). Improving Cannabis Bud Quality and Yield with Subcanopy Lighting. *HortScience*, *53*(11), 1593–1599. doi:10.21273/HORTSCI13173-18

Haykin, S. (1994). *Neural networks: a comprehensive foundation*. Prentice Hall PTR.

Hazarika, D., Laskar, S., Sarma, A., & Sarmah, P. K. (2006). PC-based instrumentation system for the detection of moisture content of tea leaves at its final stage. *IEEE Transactions on Instrumentation and Measurement*, *55*(5), 1641–1647. doi:10.1109/TIM.2006.881031

Heo, J., Lee, C., Chakrabarty, D., & Paek, K. Y. (2002). Growth responses of marigold and salvia bedding plants as affected by monochromic or mixture radiation provided by a light-emitting diode (LED). *Plant Growth Regulation*, *38*(3), 225–230. doi:10.1023/A:1021523832488

Hirai, D., & Sakai, A. (1999a). Cryopreservation of *in vitro*-grown axillary shoot-tip meristems of mint (*Mentha spicata* L.) by encapsulation vitrification. *Plant Cell Reports*, *19*(2), 150–155. doi:10.1007002990050725 PMID:30754740

Hirai, D., & Sakai, A. (1999b). Cryopreservation of *in vitro*-grown meristems of potato (*Solanum tuberosum* L.) by encapsulation vitrification. *Potato Research*, *42*(1), 153–160. doi:10.1007/BF02358405

Hirai, D., Shirai, K., Shirai, S., & Sakai, A. (1998). Cryopreservation of *in vitro*-grown meristems of strawberry (*Fragaria ×ananassa* Duch.) by encapsulation-vitrification. *Euphytica*, *101*(1), 109–115. doi:10.1023/A:1018386211577

Hoenecke, M. E., Bula, R. J., & Tibbitts, T. W. (1992). Importance of 'Blue' photon levels for lettuce seedlings grown under red-light-emitting diodes. *HortScience*, *27*(5), 427–430. doi:10.21273/HORTSCI.27.5.427 PMID:11537611

Hofstee, J. W., & Molema, G. J. (2002). *Machine vision based yield mapping of potatoes*. Paper number 021200, 2002 ASAE Annual Meeting. Available online at www.asabe.org

Hoogendoorn. (2020). *The next generation iSii process computer*. Retrieved from https://www.hoogendoorn.nl/en/product/isii-process-computer/

Hooper, L., Thompson, R. L., Harrison, R. A., Summerbell, C. D., Ness, A. R., Moore, H. J., Worthington, H. V., Durrington, P. N., Higgins, J. P. T., Capps, N. E., Riemersma, R. A., Ebrahim, S. B. J., & Smith, G. D. (2006). Risks and benefits of omega 3 fats for mortality, cardiovascular disease, and cancer: Systematic review. *BMJ (Clinical Research Ed.)*, *332*(7544), 752–760. doi:10.1136/bmj.38755.366331.2F PMID:16565093

Hornung, A., Khosla, R., Reich, R., Inman, D., & Westfall, D. G. (2006). Comparison of Site-Specific Management Zones: Soil-Color-Based and Yield-Based. *Agronomy Journal, 98*(2), 407–415. doi:10.2134/agronj2005.0240

HortiMax growing solutions. (2020a). *HortiMaX MultiMa*. Retrieved from https://www.hortimax.co.uk/water-and-nutrition/hortimax-multima/

HortiMax growing solutions. (2020b). *HortiMax Synopta*. Retrieved from https://www.hortimax.co.uk/operation-and-analysis/hortimax-synopta/

Hostiou, N., Fagon, J., Chauvat, S., Turlot, A., Kling-Eveillard, F., Boivin, X., & Allain, C. (2017). Impact of precision livestock farming on work and human-animal interactions on dairy farms. *RE:view*.

Hough, M., Warburton, H., Few, B., May, T., Man, L.-H., Witton, I., & Turnbull, P. J. (2003). *A growing market: The domestic cultivation of cannabis York*. Joseph Rowntree Foundation.

How, F.-W., & Smith, M. A. (2003). Effect of Light/Dark cycling on growth and anthocyanin production of *Ajuga reptens* in callus culture. *J. Agric. Res. China, 52*, 291–296.

Huang, C., & Chen, C. (2005). Physical properties of culture vessels for plant tissue culture. *Biosystems Engineering, 91*(4), 501–511. doi:10.1016/j.biosystemseng.2005.05.005

Huan, L. V. T., & Tanaka, M. (2004). Effect of red and blue light-emitting diodes (LEDs) on callus induction, callus proliferation, and protocorm-like body formation from callus in *Cymbidium orchid. Environment Control in Biology, 42*(1), 57–64. doi:10.2525/ecb1963.42.57

Huat, B.B.K., Faisal, H. A., & Maail, S. (2005). The effect ofnatural fber on the shear strength of soil. *American Journal of Applied Sciences*, 9-13.

Huché-Thélier, L., Crespel, L., Le Gourrierec, J., Morel, P., Sakr, S., & Leduc, N. (2016). Light signaling and plant responses to blue and UV radiations—Perspectives for applications in horticulture. *Environmental and Experimental Botany, 121*, 22–38. doi:10.1016/j.envexpbot.2015.06.009

Hughes, D.P., & Salathé, M. (2015). *An open access repository of images on plant health to enable the development of mobile disease diagnostics*. CoRR abs/1511.08060

Hummel, J. W. (1985). Monitoring and control of field machines. *Illinois Research, 27*(4), 8–10.

Hung, C. D., Hong, C.-H., Kim, S.-K., Lee, K.-H., Park, J.-Y., Nam, M.-W., Choi, D. H., & Lee, H.-I. (2016). LED light for in vitro and ex vitro efficient growth of economically important highbush blueberry (*Vaccinium corymbosum* L.). *Acta Physiologiae Plantarum, 38*(6), 1–9. doi:10.100711738-016-2164-0

Huong, L.T.L., Baiocco, M., Huy, B.P., Mezzetti, B., Santilocchi, R., & Rosati, P. (1999). Somatic embryogenesis in Canary Island date palm. *Plant Cell Tissue and Organ Culture, 56*(1), 1-7.

Husvéth, F., Galamb, E., Gaal, T., Dublecz, K., Wagner, L., & Pál, L. (2010). Milk production, milk composition, liver lipid contents and C18 fatty acid composition of milk and liver lipids in Awassi ewes fed a diet supplemented with protected cis-9, trans-11 and trans-10, cis-12 conjugated linoleic acid (CLA) isomers. *Small Ruminant Research, 94*(1-3), 25–31. doi:10.1016/j.smallrumres.2010.06.003

Huuskonen, J., & Oksanen, T. (2018). Soil sampling with drones and augmented reality in precision agriculture. *Computers and Electronics in Agriculture, 154*, 25–35. doi:10.1016/j.compag.2018.08.039

Hwang, S. (2017). Monitoring and Controlling System for an IoT Based Smart Home. *Int. J. Control Autom., 10*(2), 339–348. doi:10.14257/ijca.2017.10.2.28

Ilic, Z. S., Milenkovic, L., Stanojevic, L., Cvetkovic, D., & Fallik, E. (2012). Effects of the modification of light intensity by color shade nets on yield and quality of tomato fruits. *Scientia Horticulturae*, *139*, 90–95. doi:10.1016/j.scienta.2012.03.009

Imran, M., Rizwan, M., Kerkhi, S. A., & Ali, N. (2018). Bioinformatics for Crop Breeding and Sustainable Agriculture. *Journal*, *6*(4), 1–5. www. ijbpr. in

Isah, T. (2015). Adjusments to *in vitro* culture conditions and associated anomalies in plants. *Acta Biologica Cracoviensia. Series; Botanica*, *57*(2), 9–28. doi:10.1515/abcsb-2015-0026

Ishibashi, T., Kimura, S., Furukawa, T., & Sakaguchi, K. (2006). DNA repair mechanisms in UV-B tolerant plants. *Japan Agricultural Research Quarterly*, *40*(2), 107–113. doi:10.6090/jarq.40.107

Islam, M. T., Dembele, D. P., & Keller, E. R. J. (2005). Influence of explant, temperature and different culture vessels on in vitro culture for germplasm maintenance of four mint accessions. *Plant Cell, Tissue and Organ Culture*, *81*(2), 123–130. doi:10.100711240-004-3307-3

Ivan, M., Mir, P. S., Koenig, K. M., Rode, L. M., Neill, L., Entz, T., & Mir, Z. (2001). Effects of dietary sunflower seed oil on rumen protozoa population and tissue concentration of conjugated linoleic acid in sheep. *Small Ruminant Research*, *41*(3), 215–227. doi:10.1016/S0921-4488(01)00220-6

Ivanyi, I. (2005). *Relationship between leaf nutrient concentrations and yield of fibre hemp (Cannabis sativa L.). Alps-Adria Scientific Workshop*, Portoroz, Slovenia. 10.1556/CRC.33.2005.1.23

Ivonyi, I., Izsoki, Z., & Van der Werf, H. M. G. (1997). Influence of nitrogen supply and P K levels of the soil on dry matter and nutrient accumulation of fibre hemp (Cannabis sativa L.). *Journal of the International Hemp Association*, *4*(1), 82–87.

Jaczewski, A. (1927). *Mucnisto-rosjanyegriby* [Powdery mildew fungi] KarmannyiOpreditel'. *Gribov*, *5*(2), 412.

Jafari, F., Khalid, K., Yusoff, W. M. D. W., & Hassan, J. (2010). The analysis and design of multi-layer microstrip moisture sensor for rice grain. *Biosystems Engineering*, *106*(3), 324–331. doi:10.1016/j.biosystemseng.2010.04.005

James, G. L. (1988). *Principles of Farm Irrigation System Design*. John Wiley and Sons, Inc.

Jansen, M. A. K. (2002). Ultraviolet-B radiation effects on plants: Induction of morphogenic responses. *Physiologia Plantarum*, *116*(3), 423–429. doi:10.1034/j.1399-3054.2002.1160319.x

Jao, R.C., & Fang, W. (2004b). *Growth of potato plantlets in vitro is different when provided concurrent versus alternating red and blue light photoperiods*. Academic Press.

Jao, R. C., & Fang, W. (2004a). Effects of frequency and duty ratio on the growth of potato plantlets in vitro using light emitting diodes. *HortScience*, *39*(2), 375–379. doi:10.21273/HORTSCI.39.2.375

Javanmard, M., Abbas, K. A., & Arvin, F. (2009). A microcontroller-based monitoring system for batch tea dryer. *The Journal of Agricultural Science*, *1*(2), 101–106.

Jawad, H. M., Nordin, R., Gharghan, S. K., Jawad, A. M., & Ismail, M. (2017). Energy-Efficient Wireless Sensor Networks for Precision Agriculture: A Review. *Sensors (Basel)*, *17*(8), 1–45. doi:10.339017081781 PMID:28771214

Jensen, C. L. (2006). Effects of n-3 fatty acids during pregnancy and lactation. *The American Journal of Clinical Nutrition*, *83*(6), 1452–1457. doi:10.1093/ajcn/83.6.1452S PMID:16841854

Jeong, B. R., & Sivanesan, I. (2015). Direct adventitious shoot regeneration, in vitro flowering, fruiting, secondary metabolite content and anti- oxidant activity of *Scrophularia takesimensis* Nakai. *Plant Cell, Tissue and Organ Culture, 123*(3), 607–618. doi:10.100711240-015-0864-6

Jeschke, S., Brecher, C., Meisen, T., Özdemir, D., & Eschert, T. (2017). Industrial internet of things and cyber manufacturing systems. In *Industrial internet of things* (pp. 3–19). Springer. doi:10.1007/978-3-319-42559-7_1

Jiao, J., Gai, Q. Y., Fu, Y. J., Ma, W., Yao, L. P., Feng, C., & Xia, X. X. (2015). Optimization of *Astragalus membranaceus* hairy roots induction and culture conditions for augmentation production of astragalosides. *Plant Cell, Tissue and Organ Culture, 120*(3), 1117–1130. doi:10.100711240-014-0668-0

Jishi, T., Kimura, K., Matsuda, R., & Fujiwara, K. (2016). Effects of temporally shifted irradiation of blue and red LED light oncos lettuce growth and morphology. *Scientia Horticulturae, 198*, 227–232. doi:10.1016/j.scienta.2015.12.005

Jones, J. W., Antle, J. M., Basso, B., Boote, K. B., Conant, R. T., Ian Foster, I., ... Wheeler, T. R. (2017). Toward a new generation of agricultural system data, models, and knowledge products: State of agricultural systems science. *Agricultural Systems, 155*, 269–288. doi:10.1016/j.agsy.2016.09.021 PMID:28701818

Jones, T. R., Zarifi, M. H., & Mojgan, D. (2017). Miniaturized quarter-mode substrate integrated cavity resonators for humidity sensing. *IEEE Microwave and Wireless Components Letters, 27*(7), 612–614. doi:10.1109/LMWC.2017.2711527

Jordan, H., Lang, A. L., & Enfield, G. H. (1946). Effects of fertilizers on yields and breaking strengths of American hemp, Cannabis sativa. *Journal of the American Society of Agronomists, 38*(6), 551–563. doi:10.2134/agronj1946.00021962003800060010x

Jung, E. S., Lee, S., Lim, S. H., Ha, S. H., Liu, K. H., & Lee, C. H. (2013). Metabolite profiling of the short–term responses of rice leaves (*Oryza sativa* cv. Ilmi) cultivated under different LED lights and its correlations with antioxidant activities. *Plant Science, 210*, 61–69. doi:10.1016/j.plantsci.2013.05.004 PMID:23849114

Kadar, J., & Tarkany-Szucs, S. (2003). Effect of mineral fertilization on the yield of fibre hemp (Cannabis sativa L.). *Novenytermeles, 52*(2), 217–228.

Kalia, A. (2018). Nanotechnology in Bioengineering: Transmogrifying Plant Biotechnology. Omics Technologies and Bio-Engineering, 211-229.

Kamilaris, A., Gao, F., Prenafeta-Boldu, F. X., & Ali, M. I. (2016, December). Agri-IoT: A semantic framework for Internet of Things-enabled smart farming applications. In *2016 IEEE 3rd World Forum on Internet of Things (WF-IoT)* (pp. 442-447). IEEE. 10.1109/WF-IoT.2016.7845467

Kamińska, M., Gołębiewski, M., Tretyn, A., & Trejgell, A. (2018). Efficient long-term conservation of *Taraxacum pieninicum* synthetic seeds in slow growth conditions. *Plant Cell, Tissue and Organ Culture, 132*(3), 469–478. doi:10.100711240-017-1343-z

Kamińska, M., Skrzpek, E., Wilmowicz, E., Tretyn, A., & Trejgell, A. (2016). Effect of light conditions and ABA on cold storage and post-storage propagation of *Taraxacum pieninicum. Plant Cell, Tissue and Organ Culture, 127*(1), 25–34. doi:10.100711240-016-1026-1

Kanazawa, K., Hashimoto, T., Yoshida, S., Sungwon, P., & Fukuda, S. (2012). Short photoirradiation induces flavonoid synthesis and increases its production in postharvest vegetables. *Journal of Agricultural and Food Chemistry, 60*(17), 4359–4368. doi:10.1021/jf300107s PMID:22506664

Kanter, D. R., Bell, A. R., & McDermid, S. S. (2019). Precision Agriculture for Smallholder Nitrogen Management. *One Earth, 1*(3), 281–284. doi:10.1016/j.oneear.2019.10.015

Karlen, D. L., Sadler, E. J., & Busscher, W. J. (1990). Crop yield variation associated with Coastal Plain soil map units. *Soil Science Society of America Journal*, *54*(3), 859–865. doi:10.2136ssaj1990.03615995005400030041x

Kartha, K. K., & Engelmann, F. (1994). Cryopreservation and gemplasm storage. In I. K. Vasil & T. A. Thrope (Eds.), *Plant Cell and Tissue Culture* (pp. 195–230). Kluwer Acadimic Publishers. doi:10.1007/978-94-017-2681-8_9

Katerova & Todorova. (2011). Effect of enhanced UV-C irradiation on growth, some stress markers, polyamines, IAA and IAA-oxidase activity in pea plants (*Pisum sativum* L.). *Compt rend Acad bulg Sci, 64*(11), 1555–1562.

Katerova, Z. I., & Todorova, D. (2009). Endogenous polyamines lessen membrane damages in pea plants provoked by enhanced ultraviolet-C radiation. *Plant Growth Regulation*, *57*(2), 145–152. doi:10.100710725-008-9330-3

Katerova, Z., Ivanov, S., Mapelli, S., & Alexieva, V. (2009). Phenols, proline and low-molecular thiol levels in pea (*Pisum sativum*) plants respond differently toward prolonged exposure to ultraviolet-B and ultraviolet-C radiations. *Acta Physiologiae Plantarum*, *31*(1), 111–117. doi:10.100711738-008-0208-9

Kathner, J., & Zude-Sasse, M. (2015). Interaction of 3D soil electrical conductivity and generative growth in Prunusdomestica. *European Journal of Horticultural Science*, *80*(5), 231–239. doi:10.17660/eJHS.2015/80.5.5

Katke, K. (2019). Precision Agriculture Adoption: Challenges of Indian Agriculture. *International Journal of Research and Analytical Reviews*, *6*(1), 863–869.

Kawakita, E., Hashimoto, M., & Shido, O. (2006). Docosahexaenoic acid promotes neurogenesis in vitro and in vivo. *Neuroscience*, *139*(3), 991–997. doi:10.1016/j.neuroscience.2006.01.021 PMID:16527422

Kawka, B., Kwiecień, I., & Ekiert, H. (2017). Influence of culture medium composition and light conditions on the accumulation of bioactive compounds in shoot cultures of *Scutellaria lateriflora* L. (*American skullcap*) grown in vitro. *Applied Biochemistry and Biotechnology*, *183*(4), 1414–1425. doi:10.100712010-017-2508-2 PMID:28573603

Kay, J. K., Roche, J. R., Thomson, N. A., Griinari, J. M., & Shingfield, K. J. (2003). Increasing milk fat cis-9, trans-11 conjugated linoleic acid content in pasture-fed cows. *Journal of Dairy Science*, *86*(Suppl. 1), 146.

Keller, R. J., Nichols, M. E., & Lange, A. F. (2001). *Methods and apparatus for precision agriculture operations utilizing real time kinematic global positioning system systems*. U.S. Patent No. 6,199,000, issued March 6, 2001.

Kelly, M. L., Kolver, E. S., Bauman, D. E., VanAmburgh, M. E., & Muller, L. D. (1998). Effect of intake of pasture on concentrations of conjugated linoleic acid in milk of lactating cows. *Journal of Dairy Science*, *81*(6), 1630–1636. doi:10.3168/jds.S0022-0302(98)75730-3 PMID:9684170

Kendall, H., Naughton, P., Clark, B., Taylor, J., Li, Z., Zhao, C., Yang, G., Chen, J., & Frewer, L. J. (2017). Precision Agriculture in China: Exploring Awareness, Understanding, Attitudes and Perceptions of Agricultural Experts and End-Users in China. *Advances in Animal Biosciences*, *8*(2), 703–707. doi:10.1017/S2040470017001066

Kendrick, B. (1985). *The fifth kingdom*. Mycologue Publication.

Kendrick, R. E., & Kronenberg, G. H. (2012). *Photomorphogenesis in plants*. Springer Science & Business Media Kluwer.

Kennedy, G., & Burlingame, B. (2003). Analysis of food composition data on rice from a plant generic resources perspective. *Food Chemistry*, *80*(4), 589–596. doi:10.1016/S0308-8146(02)00507-1

Keskin, M., Sekerli, Y. E., Say, S. M., & Topcueri, M. (2017). Farmers' satisfaction level with GNSS-based tractor auto guidance systems in Adana province of Turkey. *13th Int Congress on Mechanization and Energy in Agriculture & Int. Workshop on Precision Agriculture*. Available online at https://www.researchgate.net/publication/320024304_Farmers'_Satisfaction_Level_with_GNSS-Based_Tractor_Auto_Guidance_Systems_in_Adana_Province_of_Turkey_Adana_Ilinde_GNSS_Esasli_Traktor_Otomatik_Dumenleme_Sistemi_Kullanan_Ciftcilerin_Memnuniyet_Duzeyi

Khaled, D. E., Novas, N., Gazquez, J. A., Garcia, R. M., & Agugliaro, F. M. (2015). Fruit and vegetable quality assessment via dielectric sensing. *Sensors (Basel)*, *15*(7), 15363–15397. doi:10.3390150715363 PMID:26131680

Khalid, K., & Abbas, Z. (1992). A microstrip sensor for determination of harvesting time for oil palm fruits (Tenera: ElaeisGuineesis). *The Journal of Microwave Power and Electromagnetic Energy*, *27*(1), 3–10. doi:10.1080/08327823.1992.11688165

Khalid, K., Hassan, J., & Daud, W. M. (1996). The effect of ionic conductivity and dipole orientation on the dielectric loss of the hevea rubber latex. *Proceedings of IEEE MTTS International Microwave Symposium*, 23–26.

Khanal, R. C., & Dhiman, T. R. (2004). Biosynthesis of conjugated linoleic acid (CLA): A review. *Pakistan Journal of Nutrition*, *3*(2), 72–81. doi:10.3923/pjn.2004.72.81

Khanal, S., Fulton, J., & Shearer, S. (2017). An overview of current and potential applications of thermal remote sensing in precision agriculture. *Computers and Electronics in Agriculture*, *139*, 22–32. doi:10.1016/j.compag.2017.05.001

Khan, T., Ullah, M. A., Garros, L., Hano, C., & Abbasi, B. H. (2018). Synergistic effects of melatonin and distinct spectral lights for enhanced production of anti-cancerous compounds in callus cultures of *Fagonia indica*. *Journal of Photochemistry and Photobiology. B, Biology*. Advance online publication. doi:10.1016/j.jphotobiol.2018.10.010 PMID:30482427

Kim, H. H., Goins, G. D., Wheeler, R. M., & Sager, J. C. (2004a). Green light supplementation for enhanced lettuce growth under red and blue light emitting diodes. *HortScience*, *39*(7), 1617–1622. doi:10.21273/HORTSCI.39.7.1617 PMID:15770792

Kim, H. H., Goins, G. D., Wheeler, R. M., & Sager, J. C. (2004b). Stomatal conductance of lettuce grown under or exposed to different light qualities. *Annali di Botanica*, *94*(5), 691–697. doi:10.1093/aob/mch192 PMID:15347557

Kim, K. B., Kim, J. H., Lee, S. S., & Noh, S. H. (2002). Measurement of grain moisture content using microwave attenuation at 10.5 GHz and moisture density. *IEEE Transactions on Instrumentation and Measurement*, *51*(1), 72–77. doi:10.1109/19.989904

Kim, S. J., Hahn, E. J., Heo, J. W., & Pae, K. K. Y. (2004c). Effects of LEDs on net photosynthetic rate, growth and leaf stomata of *Chrysanthemum* plantlets in vitro. *Scientia Horticulturae*, *101*(1-2), 143–151. doi:10.1016/j.scienta.2003.10.003

Kindler, J. (1992). Rationalizing water requirements with aid of fuzzy allocation model. *Journal of Water Resources Planning and Management*, *118*(3), 308–323. doi:10.1061/(ASCE)0733-9496(1992)118:3(308)

Kintzios S, Papanastasiou I, Tourgelis P, Papastellatos C, Georgopoulos V, Drossopoulos J (2002). The effects of light on callus growth and somatic embryogenesis from *Lavandula vera* and *Teucrium chamaedrys*:a preliminary study. *Journal of Herbs Spices & Medicinal Plants Spices & Medicinal Plants*, (2), 223-227.

Kintzios, S., & Taravira, N. (1997). Effect of genotype and light intensity on somatic embryogenesis and plant regeneration in melon (*Cucumis melo* L.). *Plant Breed., 116*, 359-362.

Kintzios, S., Katsouri, E., Peppes, D., & Koulocheri, S. (1998a). Somatic embryogenesis and in vitro secondary metabolite production from common mallow (*Malva sylvestris* L.) collected in Greece. *Acta Horticulturae*, (457), 173–178. doi:10.17660/ActaHortic.1998.457.21

Kintzios, S., Manos, C., & Makri, O. (1998b). Somatic embryogenesis from mature leaves of rosa (*Rosa sp.*). *Plant Cell Reports*, *18*(6), 467–472. doi:10.1007002990050605

Kišgeci, J. (1994). *Praise to Hemp*. Nolit. (in Serbian)

Kladivko, E. J., & Timmenga, H. J. (2019). Earthworms and agricultural management. In *Rhizosphere dynamics* (pp. 192–216). CRC Press. doi:10.1201/9780429304798-8

Knight, G., Hansen, S., Connor, M., Poulsen, H., McGovern, C., & Stacey, J. (2010). The results of an experimental indoor hydroponic cannabis growing study, using the "Screen of Green" (ScrOG) method-yield, tetrahydrocannabinol (THC) and DNA analysis. *Forensic Science International*, *202*(1-3), 36–44. doi:10.1016/j.forsciint.2010.04.022 PMID:20462712

Konstas, J., Kintzios, S., Drossopoulos, J., & Sarlis, G. (2003). The effect of light intensity and relative exposure under light on the expression of direct or indirect somatic embryogenesis from common mallow (*Malva sylvestris* L.). *Acta Horticulturae*, (597), 315–319. doi:10.17660/ActaHortic.2003.597.45

Kott, R. W., Hatfield, P. G., Bergman, J. W., Flynn, C. R., Van Wagoner, H., & Boles, J. A. (2003). Feedlot performance, carcass composition, and muscle and fat CLA concentrations of lambs fed diets supplemented with safflower seeds. *Small Ruminant Research*, *49*(1), 11–17. doi:10.1016/S0921-4488(03)00052-X

Kowalski, W. J. (2009). *Ultraviolet Germicidal Irradiation Handbook: UVGI for Air and Surface Disinfection*. Springer. doi:10.1007/978-3-642-01999-9

Kozai, T., & Smith, M. A. L. (1995) Environmental control in plant tissue culture-general overview. In Automation and Environmental Control in Plant Tissue Culture. Kluwer Academic Publishers. doi:10.1007/978-94-015-8461-6_14

Kozai, T. (2013). Resource use efficiency of closed plant production system with artificial light: Concept, estimation and application to plant factory. *Proc Japn Acad Series B*, *89*(10), 447–461. doi:10.2183/pjab.89.447 PMID:24334509

Kozai, T. (2016). Why LED lighting for urban agriculture? In T. Kozai, K. Fujiwara, & E. Runkle (Eds.), *LED lighting for urban agriculture* (pp. 3–18). Springer. doi:10.1007/978-981-10-1848-0_1

Kraszewski, A. W. (2005). Recent Developments in Electromagnetic Aquametry. In K. Kupfer (Ed.), *Electromagnetic Aquametry* (pp. 1–12). Springer. doi:10.1007/3-540-26491-4_1

Krewzaler, F., & Hahlbrock, K. (1973). Flavonoid glycosides from illuminated cell suspension cultures of *Petroselinum hortense*. *Phytochemistry*, *12*(5), 1149–1152. doi:10.1016/0031-9422(73)85031-9

Krishna, K. R. (2018). *Agricultural drones: a peaceful pursuit*. Apple Academic Press. doi:10.1201/9781315195520

Krishnan, M., Foster, Ch. A., Strosser, R. P., Glancey, J. L., & Sun, J.-Q. (2006). Adaptive modeling and control of a manure spreader for precision agriculture. *Computers and Electronics in Agriculture*, *52*(1–2), 1–10. doi:10.1016/j.compag.2005.11.005

Ku, K.-L., Chang, P.-S., Cheng, Y.-C., & Lien, C.-Y. (2005). Production of stilbenoids from the callus of *Arachis hypogaea*: A novel source of anticancer compound piceatannol. *Journal of Agricultural and Food Chemistry*, *53*(10), 3877–3881. doi:10.1021/jf050242o PMID:15884811

Kumar, D., & Kalita, P. (2017). Reducing postharvest losses during storage of grain crops to strengthen food security in developing countries. *Foods*, *6*(1), 1–22. doi:10.3390/foods6010008 PMID:28231087

Kumar, H. S., & Kusuma, S. (2016). Automated Irrigation System Based on Wireless Sensor Network and GPRS Module. *Int. Res. J. Eng. Technol.*, *3*(4), 148–151.

Kumari, S., Singh, T. P., & Prasad, S. (2019). Climate Smart Agriculture and Climate Change. *International Journal of Current Microbiology and Applied Sciences*, *8*(3), 1112–1137. doi:10.20546/ijcmas.2019.803.134

Kuo, T. C. Y., Chen, C. H., Chen, S. H., Lu, I. H., Chu, M. J., Huang, L. C., Lin, C. Y., Chen, C. Y., Lo, H. F., Jeng, S. T., & Chen, L. F. O. (2015). The effect of red light and far-red light conditions on secondary metabolism in agarwood. *BMC Plant Biology*, *15*(1), 139. doi:10.118612870-015-0537-y PMID:26067652

Kutter, T., Tiemann, S., Siebert, R., & Fountas, S. (2011). The role of communication and cooperation in the adoption of precision farming. *Precision Agriculture*, *12*(1), 2–17. doi:10.100711119-009-9150-0

Kwak, S. M.; Myung, S. K.; Lee, Y. J., & Seo, H.G. (2012). Efficacy of Omega-3 Fatty Acid Supplements (Eicosapentaenoic Acid and Docosahexaenoic Acid) in the Secondary Prevention of Cardiovascular Disease: A Meta-analysis of Randomized, Double-blind, Placebo-Controlled Trials. *Archives of Internal Medicine*.

Kwasnewska, E., Dziedzic, E., & Pawlowska, B. (2017). Integration of cryopreservation and tissue culture for germplasm conservation and ropagation of Rosa pomifera 'Karpatia '. *Notulae Botanicae Horti Agrobotanici Cluj-Napoca*, *45*(1), 208–214. doi:10.15835/nbha45110566

Labuza, T. P., &Altunakar, B. (2020). Water activity prediction and moisture sorption isotherms. *Water activity in foods: Fundamentals and applications*, 161-205.

Lakhiar, I. A., Jianmin, G., Syed, T. N., Chandio, F. A., Buttar, N. A., & Qureshi, W. A. (2018). Monitoring and Control Systems in Agriculture Using Intelligent Sensor Techniques: A Review of the Aeroponic System. Journal of Sensors Volume 2018. *Article ID*, *8672769*, 18. doi:10.1155/2018/8672769

Larsen, W. E., Tyler, D. A., & Nielsen, G. A. (1988). Field navigation using the global positioning system (GPS). Am. Soc. Agric. Eng.

Larson, W., Lamb, J., Khakural, B., Ferguson, R., & Rehm, G. (1997). Potential of site-specific management for nonpoint environmental protection. In F. Pierce & E. Sadler (Eds.), *The State of Site-Specific Management for Agriculture* (pp. 337–367). ASA-CSSA-SSSA. doi:10.2134/1997.stateofsitespecific.c15

Lata, H., Moraes, R. M., Bertoni, B., & Pereira, A. M. S. (2010). *In vitro* germplasm conservation of *Podophyllum peltatum* L. under slow growth conditions. *In Vitro Cellular & Developmental Biology. Plant*, *46*(1), 22–27. doi:10.100711627-009-9243-5

Latorre, B. A., Rojas, S., Diaz, G. A., & Chuaqui, H. (2012). Germicidal effect of UV light on epiphytic fungi isolated from bleberry. *CienInvAgr*, *39*(3), 473–480.

Laura, J., & denHartigh. (2019). Conjugated Linoleic Acid Effects on Cancer, Obesity, and Atherosclerosis: A Review of Pre-Clinical and Human Trials with Current Perspectives. *Nutrients*, *2019*(11), 370.

Lavola, A., Aphalo, P. J., Lahti, M., & Julkunen-Tiitto, R. (2003). Nutrient availability and the effect of increasing UV-B radiation on secondary plant compounds in Scots pine. *Environmental and Experimental Botany*, *49*(1), 49–60. doi:10.1016/S0098-8472(02)00057-6

Lavric, A., & Popa, V. (2017, July). Internet of things and LoRa™ low-power wide-area networks: a survey. In *2017 International Symposium on Signals, Circuits and Systems (ISSCS)* (pp. 1-5). IEEE. 10.1109/ISSCS.2017.8034915

Lawless, F., Stanton, C., Escop, P. L., Devery, R., Dillon, P., & Murphy, J. J. (1999). Influence of breed on bovine milk cis-9, trans-11-conjugated linoleic acid content. *Livestock Production Science*, *62*(1), 43–49. doi:10.1016/S0301-6226(99)00053-6

Lawrance, C. J., Rickson, R. J., & Clark, J. E. (1996). The Effect of Grass Roots on the Shear Strength of Colluvial Soils in Nepal. In Advances in Hillslope Processes. John Wiley and Sons.

Lawson, R. E., Moss, A. R., & Givens, D. I. (2001). The role of dairy products in supplying conjugated linoleic acid to man's diet: A review. *Nutrition Research Reviews*, *14*(1), 153–172. doi:10.1079/095442201108729178 PMID:19087420

Leamer, R. W., Myers, V. I., & Silva, L. F. (1973). A spectroradiometer for field use. *The Review of Scientific Instruments*, *44*(5), 611–614. doi:10.1063/1.1686195

Lefsrud, M. G., Kopsell, D. A., & Sams, C. E. (2008). Irradiance from distinct wavelength light emitting diodes affect secondary metabolites in kale. *HortScience*, *43*(7), 2243–2244. doi:10.21273/HORTSCI.43.7.2243

Leggett, T. (2006). Review of the world cannabis situation. *Bulletin on Narcotics*, *58*, 1–155. PMID:19066071

Lemberger, L. (1980). Potential therapeutic usefulness of marihuana. *Annual Review of Pharmacology and Toxicology*, *20*(1), 151–172. doi:10.1146/annurev.pa.20.040180.001055 PMID:6104468

Leonard, E. (2014). *Precision Ag Down Under*. Available online at https://www.precisionag.com/in-field-technologies/guidance/precision-ag-down-under/

Lesch, S. M., Rhoades, J. D., Lund, L. J., & Corwin, D. L. (1992). Mapping soil salinity using calibrated electromagnetic measurements. *Soil Science Society of America Journal*, *56*(2), 540–548. doi:10.2136ssaj1992.03615995005600020031x

Li, Z., Taylor, J., Frewer, L., Zhao, C., Yang, G., Liu, Z., & Yu, C. (2019). A comparative review on the state and advancement of site-specific crop management in the UK and China. *Front. Agric. Sci. Eng.*, 1–21. doi:10.15302/J-FASE-2018240

Lian, M. L., Murthy, H. N., & Paek, K. Y. (2002). Effects of light emitting diodes (LED) on the in vitro induction and growth of bulblets of *Lilium* oriental hybrid 'Pesaro'. *Scientia Horticulturae*, *94*(3-4), 365–370. doi:10.1016/S0304-4238(01)00385-5

Liao, M. (2017). *XairCraft Launched in Japan Targeting Global Precision Farming*. Available online at https://globenewswire.com

Li, H., Xu, Z., & Tang, C. (2010). Effect of light-emitting diodes on growth and morphogenesis of upland cotton (*Gossypium hirsutum* L.) plantlets in vitro. *Plant Cell, Tissue and Organ Culture*, *103*(2), 155–163. doi:10.100711240-010-9763-z

Lingyan, R., Yanning, Z., Wei, W., & Qilin, Z. (2017). A hyperspectral image classification framework with spatial pixel pair features. *Sensors (Basel)*, *17*(10), 2421. doi:10.339017102421 PMID:29065535

Linsley, C. M., & Bauer, F. C. (1929). Illinois Agricultural Experiment Station. *Circular, 346.*

Lin, Y., Li, J., Li, B., He, T., & Chun, Z. (2010). Effects of light quality on growth and development of protocorm-like bodies of *Dendrobium officinale* in vitro. *Plant Cell, Tissue and Organ Culture*, *105*(3), 329–335. doi:10.100711240-010-9871-9

Lisson, S. N., Mendham, N. J., & Carberry, P. S. (2000). Development of a hemp (Cannabis sativa L.) simulation model 3. The effect of plant density on leaf appearance, expansion and senescence. *Australian Journal of Experimental Agriculture*, *40*(3), 419–423. doi:10.1071/EA99060

Lisson, S., & Mendham, N. (1998). Response of fibre hemp (Cannabis sativa L.) to varying irrigation regimes. *Journal of the International Hemp Association*, *5*(1), 9–15.

Li, T., Ji, Y. H., Zhang, M., & Li, M. Z. (2017). Determining optimal CO_2 concentration of greenhouse tomato based on PSO-SVM. *Applied Engineering in Agriculture*, *33*(2), 157–166. doi:10.13031/aea.11578

Liu, A. R., Chen, S. C., Lin, X. M., Wu, S. Y., Xu, T., Cai, F. M., & Raesh, J. (2010). Endophytic Pestalotiopsis species spp. associated with plants of Palmae, Rhizophoraceae, Planchonellae and Podocarpaceae in Hainan, China. *African Journal of Microbiological Research*, *4*, 2661–2669.

Liu, C. Z., Guo, C., Wang, F. O., & Ouyang, F. (2003). Factors influencing artemisinin production from shoot cultures of *Artemisia annua* L. *World Journal of Microbiology & Biotechnology*, *19*(5), 535–538. doi:10.1023/A:1025158416832

Liu, S., Zhu, Y., Yue, E., & Guo, X. (2020). Key Technology and Application Platform of Agricultural Economic Spatial Information Service—Chinese Agricultural Economic Electronic Map. In *China's e-Science Blue Book 2018* (pp. 269–289). Springer. doi:10.1007/978-981-13-9390-7_16

Li, X. L., Xie, C. Q., He, Y., Qiu, Z. J., & Zhang, Y. C. (2012). Characterizing the moisture content of tea with diffuse reflectance spectroscopy using wavelet transform and multivariate analysis. *Sensors (Basel)*, *12*(7), 9847–9861. doi:10.3390120709847 PMID:23012574

Li, X., & Zhu, W. (2011). Apple grading method based on features fusion of size, shape and color. *Procedia Engineering*, *15*, 2885–2891. doi:10.1016/j.proeng.2011.08.543

Lock, A. L., & Bauman, D. E. (2004). Modifying milk fat composition of dairy cows to enhance fatty acids beneficial to human health. *Lipids*, *39*(12), 1197–1206. doi:10.100711745-004-1348-6 PMID:15736916

Lohr, S. (2012). The age of big data. *New York Times*, *11*(2012).

Lopez-Luc, G. (2013). An effective *in vitro* slow growth protocol for conservation of the orchid *Epidendrum chlorocorymbos* Schltr. *Tropical and Subtropical Agroecosystems*, *16*, 61–68.

Lorenzini, G. (2002). Air temperature effect on spray evaporation in sprinkler irrigation. *Irrigation and Drainage*, *51*(4), 301–309. doi:10.1002/ird.68

Loughrin, J. H., & Kasperbauer, M. J. (2001). Light reflected from colored mulches affects aroma and phenol content of sweet basil (Ocimumbasilicum L.) leaves. *Journal of Agricultural and Food Chemistry*, *49*(3), 1331–1335. doi:10.1021/jf0012648 PMID:11312860

Lowenberg-DeBoer, J. (1999). Precision Agriculture in Argentina. *Modern Agriculture*, *2*(2), MA13–MA15.

Loyall, L., Uchida, K., Furuya, M., & Frohnmeyer, H. (2000). Glutathione and a UV light induced glutathione S-transferase are involved in signaling to chalcone synthase in cell cultures. *The Plant Cell*, *12*, 1939–1950. PMID:11041888

Ludden, P. A., Kucukb, O., Rule, D. C., & Hess, B. W. (2009). Growth and carcass fatty acid composition of beef steers fed soybean oil for increasing duration before slaughter. *Meat Science*, *82*(2), 185–192. doi:10.1016/j.meatsci.2009.01.009 PMID:20416764

Luellen, W. R. (1985). Fine-tuned fertility: Tomorrow's technology here today. *Crops Soils*, *38*, 18–22.

Lumsden, P., Nicholas, J., & Davies, W. (1994). *Physiology, Growth, and Development of Plants in Culture*. Kluwer Academic Publishers. doi:10.1007/978-94-011-0790-7

Luo, L., Hou, X., Cai, W., & Guo, B. (2020). Incremental route inference from low-sampling GPS data: An opportunistic approach to online map matching. *Information Sciences*, *512*, 1407–1423. doi:10.1016/j.ins.2019.10.060

Lynch, P. T., Harris, W. C., & Chartier-Hollis, J. M. (1996). The cryopreservation of shoot tips of Rosa multiflora. *Plant Growth Regulation*, *20*(1), 43–45. doi:10.1007/BF00024056

MacDonald, J. M., & McBride, W. D. (2009). The Transformation of U.S. Livestock Agriculture: Scale, Efficiency, and Risks. *Economic Information Bulletin, 43*. Available online at https://www.ers.usda.gov/webdocs/publications/44292/10992_eib43.pdf?v=0

Machado, J., de Fátima Padilha, M. D. R., de Lira, F. P., de Oliveira, J. G., da Silva, R. S., & Caetano, M. B. C. (2017). Agricultura de Precisão e abertura de novas fronteiras no Brasil| Precisionagricultureandopening new frontiers in Brazil. *Revista Geama, 4*(1), 49–53.

MacLean, C. H., Newberry, S. J., Mojica, W. A., Khanna, P., Issa, A. M., Suttorp, M. J., Lim, Y. W., Traina, S. B., Hilton, L., Garland, R., & Morton, S. C. (2006). Effects of omega-3 fatty acids on cancer risk: A systematic review. *Journal of the American Medical Association, 295*(4), 403–415. doi:10.1001/jama.295.4.403 PMID:16434631

Maes, W. H., & Steppe, K. (2019). Perspectives for remote sensing with unmanned aerial vehicles in precision agriculture. *Trends in Plant Science, 24*(2), 152–164. doi:10.1016/j.tplants.2018.11.007 PMID:30558964

Magruder, J. R. (2018) An Assessment of Experimental Evidence on Agricultural Technology Adoption in Developing Countries. *Annu. Rev. Resour. Econ., 10*, 299–316. doi:10.1146/annurev-resource100517-023202

Mahmud, M. S. (2019). *Development of a machine vision system for strawberry powdery mildew disease detection* (MS thesis). Dalhousie University, Halifax, Nova Scotia, Canada.

Majada, J. P., Sierra, M. I., & Sanchez-Tames, R. (2001). Air exchange rate affects the in vitro developed leaf cuticle of carnation. *Scientia Horticulturae, 87*(1), 121–130. doi:10.1016/S0304-4238(00)00162-X

Majumdar, S., & Jayas, D. S. (2000). Classification of cereal grains using machine vision: IV. Combined morphology, color, and texture models. *Transactions of the ASAE. American Society of Agricultural Engineers, 43*(6), 1689. doi:10.13031/2013.3069

Mallery, M. (2011). Marijuana National Forest: Encroachment on California Public Lands for Cannabis Cultivation. *Berkeley Undergraduate Journal, 23*(2), 51.

Mallon, R., Barros, P., Luzardo, A., & Gonzalez, M.L. (2007). Encapsulation of moss buds: an efficient method for the *in vitro* conservation and regeneration of the endangered moss *Splachnum ampullaceum. Plant Cell, Tissue and Organ Culture, 88*, 41–49.

Manglia, M. (2010). *Precision Agriculture at a Glance.* Agropedia. Retrieved from: http://agropedia.iitk.ac.in/content/precision-agriculture-glance

Manivannan, A., Soundararajan, P., Halimah, N., Ko, C. H., & Jeong, B. R. (2015). Blue LED light enhances growth, phytochemical contents, and antioxidant enzyme activities of *Rehmannia glutinosa* cultured in vitro. *Horticulture, Environment and Biotechnology, 56*(1), 105–113. doi:10.100713580-015-0114-1

Mao, H. L., Wang, J. K., Lin, J., & Liu, J. X. (2012). Fatty Acid Profiles and Stearoyl-CoA Desaturase Gene Expression in Longissimusdorsi Muscle of Growing Lambs Influenced by Addition of Tea Saponins and Soybean Oil. *Journal of Animal Science, 25*(5), 648–652. PMID:25049609

Marino, S., & Alvino, A. (2014). Proximal sensing and vegetation indices for site-specific evaluation on an irrigated crop tomato. *European Journal of Remote Sensing, 47*(1), 271–283. doi:10.5721/EuJRS20144717

Marti, G., Schnee, S., Andrey, Y., Simões-Pires, C., Carrupt, P., & Gindro, J. K. (2014). Study of Leaf Metabolome Modifications Induced by UV-C Radiations in Representative Vitis, Cissus and Cannabis Species by LC-MS Based Metabolomics and Antioxidant Assays. *Molecules (Basel, Switzerland), 19*(9), 14004–14021. doi:10.3390/molecules190914004 PMID:25197936

Martinez, D., & Revilla, M. A. (1999). Cold acclimation and thermal transitions in the cryopreservation of hop shoot tips. *Cryo Letters*, *19*, 333–342.

Massa, G. D., Kim, H. H., Wheeler, R. M., & Mitchell, C. A. (2008). Plant productivity in response to LED lighting. *HortScience*, *43*(7), 1951–1956. doi:10.21273/HORTSCI.43.7.1951

Massey, R. E., Myers, D. B., Kitchen, N. R., & Sudduth, K. A. (2008). Profitability maps as an input for site-specific management decision making. *Agronomy Journal*, *100*(1), 52–59. doi:10.2134/agronj2007.0057

Matsumoto, T., & Sakai, A. (2000). Cryopreservation of in vitro-cultured axillary shoot tips of Vitis by vitrification. *Acta Horticulturae*, (538), 177–181. doi:10.17660/ActaHortic.2000.538.28

Matsumoto, T., & Sakai, A. (2003). Cryopreservation of axillary shoot tips of in vitro-grown grape (Vitis) by a two-step vitrification protocol. *Euphytica*, *131*(3), 299–304. doi:10.1023/A:1024024909864

Matsumoto, T., Sakai, A., Takahashi, C., & Yamada, K. (1995). Cryopreservation of *in vitro*-grown apical meristems of wasabi (*Wasabijaponica*) by encapsulation-vitrification method. *Cryo Letters*, *16*, 189–196.

Matsumoto, T., Sakai, A., & Yamada, K. (1994). Cryopreservation of *in vitro*-grown apical meristems of wasabi (*Wasabia japonica*) by vitrification and subsequent high plant regeneration. *Plant Cell Reports*, *13*(8), 442–446. doi:10.1007/BF00231963 PMID:24194022

Matsuura, H. N., de Costa, F., Yendo, A. C. A., & Fett-Neto, A. G. (2013). Photoelicitation of bioactive secondary metabolites by ultraviolet radiation: mechanisms, strategies, and applications. In S. Chandra, H. Lata, & A. Varma (Eds.), *Biotechnology for Medicinal Plants* (pp. 171–190). Springer-Verlag. doi:10.1007/978-3-642-29974-2_7

Maxted, N., & Kell, S. P. (2003). Biodiversity and conservation / Plant diversity, conservation and use. Encyclopedia of Applied Plant Sciences, 25-48.

Mayne, P. W. (2006). In-situ Test Calibrations for Evaluating Soil Parameters. In T. S. Tan, K. K. Phoon, D. W. Hight, & S. Leroueil (Eds.), *Singapore Workshop* (pp. 1–56). 10.1201/NOE0415426916.ch2

Mazereeuw, G., Lanctôt, K. L., Chau, S. A., Swardfager, W., & Herrmann, N. (2012). Effects of omega-3 fatty acids on cognitive performance: A meta-analysis. *Neurobiology of Aging*, *33*(7), 17–29. doi:10.1016/j.neurobiolaging.2011.12.014 PMID:22305186

McBratney, A., Whelan, B., Ancev, T., & Bouma, J. (2005). Future Directions of Precision Agriculture. *Precision Agriculture*, *6*(1), 7–23. doi:10.100711119-005-0681-8

McBride, W. D., & Daberkow, S. D. (2003). Information and the Adoption of Precision Farming Technologies. *Journal of Agribusiness*, *21*(1), 21–38.

McCann, I. R., & Stark, J. C. (1993). *Method and apparatus for variable application of irrigation water and chemicals.* U.S. Patent No. 5,246,164, issued September 21, 1993.

McEntee, P. J., Bennett, S. J., & Belford, R. K. (2020). Mapping the spatial and temporal stability of production in mixed farming systems: An index that integrates crop and pasture productivity to assist in the management of variability. *Precision Agriculture*, *21*(1), 77–106. doi:10.100711119-019-09658-6

McPartland, J. M. (2000). *Cannabis pathogens XI: Septoria spp. on Cannabis sativa, sensustricto*. Academic Press.

McPartland, J. M., & Hughes, S. (1994). Cannabis pathogens VII: A new species, Schiffnerula cannabis. *Mycologia*, *86*(6), 867–869. doi:10.1080/00275514.1994.12026491

Mekala, M. S., & Viswanathan, P. (2017, August). A Survey: Smart agriculture IoT with cloud computing. In *2017 international conference on microelectronic devices, circuits and systems (ICMDCS)* (pp. 1-7). IEEE.

Mekki, M., Abdallah, O., Amin, M. B. M., Eltayeb, M., Abdalfatah, T., & Babiker, A. (2015). Greenhouse monitoring and control system based on wireless Sensor Network. *Proc. - 2015 Int. Conf. Comput. Control. Networking, Electron. Embed. Syst. Eng. ICCNEEE*, 384–387. 10.1109/ICCNEEE.2015.7381396

Melsted, S. W. (1967). The philosophy of soil testing. In *Soil Testing and Plant Analysis Part I.* SSSA Special Publication Series No. 2. Soil Science Society of America, Inc.

Memon, M. U., & Lim, S. J. (2018). Microfluidic high-Q circular substrate-integrated waveguide (SIW) cavity for radio frequency (RF) chemical liquid sensing. *Sensors (Basel)*, *18*(1), 1–13. PMID:29316646

Méndez López, J. C. (2017). *Condicionantes y posibilidades en el uso de las TIC. Estudio de caso de dos cooperativas de Chinamperos de Xochimilco.* Beneméritca Universidad Autónoma de Puebla.

Mengxi, L., Zhigang, X., Yang, Y., & Yijie, F. (2011). Effects of different spectral lights on Oncidium PLBs induction, proliferation, and plant regeneration. *Plant Cell, Tissue and Organ Culture*, *106*(1), 1–10. doi:10.100711240-010-9887-1

Menin, B., Moglia, A., Comino, C., Hakkert, J. C., Lanteri, S., & Beekwilder, J. (2013). In vitro callus-induction in globe artichoke (*Cynara cardunculus* L. var. scolymus) as a system for the production of caffeoylquinic acids. *The Journal of Horticultural Science & Biotechnology*, *88*(5), 537–542. doi:10.1080/14620316.2013.11513003

Mensink, R. P. (2005). Effects of stearic acid on plasma lipid and lipoproteins in humans. *Lipids*, *40*(12), 1201–1205. doi:10.100711745-005-1486-x PMID:16477803

Mercier, J., Baka, M., Reddy, B., Corcuff, R., & Arul, J. (2001). Shortwave Ultraviolet Irradiation for Control of Decay Caused by Botrytis cinerea in Bell Pepper: Induced Resistance and Germicidal Effects. *Journal of the American Society for Horticultural Science*, *126*(1), 128–133. doi:10.21273/JASHS.126.1.128

Mikula, K., Izydorczyk, G., Skrzypczak, D., Mironiuk, M., Moustakas, K., Witek-Krowiak, A., & Chojnacka, K. (2019). Controlled release micronutrient fertilizers for precision agriculture–A review. *The Science of the Total Environment.* PMID:31935544

Milanes-Murcia, M. E. (2017). Cannabis and Water Management: International Regulation and the Legal Framework of the European Union. *University of Pennsylvania Law Review*, *49*(1), 117–153.

Minango, P., Iano, Y., Monteiro, A. C. B., França, R. P., & de Oliveira, G. G. (2019). *Classification of Automatic Skin Lesions from Dermoscopic Images Utilizing Deep Learning. SET International Journal of Broadcast Engineering.* doi:10.18580etijbe.2019.14

Mir, P. S., McAllister, T. A., Zaman, S., Jones, S. D. M., & He, M. L. (2003). Effect of dietary sunflower oil and vitamin E on beef cattle performance, carcass characteristics and meat quality. *Canadian Journal of Animal Science*, *83*(1), 53–66. doi:10.4141/A02-014

Mitchell, S., Weersink, A., & Erickson, B. (2018). Adoption of precision agriculture technologies in Ontario crop production. *Canadian Journal of Plant Science*, *98*(6), 1384–1388. doi:10.1139/cjps-2017-0342

Mizukami, Y., Sawai, Y., & Yamaguchi, Y. (2006). Moisture content measurement of tea leaves by electrical impedance and capacitance. *Biosystems Engineering*, *93*(3), 293–299. doi:10.1016/j.biosystemseng.2005.12.009

Moges, A. D., Karam, N. S., & Shibli, R. A. (2003). Slow growth in vitro preservation of African violet (Saintpaulia ionantha Wendl) shoot tips. *Advanced HortScience*, *17*, 1–8.

Mogili, U. R., & Deepak, B. B. V. L. (2018). Review on application of drone systems in precision agriculture. *Procedia Computer Science*, *133*, 502–509. doi:10.1016/j.procs.2018.07.063

Mohammadi, A., Khalid, K., Bafti, P. S., & Homaiee, M. (2012). Development of microcontroller-based microwave system to measure solid content of Hevea rubber latex. *Measurement*, *45*(3), 571–579. doi:10.1016/j.measurement.2011.10.007

Molano Bernal, L. C. (2017). *Estudio del uso de las tecnologías de la información y la comunicación (TIC) como herramientas de transferencia de tecnología en el sector agropecuario colombiano: caso Corpoica*. Universitat Politècnica de València. Retrieved from https://riunet.upv.es/handle/10251/90228

Molnar, J. J., & Kinnucan, H. (2019). Introduction: The biotechnology revolution. In Biotechnology and the new agricultural revolution (pp. 1-18). CRC Press.

Moloney, A. P., Mooney, M. T., Kerry, J. P., & Troy, D. J. (2001). Producing tender and flavoursome beef with enhanced nutritional characteristics. *The Proceedings of the Nutrition Society*, *60*(2), 221–229. doi:10.1079/PNS200077 PMID:11681638

Moncada, K. M. (2017). *Introduction to agronomy: food, crops, and environment*. Delmar Cengage Learning.

Mondal, P., & Basu, M. (2009). Adoption of precision agriculture technologies in India and in some developing countries: Scope, present status and strategies. *Progress in Natural Science*, *19*(6), 659–666. doi:10.1016/j.pnsc.2008.07.020

Monette, P. L. (1986). Cold storage of kiwi shoot tips *in vitro. HortScience*, *21*, 1203–1205.

Montgomery, P., & Richardson, A. J. (2008). Omega-3 fatty acids for bipolar disorder. *Cochrane Database of Systematic Reviews*, (2).

Montzka, S. A., Dlugokencky, E. J., & Butler, J. H. (2011). Non-CARBON DIOXIDE greenhouse gases and climate change. *Nature*, *476*(7358), 43–50. doi:10.1038/nature10322 PMID:21814274

Morel, G. (1975). Merstim culture techniques for the long-term storage of cultivated plants. In O. H. Frankel & J. G. Hawkes (Eds.), *Crop Genetic Resources for Today and Tomorrow* (pp. 327–332). Cambridge Univ. Press.

Morris, K. (1997). The cannabis remedy-wonder worker or evil weed? *Lancet*, *350*(9094), 1828. doi:10.1016/S0140-6736(05)63651-3 PMID:9454053

Morrow, R. C. (2008). LED lighting in horticulture. *HortScience*, *43*(7), 1949–1950. doi:10.21273/HORTSCI.43.7.1947

Muangprathub, J., Boonnam, N., Kajornkasirat, S., Lekbangpong, N., Wanichsombat, A., & Nillaor, P. (2019). IoT and agriculture data analysis for smart farm. *Computers and Electronics in Agriculture*, *156*, 467–474. doi:10.1016/j.compag.2018.12.011

Mubbarakh, S. A., Izhar, N. A., Rajasegar, A., & Subramaniam, S. (2014). Establishment of encapsulation-dehydration technique for *in vitro* fragmented explants of *Rosa hybrida* L. cv. Helmut Schmidt. *Emirates Journal of Food and Agriculture*, *26*(6), 565–276. doi:10.9755/ejfa.v26i6.18024

Mukherjee, A., & Lal, R. (2014). Comparison of soil quality index using three methods. *PLoS One*, *9*(8), e105981. Advance online publication. doi:10.1371/journal.pone.0105981 PMID:25148036

Mulla, D. J. (1989). Soil spatial variability and methods of analysis: I. Kriging soil fertility patterns. In C. E. Whitman, J. F. Parr, R. I. Papendick, & R. E. Meyer (Eds.), *Rainfed Agriculture in the Near East Region: Soil, Water and Crop/Livestock Management Systems* (pp. 185–204). USDA/USAID.

Mulla, D. J. (1991). Using geostatistics and GIS to manage spatial patterns in soil fertility. In G. Kranzler (Ed.), *Automated Agriculture for the 21st Century* (pp. 336–345). American Society of Agricultural Engineers.

Mulla, D. J. (1993). Mapping and managing spatial patterns in soil fertility and crop yield. In P. Robert, W. Larson, & R. Rust (Eds.), *Soil Specific Crop Management* (pp. 15–26). American Society of Agronomy. doi:10.2134/1993.soilspecificcrop.c2

Mulla, D. J., Bhatti, A. U., Hammond, M. W., & Benson, J. A. (1992). A comparison of winter wheat yield and quality under uniform versus spatially variable fertilizer management. *Agriculture, Ecosystems & Environment*, *38*(4), 301–311. doi:10.1016/0167-8809(92)90152-2

Mulla, D. J., Gowda, P., Koskinen, W. C., Khakural, B. R., Johnson, G., & Robert, P. C. (2002). Modeling the effect of precision agriculture: Pesticide losses to surface waters. In *Terrestrial Field Dissipation Studies. ACS Symposium Series No*. 842. ACS. 10.1021/bk-2002-0842.ch020

Mulla, D. J., & Hammond, M. W. (1988). Mapping of soil test results from large irrigation circles. *Proc. 39th Annual Far West Regional Fertilizer Conference.*

Mulla, D. J., & McBratney, A. B. (2000). Soil spatial variability. In M. E. Sumner (Ed.), *Handbook of Soil Science* (pp. A321–A352). CRC Press.

Mulla, D. J., Perillo, C. A., & Cogger, C. G. (1996). A site-specific farm-scale GIS approach for reducing groundwater contamination by pesticides. *Journal of Environmental Quality*, *25*(3), 419–425. doi:10.2134/jeq1996.00472425002500030006x

Mulla, D., & Khosla, R. (2016). *Historical evolution and recent advances in precision farming. Soil specific farming.* CRS Press. https://www. taylorfrancis. com/books/9781482245349

Mun, H. K., You, K. Y., & Dimon, M. N. (2013). Broken rice detection based on microwave measurement technique using microstrip wide-ring sensor and microstrip coupled-line sensor. *Australian Journal of Crop Science*, *7*(13), 2079–2090.

Murielle, G., Veylon, G., Bernard, A., Vanel, Q., & Stokes, A. (2014). Influence of plant root system morphology and architectural traits on soil shear resistance. *Plant and Soil*, *377*(1-2), 43–61. doi:10.100711104-012-1572-1

N` Nan, O., Hocher, V., & Verdeil, J. L. (2008). Cryopreservation by encapsulation – dehydration of plumules of coconut (*Cocos nucifera* L.). *Cryo Letters*, *29*, 339–350. PMID:19137197

Nagel Amaro, J., & Martínez Vergara, C. (2016). *Acceso y uso de TICs y necesidades de información de los pequeños agricultores de las Regiones de O'Higgins y Maule*. Fundación para la Innovación Agraria.

Naliwaiko, K., Araújo, R. L., da Fonseca, R. V., Castilho, J. C., Andreatini, R., Bellissimo, M. I., Oliveira, B. H., Martins, E. F., Curi, R., Fernandes, L. C., & Ferraz, A. C. (2004). Effects of fish oil on the central nervous system: A new potential antidepressant? *Nutritional Neuroscience*, *7*(2), 91–99. doi:10.1080/10284150410001704525 PMID:15279495

National Institute of Health. (2005). *Omega-3 fatty acids, fish oil, alpha-linolenic acid.* Author.

Navulur, S., Sastry, A. S. C. S., & Giri Prasad, M. N. (2017). Agricultural Management through Wireless Sensors and Internet of Things. *Iranian Journal of Electrical and Computer Engineering*, *7*(6), 3492–3499. from. doi:10.11591/ijece.v7i6.pp3492-3499

Nelson, S. O. (1991). Dielectric properties of agricultural products – measurements and applications. *IEEE Transactions on Electrical Insulation*, *26*(5), 845–869. doi:10.1109/14.99097

Nelson, S. O. (2015). *Dielectric Properties of Agricultural Materials and Their Applications*. Elsevier.

Nelson, S. O., Bartley, P. G., & Lawrence, K. C. (1997). Measuring RF and microwave permittivities of adult rice weevils. *IEEE Transactions on Instrumentation and Measurement*, *46*(4), 941–946. doi:10.1109/19.650805

Nelson, S. O., Forbus, W. R. Jr, & Lawrence, K. C. (1994). Permittivities of fresh fruits and vegetables at 0.2 to 20 GHz. *Journal of Microwave & Electromagnetic Energy*, *29*(2), 81–93. doi:10.1080/08327823.1994.11688235 PMID:8083786

Nelson, S. O., Forbus, W. R. Jr, & Lawrence, K. C. (1995). Assessment of microwave permittivity for sensing peach maturity. *Transactions of the ASAE. American Society of Agricultural Engineers*, *38*(2), 579–585. doi:10.13031/2013.27869

Nelson, S. O., Guo, W. C., Trabelsi, S., & Kays, S. J. (2007). Dielectric spectroscopy ofwatermelons for quality sensing. *Measurement Science & Technology*, *18*(7), 1887–1892. doi:10.1088/0957-0233/18/7/014

Nelson, S. O., & Stetson, L. (1974). Possibilities for controlling inserts with microwave and lower frequency RF energy. *IEEE Transactions on Microwave Theory and Techniques*, *22*(12), 1303–1305. doi:10.1109/TMTT.1974.1128482

Nemali, K. S., & Van Iersel, M. W. (2006). An automated system for controlling drought stress and irrigation in potted plants. *Scientia Horticulturae*.

Nganga, W. B., Ng'etich, K. O., Macharia, M. J., Kiboi, N. M., Adamtey, N., & Ngetich, K. F. (2020). Multi-influencing-factors' evaluation for organic-based soil fertility technologies out-scaling in Upper Tana Catchment in Kenya. *Scientific American*, *7*, e00231.

Ng, S., & Ng, N. (1991). Reduced – growth storage of germplasm. In J. Dodds (Ed.), *In Vitro Methods for Conservation of Plant Genetic Resources* (pp. 11–39). Chapman and Hall. doi:10.1007/978-94-011-3072-1_2

Nguyen, L., Hai, F. I., Yang, S., Kang, J., Leusch, F., Roddick, F., Price, W. E., & Nghiem, L. D. (2014). Removal of pharmaceuticals, steroid hormones, phytoestrogens, UV-filters, industrial chemicals and pesticides by *Trametes versicolor*: Role of biosorption and biodegradation. *International Biodeterioration & Biodegradation*, *88*(March), 169–175. doi:10.1016/j.ibiod.2013.12.017

Nhut, D. T., Hong, L. T. A., Watanabe, H., Goi, M., & Tanaka, M. (2000). Growth of banana plantlets cultured in vitro under red and blue lightemitting diode (LED) irradiation source. *Acta Horticulturae*, (575), 7–23.

Nhut, D. T., Huy, N. P., Tai, N. T., Nam, N. B., Luan, V. Q., Hien, V. T., Tung, H. T., Vinh, B. T., & Luan, T. C. (2015). Light-emitting diodes and their potential in callus growth, plantlet development and saponin accumulation during somatic embryogenesis of *Panax vietnamensis* Ha et Grushv. *Biotechnology, Biotechnological Equipment*, *29*(2), 299–308. doi:10.1080/13102818.2014.1000210 PMID:26019644

Nhut, D. T., & Nam, N. B. (2010). Light emitting diodes (LEDs): an artificial lighting source for biological studies. *Proceedings of the 3rd international conference of the development of BME in Vietnam*. 10.1007/978-3-642-12020-6_33

Nhut, T. D., Loan, P. T. L., Thao, P. T. L., Suong, T. T. N., & Lien, H. V. (2006). Effects of some artificial lighting sources on the photomorphogenesis of callus induced embryo and the plantlet regeneration of *Phalaenopsis amabilis* and its applications in micropropagation. *Proceedings of International Workshop on Agricultural Biotechnology*, 20–21.

Niino, T., & Sakai, A. (1992). Cryopreservation of alginate coated *in vitro* – grown shoot tips of apple, Pear and mulberry. *Plant Science*, *87*(2), 199–206. doi:10.1016/0168-9452(92)90151-B

Niino, T., Tashino, K., Suzuki, M., Ohuchi, S., Magoshi, J., & Akihama, T. (1997). Cryopreservation of *in vitro* grown shoot tips of cherry and sweet cherry by one – step vetrification. *Scientia Horticulturae*, *70*(2-3), 155–163. doi:10.1016/S0304-4238(97)00062-9

Niino, T., Yamamoto, S., Matsumoto, T., Engelmann, F., Valle Arizaga, M., & Tanaka, D. (2019). Development of V and D cryo-plate methods as effective protocols for cryobanking. *Acta Horticulturae*, (1234), 249–262. doi:10.17660/ActaHortic.2019.1234.33

Nishio, J. N. (2000). Why are higher plants green? Evolution of the higher plant photosynthetic pigment complement. *Plant, Cell & Environment, 23*(6), 539–548. doi:10.1046/j.1365-3040.2000.00563.x

Niu, Q., Fratta, D., & Wang, Y.-H. (2015). Precision agriculture - a worldwide overview. *Journal of Hydrology (Amsterdam), 522*(20150806), 475–487. Retrieved from https://linkinghub.elsevier.com/retrieve/pii/S002216941401066X

Noci, F., Kiely, P. O., Monahan, F. J., Stanton, C., & Moloney, A. P. (2005). Conjugated linoleic acid concentration in M. Longissimusdorsi from heifers offered sunflower oil-based concentrates and conserved forages. *Meat Science, 69*(3), 509–518. doi:10.1016/j.meatsci.2004.09.006 PMID:22062990

Norton, G. W., & Swinton, S. M. (2001). Precision Agriculture: Global Prospects and Environmental Implications. In G. H. Peters & P. Pingali (Eds.), *Tomorrow's Agriculture: Incentives, Institutions, Infrastructure and Innovations*. Ashgate.

NRCS East National Technology Support Center. (2011). *Soil Quality for Environmental Health*. Retrieved from http://soilquality.org/home.html

Nudda, A., Mc Guire, M. A., Battacone, G., & Pulina, G. (2005). Seasonal variation in conjugated linoleic acid and vaccenic acid in milk fat of sheep and its transfer to cheese and ricotta. *Journal of Dairy Science, 88*(4), 1311–1319. doi:10.3168/jds.S0022-0302(05)72797-1 PMID:15778298

Nuernberg, K., Nuernberg, G., Ender, K., Lorenz, S., Winkler, K., & Rickert, R. (2002). n-3 fatty acids, and conjugated linoleic acids of longissimus muscle in beef cattle. *European Journal of Lipid Science and Technology, 104*, 463–471.

Nyfors, E., & Vainikainen, P. (1989). *Industrial Microwave Sensors*. Artech House, Inc.

O'Connor, M., Bell, T., Elkaim, G., & Parkinson, B. 1996. Automatic steering of farm vehicles using GPS. In P. C. Robert, R. H. Rust, & W. E. Larson (Eds.), *Precision Agriculture. Proc. 3rd Intl. Conf.* ASA-CSSA-SSSA. 10.2134/1996.precisionagproc3.c91

O'Loughlin, C., & Ziemer, R. R. (1982). The importance of root strength and deterioration rates upon edaphic stability in steepland forests. In Carbon uptake and allocation in subalpine ecosystems as a key to management. Oregon State University.

Ojha, T., Misra, S., & Raghuwanshi, N. S. (2015). Wireless sensor networks for agriculture: The state-of-the-art in practice and future challenges. *Computers and Electronics in Agriculture, 118*, 66–84. doi:10.1016/j.compag.2015.08.011

Okamura, S., & Tomita, F. (1994). Microwave moisture sensing system in drying process for green tea production. *Proceedings of Instrumentation and Measurement Technology Conference*, 1253-1256. 10.1109/IMTC.1994.351828

Okamura, S., Zhang, Y., & Tsukamoto, N. (2007). A new microstripline-type moisture sensor for heavily wet tea leaves. *Measurement Science & Technology, 18*(4), 1022–1028. doi:10.1088/0957-0233/18/4/009

Operstein, V., & Frydman, S. (2000). The influence of vegetation on soil strength. *Ground Improvement, 4*(2), 81–89. doi:10.1680/grim.2000.4.2.81

Orlikowska, T. (1992). Effect of *in vitro* storage at 4° C on survival and proliferation of two apple rootstocks. *Plant Cell, Tissue and Organ Culture, 31*(1), 1–7. doi:10.1007/BF00043468

Ortlip, E. W. (1986). *Method and apparatus for spreading fertilizer*. U.S. Patent No. 4,630,773, issued December 23, 1976, Assignee: Soil Teq.

Ozudogru, E. A., Previati, A., & Lambardi, M. (2010). *In vitro* conservation and cryopreservation of ornamental plants. In S. M. Jain & S. J. Ochatt (Eds.), *Protocols for InVitro Propagation of Ornamental Plants* (pp. 303–324). Humana Press, Springer. doi:10.1007/978-1-60327-114-1_28

Ozudogru, E. A., Silva, D. P. C., Kaya, E., Dradi, G., Paiva, R., & Lambardi, M. (2013). *In vitro* conservation and cryopreservation of *Nandina domestica*, an outdoor ornamental shrub. *Notulae Botanicae Horti Agrobotanici Cluj-Napoca, 41*(2), 638–645. doi:10.15835/nbha4129335

Padilha, R., Iano, Y., Monteiro, A. C. B., Arthur, R., & Estrela, V. V. (2018, October). Betterment Proposal to Multipath Fading Channels Potential to MIMO Systems. In *Brazilian Technology Symposium* (pp. 115-130). Springer.

Pair, C.H. (1960). *Sprinkler Irrigation.* Soil and Water Conservation Research Division, Agricultural Research Service. Leaflet-No. 476.

Paliwal, J., Visen, N., Jayas, D., & White, N. (2003). Comparison of a neural network and a non-parametric classifier for grain kernel identification. *Biosystems Engineering, 85*(4), 405–413. doi:10.1016/S1537-5110(03)00083-7

Palmer, N. (2012). *Las TIC en agricultura. Conectando a pequeños agricultores con el conocimiento, las redes y las instituciones.* Retrieved December 3, 2019, from http://www.fao.org/3/ap561s/ap561s.pdf

Palmer, R. J. (1991). Progress report of a local positioning system. In G. Kranzler (Ed.), *Automated Agriculture for the 21st Century* (pp. 403–408). American Society of Agricultural Engineers.

Panda, S. S., Ames, D. P., & Panigrahi, S. (2010). Application of vegetation indices for agricultural crop yield prediction using neural network techniques. *Remote Sensing, 2*(3), 673–696. doi:10.3390/rs2030673

Pandey, N., & Pandey-Rai, S. (2014a). Short-term UV-B radiation-mediated transcriptional responses and altered secondary metabolism of in vitro propagated plantlets of *Artemisia annua* L. *Plant Cell, Tissue and Organ Culture, 116*(3), 371–385. doi:10.100711240-013-0413-0

Pandey, N., & Pandey-Rai, S. (2014b). Modulations of physiological responses and possible involvement of defense-related secondary metabolites in acclimation of *Artemisia annua* L. against short-term UV-B radiation. *Planta, 240*(3), 611–627. doi:10.100700425-014-2114-2 PMID:25023630

Pandino, G., Meneghini, M., Tavazza, R., Lombardo, S., & Mauromicale, G. (2017). Phytochemicals accumulation and antioxidant activity in callus and suspension cultures of *Cynara scolymus* L. *Plant Cell, Tissue and Organ Culture, 128*(1), 223–230. doi:10.100711240-016-1102-6

Panis, B. (1995). *Cryopreservation of Banana (Musa spp) Germplasm* (PhD Thesis). Katholieke University, Leuven, Belgium.

Panis, B., Piette, B., & Swennen, R. (2005). Droplet vitrification of apical meristems: A cryopreservation protocol applicable to all *Musaceae. Plant Science, 168*(1), 45–55. doi:10.1016/j.plantsci.2004.07.022

Panis, B., Totte, N., Nimmen, K., Withers, L., & Swennen, R. (1996). Cryopreservation of banana (*Musa spp*) meristem cultures after preculture on sucrose. *Plant Science, 121*(1), 95–106. doi:10.1016/S0168-9452(96)04507-4

Papendick, R., & Parr, J. (1992). Soil quality—The key to a sustainable agriculture. *American Journal of Alternative Agriculture, 2–3*(1-2), 2–3. Advance online publication. doi:10.1017/S0889189300004343

Paris, J. O., Gontijo, I., Partelli, F. L., & Facco, A. G. (2020). Variability and spatial correlation of soil micronutrients and organic matter with macadamia nut production. *Revista Brasileira de Engenharia Agrícola e Ambiental, 24*(1), 31–36. doi:10.1590/1807-1929/agriambi.v24n1p31-36

Pariza, M. W., Park, Y., & Cook, M. E. (2001). The biologically active isomers of conjugated linoleic acid. *Progress in Lipid Research, 40*(4), 283–298. doi:10.1016/S0163-7827(01)00008-X PMID:11412893

Park, S. Y., Lee, J. G., Cho, H. S., Seong, E. S., Kim, H. Y., Yu, C. Y., & Kim, J. K. (2013). Metabolite profiling approach for assessing the effects of colored light–emitting diode lighting on the adventitious roots of ginseng (*Panax ginseng* C. A Mayer). *Plant Omics*, *6*, 224–230.

Park, S., Im, J., Jang, E., & Rhee, J. (2016). Drought assessment and monitoring through blending of multi-sensor indices using machine learning approaches for different climate regions. *Agricultural and Forest Meteorology*, *216*, 157–169. doi:10.1016/j.agrformet.2015.10.011

Park, Y., Albright, K. J., Liu, W., Storkson, J. M., Cook, M. E., & Pariza, M. W. (1997). Effect of conjugated linoleic acid on body composition in mice. *Lipids*, *32*(8), 853–858. doi:10.100711745-997-0109-x PMID:9270977

Park, Y., McGuire, M. K., Behr, R., McGuire, M. A., Evans, M. A., & Schultz, T. D. (1999). High-fat dairy product consumption increases delta 9c, 11t-18: 2 (rumenic acid) and total lipid concentrations of human milk. *Lipids*, *34*(6), 543–549. doi:10.100711745-999-0396-2 PMID:10405966

Parra, L., Rocher, J., García, L., Lloret, J., Tomás, J., Romero, O., Rodilla, M., Falco, S., Sebastiá, M. T., Mengual, J., González, J. A., & Roig, B. (2018). Design of a WSN for smart irrigation in citrus plots with fault-tolerance and energy-saving algorithms. *Network Protocols and Algorithms.*, *10*(2), 95–115. doi:10.5296/npa.v10i2.13205

Patil, A., Beldar, M., Naik, A., & Deshpande, S. (2016). Smart farming using Arduino and data mining. *Proc. 10th INDIACom; 2016 3rd Int. Conf. Comput. Sustain. Glob. Dev. INDIACom 2016*, 1913–1917.

Patil, P., & Desai, B. L. (2013). Intelligent Irrigation Control System by Employing Wireless Sensor Network. *International Journal of Computers and Applications*, *79*(11), 2–5.

Patil, V. C., Nadagouda, B. T., & Al-Gaadi, K. A. (2012). Spatial Variability and Precision Nutrient Management in Sugarcane. *Photonirvachak (Dehra Dun)*, *41*(1), 183–189. doi:10.100712524-011-0181-4

Patterson, R. E., Flatt, S. W., Newman, V. A., Natarajan, L., Rock, C. L., Thomson, C. A., Caan, B. J., Parker, B. A., & Pierce, J. P. (2010). Marine Fatty Acid Intake is Associated with Breast Cancer Prognosis. *The Journal of Nutrition*, *141*(2), 201–206. doi:10.3945/jn.110.128777 PMID:21178081

Paulet, F., Engelmann, F., & Glaszmann, J. C. (1993). Cryopreservation of apices of *in vitro* plantlets of sugarcane (*Saccharum* sp. hybrids) using encapsulation/ dehydration. *Plant Cell Reports*, *12*(9), 525–529. doi:10.1007/BF00236101 PMID:24196115

Paul, S. S., Coops, N. C., Johnson, M. S., Krzic, M., Chandna, A., & Smukler, S. M. (2020). Mapping soil organic carbon and clay using remote sensing to predict soil workability for enhanced climate change adaptation. *Geoderma*, *363*, 114177. doi:10.1016/j.geoderma.2020.114177

Paustian, M., & Theuvsen, L. (2017). Adoption of precision agriculture technologies by German crop farmers. *Precision Agriculture*, *18*(5), 701–716. doi:10.100711119-016-9482-5

Pawłowska, B. (2011). The effect of BA and GA3 on shoot multiplication in *in vitro* cultures of Polish wild roses. *Folia Horticulturae*, *23*(2), 145–149. doi:10.2478/v10245-011-0022-5

Pawłowska, B., & Bach, A. (2011). Cryopreservation by encapsulation dehydration of *in vitro* grown shoot buds of *Rosa* 'New Dawn'. *Acta Horticulturae*, (908), 303–307. doi:10.17660/ActaHortic.2011.908.40

Pawłowska, B., & Szewczyk-Taranek, B. (2015). Efficient cryopreservation by droplet vitrification of pentaploid roses and the phenotype of regenerated plants. *Acta Societatis Botanicorum Poloniae*, *84*(4), 439–442. doi:10.5586/asbp.2015.038

Pejic, B., Sikora, V., Milic, S., Mackic, K., Koren, A., & Bajic, I. (2018). *Effect of drip irrigation on yield and evapotranspiration of fibre hemp (Cannabis sativa L.)*. Academic Press.

Peng, H., Jia, Y., Zhan, C., & Xu, W. (2020). Topographic controls on ecosystem evapotranspiration and net primary productivity under climate warming in the Taihang Mountains, China. *Journal of Hydrology (Amsterdam)*, *581*, 124394. doi:10.1016/j.jhydrol.2019.124394

Peng, Y. S., Brown, M. A., Wua, J. P., & Liu, Z. (2010). Different oilseed supplements alter fatty acid composition of different adipose tissues of adult ewes. *Meat Science*, *85*(3), 542–549. doi:10.1016/j.meatsci.2010.03.003 PMID:20416838

Pérez-Tornero, O., Egea, J., Olmos, E., & Burgos, L. (2001). Control of hyperhydricity in micropropagated apricot cultivars. *In Vitro Cellular & Developmental Biology – Plant, 37*(2), 250–254.

Perez-Tornero, O., Egea, J., Olmos, E., & Burgos, L. (2001). Control of hyperhydrcity in micropropagated apricot cultivars. *In Vitro Cellular & Developmental Biology. Plant*, *37*(2), 250–254. doi:10.100711627-001-0044-8

Perica, M. M., & Delas, I. (2011). Essential fatty acids and psychiatric disorders". Nutrition in clinical practice. *Official Publication of the American Society for Parenteral and Enteral Nutrition*, *26*(4), 409–425. doi:10.1177/0884533611411306

Pertwee, R. G. (2014). *Handbook of cannabis 781.* doi:10.1093/acprof:oso/9780199662685.001.0001

Peters, J., Taute, W., Döscher, C., Meier, R., Höft, M., Knöchel, R., & Breitkreutz, J. (2018). From laboratory-to pilot-scale: Moisture monitoring in fluidized bed granulation by a novel microwave sensor using multivariate calibration approaches. *Drug Development and Industrial Pharmacy*, *44*(6), 961–968. doi:10.1080/03639045.2018.1425427 PMID:29308682

Phocaides, A. (2007). *Handbook om pressurized irrigation technique. Food and Agriculture Organization of the United Nations.*

Pichu, R., & Alla, M. (2011). Cation ratio of soil structural stability (CROSS). *Soil Research (Collingwood, Vic.)*, *49*(3), 280. doi:10.1071/SR10105

Pierpaoli, E., Carlia, G., Pignattia, E., & Canavaria, M. (2013). Drivers of Precision Agriculture Technologies Adoption: A Literature Review. *Procedia Technology*, *8*, 61–69. doi:10.1016/j.protcy.2013.11.010

Pineda, M., Pérez-Bueno, M. L., & Barón, M. (2018). Detection of bacterial infection in melon plants by classification methods based on imaging data. *Frontiers in Plant Science*, *9*, 164. doi:10.3389/fpls.2018.00164 PMID:29491881

Pinter, P. J. Jr, Jackson, R. D., Idso, S. B., & Reginato, R. J. (1981). Multidate spectral reflectance as predictors of yield in water stressed wheat and barley. *International Journal of Remote Sensing*, *2*(1), 43–48. doi:10.1080/01431168108948339

Piper, C. S. (2017). *Soil and plant analysis*. Scientific Publishers.

Playán, E., Salvador, R., Bonet, L., Camacho, E., Intrigliolo, D. S., Moreno, M. A., ... Sánchez-de-Ribera, A. (2018). Assessing telemetry and remote control systems for water users associations in Spain. *Agricultural Water Management*, *202*, 89–98. doi:10.1016/j.agwat.2018.02.015

Plumecocq, G., Debril, T., Duru, M., Magrini, M. B., Sarthou, J. P., & Therond, O. (2018). The plurality of values in sustainable agriculture models. *Ecology and Society*, *23*(1), art21. doi:10.5751/ES-09881-230121

Pongnumkul, S., Chaovalit, P., & Surasvadi, N. (2015). Applications of smartphone-based sensors in agriculture: A systematic review of research. *Journal of Sensors*.

Pooja, S., Uday, D. V., Nagesh, U. B., & Talekar, S. G. (2017). Application of MQTT protocol for real time weather monitoring and precision farming. *2017 International Conference on Electrical, Electronics, Communication, Computer, and Optimization Techniques (ICEECCOT)*, 1-6. 10.1109/ICEECCOT.2017.8284616

Potter, G. R. (2006). *Weed, Need and Greed: Domestic Marijuana Production and the UK Cannabis Market* (PhD thesis). University of Sheffield.

Potter, D. J. (2014). A review of the cultivation and processing of cannabis (Cannabis sativa L.) for production of prescription medicines in the UK. *Drug Testing and Analysis*, *6*(1-2), 31–38. doi:10.1002/dta.1531 PMID:24115748

Potter, D. J., & Duncombe, P. (2012). The effect of electrical lighting power and irradiance on indoor-grown cannabis potency and yield. *Journal of Forensic Sciences*, *57*(3), 618–622. doi:10.1111/j.1556-4029.2011.02024.x PMID:22211717

Poudel, P. R., Kataoka, I., & Mochioka, R. (2008). Effect of red-and blue-light-emitting diodes on growth and morphogenesis of grapes. *Plant Cell, Tissue and Organ Culture*, *92*(2), 147–153. doi:10.100711240-007-9317-1

Poulson, C. S., Dhiman, T. R., Ure, A. L., Cornforth, D., & Olson, K. C. (2004). *Conjugated linoleic acid content of beef from cattle fed diets containing high grain, CLA, or raised on forages*. Livestock Production. doi:10.1016/j.livprodsci.2004.07.012

Prathibha, S.R., Hongal, A., & Jyothi, M.P. (2017). IOT Based Monitoring System in Smart Agriculture. *Proc. - 2017 Int. Conf. Recent Adv. Electron. Commun. Technol. ICRAECT*, 81–84.

Prathibha, S. R., Hongal, A., & Jyothi, M. P. (2017, March). IoT based monitoring system in smart agriculture. In *2017 International Conference on Recent Advances in Electronics and Communication Technology (ICRAECT)* (pp. 81-84). IEEE. 10.1109/ICRAECT.2017.52

Preece, J. E., & Sutter, E. G. (1990). Acclimatization of micropropagated plants to the greenhouse and field. In P. C. Debergh & R. H. Zimmerman (Eds.), *Micropropagation: Technology and Application* (pp. 71–93). Kluwer Academic Publishers.

Priva, B. V. (2019a). *Maximizer|Priva.* Retrieved from http://www.priva-asia.com/en/products/maximizer/

Priva, B. V. (2019b). *Priva Connext.* Retrieved from https://www.priva.com/horticulture/products/connext

Pruski, K. (2001). *Micropropagation technology in early phases of commercial seed potato production* (PhD Thesis). Wageningen University, Wageningen, The Netherlands.

Public-Private Analytic Exchange Program. (2018). *Threats to Precision Agriculture (2018).* Available online at https://www.dhs.gov/sites/default/files/publications/2018%20AEP_Threats_to_Precision_Agriculture.pdf

Purba, Paengkoum, & Paengkoum. (2019). *The link between supplementary tannin level and conjugated linoleic acid (CLA) formation in ruminants: A meta-analysis A meta-analysis of tannin's level prediction on ruminalbiohydrogenation.* bioRxiv preprint first posted online Apr. 17, 2019; doi:10.1101/612523

Quisumbing, A. R. (2003). Household Decisions, Gender, and Development A Synthesis of Recent Research. An earlier, expanded version of Chapter 8, "Gender differentials in farm productivity: Implications for household efficiency and agricultural policy." Food Policy, 20, 407–423.

Rademacher, W. (2000). Growth retardants: Effects on gibberellin biosynthesis and other metabolic pathways. *Annual Review of Plant Physiology and Plant Molecular Biology*, *51*(1), 501–531. doi:10.1146/annurev.arplant.51.1.501 PMID:15012200

Radwan, M. M., ElSohly, M. A., El-Alfy, A. T., Ahmed, S. A., Slade, D., Husni, A. S., Manly, S. P., Wilson, L., Seale, S., Cutler, S. J., & Ross, S. A. (2015). Isolation and pharmacological evaluation of minor cannabinoids from highpotency Cannabis sativa. *Journal of Natural Products*, *78*(6), 1271–1276. doi:10.1021/acs.jnatprod.5b00065 PMID:26000707

Raes, K., de Smet, S., & Demeyer, D. (2004). Effect of dietary fatty acids on incorporation of long chain polyunsaturated fatty acids and conjugated linoleic acid in lamb, beef and pork meat: A review. *Animal Feed Science and Technology*, *113*(1-4), 199–221. doi:10.1016/j.anifeedsci.2003.09.001

Rahman, L. (2016). *Precision agriculture in the World and its prospect in Bangladesh Research in Agriculture, Livestock and Fisheries*. Academic Press.

Rains, C. R., & Thomas, D. L. (2009). Precision farming: An introduction. The University of Georgia Bulletin, 1186, 12.

Rai, R., Meena, R. P., Smita, S. S., Shukla, A., Rai, S. K., & Pandey-Rai, S. (2011b). UV-b and UV-C pre-treatments induce physilogical changes and artemisinin biosynthesis in *Artemisia annua* L.- An antimalarial plant. *Journal of Photochemistry and Photobiology. B, Biology*, *105*(3), 216–225. doi:10.1016/j.jphotobiol.2011.09.004 PMID:22019553

Rai, R., Pandey, S., & Rai, S. P. (2011a). Arsenic-induced changes in morphological, physiological, and biochemical attributes and artemisinin biosynthesis in *Artemisia annua*, an antimalarial plant. *Ecotoxicology (London, England)*, *20*(8), 1900–1913. doi:10.100710646-011-0728-8 PMID:21710305

Rajvanshi, A. K. (2015). *Is precision agriculture the solution to India's farming crisis" Business Standards, online article*. Retrieved from: https://www.business-standard.com/article/news-ians/is-precision-agriculture-the-solution-to-india-s-farming-crisis-comment-special-to-ians-113101500127_1.html

Ramawat, K. G., & Mathur, M. (2007). Factors affecting production of secondary metabolites. In K. G. Ramawat & J. M. Merillon (Eds.), *Biotechnology secondary metabolites* (pp. 59–101). Sci Pub. doi:10.1201/b10756-4

Ramsay, M. M., Jacskon, A. D., & Porley, R. D. (2000). A Pilot study for *ex situ* conservation of UK Bryophytes. *Proceedings of EuroGard 2000—II European Botanic Garden Congress*, 52–57.

Ranalli, P. (1999). Agronomical and physiological advances in hemp crops. In P. Ranalli (Ed.), *Advances in Hemp Research* (pp. 61–84). Food Products Press, The Haworth Press Inc. doi:10.1201/9781498705820-11

Rao, N. K. (2004). Plant genetic resources: Advancing conservation and use through biotechnology. *African Journal of Biotechnology*, *3*(2), 136–145.

Rao, N., & Han, A. T. (2014). Research on non-destructive measuring moisture content for packaged tea based on the LC resonant sensor. *Proceedings of the 26th Chinese Control and Decision Conference*, 2476-2480.

Rao, R. S., & Ravishankar, G. A. (2002). Plant cell cultures: Chemical factories of secondary metabolites. *Biotechnology Advances*, *20*(2), 101–153. doi:10.1016/S0734-9750(02)00007-1 PMID:14538059

Rasaei, Z., Rossiter, D. G., & Farshad, A. (2020). Rescue and renewal of legacy soil resource inventories in Iran as an input to digital soil mapping. *Geoderma Regional*, *21*, e00262. doi:10.1016/j.geodrs.2020.e00262

Raun, W. R., Solie, J. B., Johnson, G. V., Stone, M. L., Mullen, R. W., Freeman, K. W., Thomason, W. E., & Lukina, E. V. (2002). Improving nitrogen use efficiency in cereal grain production with optical sensing and variable rate application. *Agronomy Journal*, *94*(4), 815–820. doi:10.2134/agronj2002.8150

Raviv, M., & Lieth, J. H. (2007). *Soilless culture: Theory and practice* (1st ed.). Elsevier B.V.

Raza, M. H., Khan, G. A., Shahbaz, B., & Saleem, M. F. (2020). Effectiveness of information and communication technologies as information source among farmers in pakistan. *Pakistan Journal of Agricultural Sciences*, *57*(1), 281–288. doi:10.21162/PAKJAS/19.8292

Razdan, M. K., & Cockling, E. C. (1997). Biotechnology in conservation of genetic resources. In M. K. Razdan & E. C. Cockling (Eds.), *Conservation of Plant Genetic Resources in vitro, General Aspects* (pp. 1–25). Science Publishers Inc.

Read, P. E., & Preece, J. E. (2003). Environmental management for optimizing micropropagation. *Acta Horticulturae*, (616), 129–133.

Realini, C. E., Duckett, S. K., Brito, G. W., Dalla Rizza, M., & De Mattos, D. (2004). Effect of pasture vs. concentrate feeding with or without antioxidants on carcass characteristics, fatty acid composition, and quality of Uruguayan beef. *Meat Science*, *66*(3), 567–577. doi:10.1016/S0309-1740(03)00160-8 PMID:22060866

Reed, B. (1993). Improved survival of *in vitro* stored rubus germplasm. *Journal of the American Society for Horticultural Science*, *118*(6), 890–895. doi:10.21273/JASHS.118.6.890

Reed, B. (1999). *In vitro* storage conditions for mint germplasm. *HortScience*, *34*(2), 350–352. doi:10.21273/HORTSCI.34.2.350

Reed, C. A. (Ed.). (2011). *Origins of agriculture*. Walter de Gruyter.

Reichardt, M., & Jürgens, C. (2009). Adoption and future perspective of precision farming in Germany: Results of several surveys among different agricultural target groups. *Precision Agriculture*, *10*(1), 73–94. doi:10.100711119-008-9101-1

Reid, J. F., & Searcy, S. W. (1987). Vision-based guidance of an agriculture tractor. IEEE Control Syst. Mag., 7, 39–43. doi:10.1109/MCS.1987.1105271

Reinhoud, P. J., Schrynemakers, E. W. M., Iren, F. V., & Kijne, J. W. (1995). Vitrification and a heat-shock treatment improve cryopreservation of tobacco cell suspensions compared to two-step freezing. *Plant Cell, Tissue and Organ Culture*, *42*(3), 261–267. doi:10.1007/BF00029997

Reisinger, P., Lehoczky, E., & Komives, T. (2005). Competitiveness and precision management of the noxious weed Cannabis sativa L. in winter wheat. *Communications in Soil Science and Plant Analysis*, *36*(4-6), 629–634. doi:10.1081/CSS-200043303

Rhoades, J. D., & Corwin, D. L. (1981). Determining soil electrical conductivity-depth relations using an inductive electromagnetic soil conductivity meter. *Soil Science Society of America Journal*, *45*(2), 255–260. doi:10.2136ssaj1981.03615995004500020006x

Richardson, A. J. (2006). Omega-3 fatty acids in ADHD and related neurodevelopmental disorders. *International Review of Psychiatry (Abingdon, England)*, *18*(2), 155–172. doi:10.1080/09540260600583031 PMID:16777670

Richardson, A. J., & Montgomery, P. (2005). The Oxford-Durham study: A randomized, controlled trial of dietary supplementation with fatty acids in children with developmental coordination disorder. *Pediatrics*, *115*(5), 1360–1366. doi:10.1542/peds.2004-2164 PMID:15867048

Ristorto, G., Gallo, R., Gasparetto, A., Scalera, L., Vidoni, R., &Mazzetto, F. (2017). *A mobile laboratory for orchard health status monitoring in precision farming*. Academic Press.

Robert, P. C. (1982). *Evaluation of some remote sensing techniques for soil and crop management* (PhD dissertation). University of Minnesota.

Roca, W. M., Reyes, R., & Beltran, J. (1984). Effect of various factors on minimal growth in tissue culture storage of cassava germplasm. *Proc. 6th Symp. Int. Soc. For Tropical Root Crops*, 441-446.

Rodrigues, P. H. V., Arruda, F., & Forti, V. A. (2018). Slow growth *in vitro* conservation of *Heliconia champneianacv.* Splash under different light spectra. *Scientia Horticulturae*, *75*, 163–166.

Rojas-Moraleda, R., Valous, N. A., Gowen, A., Esquerre, C., Härtel, S., Salinas, L., & O'Donnell, C. (2017). A frame-based ANN for classification of hyperspectral images: Assessment of mechanical damage in mushrooms. *Neural Computing & Applications*, *28*(1), 969–981. doi:10.100700521-016-2376-7

Romano, A., & Martins-Loução, M. A. (1999). In vitro cold storage of cork oak shoot cultures. *Plant Cell, Tissue and Organ Culture*, *591*(2), 155–157. doi:10.1023/A:1006465706905

Rosman, R., Dimon, M. N., & You, K. Y. (2018). Correlation between reflection coefficient, dielectric properties and brix level of Malaysian oranges at microwave frequencies. *Indonesian Journal of Electrical Engineering and Computer Science*, *10*(3), 853–858. doi:10.11591/ijeecs.v10.i3.pp853-858

Roxas, N. J. L., Tashiro, Y., Miyazaki, S., Isshiki, S., & Takeshita, A. (1995). Meiosis and Pollen Fertility in Higo Chrysanthemum *Dendranthema grandiflorum* (Ramat.) Kitam. *Journal of the Japanese Society for Horticultural Science*, *64*(1), 161–168. doi:10.2503/jjshs.64.161

Ruggiero, C., Lattanzio, F., Lauretani, F., Gasperini, B., Andres-Lacueva, C., & Cherubini, A. (2009). Omega-3 polyunsaturated fatty acids and immune-mediated diseases: Inflammatory bowel disease and rheumatoid arthritis. *Current Pharmaceutical Design*, *15*(36), 4135–4148. doi:10.2174/138161209789909746 PMID:20041815

Rusliyadi, M., & Jamil, A. B. (2021). Food Security Policy Analysis of Household Level: Case of Food Security Village Programme in Indonesia. In Ordóñez de Pablos, P., & Lytras, M. D. (Ed.), Global Challenges and Strategic Disruptors in Asian Businesses and Economies (pp. 60-75). IGI Global. http://doi:10.4018/978-1-7998-4787-8.ch004

Rusliyadi, M., Jamil, A. B., Kumalasari, R. T., & Rouf, A. A. (2021). Institutional Role and Asisstance Service Analysis on Food Security Policy Case in Indonesia. In Ordóñez de Pablos, P., & Lytras, M. D. (Ed.), Global Challenges and Strategic Disruptors in Asian Businesses and Economies (pp. 43-59). IGI Global. http://doi:10.4018/978-1-7998-4787-8.ch003

Rusliyadi, M., Jamil, A. B., Kumalasari, R. T., & Rouf, A. A. (2021). The Institutional Impact and Community Perception Implementation of Food Security Policy on Poverty Reduction: Case of Food Self-Sufficiency Village Programme, Indonesia. In Ordóñez de Pablos, P., & Lytras, M. D. (Ed.), Global Challenges and Strategic Disruptors in Asian Businesses and Economies (pp. 29-42). IGI Global. http://doi:10.4018/978-1-7998-4787-8.ch002

Rusliyadi, M., Mohd, H., Jamil, A., & Anshari, M. (2020). Analysis of Food Security Policy by Participatory Poverty Assessment (PPA) Effort: Case in Indonesia. In P. Ordoñez de Pablos, M. Almunawar, & M. Abduh (Eds.), Economics, Business, and Islamic Finance in ASEAN Economics Community (pp. 105-127). Hershey, PA: IGI Global. doi:10.4018/978-1-7998-2257-8.ch006

Rusliyadi, M., Bin Hj. Mohd. Jamil, A., Maseleno, A., Tri Kumalasari, R., & Othman, M. (2019). Agricultural extension policy, agricultural growth and poverty reduction in Indonesia. *IACSIT International Journal of Engineering and Technology*, *7*(4), 5539–5550. doi:10.14419/ijet.v7i4.13337

Rusliyadi, M., & Jamil, A. B. (2020). Participatory Poverty Assessment Effort in Food Security and Extension Policy: Evidence From Indonesia. In V. Erokhin & T. Gao (Eds.), *Handbook of Research on Globalized Agricultural Trade and New Challenges for Food Security* (pp. 481–499). IGI Global., doi:10.4018/978-1-7998-1042-1.ch024

Rusliyadi, M., Jamil, A., Mohd, B., & Kumalasar, R. T. (2019). Analysis of household food security policy: Case of food security village programme, Indonesia. *Asian Journal of Agriculture and Rural Development*, *9*(1), 19–32.

Rusliyadi, M., & Libin, W. (2018). Agriculture development programs for poverty reduction evidences from Indonesia and china - comparative study case. *Asian Journal of Agriculture and Rural Development*, *8*(2), 104–118.

Rust, R. H. (1985). Computerized soil maps benefit crop management. *Farming with Pride Magazine*, 4–10.

Saha, A. K., Saha, J., Ray, R., Sircar, S., Dutta, S., Chattopadhyay, S. P., & Saha, H. N. (2018, January). IoT-based drone for improvement of crop quality in agricultural field. In *2018 IEEE 8th Annual Computing and Communication Workshop and Conference (CCWC)* (pp. 612-615). IEEE. 10.1109/CCWC.2018.8301662

Sakai, A., & Engelmann, F. (2007). Vitrification, encapsulation–vitrification and droplet–vitrification: A review. *Cryo Letters*, *28*, 151–172. PMID:17898904

Sakai, A., Hirai, D., & Niino, T. (2008). Development of PVS-based vitrification and encapsulation–vitrification protocols. In B. M. Reed (Ed.), *Plant cryopreservation: a practical guide* (pp. 33–58). Springer. doi:10.1007/978-0-387-72276-4_3

Sakai, A., Kobayashi, S., & Oiyama, I. (1990). Cryopreservation of nucellar cells of navel orange (Citrus sinensis Osb. var. brasiliensis Tanaka) by vitrification. *Plant Cell Reports*, *9*(1), 30–33. doi:10.1007/BF00232130 PMID:24226373

Sakai, A., & Yoshida, S. (1976). Survival of plant tissue at super low temperature VI. Effects of cooling and re- warming rates on survival. *Plant Physiology*, *42*(12), 1695–1701. doi:10.1104/pp.42.12.1695 PMID:16656707

Salic, A., &Zelic, B. (2018). Introduction to environmental engineering. *Physical Sciences Reviews, 3.*

Sallanon, H., & Maziere, Y. (1992). Influence of growth room and vessel humidity on the in vitro development of rose plants. *Plant Cell, Tissue and Organ Culture*, *30*(2), 121–125. doi:10.1007/BF00034305

Sandri, D., Pereira, J. A., & Vargas, R. B. (2014). Production costs and profitability of watermelon under different water depths and irrigation systems. Irriga. *Botucatu*, *19*(3), 414–429. doi:10.15809/irriga.2014v19n3p414

Sankari, H., & Mela, T. J. N. (1998). Plant development and stem yield of non-domestic fibre hemp (Cannabis sativa L.) cultivars in long-day growth conditions in Finland. *Journal Agronomy & Crop Science*, *181*(3), 153–159. doi:10.1111/j.1439-037X.1998.tb00411.x

Sarkar, D., & Naik, P. S. (1998). Factors affecting minimal growth conservation of potato microplants *in vitro. Euphytic.*, *102*(2), 275–280. doi:10.1023/A:1018309300121

Savakar, D. G., & Anami, B. S. (2015). Grading of bulk food grains and fruits using computer vision. *Journal of Agricultural Engineering and Biotechnology*, *3*(1), 1–10. doi:10.18005/JAEB0301001

Savary, S., Ficke, A., Aubertot, J.-N., & Hollier, C. (2012). *Crop losses due to diseases and their implications for global food production losses and food security*. Springer. doi:10.100712571-012-0200-5

Sawyer, J. E. (1994). Concepts of variable rate technology with considerations for fertilizer application. *Journal of Production Agriculture*, *7*(2), 195–201. doi:10.2134/jpa1994.0195

Say, S. M., Keskin, M., Sehri, M., & Sekerli, Y. E. (2017) *Adoption of precision agriculture technologies in developed and developing countries*. International Science and Technology Conference, Berlin, Germany. Available online at http://tojsat.net/journals/tojsat/articles/v08i01/v08i01-02.pdf

Scheeder, M. R. (2004). MarkanterZusatznutzenmitfunktionellenFettsauren. In M. Kreuzer, C. Wenk, & T. Lanzini (Eds.), *Lipide in Fleisch, Milch und Ei - Herausforderungfu¨ r die Tierernaˇhrung* (pp. 52–68). ETH Zurich.

Scheifele, G. (1999). *Determining the feasibility and potential of field production of low THC industrial hemp (Cannabis sativa) for fibre and seed grain in Northern Ontario*. Ontario Ministry of Agriculture and Food.

Scherer, T. F., Kranz, W., Pfost, P., Wermer, H., Wright, J. A., & Yonts, C. D. (1999). Sprinkler irrigation systems. WCAinfoNET.

Schmid, A. M., Sieber, R., & Bee, G. (2006). Conjugated linoleic acid in meat and meat products: A review. *Meat Science*, *73*(1), 29–41. doi:10.1016/j.meatsci.2005.10.010 PMID:22062051

Schmitt, M. A., Walker, W. G., & Fairchild, D. (1986). Computerized fertilizer application by soil type. In *Proc. Great Plains Soil Fertility Workshop*. Kansas State Univ.

Schmitz, A., & Moss, C. B. (2015). Mechanized Agriculture: Machine Adoption, Farm Size, and Labor Displacement. *AgBioForum*, *18*(3), 278–296.

Schueller, J. K. (1991). In-field site-specific crop production. In G. Kranzler (Ed.), *Automated Agriculture for the 21st Century* (pp. 291–292). American Society of Agricultural Engineers.

Schueller, J. K. (1992). A review and integrating analysis of spatially-variable control of crop production. *Fertilizer Research*, *33*(1), 1–34. doi:10.1007/BF01058007

Schueller, J. K., & Bae, Y. H. (1987). Spatially attributed automatic combine data acquisition. *Computers and Electronics in Agriculture*, *2*(2), 119–127. doi:10.1016/0168-1699(87)90022-6

Schueller, J. K., & Wang, M.-W. (1994). Spatially-variable fertilizer and pesticide application with GPS and DGPS. *Computers and Electronics in Agriculture*, *11*(1), 69–83. doi:10.1016/0168-1699(94)90053-1

Schuerger, A. C., Brown, C. S., & Stryjewski, E. C. (1997). Anatomical features of pepper plants (*Capsium annuum* L.) grown under red light emitting diodes supplemented with blue or far-red light. *Annali di Botanica*, *79*(3), 273–282. doi:10.1006/anbo.1996.0341 PMID:11540425

Schultes, R. E., Klein, W. M., Plowman, T., & Lockwood, T. E. (1974). Cannabis: An example of taxonomic neglect. *Botanical Museum Leaflets, Harvard University*, *23*, 337–367.

Schumacher, J., & Tudzynski, P. (2011). Morphogenesis and Pathogenicity in Fungi. *Current Genetics*, *22*, 225–241.

Sciortino, M., De Felice, M., De Cecco, L., & Borfecchia, F. (2020). Remote sensing for monitoring and mapping Land Productivity in Italy: A rapid assessment methodology. *Catena*, *188*, 104375. doi:10.1016/j.catena.2019.104375

Scollan, N. D., Enser, M., Gulati, S. K., Richardson, I., & Wood, J. D. (2003). Effects of including a ruminally protected lipid supplement in the diet on the fatty acid composition of beef muscle. *British Journal of Nutrition*, *90*(3), 709–716. doi:10.1079/BJN2003933 PMID:13129478

Scoones, I. (2017). *Free-Book: Pathways To Sustainable Agriculture*. Academic Press.

Scottez, C., Chevreau, E., Godard, N., Arnaud, Y., Duron, M., & Dereuddre, J. (1992). Cryopreservation of cold-acclimated shoot tips of pear *in vitro* cultures after encapsulation-dehydration. *Cryobiology*, *29*(6), 691–700. doi:10.1016/0011-2240(92)90073-B

Seabrook, J. E. A., Coleman, S., & Levy, D. (1993). Effect of photoperiod on *in vitro* tuberization of potato (*Solanum tuberosum* L.). *Plant Cell, Tissue and Organ Culture*, *34*(1), 43–51. doi:10.1007/BF00048462

Searcy, S. W., Schueller, J. K., Bae, Y. H., Borgelt, S. C., & Stout, B. A. (1989a). Mapping of spatially-variable yield during grain combining. *Transactions of the ASAE. American Society of Agricultural Engineers*, *32*(3), 826–829. doi:10.13031/2013.31077

Secher, N. J. (2007). Does fish oil prevent preterm birth? *Journal of Perinatal Medicine*, *35*(Suppl 1), S25–S27. PMID:17302537

Shabbir, F., & Tahir, Z. (2014). Impact of Electric power on Agriculture industrialization and value chain. *Proceeding of international Conference entitled "emerging horizons of Agricultural extension for sustainable rural development.*

Shafi, U., Mumtaz, R., García-Nieto, J., Hassan, S. A., Zaidi, S. A. R., & Iqbal, N. (2019). Precision Agriculture Techniques and Practices: From Considerations to Applications. *Sensors (Basel)*, *19*(17), 3796. doi:10.339019173796 PMID:31480709

Shakhatreh, H., Sawalmeh, A. H., Al-Fuqaha, A., Dou, Z., Almaita, E., Khalil, I., Othman, N. S., Khreishah, A., & Guizani, M. (2019). Unmanned aerial vehicles (UAVs): A survey on civil applications and key research challenges. *IEEE Access: Practical Innovations, Open Solutions*, *7*, 48572–48634. doi:10.1109/ACCESS.2019.2909530

Shalloo, L., O'Donovan, M., Leso, L., Werner, J., Ruelle, É., Geoghegan, A., ... O'Leary, N. (2018, September). *Simulation of the variation in grass growth, milk production and residency time in a 40-hectare farm composed of either 6 or 40 paddocks*. Academic Press.

Shama, G., & Alderson, P. (2005). UV hormesis in fruits: A concept ripe for commercialization. *Trends in Food Science & TechnologyVolume*, *16*(4), 128–136. doi:10.1016/j.tifs.2004.10.001

Shamshiri, R. R., Hameed, I. A., Balasundram, S. K., Ahmad, D., Weltzien, C., & Yamin, M. (2018). *Fundamental research on unmanned aerial vehicles to support precision agriculture in oil palm plantations*. Agricultural Robots-Fundamentals and Application.

Shang, Y., KamrulHasan, M., Ahammed, G. J., Li, M., Yin, H., & Zhou, J. (2019). Applications of Nanotechnology in Plant Growth and Crop Protection: A Review. *Molecules (Basel, Switzerland)*, *24*, 2558. doi:10.3390/molecules24142558

Shankar, S., Rao, M., Shetty, P., Thombre, J., & Manek, H. (2020). A data analytics framework for decision-making in agriculture. In Advances in Data Sciences, Security and Applications (Vol. 612, pp. 85–98). Springer Nature Singapore. doi:10.1007/978-981-15-0372-6_7

Shannon, D. K., Clay, D. E., & Sudduth, K. A. (2018). An introduction to precision agriculture. *Precision Agriculture Basics,* 1-12.

Sheibani, S., Asgharipour, M. R., Ghanbari, A., & Abolpour, B. (2020). Modeling the yield spatial distribution based on the basic production factors under uncertainty conditions in Kamin, Iran. *Arabian Journal of Geosciences*, *13*(2), 1–11. doi:10.100712517-019-5015-7

Sheil, D., & Liswanti, N. (2006). Scoring the importance of tropical forest landscapes with local people: Patterns and insights. *Environmental Management*, *38*(1), 126–136. doi:10.100700267-005-0092-7 PMID:16508796

Shekhar, C. (2020). Spatial Variability of Ground Water Quality for Irrigation of Mansa District, Punjab. *International Journal of Agriculture Sciences*.

Shelton, B. G., Kirkland, K. H., Flanders, W. D., & Morris, G. K. (2002). Profiles of Airborne Fungi in Buildings and Outdoor Environments in the United States. *Applied and Environmental Microbiology*, *68*(4), 1743–1753. doi:10.1128/AEM.68.4.1743-1753.2002 PMID:11916692

Shibli, R. A. (1991). *In vitro* acclimatization of *(Chrysanthemum morifolium* Ramat.) after *in vitro* water stress. *Plant Cell, Tissue and Organ Culture*, *11*, 15–23.

Shibli, R. A. (2000). Cryopreservation of black iris (*Iris nigricans*) somatic embryos by encapsulation – dehydration. *Cryo Letters*, *21*, 39–46. PMID:12148063

Shibli, R. A., Ajlouni, M. M., Jaradat, A., Al Janabi, S., & Shatanawi, M. (1997). Micropropagation of wild pear (*Pyrus syrica*). *Scientia Horticulturae*, *69*(1-4), 237–242. doi:10.1016/S0304-4238(96)00972-7

Shibli, R. A., Shatnawi, M. A., Ajlouni, M. M., Jaradat, A., & Adham, Y. (1999). Slow growth *in vitro* conservation of bitter almond (*Amygdalus communis* L.). *Advances in Horticultural Science*, *13*, 133–134.

Shibli, R. A., Smith, M. A. L., & Spomer, L. A. (1992). Osmotic adjustment and growth responses of three *(Chrysanthemum morifolium* Ramat.) cultivars to osmotic stress induced *in vitro*. *Journal of Plant Nutrition*, *15*(9), 1373–1381. doi:10.1080/01904169209364404

Shin, S. K., Murthy, H. N., Heo, J. W., & Paek, K. Y. (2004). Induction of betalain pigmentation in hairy roots of red beet under different radiation sources. *Biologia Plantarum*, *47*(1), 149–152. doi:10.1023/A:1027313805930

Shin, S. K., Murthy, N. H., Heo, W. J., Hahn, J. E., & Paek, Y. K. (2008). The effect of light quality on the growth and development of in vitro cultured Doritaenopsis plants. *Acta Physiologiae Plantarum*, *30*(3), 339–343. doi:10.100711738-007-0128-0

Shohael, A. M., Ali, M. B., Yu, K. W., Hahn, E. J., & Paek, K. Y. (2006). Effect of temperature on secondary metabolites production and antioxidant enzyme activities in Eleutherococcus senticosus somatic embryos. *Plant Cell, Tissue and Organ Culture*, *85*(2), 219–228. doi:10.100711240-005-9075-x

Shonk, J. L., & Gaultney, L. D. (1988). Spectroscopic sensing for the determination of organic matter content. Am. Soc. Agric. Engin..

Shukla, M.R., Amritpal, S.S., Kevin, P., Praveen, K.S., & Maxwell, A.P.J. (2017). Application of 3D printing to prototype and develop novel plant tissue culture systems. *Plant Methods*., 13:6 1:10.

Siddique, A., Prabhu, B., Chaskar, A., & Pathak, R. (2019). A review on intelligent agriculture service platform with lora based wireless sensor network. *Life, 100*, 7000.

Sigrimis, N., Arvanitis, K. G., Pasgianos, G. D., & Ferentinos, K. (2001). Hydroponics water management using adaptive scheduling with an on-line optimiser. *Computers and Electronics in Agriculture*, *31*(1), 31–46. doi:10.1016/S0168-1699(00)00172-1

Silva, C. B., do Vale, S. M. L. R., Pinto, F. A., Müller, C. A., & Moura, A. D. (2007). The economic feasibility of precision agriculture in MatoGrossodoSul State, Brazil: A case study. *Precision Agriculture*, *8*(6), 255–265. doi:10.100711119-007-9040-2

Silva, S. T., Bertolucci, S. K. V., da Cunha, S. H. B., Lazzarini, L. E. S., Tavares, M. C., & Pinto, J. E. B. P. (2017). Effect of light and natural ventilation systems on the growth parameters and carvacrol content in the in vitro cultures of *Plectranthus amboinicus* (Lour.) Spreng. *Plant Cell, Tissue and Organ Culture*, *129*(3), 501–510. doi:10.100711240-017-1195-6

Silva, T. L., & Scherwinski-Pereira, J. E. (2011). *In vitro* Conservation of *Piper aduncum* and *Piper hispidinervum*under slow-growth conditions. *Pesquisa Agropecuária Brasileira*, *46*(4), 384–389. doi:10.1590/S0100-204X2011000400007

Singh, H., & Huat, B. B. K. (2004). Origin, formation and occurrence of tropical residual soils. In Tropical Residual Soil Engineering. Taylor and Francis. doi:10.1201/9780203024621.ch1

Singh, P., & Saikia, S. (2017). Arduino-based smart irrigation using water flow sensor, soil moisture sensor, temperature sensor and ESP8266 WiFi module. *IEEE Reg. 10 Humanit. Technol. Conf. 2016, R10-HTC 2016 - Proc.*

Singh, A. K., Varshney, R., Sharma, M., Agarwal, S. S., & Bansal, K. C. (2006). Regeneration of plants from alginate-encapsulated shoot tips of *Withania somnifera* (L.)Dunal, a medicinally important plant species. *Journal of Plant Physiology*, *163*(2), 220–223. doi:10.1016/j.jplph.2005.06.013 PMID:16399013

Singh, K. R., & Chaudhury, S. (2016). Efficient technique for rice grain classification using back-propagation neural network and wavelet decomposition. *IET Computer Vision*, *10*(8), 780–787. doi:10.1049/iet-cvi.2015.0486

Singh, S. K., Rai, M. K., Asthana, P., & Sahoo, L. (2010). Alginate-encapsulation of nodal segments for propagation, short-term conservation and germplasm exchange and distribution of *Eclipta alba* (L.). *Acta Physiologiae Plantarum*, *32*(3), 607–610. doi:10.100711738-009-0444-7

Sinija, V. R., & Mishra, H. N. (2011). FTNIR spectroscopic method for determination of moisture content in green tea granules. *Food and Bioprocess Technology*, *4*(1), 136–141. doi:10.100711947-008-0149-8

Smetanska, I. (2008). Production of secondary metabolites using plant cell cultures. *Advances in Biochemical Engineering/Biotechnology, 111*, 187–228. doi:10.1007/10_2008_103 PMID:18594786

Smith, K. (2013). *The science of photobiology*. Springer Science & Business Media.

Smith, M. A. L., & Spomer, L. (1995). Vessels, gels, liquid media, and support systems. In J. Aitken-Chistie (Ed.), *Automation and Environmental Control in Plant Tissue Culture* (pp. 371–404). Kluwer Academic Publishers. doi:10.1007/978-94-015-8461-6_16

Soehne, W. (1956). Some principles of soil mechanics as applied to agricultural engineering. *Inst. Agric. Eng. English Translation, 53*, 11-27.

Solie, J. B., Raun, W. R., Whitney, R. W., Stone, M. L., & Ringer, J. D. (1996). Optical sensor-based field element size and sensing strategy for nitrogen. *Transactions of the ASAE. American Society of Agricultural Engineers, 39*(6), 1983–1992. doi:10.13031/2013.27700

Soltani, A., Alimagham, S. M., Nehbandani, A., Torabi, B., Zeinali, E., Zand, E., Ghassemi, S., Vadez, V., Sinclair, T. R., & van Ittersum, M. K. (2020). Modeling plant production at country level as affected by availability and productivity of land and water. *Agricultural Systems, 183*, 102859. doi:10.1016/j.agsy.2020.102859

Son, S. H., Chun, Y. W., & Hall, R. B. (1991). Cold storage of in vitro cultures of hybrid poplar shoots (*Populus alba* L. x *P. grandidentata* Michx.). *Plant Cell, Tissue and Organ Culture, 27*(2), 161–168. doi:10.1007/BF00041285

Sophocleous, M. (2015). *Thick-film underground sensors* (Doctoral dissertation). University of Southampton.

Sophocleous, M. (2017). Electrical Resistivity Sensing Methods and Implications. In Electrical Resistivity and Conductivity. doi:10.5772/67748

Sophocleous, M. (2016). *Thick-Film Underground Sensors*. LAP LAMPERT Academic Publishing.

Sophocleous, M., Garcia-Breijo, E., Atkinson, J. K., & Georgiou, J. (2019). Modeling & Performance Comparison of Screen-Printed, Impedance Spectroscopy Probes for Harsh Environments. *IEEE Sensors Journal*. Advance online publication. doi:10.1109/JSEN.2019.2951200

Sophocleous, M., & Georgiou, J. (2017). Precision agriculture: Challenges in sensors and electronics for real-time soil and plant monitoring. *2017 IEEE Biomedical Circuits and Systems Conference (BioCAS)*, 1–4. 10.1109/BIOCAS.2017.8325180

Stafford, J. V. (2000). Implementing precision agriculture in the 21st century. *Journal of Agricultural Engineering Research, 76*(3), 267–275. doi:10.1006/jaer.2000.0577

Stafford, J. V., Ambler, B., & Smith, M. P. (1991). Sensing and mapping grain yield variation. In G. Kranzler (Ed.), *Automated Agriculture for the 21st Century* (pp. 356–365). American Society of Agricultural Engineers.

Stafford, J. V., & Miller, P. C. H. (1993). Spatially selective application of herbicide to cereal crops. *Computers and Electronics in Agriculture, 9*(3), 217–229. doi:10.1016/0168-1699(93)90040-8

Stanton, C., Murphy, J., McGrath, E., & Devery, R. (2003). *Animal feeding strategies for conjugated linoleic acid enrichment of milk*. Academic Press.

Stasiniewicz, T., Strzetelski, J., Kowalczyk, J., Osieglowski, S., & Pustkowiak, H. (2000). Performance and meat quality of fattening bulls fed complete feed with rapeseed oil cake or linseed. *Journal of Animal and Feed Sciences, 9*(2), 283–296. doi:10.22358/jafs/68046/2000

Stearns, E. I. (1987). Commission Internationale de l'Eclairage, Colorimetry (2nd ed.). Publication CIE No. 15.2.

Steeneveld, W., Tauer, L. W., Hogeveen, H., & Oude Lansink, A. G. J. M. (2012). Comparing technical efficiency of farms with an automatic milking system and a conventional milking system. *Journal of Dairy Science*, *95*(12), 7391–7398. doi:10.3168/jds.2012-5482 PMID:23021754

Stehr, N. J. (2015). Drones: The Newest Technology for Precision Agriculture. *Natural Sciences Education*, *44*(1), 89–91. doi:10.4195/nse2015.04.0772

Stemeroff, J. (1917). *Irrigation Management Strategies for Medical Cannabis in Controlled Environments* (PhD thesis). University of Guelph.

Story, D., Kacira, M., Kubota, C., Akoglu, A., & An, L. (2010). Lettuce calcium deficiency detection with machine vision computed plant features in controlled environments. *Computers and Electronics in Agriculture*, *74*(2), 238–243. doi:10.1016/j.compag.2010.08.010

Struik, P. C., Amaducci, S., Bullard, M. J., Stutterheim, N. C., Venturi, G., & Cromack, H. T. H. (2000). Agronomy of fibre hemp (Cannabis sativa L.) in Europe. *Industrial Crops and Products*, *11*(2-3), 107–118. doi:10.1016/S0926-6690(99)00048-5

Strzetelski, J., Kowalczyk, J., Osiegowski, S., Stasiniewicz, T., Lipiarska, E., & Pustkowiak, H. (2001). Fattening bulls on maize silage and concentrate supplemented with vegetable oils. *Journal of Animal and Feed Sciences*, *10*(2), 259–271. doi:10.22358/jafs/67982/2001

Subaih, W. S., Shatnawi, M. A., & Shibli, R. A. (2007). Cryopreservation of date palm (*Phoenix dactylifera*) embryogenic callus by encapsulation-dehydration, vitrification and encapsulation-vitrification. *J. J. of Agric. Sci.*, *3*(2), 156–171.

Suciu, G., Istrate, C., & Dițu, M. (2019). *Secure smart agriculture monitoring technique through isolation*. Available online at https://ieeexplore.ieee.org/document/8766433/authors#authors

Sudduth, K. A., Hummel, J. W., & Birrell, S. J. (1997). Sensors for Site-Specific Management.

Sudduth, K. A., Hummel, J. W., & Cahn, M. D. (1991). Soil organic matter sensing: A developing science. In G. A. Kranzler (Ed.), *Automated Agriculture for the 21st Century* (pp. 307–316). ASAE.

Sulecki, J. (2018). *Association seeks definitive definition of "precision agriculture"—What's your vote?* https://www.precisionag.com/events/association-seeks-definitive-definition-of-precision-agriculturewhats-your-vote

Sulkava, M., Sepponen, A. M., Yli-Heikkilä, M., & Latukka, A. (2015). Clustering of the self-organizing map reveals profiles of farm profitability and upscaling weights. *Neurocomputing*, *147*, 197–206. doi:10.1016/j.neucom.2013.09.063

Suma, N., Samson, S. R., Saranya, S., Shanmugapriya, G., &Subhashri, R. (2017). IoT based smart agriculture monitoring system. *International Journal on Recent and Innovation Trends in Computing and Communication, 5*(2), 177-181.

Summart, J., Panichajakul, S., Prathepha, P., & Thanonkeo, P. (2008). Callus Induction and Influence of Culture Condition and Culture Medium on Growth of Thai Aromatic Rice, Khao Dawk Mali 105. *Cell CultureWorld Applied Sciences Journal*, *5*(2), 246–251.

Sun, D. W. (2000). Inspecting pizza topping percentage and distribution by a computer vision method. *Journal of Food Engineering*, *44*(4), 245–249. doi:10.1016/S0260-8774(00)00024-8

Sun, D. W. (Ed.). (2016). *Computer vision technology for food quality evaluation*. Academic Press.

Suthaparan, A., Stensvand, A., Solhaug, K. A., Torre, S., Mortensen, L. M., Gadoury, D. M., Seem, R. C., & Gislerød, H. R. (2012). Suppression of powdery mildew (Podosphaerapannosa) in greenhouse roses by brief exposure to supplemental UV-B radiation. *Plant Disease*, *96*(11), 1653–1660. doi:10.1094/PDIS-01-12-0094-RE PMID:30727451

Swan, T. W., Ohare, D., Gill, R. A., & Lynch, P. T. (1999). Influence of preculture conditions on the post – thaw recovery of suspension cultures of Jerusalem artichoke (*Helianthus tuberosus* L.). *Cryo Letters*, *20*, 325–336.

Swarner, L. R., Langley, M. N., Maletic, J. T., Phelan, J. T., Lawhon, L. F., Shockley, D. G., & Brownscombe, R. H. (1959). *Irrigation on Western Farms*. Bulletin No. 199. United States Department of the Interior, Bureau of Reclamation. United States Department of Agriculture, Soil Conservation Service Agriculture Information.

Swinton, S. M., & Lowenberg-Deboer, J. (2001, June). Global adoption of precision agriculture technologies: Who, when and why. In *Proceedings of the 3rd European Conference on Precision Agriculture* (pp. 557-562). Citeseer.

Swinton, S. M., & Lowenberg-DeBoer, J. (1998). Evaluating the Profitability of Site-Specific Farming. *Journal of Production Agriculture*, *11*(4), 439–446. doi:10.2134/jpa1998.0439

Syvitski, J. P. M., Overeem, I., Brakenridge, G. R., & Hannon, M. (2012). Floods, floodplains, delta plains - A satellite imaging approach. *Sedimentary Geology*, *267–268*, 1–14. doi:10.1016/j.sedgeo.2012.05.014

Szopa, A., & Ekiert, H. (2016). The importance of applied light quality on the production of lignans and phenolic acids in *Schisandra chinensis* (Turcz.) Baill.cultures in vitro. *Plant Cell, Tissue and Organ Culture*, *127*(1), 115–121. doi:10.100711240-016-1034-1

Szopa, A., Ekiert, H., Szewczyk, A., & Fugas, E. (2012). The production of therapeutically important phenolic acids and furanocoumarins in in vitro cultures of *Ruta graveolens* and *Ruta graveolens* ssp. divaricata cultured under different light conditions. *Plant Cell, Tissue and Organ Culture*, *110*, 329–336. doi:10.100711240-012-0154-5

Szumacher-Strabel, M., Potkanski, A., Cieslak, A., Kowalczyk, J., & Czauderna, M. (2001). The effects of different amounts and types of fat on the level of conjugated linoleic acid in the meat and milk of sheep. *Journal of Animal and Feed Sciences*, *10*(Suppl. 2), 103–108. doi:10.22358/jafs/70041/2001

Tagarakis, A. C., van Evert, F., Kempenaar, C., Ljubicic, N., Milic, D., Crnojevic-Bengin, V., & Crnojevic, V. (2018). Opportunities for precision agriculture in Serbia. *Proceedings of the 14th International Conference on Precision Agriculture*. Available online at https://ispag.org/proceedings/?action=download&item=5042

Tahtamouni, R. W., & Shibli, R. A. (1999). Preservation at low temperature and cryopreservation in wild pear (*Pyrus syriaca*). *Advances in Horticultural Science*, *13*, 156–160.

Takacs-Gyorgy, K. (2009). Importance of precision farming in improving the environment. *Žemes Ukio Mokslai*, *16*(3–4), 217–223. Available online at http://mokslozurnalai.lmaleidykla.lt/publ/1392-0200/2009/3-4/217-223.pdf

Tamirat, T. W., & Pedersen, S. M. (2019). Precision irrigation and harvest management in orchards: an economic assessment. *Journal of Central European Agriculture*, *20*(3),1009-1022. DOI: /. doi:10.5513/JCEA01/20.3.2160

Tanaka, K., Fujiwara, K., & Kozai, T. (1992). Effects of relative humidity in the culture vessels on the evaporation and net photosynthetic rates of potato plantlets in vitro. *Acta Horticulturae*, (319), 59–64. doi:10.17660/ActaHortic.1992.319.3

Taner, A., Tekgüler, A., & Sauk, H. (2015). Yapay sinir ağları ile makarnalık buğday çeşitlerinin sınıflandırılması. *Anadolu Tarım Bilimleri Dergisi*, *30*(1), 51–59. doi:10.7161/anajas.2015.30.1.51-59

Tariq, U., Ali, M., & Abbasi, B. H. (2014, January 5). Morphogenic and biochemical variations under different spectral lights in callus cultures of *Artemisia absinthium* L. *Journal of Photochemistry and Photobiology. B, Biology*, *130*, 264–271. doi:10.1016/j.jphotobiol.2013.11.026 PMID:24362323

Tarmizi, A., Marziah, A., & Halim, A. (1993). Effects of various concentrations of sucrose on growth and proline accumulation in oil palm polyembryogenic cultures. *Biotechnology in Agric.*, 365 – 368.

Taulavuoria, K., Hyökya, V., Oksanena, J., Taulavuoria, E., & Tiittob, R. J. (2016). Species-specific differences in synthesis of flavonoids and phenolic acids under increasing periods of enhanced blue light. *Environmental and Experimental Botany*, *121*, 145–150. doi:10.1016/j.envexpbot.2015.04.002

Tein, S. Y., Then, Y. L., & You, K. Y. (2017). Tea leaves moisture measurement and prediction using RF waveguide antenna. *Proceedings of 2017 Asia Pacific Microwave Conference*, 670-673. 10.1109/APMC.2017.8251535

Tejada-Castro, M., Delgado-Vera, C., Garzón-Goya, M., Sinche-Guzmam, A., & Cárdenas-Rosales, X. (2019). Trends in the Use of Webapps in Agriculture: A Systematic Review. In R. Valencia García, G. Alcaraz Mármol, J. del Cioppo-Morstadt, N. Vera Lucio, & M. Bucaram Leverone (Eds.), *ICT for Agriculture and Environment* (pp. 130–142). Springer Nature Switzerland. doi:10.1007/978-3-030-10728-4_14

Tey, Y. S., & Brindal, M. (2012). Factors influencing the adoption of precision agricultural technologies: A review for policy implications. *Precision Agriculture*, *13*(6), 713–730. doi:10.100711119-012-9273-6

Thakur, S. K. P., & Kolmes, W. S. (2001). Permittivity of rice grain from electromagnetic scattering. *4th International Conference on Electromagnetic Wave Interaction with Water and Moist Substance*, 203-210.

Thenkabail, P. S., Lyon, J. G., & Huete, A. (2010). *Hyperspectral remote sensing of vegetation and agricultural crops: Knowledge gain and knowledge gap after 40 years of research.* CRC Press.

Thessler, S., Kooistra, L., Teye, F., Huitu, H., & Bregt, A. K. (2011). Geosensors to Support Crop Production: Current Applications and User Requirements. *Sensors (Basel)*, *11*(7), 6656–6684. doi:10.3390110706656 PMID:22163978

Thompson, Stafford, & Miller. (1990). *Selective application of herbicides to UK cereal crops.* Paper, American Society of Agricultural Engineers 90-1629.

Thompson, J. F., Stafford, J. V., & Miller, P. C. H. (1991). Potential for automatic weed detection and selective herbicide application. *Crop Protection (Guildford, Surrey)*, *10*(4), 254–259. doi:10.1016/0261-2194(91)90002-9

Thorne, C. R. (1990). Effects of vegetation on riverbank erosion and stability. In Vegetation and Erosion. John Wiley and Sons Inc.

Tisserat, B. (1979). Propagation of Date Palm (*Phoenix dactylifera* L.) *in vitro. Journal of Experimental Botany*, *30*(119), 1275–1283. doi:10.1093/jxb/30.6.1275

Tisserat, B. H. (1981). Cryogenic preservation and regeneration of date palm tissue. *HortScience*, *16*(1), 47–48.

Togola, A., Seck, P., Glitho, I., Diagne, A., Adda, C., Toure, A., & Nwilene, F. (2013). Economic losses from insect pest infestation on rice stored on-farm in Benin. *Journal of Applied Sciences (Faisalabad)*, *13*(2), 278–285. doi:10.3923/jas.2013.278.285

Toivonen, L., Laakso, S., & Rosenguist, H. (1992). The effect of temperature on growth, indole alkaloid accumulation and lipid composition of *Catharanths roseus* cell suspension culture. *Plant Cell Reports*, *11*, 390–394. PMID:24201539

Toscano, P., Castrignano, A., Di Gennaro, S. F., Vonella, A. V., Ventrella, D., & Matese, A. (2019). A Precision Agriculture Approach for Durum Wheat Yield Assessment Using Remote Sensing Data and Yield Mapping. *Agronomy (Basel)*, *9*(8), 437. doi:10.3390/agronomy9080437

Towill, L. E. (1991). Cryopreservation. In J. H. Dodds (Ed.), *In vitro methods for conservation of plant genetic resources* (pp. 41–70). Chapman and Hall. doi:10.1007/978-94-011-3072-1_3

Trabelsi, S., Kraszewski, A. W., & Nelson, S. O. (1999). Unified calibration method for nondestructive dielectric sensing of moisture content in granular materials. *Electronics Letters*, *35*(16), 1346–1347. doi:10.1049/el:19990890

Transhel, V., Gutner, L., & Khokhryakov, M. (1933). A list of fungi found on new cultivated textile plants. Moscow Inst. Nov. Lubian. Syria. *Trudy*, *4*, 127–140.

Trejgell, A., Kamińska, M., & Tretyn, A. (2015). In vitro slow growth storage of Senecio macrophyllus shoots. *Acta Physiologiae Plantarum*, *37*(11), 234. doi:10.100711738-015-1983-8

Tsiplakou, E., Kominakis, A., & Zervas, G. (2008). The interaction between breed and diet on CLA and fatty acids content of milk fat of four sheep breeds kept indoors or at grass. *Small Ruminant Research*, *74*(1-3), 179–187. doi:10.1016/j.smallrumres.2007.06.008

Tyler, D. A. (1993). Positioning technology (GPS). In P. C. Robert, R. H. Rust, & W. E. Larsen (Eds.), *Proc. Soil Specific Crop Management* (pp. 159–166). American Society of Agronomy.

Tzounis, A., Katsoulas, N., Bartzanas, T., & Kittas, C. (2017). Internet of Things in agriculture, recent advances and future challenges. *Biosystems Engineering*, *164*, 31–48. doi:10.1016/j.biosystemseng.2017.09.007

Ulrich, J. M., Finkle, B. J., & Tisserat, B. H. (1982). Effects of Cryogenic treatment on plantlet production from frozen and unfrozen date palm callus. *Plant Physiology*, *69*(3), 624–627. doi:10.1104/pp.69.3.624 PMID:16662261

UN Economic and Social Council (ECOSOC). (1988). *United Nations Convention Against Illicit Traffic in Narcotic Drugs and Psychotropic Substances*. Available at: https://www.refworld.org/docid/49997af90.html

UN General Assembly. (1971). *Convention on Psychotropic Substances*. Available at: https://www.refworld.org/docid/3b00f1ad4b.html

UN General Assembly. (1972). *Protocol Amending the Single Convention on Narcotic Drugs*. Available at: https://www.refworld.org/docid/3b00f1c118.html

United Nations Conference on Environment and Development (UNCED). (1992). *Convention on Biological Diversity*. UNCED.

United States Department of Agriculture (USDA). (2005). *Dietary guidelines for Americans*. Available from http://www.health.gov/dietaryguidelines/dga2005/document/

Van Dam, J. E. G. (1995). Potentials of hemp as an industrial fibre crop. *Proceedings of First Bioresource Hemp Symposium, nova-Institute*, 405-422.

Van der Werf, H. M. G., & Van den Berg, W. (1995). Nitrogen fertilization and sex expression affect size variability of fibre hemp (Cannabis sativa L.). *Oecologia*, *103*(4), 462–470. doi:10.1007/BF00328684 PMID:28306994

Van der Werf, H. M. G., Van Geel, W. C. A., Van Gils, L. J. C., & Haverkort, A. J. (1995). Nitrogen fertilization and row width affect self-thinning and productivity of fibre hemp (Cannabis sativa L.). *Field Crops Research*, *42*(1), 27–37. doi:10.1016/0378-4290(95)00017-K

van der Werf, H. M. G., Wijlhuizen, M., & De Schutter, J. (1995). Plant density and self-thinning affect yield and quality of fibre hemp (Cannabis sativa L.). *Field Crops Research*, *40*(3), 153–164. doi:10.1016/0378-4290(94)00103-J

Vanhove, W., Van Damme, P., & Meert, N. (2011). Factors determining yield and quality of illicit indoor cannabis (Cannabis spp.) production. *Forensic Science International*, *212*(1-3), 158–163. doi:10.1016/j.forsciint.2011.06.006 PMID:21737218

Vasta, V., Nuddab, A., Cannas, A., Lanza, M., & Priolo, A. (2008). Alternative feed resources and their effects on the quality of meat and milk from small ruminants. *Animal Feed Science and Technology*, *147*(1-3), 223–246. doi:10.1016/j.anifeedsci.2007.09.020

Velázquez-López, N., Sasaki, Y., Nakano, K., Mejía-Muñoz, J. M., & Romanchik, K. E. (2011). Detection of powdery mildew disease on rose using image processing with Open CV. *Revista Chapingo Serie Horticultura, 17*(2), 151–160. doi:10.5154/r.rchsh.2011.17.017

Vemuri, V. K., & Makriyannis, A. (2015). Medicinal chemistry of cannabinoids. *Clinical Pharmacology and Therapeutics, 97*(6), 553–558. doi:10.1002/cpt.115 PMID:25801236

Vera, C. L., Malhi, S. S., Raney, J. P., & Wang, Z. H. (2004). The effect of N and P fertilization on growth, seed yield and quality of industrial hemp in the Parkland region of Saskatchewan. *Canadian Journal of Plant Science, 84*(4), 939–947. doi:10.4141/P04-022

Veris. (2014). *Veris Technologies.* Retrieved from https://www.veristech.com/the-sensors/u-series

Vermeulen, P., Flémal, P., Pigeon, O., Dardenne, P., Fernández Pierna, J. A., & Baeten, V. (2017). Assessment of pesticide coating on cereal seeds by near infrared hyperspectral imaging. *Journal of Spectral Imaging, 6*, 1–7. doi:10.1255/jsi.2017.a1

Veroustraete, F. (2015). The Rise of the Drones in Agriculture. *EC Agriculture, 2*(2), 325–327.

Vidoni, R., Gallo, R., Ristorto, G., Carabin, G., Mazzetto, F., Scalera, L., & Gasparetto, A. (2017, November). ByeLab: an agricultural mobile robot prototype for proximal sensing and precision farming. In *ASME 2017 International Mechanical Engineering Congress and Exposition.* American Society of Mechanical Engineers Digital Collection. 10.1115/IMECE2017-71216

Visen, N., Paliwal, J., Jayas, D., & White, N. (2002). Ae—automation and emerging technologies: Specialist neural networks for cereal grain classification. *Biosystems Engineering, 82*(2), 151–159. doi:10.1006/bioe.2002.0064

Voronin, B., Mitin, A., & Pichugin, O. (2019, June). The complexity of digitization of agriculture in Russia. In *International Scientific and Practical Conference "Digital agriculture-development strategy" (ISPC 2019).* Atlantis Press. 10.2991/ispc-19.2019.29

Vuran, M. C., Salam, A., Wong, R., & Irmak, S. (2018). Internet of underground things in precision agriculture: Architecture and technology aspects. *Ad Hoc Networks, 81*, 160–173. doi:10.1016/j.adhoc.2018.07.017

Wachira, A. M., Sinclair, L. A., Wilkinson, R. G., Enser, M., Wood, J. D., & Fisher, A. V. (2002). Effects of dietary fat source and breed on the carcass composition, n-3 polyunsaturated fatty acid and conjugated linoleic acid content of sheep meat and adipose tissue. *British Journal of Nutrition, 88*(6), 697–709. doi:10.1079/BJN2002727 PMID:12493092

Waide, P. (2010). *Information paper. Phase out of incandescent lamps. Implications for international supply and demand for regulatory compliant lamps.* https://www.iea.org/publications/freepublications/publication/phase_out.pdf

Walburg, G., Bauer, M., Daughtry, C. S. T., & Housley, T. L. (1982). Effects of nitrogen nutrition on the growth, yield and reflectance characteristics of corn canopies. *Agronomy Journal, 74*(4), 677–683. doi:10.2134/agronj1982.00021962007400040020x

Waldron, L. J. (1977). The shear resistance of root-permeated homogeneous and stratified soil. *Soil Science Society of America Journal, 41*(5), 843–849. doi:10.2136ssaj1977.03615995004100050005x

Waldron, L. J., & Dakessian, S. (1982). Effects of Grass, Legume, and Tree Roots on Soil Shearing Resistance. *Soil Science Society of America Journal, 46*(5), 894–899. doi:10.2136ssaj1982.03615995004600050002x

Wall, R., Ross, R. P., Fitzgerald, G. F., & Stanton, C. (2010). Fatty acids from fish: The anti-inflammatory potential of long-chain omega-3 fatty acids. *Nutrition Reviews, 68*(5), 280–289. doi:10.1111/j.1753-4887.2010.00287.x PMID:20500789

Wanas, W. (1999). The application of *in vitro* technique for the genetic conservation of some *Citrus spp. Anal. of Agric. Sci., 44*(2), 653–672.

Wang, H., Ma, L. G., Li, J. M., Zhao, H. Y., & Deng, X. W. (2001). Direct interaction of Arabidopsis cryptochomes with COP1 in light control development. *Science, 294*(5540), 154–158. doi:10.1126cience.1063630 PMID:11509693

Wang, P., & Hu, C. (1982). In vitro mass suberisation and virus-free seed potato production in Taiwan. *American Potato Journal, 59*(1), 33–37. doi:10.1007/BF02854881

Wang, Q. C., Batuman, Ö., Bar-Joseph, P., Li, M., & Gafny, R. (2002). Cryopreservation of *in vitro*-grown shoot tips of 'Troyer' citrange [*Poncirus trifoliata* (L.) Raf. ×*Citrus sinesis* (L.) Osbeck.] by encapsulation-vitrification. *Euphytica, 128*(1), 135–142. doi:10.1023/A:1020683305690

Wang, Q., Gafny, R., Sahae, N., Sela, I., Mawassi, M., Tanne, E., & Perl, A. (2002). Cryopreservation of grapevine (*Vitis vinifera* L.) embryogenic cell suspension by encapsulation – dehydration and subsequent plant regeneration. *Plant Science, 162*(4), 551–558. doi:10.1016/S0168-9452(01)00594-5

Wang, S., & Tang, J. (2001). Radio frequency and microwave alternative treatments for insect control in nuts: A review. *Asian Agricultural Engineering, 10*(3 & 4), 105–120.

Wang, X., & Qi, Q. (2011). Design and realization of precision agriculture information system based on 5S. *Proc. - 19th Int. Conf. Geoinformatics, Geoinformatics*, 2–5.

Wang, X., Yang, W., Wheaton, A., Cooley, N., & Moran, B. (2010). Wheaton, Cooley, A.N., and Moran, B., Efficient registration of optical and IR images for automatic plant water stress assessment. *Computers and Electronics in Agriculture, 74*(2), 230–237. doi:10.1016/j.compag.2010.08.004

West, J. (2018). A Prediction Model Framework for Cyber-Attacks to Precision Agriculture Technologies. *Journal of Agricultural & Food Information*, 1-24. Doi:10.1080/10496505.2017.1417859

West, J. S., Canning, G. G., Perryman, S. A., & King, K. (2017). Novel Technologies for the detection of Fusarium head blight disease and airborne inoculum. *Tropical Plant Pathology, 42*(3), 203–209. doi:10.100740858-017-0138-4 PMID:32714500

Whelan, B. M., McBratney, A. B., & Stein, A. (2003). On-farm field experiments for precision agriculture. *Precision Agriculture, Proceedings of the 4th European Conference on Precision Agriculture*, 731–737.

Whitlock, L., Schingoethe, D.J., Hippen, A.R., Kalscheur, K.F., Baer, R.J., & Ramaswamy, N., & Kasperson, K.M. (2002). Fish oil and extruded soybeans fed in combination increase CLA in milk of dairy cows more than when fed separately. *Journal of Dairy Science, 85*, 234–243. doi:10.3168/jds.S0022-0302(02)74072-1 PMID:11860116

Whitney, J. D., Ling, Q., Miller, W. M., & Wheaton, T. A. (2001). A DGPS yield monitoring system for Florida citrus. *Applied Engineering in Agriculture, 17*(2), 115–119. doi:10.13031/2013.5451

Wiegand, C. L., Richardson, A. J., Escobar, D. E., & Gerbermann, A. H. (1991). Vegetation indices in crop assessment. *Remote Sensing of Environment, 35*(2-3), 105–119. doi:10.1016/0034-4257(91)90004-P

Willatt, S. T., & Sulistyaningsih, N. (1990). Effect of plantroots on soil strength. *Soil & Tillage Research, 16*(4), 329–336. doi:10.1016/0167-1987(90)90068-O

Williamson, J., Jennings, D., Guo, W., Pharr, M., & Ehrenshaft, M. (2002). Sugar alcohols, salt stress and fungal resistance: Polyols, multifunctional plant protection. *Journal of the American Society for Horticultural Science, 127*(4), 463–467. doi:10.21273/JASHS.127.4.467

Wilson, C. L., & Wisniewski, M. E. (1989). Biological control of postharvest diseases of fruits and vegetables: An emerging technology. *Annual Review of Phytopathology*, *27*(1), 425–441. doi:10.1146/annurev.py.27.090189.002233

Wilson, S. B., Rajapakse, N. C., & Young, R. E. (2000). Media composition and light affect storability and post storage recovery of micropropagated hosta plantlets. *HortScience*, *34*, 1158–1162.

Window, M. (2019). *Security in Precision Agriculture Vulnerabilities and risks of agricultural systems* (M Sc. Thesis). Lulea University of Technology Department of Computer Science, Electrical and Space Engineering. Available online at http://ltu.diva-portal.org/smash/get/diva2:1322203/FULLTEXT02.pdf

Wiseman, L., Sanderson, J., & Robb, L. (2018). Rethinking Ag data ownership. *Farm Policy Journal*, *15*(1), 71–77.

Withers, L. (1982). Institutes working on tissue culture for genetic conservation. IBGR Secretariat.

Withers, L. A. (1991). Biotechnology and plant genetic resources conservation. In Plant Genetic Resources, Conservation and Management Concepts and Application. International Board for Plant genetic Resources.

Withers, L. A., & Engelmann, F. (1998). *In vitro* conservation of plant genetic resources. In A. Altman (Ed.), *Agricultural Biotechnology* (pp. 57–87). Marcel Dekker.

Wixted, A. J., Kinnaird, P., Larijani, H., Tait, A., Ahmadinia, A., & Strachan, N. (2016, October). Evaluation of LoRa and LoRaWAN for wireless sensor networks. In *2016 IEEE SENSORS* (pp. 1-3). IEEE.

Wolf, J., Bhandari, S., & Raheja, A. (2017). *Unmanned Aerial Vehicles for Precision Agriculture in Orchard Crops*. Academic Press.

Wolfert, S., Ge, L., Verdouw, C., & Bogaardt, M. J. (2017). Big data in smart farming–a review. *Agricultural Systems*, *153*, 69–80. doi:10.1016/j.agsy.2017.01.023

Wollenhaupt, N. C., Mulla, D. J., & Gotway, C. A. (1997). Soil sampling and interpolation techniques for mapping spatial variability of soil properties. In F. J. Pierce & E. J. Sadler (Eds.), *The State of Site-Specific Management for Agriculture* (pp. 19–54). ASA/CSSA/SSSA. doi:10.2134/1997.stateofsitespecific.c2

Wright, R. D., & Niemiera, A. X. (1987). Nutrition of container-grown woody nursery crops. *Horticultural Reviews*, *9*, 75–101.

Wspanıaly, P., & Moussa, M. (2016). Early powdery mildew detection system for application in greenhouse Automation. *Computers and Electronics in Agriculture*, *127*, 487–494. doi:10.1016/j.compag.2016.06.027

Wu T.H., Kokesh, C.M., Trenner,B. R. & Fox, P. J. (2014). Use of live poles for stabilization of a shallow slope failure. *Journal of Geotechnical and Geoenvironmental Engineering*, *140*(10), 1-13.

Wu, T. H. (1995). Slope stabilization. In Slope stabilization and erosion control: a bioengineering approach. Academic Press.

Wu, T. H. (1984). *Effect of Vegetation on Slope Stability. In soil reinforcement and moisture effects on slope stability.* Transportation Research Board.

Wu, T. H., Mckinnell, W. P., & Swanson, D. N. (1979). Strength of tree roots and landslides on Prince of Wales Island, Alaska. *Canadian Geotechnical Journal*, *16*(1), 19–33. doi:10.1139/t79-003

Xu, G., Wolf, S., & Kafkafi, U. (2001). Effect of varying nitrogen form and concentration during growing season on sweet pepper flowering and fruit yield. *Journal of Plant Nutrition*, *24*(7), 1099–1116. doi:10.1081/PLN-100103806

Yadav, V. K., Yadav, M. R., Kumar, R., Yadav, B., & Verma, A. K. (2017). Potentials and Prospects of Precision Farming in Indian Agriculture. *Marumegh, 2*(2), 47-53. Available online at www.marumegh.com

Yagihara, S., Oyama, M., Inoue, A., Asano, M., Sudo, S., & Shinyashiki, N. (2007). Dielectric relaxation measurement and analysis of restricted water structure in rice kernels. *Measurement Science & Technology, 18*(4), 983–990. doi:10.1088/0957-0233/18/4/004

Yamamoto, S., Fukui, K., Rafique, T., Khan, N., Castillo Martinez, C., Sekizawa, K., & Niino, T. (2012). Cryopreservation of in vitro-grown shoot tips of strawberry by the vitrification method using aluminium cryo-plates. *Plant Genetic Resources, 10*(1), 14–19. doi:10.1017/S1479262111000906

Yamamoto, S., Fukui, K., Rafique, T., Sekizawa, K., Koyama, A., Ichihashi, I., & Niino, T. (2012). Development of an effective cryopreservation protocol using aluminum cryo-plates for in vitro-grown shoot tips of mulberries (Morus spp.) originated from the tropics and subtropics. *Sanshi-Konchu Biotec, 81*(1), 57–62.

Yeh, H., & Chung, J. P. (2009). High-brightness LEDs-energy efficient lighting sources and their potential in indoor plant cultivation. *Renewable & Sustainable Energy Reviews, 13*(8), 2175–2180. doi:10.1016/j.rser.2009.01.027

Yirga, C., & Hassan, R. M. (2010). Social costs and incentives for optimal control of soil nutrient depletion in the central highlands of Ethiopia. *Agricultural Systems, 103*(3), 153–160. doi:10.1016/j.agsy.2009.12.002

Yorio, N. C., Goins, G. D., Kagie, H. R., Wheeler, R. M., & Sager, J. C. (2001). Improving spinach, radish, and lettuce growth under red light-emitting diodes (LEDs) with blue light supplementation. *HortScience, 36*(2), 380–383. doi:10.21273/HORTSCI.36.2.380 PMID:12542027

Yost, M. A., Sudduth, K. A., Walthall, C. L., & Kitchen, N. R. (2019, December 22). Public–private collaboration toward research, education and innovation opportunities in precision agriculture. *Precision Agriculture, 20*(1), 4–18. doi:10.100711119-018-9583-4

You, K. Y., Salleh, J., Abbas, Z., & You, L. L. (2011). Cylindrical slot antennas for monitoring the quality of milled rice. *Progress in Electromagnetics Research Symposium Proceedings*, 1370-1373.

You, K. Y. (2015). *RF Coaxial Slot Radiators: Modeling, Measurements, and Applications*. Artech House.

You, K. Y. (2017). Materials characterization using microwave waveguide system. In S. K. Goudos (Ed.), *Microwave Systems and Applications* (pp. 341–358). InTech. doi:10.5772/66230

You, K. Y., Lee, C. Y., Chan, K. S., Lee, K. Y., Cheng, E. M., & Lee, Y. S. (2018). Monitoring moisture content for various kind of tea leaves in drying processes using RF reflectometer-sensor system. *Instruments, 2*(18), 1–8. doi:10.3390/instruments2030018

You, K. Y., Lee, C. Y., Then, Y. L., Chen Chong, S. H., You, L. L., Abbas, Z., & Cheng, E. M. (2013). Precise moisture monitoring for various soil types using handheld microwave-sensor meter. *IEEE Sensors Journal, 13*(7), 2563–2570. doi:10.1109/JSEN.2013.2253318

You, K. Y., Mun, H. K., You, L. L., Jamaliah, S., & Abbas, Z. (2013). Small and slim coaxial probe for single rice grain moisture sensing. *Sensors (Basel), 13*(3), 3652–3663. doi:10.3390130303652 PMID:23493127

You, K. Y., You, L. L., Yue, C. S., Mun, H. K., & Lee, C. Y. (2017). Physical And Chemical Characterization Of Rice Using Microwave And Laboratory Methods. In S. F. Amanullah (Ed.), *Rice* (pp. 81–99). InTech.

You, K. Y., Zulkifly, A., & Khalid, K. (2010). Application of microwave moisture sensor for determination of oil palm fruit ripeness. *Measurement Science Review, 10*(1), 7–14.

You, K. Y., Zulkifly, A., Khalid, K., & Abdul Rahman, M. Z. (2010). Improved dielectric model for polyvinyl alcohol-water hydrogel at microwave frequencies. *American Journal of Applied Sciences*, *7*(2), 270–276. doi:10.3844/ajassp.2010.270.276

You, K. Y., Zulkifly, A., You, L. L., Lee, K. Y., & Cheng, E. M. (2014). Palm oil moisture based on dielectric properties at microwave frequencies. *International Journal of Microwave and Optical Technology*, *9*(14), 294–302.

Younas, M., Drouet, S., Nadeem, M., Giglioli-Guivarc'h, N., Hano, C., & Abbasi, B. H. (2018). Differential accumulation of silymarin induced by exposure of *Silybum marianum* L. callus cultures to several spectres of monochromatic lights. *Journal of Photochemistry and Photobiology. B, Biology*, *184*, 61–70. doi:10.1016/j.jphotobiol.2018.05.018 PMID:29803074

Yousefi, M. R., & Razdari, A. M. (2014). Application of GIS and GPS in Precision Agriculture (A Review). *International Journal of Advanced Biological and Biomedical Research*, *2*(4), 473–476.

You, T. S., & Nelson, S. O. (1988). Microwave dielectric properties of rice kernels. *The Journal of Microwave Power and Electromagnetic Energy*, *23*(3), 150–158. doi:10.1080/08327823.1988.11688050

Yu, K. W., Murthy, H. N., Hahn, E. J., & Paek, K. Y. (2005). Ginsenoside production by hairy root cultures of Panax ginseng: Influence of temperature and light quality. *Biochemical Engineering Journal*, *23*(1), 53–56. doi:10.1016/j.bej.2004.07.001

Yu, X. R., Su, Y., Wang, Y., Zheng, T. S., Zhao, S. M., & Huang, S. Z. (1998). Study on microwave drying of grain. *Proceedings of the 7th International Working Conference on Stored-product Protection*, 1096-1101.

Yu, X., Wu, P., Han, W., & Zhang, Z. (2013). A survey on wireless sensor network infrastructure for agriculture. *Computer Standards & Interfaces*, *35*(1), 59–64. doi:10.1016/j.csi.2012.05.001

Zadra, A., & Robert, G. (2012). Dream recall frequency: Impact of prospective measures and motivational factors. *Consciousness and Cognition*, *21*(4), 1695–1702. doi:10.1016/j.concog.2012.08.011 PMID:23092675

Zaman, Q. U., Schumann, A. W., & Miller, W. M. (2005). Variable rate nitrogen application in Florida citrus based on ultrasonically-sensed tree size. *Applied Engineering in Agriculture*, *21*(3), 331–335. doi:10.13031/2013.18448

Zarco-Tejada, J. P., Hubbard, N., & Loudjani, P. (2014). *Precision Agriculture: An opportunity for EU farmers - Potential Support with the CAP 2014-2020*. Academic Press.

Zarifi, M. H., Gholidoust, A., Abdolrazzaghi, M., Shariaty, P., Hashisho, Z., & Daneshmand, M. (2018). Sensitivity enhancement in planar microwave active-resonator using metal organic framework for CO_2 detection. *Sensors and Actuators. B, Chemical*, *255*, 1561–1568. doi:10.1016/j.snb.2017.08.169

Zhang, Y. (2011). *Design of the node system of wireless sensor network and its application in digital agriculture*. Paper presented at the 2011 International Conference on Computer Distributed Control and Intelligent Environmental Monitoring. 10.1109/CDCIEM.2011.371

Zhang, M., & Meng, Q. (2011). Automatic citrus canker detection from leaf images captured in field. *Pattern Recognition Letters*, *32*(15), 2036–2046. doi:10.1016/j.patrec.2011.08.003

Zhang, M., Qin, Z., Liu, X., & Ustin, S. L. (2003). Detection of stress in tomatoes induced by late blight disease in California, USA, using hyperspectral remote sensing. *International Journal of Applied Earth Observation and Geoinformation*, *4*(4), 295–310. doi:10.1016/S0303-2434(03)00008-4

Zhang, N., Wang, M., & Wang, N. (2002). Precision agriculture - a worldwide overview. *Computers and Electronics in Agriculture*, *36*(2-3), 113–132. doi:10.1016/S0168-1699(02)00096-0

Zhang, Q., Reid, J. F., & Noguchi, N. (1999). Agricultural vehicle navigation using multiple guidance sensors. *Proc. Intl. Conf. Field Service Robotics.*

Zhang, Y. J., & Okamura, S. A. (2000). Moisture content measurement for green tea using phase shifts at two microwave frequencies. *Subsurface Sensing Technologies and Applications*, *1*(2), 489–496. doi:10.1023/A:1026571800893

Zhang, Y. J., & Okamura, S. A. (2006). Density independent method for high moisture content measurement using microstrip transmission line. *Journal of Microwave & Electromagnetic Energy*, *40*(2), 110–118. doi:10.1080/08327823.2005.11688527 PMID:17278794

Zhang, Z. Z., Che, X. N., Pan, Q. H., Li, X. X., & Duan, C. Q. (2013). Transcriptional activation of flavan-3-ols biosynthesis in grape berries by UV irradiation depending on developmental stage. *Plant Science*, *208*, 64–74. doi:10.1016/j.plantsci.2013.03.013 PMID:23683931

Zheng, F., & Schreier, H. (1988). Quantification of soil patterns and field soil fertility using spectral reflection and digital processing of aerial photographs. *Fertilizer Research*, *16*(1), 15–30. doi:10.1007/BF01053311

Zheng, L., Zhang, J., & Wang, Q. (2009). Mean-shift-based color segmentation of images containing green vegetation. *Computers and Electronics in Agriculture*, *65*(1), 93–98. doi:10.1016/j.compag.2008.08.002

Zhou, J., & Zhang, B. (Eds.). (2019). *Agricultural Robots: Fundamentals and Applications*. BoD–Books on Demand. doi:10.5772/intechopen.74631

Zimmermann, U. (1978). Physics of turgor and osmo – regulation. *Annual Review of Plant Physiology*, *29*(1), 121–148. doi:10.1146/annurev.pp.29.060178.001005

Zimmerman, R. H. (1995). Environmental effects and their control in plant tissue culture review. *Acta Horticulturae*, (393), 11–14. doi:10.17660/ActaHortic.1995.393.1

Ziv, M. (1990). Vitrification: morphological and physiological disorders of in vitro plants. In P. C. Debergh & R. H. Zimmerman (Eds.), *Micropropagation: Technology and Application* (pp. 45–69). Kluwer Academic Publishers.

Zude-Sasse, M., Fountas, S., Gemtos, T. A., & Abu-Khalaf, N. (2016). Applications of precision agriculture in horticultural crops. *European Journal of Horticultural Science*, *81*(2), 78–90. doi:10.17660/eJHS.2016/81.2.2

Zu, Y. G., Wei, X. X., Yu, J. H., Li, D. W., Pang, H. H., & Tong, L. (2011). Responses in the physiology and biochemistry of Korean pine (*Pinus koraiensis*) under supplementary UV-B radiation. *Photosynthetica*, *49*(3), 448–458. doi:10.100711099-011-0057-4

About the Contributors

Sherine Mohamed Abd El-Kader has her MSc, & PhD degrees from the Electronics & Communications Dept. & Computers Dept., Faculty of Engineering, Cairo University, at 1998 & 2003. Dr. Abd El-kader is a Professor in Computers & Systems Dept., at the Electronics Research Institute (ERI) since April 2014. She is currently the head of computers and systems Dept., at ERI since 2017. Also she is the head of Technology, Innovation and Commercialization Office since 2018. She was the head of Technology Innovation Support Center (TISC) at ERI from 2013 to 2018. She supervised more than 20 MSc and PhD students. She has published more than 50 papers, 6 book chapters in computer networking area. She is working in many hot topics such as; IOT, 5G, cognitive radio, Wi-MAX, Wi-Fi, QoS, Wireless sensors Networks, Ad-Hoc Networking, real time traffics, and localization algorithms. She was an Associate Prof., at Faculty of Engineering, Akhbar El Yom Academy from 2007 till 2009. She was heading the Internet and Networking unit at ERI from 2003 till 2014. She also was heading the information and decision making support center at ERI from 2009 to 2014. She supervised many automation and web projects for ERI.

Basma Mamdouh Mohammad El-Basioni is a researcher in Computers and Systems Department at the Electronics Research Institute (ERI) in Egypt. She received her M.Sc. and Ph.D. degrees, both in Electronics and Communications Engineering, from Al-Azhar University, Cairo, Egypt, in 2011 and 2017 respectively. Her current research interests focus on IoT technology and communication networks especially WSN in different applications. Dr. Basma has published more than twelve paper and chapter, and has two copyrights and a patent application; she is a reviewer for some international publishers. She has a good background on Intellectual Property (IP) and she is the supervisor of Technology and Innovation Support Center (TISC) at ERI.

* * *

Suhail Abdullah is a bachelor in computer and communications (AUC) Iraq. M. Eng. Electronics and telecommunications (UTM) Malaysia. PhD candidate electronics and telecommunications (UTM) Malaysia.

Waleed Abobatta is Dr. at (Citrus research Department), Horticulture Research Institute, Agriculture Research Center, he's a member of Scientific Committee for Greenhouses Plantation (A R C), and member of Scientific Team for National Campaign for Navel Orange improvement, Abobatta has published significantly more than 11 books and research papers with the international journals since 2014. Abobatta has been serving as an editorial board member in many journals.

Rangel Arthur holds a degree in Electrical Engineering from the Paulista State University Júlio de Mesquita Filho (1999), a Master's degree in Electrical Engineering (2002) and a Ph.D. in Electrical Engineering (2007) from the State University of Campinas. Over the years from 2011 to 2014 he was Coordinator and Associate Coordinator of Technology Courses in Telecommunication Systems and Telecommunication Engineering of FT, which was created in its management. From 2015 to 2016 he was Associate Director of the Technology (FT) of Unicamp. He is currently a lecturer and advisor to the Innovation Agency (Inova) of Unicamp. He has experience in the area of Electrical Engineering, with emphasis on Telecommunications Systems, working mainly on the following topics: computer vision, embedded systems and control systems.

Virginia Barradas graduated from Computer Science, from the Universidad Veracruzana with a Master's Degree in Computational Sciences from the Rosenblueth Foundation, a Master's Degree in University Teaching from the Universidad de Xalapa and a PHD in Education. Full-time academic of the Master's Degree and Engineering in Computer Systems of the Instituto Tecnológico Superior de Xalapa (19 years), as well as a professor at the Universidad Veracruzana (24 years), PRODEP Profile and member of IEEE and WIE. Her current interests reside in topics on Educative technology, Software Project Management, User Experience, Agile Project Management, as well as various topics of Data Science.

Canset Koçer Baykara was born in Ankara, Turkey in 1988. I graduated from Eskişehir Osmangazi University in 2012. First I started to work in Amasya University as a research assistant and then pass to Erciyes University at the same position in 2013-2014. Since 2014 I have been working in Turkish Grain Board (TMO) as electrical elevtronics engineer. In this period, I have taken my master degree from Mechetronics Engineering Department of Erciyes University. I am married and have a fifteen months baby.

Rabaie Benameur, Master student, is affiliated with the Research Laboratory in Industrial Computing and Networks (RIIR), University of Oran 1 Ahmed Benbella, Algeria. His main research interests include wireless sensor networks, their security, routing and management, intrusion detection, and MAC protocols design issues, IA Techniques.

Amine Dahane, PhD, is affiliated with the Research Laboratory in Industrial Computing and Networks (RIIR), University of Oran 1 Ahmed Benbella, Algeria. His main research interests include wireless sensor networks, their security, routing and management, intrusion detection, and MAC protocols design issues. He is a reviewer of several journals and participates regularly at professional conferences. He received his master's degree in Computer Systems and Networks from the University of Bechar and his PhD degree in Electronics from the University of Sciences and Technology of Oran (USTO, Algeria). He is currently a lecturer and researcher at the institute of Sciences and Applied Techniques of the University of Oran 1, Ahmed Benbella (Algeria) where he teaches Instrumentation, Informatics, networking, Programming languages.

M. El-Hagarey has a PhD at irrigation and drainage, 2010, He works as a associate prof. at the Desert Research Center, Soil Conservation and Water Resources department, Irrigation and Drainage unit. Recently, he is awarded Kashida Prize for Scientific Excellence in The Field of Agricultural and Bioengineering,2019, He is the winner of "ICID WatSave Awards 2016"under 'WatSave Young Professional Award' Also, he is the winner of TWAS-ARO Young Arab Scientist (YAS) Prize 2016 with the topic "The Implementation of Food Security Research for the Arab Region Sustainability", His work as vice president of Halayb and the Shalateen research station and the president of Al-Kantra Shark research station. He records some patents in hydraulics devices, especially in fluid mechanics and irrigation sciences. Dr. ElHagarey contributed in many international and national scientific projects, having many of international publications (Article researches, Books, Bulletins, Reports, posters, and journals), and works in Africa, Asia and Europe in the scientific research fields. He contributed in scientific activity by American scientists; he has been a member of international and national Societies. Dr. ElHagarey is a consultant in many companies and committees. Recently he has been an Accreditation and Quality Assurance Bureau member in Desert Research Centre, Provisional Member of the Working Group on Water Saving in Irrigated Areas (WG-WATS, ICID), in addition to have the training courses of Leadership development courses at the National Defence College, Nasser Higher Military Academy, Ministry of Defence.

Eman Hassan Elsabaawy is a Researcher in National Research Center, Agricultural and Biological Research Division, Animal production department. She obtained a PhD in ruminant nutrition at the University of Cairo (2016), an MSc in animal nutrition at the University of Cairo (2008), and BSc in agriculture at Cairo University (2000). Her filed of interest is improve quality of meat through functional food and increase efficiency of feed utilization and decrease environmental pollution and interest in the areas of protein and energy nutrition of lactating cows.

Aitazaz A. Farooque is working as an Associate Professor & Industry Research Chair - Precision Agriculture (PA) at the Faculty of Sustainable Design Engineering, University of Prince Edward Island. Dr. Farooque's research focuses on fundamental understanding and development of state of the art PA technologies for Eastern Canada's agriculture industry. Development of innovative and novel PA systems utilizes knowledge of engineering design, development and management, instrumentation, design and evaluation of sensors and controllers, development of hardware and software for automation of machines to sense targets in real-time for spot application of agrochemicals on an as-needed basis to improve farm profitability while maintaining environmental sustainability. Dr. Farooque is actively working on machine vision, application of multispectral and thermal imagery using drone technology, delineation of management zones for site-specific fertilization, electromagnetic induction methods, remote sensing, and digital photography technique for mapping, bio-systems modeling, hydrological modeling, artificial neural network, deep learning, analog and digital sensor integration into agricultural equipment for real-time soil, plant, and yield mapping. Dr. Farooque has been developing and evaluating the variable rate technologies for potential environmental risks. Dr. Farooque has been very successful in securing research funding from Natural Science and Engineering Council of Canada (NSERC), Provincial and Federal Governments and agriculture industry. Dr. Farooque was selected as one of the Top 125 researchers of Canada to receive NSERC Discovery Accelerator Supplement Research Grant. He has been supervising undergraduate and graduate students, research assistants, and post-doctoral fellow at UPEI and other collaborating institutions. He is also serving as an adjunct professor at couple of national and international institutions. Dr. Farooque highlighted and published the significance of his research in peer-reviewed journals, conference proceedings, workshops, industry meetings and farmer's field days. He is a member of various professional organizations, industry and university committees. He is also serving on editorial boards of few peer-reviewed journals. He had organized a couple of international conferences and various research seminars in Canada and internationally. Based on his student centered approach to teaching by going above and beyond for students and encouraging students to push the boundaries of their learning, he has been awarded with Dalhousie University Teaching Impact Award 2015. He also was awarded with the Engineers PEI Award for Engineering Excellence 2018 for excellence in the practice of engineering, the teaching of engineering and engineering research. He is also a recipient of the 2019 Presidential Award for Combined Achievement (Teaching, research and service) from the University of Prince Edward Island. He is a Professional Engineer and registered in both PEI and Nova Scotia. http://www.islandscholar.ca/people/afarooque https://scholar.google.ca/citations?user=GnIErVgAAAAJ.

Reinaldo França is B.Sc. in Computer Engineering in 2014. Currently, he is an Ph.D. degree candidate by Department of Semiconductors, Instruments and Photonics, Faculty of Electrical and Computer Engineering at the LCV-UNICAMP working with technological and scientific research as well as in programming and development in C / C ++, Java and .NET languages. His main topics of interest are simulation, operating systems, software engineering, wireless networks, internet of things, broadcasting and telecommunications systems.

Sawsan M. Gad is a Professor on Animal Nutrition, National Research Center, Egypt. She Interest on: Ruminant nutrition and feed quality evaluation and improving nutritional values of feeds through chemical and biological treatments. Using enzymes, probiotics or prebiotics as feed additives to improve nutritional quality of raw materials, increase efficiency of feed utilization and decrease environmental pollution and Producing meat and milk of high quality through feed manipulation. Dr. Sawsan supervised many M.Sc. and Ph D. thesis and published a lot of scientific research papers in National and International Journals on her interest points.

Rosa Maria Gonzalez-Amaro is a PhD in Ecology and sustainable development with experience in native maize conservation. Her research interests are plant genetic resources, agroecology and food sufficiency. She has participated in national and international forums and has published in scientific journals and digital media. She is currently a researcher at CONACYT, Institute of Ecology, participating in the project "Strategies for the conservation of corn and food sufficiency".

Miguel Hidalgo-Reyes has a PhD in Computer Science with experience in research and technological development. His research interests are data mining, usability in software development and information visualization. He has participated in national and international forums and has published in scientific journals and digital media. He is currently a CONACYT chair at the Institute of Ecology, participating in the Big Data and Sustainable Development project.

Yuzo Iano. B.Sc. (1972), M.Sc. (1974) and Ph.D. degrees (1986) in Electrical Eng. at UNICAMP, Brazil. Since then he has been working in the technological production field, with 1 patent granted, 8 filed patent applications and 36 projects completed with research and development agencies. He has supervised 29 doctoral theses, 49 master's dissertations, 74 undergraduate and 48 scientific initiation works. He has participated in 100 master's examination boards, 50 doctoral degrees, author of 2 books and more than 250 published articles. He is currently Professor at UNICAMP, Editor-in-Chief of the SET International Journal of Broadcast Engineering and General Chair of the Brazilian Symposium on

Technology (BTSym). He has experience in Electrical Engineering, with knowledge in Telecommunications, Electronics and Information Technology, mainly in the field of audio-visual communications and multimedia.

Muzaffer Kanaan received his B.S. degree in from the Eastern Mediterranean University (Famagusta, North Cyprus), M.S. degree from the New Jersey Institute of Technology (Newark, NJ, USA), and Ph.D. degree from the Worcester Polytechnic Institute (Worcester, MA, USA) in 2008, all in Electrical and Computer Engineering. He worked as a Distinguished Member of the Technical Staff at Verizon Laboratories in the 1997-2008 timeframe, where he carried out research and development activities in wireless and fiber optic communications. Since 2009, he has been with the Department of Mechatronics Engineering at Erciyes University (Kayseri, Turkey), where he is currently an Associate Professor and Department Vice-Chair. In 2018, he founded a technology start-up, Sanotech Corporation, which specializes in wearable technology and precision agriculture applications. He is the holder of 5 US patents, with another two pending.

Bouabdellah Kechar is Professor in the Department of Computer Scienc at Oran 1 Ahmed Ben Bella University, Algeria where he received his Ph.D in 2010. He has authored several journal publications, refereed conferenc publications and book chapters. He has been a member of the technica program and organizing committees of several internationa IEEE/ACM/ELSEVIER conferences and workshops. He also serves as a referee of renowned journals such as IEEE Transactions on computers IEEE Communications Magazine, IEEE Systems Journal and South African Journal of Science. His current research interests include IoT and big data technologies, mobile wireless sensor/actuator networks, Zigbee/IEEE 802.15.4 technologies deployment, Vehicular ad-hoc networks, vehicula sensor networks and heterogeneous wireless networks, with specia emphasis on radio resource management techniques, performance modelling, provisioning QoS and practical societal and industria applications. He has supervised and co-supervised several undergraduat graduate students in these areas. He has co-founded in 2011 the laboratory of industrial computing and networking (RIIR). He is currently head of research team on wireless sensor networks and their societal and industria applications.

Ana Carolina Monteiro is a Ph.D. student at the Faculty of Electrical and Computer Engineering (FEEC) at the State University of Campinas - UNICAMP, where she develops research projects regarding health software with emphasis on the development of algorithms for the detection and counting of blood cells through processing techniques. digital images. These projects led in 2019 to a computer program registration issued by the INPI (National Institute of Industrial Property). She holds a Master's degree in Electrical Engineering from the State University of Campinas - UNICAMP (2019) and graduated in Biomedicine from the University Center Amparense - UNIFIA with a degree in Clinical Pathology - Clinical Analysis (2015). In 2019, he acquired a degree in Health Informatics. Has experience in the areas of Molecular Biology and management with research animals. Since 2017, she has been a researcher at the FEEC/UNICAMP Visual Communications Laboratory (LCV) and has worked at the Brazilian Technology Symposium (BTSym) as a member of the Organizational and Executive Committee and as a member of the Technical Reviewers Committee. In addition, she works as a reviewer at the Health magazines of the Federal University of Santa Maria (UFSM - Brazil), Medical Technology Journal MTJ (Algeria) and Production Planning & Control (Taylor & Francis). Interested in digital image processing, hematology, clinical analysis, cell biology, medical informatics, Matlab and teaching.

Muhamad Rusliyadi is Lecturer at the Polytechnic of Agricultural Development Yogyakarta-Magelang, Agricultural Extension and Human Resource Development Agency, Ministry of Agriculture, Indonesia.

Man Seng Sim was born in Johor, Malaysia in 1993. In 2017, he received his Bachelor of Science (Applied Physics) with Honours from Universiti Tun Hussein Onn Malaysia (UTHM). In 2019, he received his Master of Philosophy in Electrical Engineering from Universiti Teknologi Malaysia (UTM). He is currently pursuing his Doctor of Philosophy in Electrical Engineering in UTM. His research interest is in electromagnetic metamaterials, microwave characterisation techniques and microwave sensors.

Marios Sophocleous is a Special Scientist at the Holistic Electronic Research Laboratory of the University of Cyprus since 2016. He obtained his Masters of Engineering degree from the University of Southampton in Mechatronics and his PhD in Thick-Film Underground Sensors in 2016 from the same University. He won two scholarships being rated in the top 10 students of the year during his undergraduate studies where he won the Jim Graham prize for the best experimental project of the year for his individual project. He further won two prestigious partial PhD scholarships from the University of Southampton for his PhD work. During his time at the University of Cyprus, he won one prestigious PostDoctoral Scholarship from the University of Cyprus during when he co-supervised a team of undergraduate students ranked in the top 4 in the world for the CASS Student competition 2018. He has successfully obtained more than 200 thousand euros of funding up to date setting up the first Thick Film Unit in the country. He has been nominated for the Outstanding Young Engineer Award of the IEEE Instrumentation and Measurement Society in 2019 whilst he is also a nominee for the Young Scientist Award 2020 in Cyprus. Marios has already co-authored more than 25 peer-reviewed journal and conference publications in very high impact journals and major conferences, 3 book chapters and 1 magazine article. Only 4 years after his PhD, he has an h-index and i-10 index of 9. Marios has served as a technical and/or organizing committee member in major conferences whilst he is a member of the Editorial Advisory Board of Sensors and Actuators A: Physical Journal and Microelectronics International Journal. Furthermore, he is an active member of the IEEE Sensors Council and of the IEEE community in general.

Akalpita Tendulkar is a major in Life Sciences research, focusing on the utilization of biotechnological techniques in environmental science. Have a three plus years of experience as Lecturer with a demonstrated history of working in the education management industry. Skilled in Molecular Biology, Polymerase Chain Reaction (PCR), and Environmental Biotechnology. Strong education professional with a Doctor of Philosophy (PhD) in Biotechnology from Malaysia University of Science and Technology.

Kok Yeow You was born in 1977. He obtained his B.Sc. Physics (Honours) degree in Universiti Kebangsaan Malaysia (UKM) in 2001. He pursued his M.Sc. in Microwave at the Faculty of Science in 2003 and his Ph.D. in Wave Propagation at the Institute for Mathematical Research in 2006 in Universiti Putra Malaysia (UPM), Malaysia. Recently, he is a senior lecturer at the School of Electrical Engineering, Faculty of Engineering, Universiti Teknologi Malaysia (UTM), Malaysia. His main personnel research interest is in the theory, simulation, and instrumentation of electromagnetic wave propagation at microwave frequencies focusing on the development of microwave passive devices and sensors for medical and agricultural applications.

Index

K

M

N

O

P

R

S

www.ingramcontent.com/pod-product-compliance
Lightning Source LLC
LaVergne TN
LVHW080006210926
848481LV00011B/2959